Fundamentals of VHDL for FPGA Programming Using Vivado

IEEE Press

445 Hoes Lane

Piscataway, NJ 08854

Fundamentals of VHDL for FPGA Programming Using Vivado

Majid Pakdel

Electrical, Automation, and Instrumentation Expert at MISCO

IEEE Press
WILEY

Published by John Wiley & Sons, Inc., Hoboken, New Jersey.
Published simultaneously in Canada.

For general information on our other products and services or for technical support, please contact our Customer Care Department within the United States at (800) 762-2974, outside the United States at (317) 572-3993 or fax (317) 572-4002.

Wiley also publishes its books in a variety of electronic formats. Some content that appears in print may not be available in electronic formats. For more information about Wiley products, visit our web site at www.wiley.com.

Library of Congress Cataloging-in-Publication Data applied for:

Hardback ISBN: 9781394343096
ePDF ISBN: 9781394343119
ePub ISBN: 9781394343102
oBook ISBN: 9781394343126

Cover Design: Wiley
Cover Image: © whiteMocca/Shutterstock

To all the aspiring engineers and innovators who dare to dream beyond the boundaries of conventional design, this book is dedicated to you. May the knowledge within these pages empower you to harness the power of VHDL and FPGA programming, inspiring you to create solutions that shape the future.

To the dedicated staff at Wiley, whose guidance and support have been invaluable throughout this journey, thank you for believing in this work.

To my family, whose love and patience have been my unwavering foundation, I am forever grateful.

Lastly, in loving memory of my mother, whose absence has left a profound void in my life. I hope she forgives me, and may her soul rest in eternal peace.

Contents

About the Author

Majid Pakdel received his Master's degree in Electrical Power Engineering from Isfahan University of Technology, Iran, in 2007, his PhD in Electrical Power Engineering from the University of Zanjan in 2018, and his second Master's degree in Computer Engineering – Artificial Intelligence and Robotics from Malek Ashtar University of Technology, Tehran, in 2023. He has published over 20 papers and books in the fields of electrical engineering and computer science.

Preface

As technology continues to evolve at an unprecedented pace, the ability to effectively design and implement digital systems has become increasingly essential. Field-programmable gate arrays (FPGAs) represent a pivotal advancement in this domain, providing unparalleled flexibility and performance for a wide variety of applications – ranging from embedded systems to high-speed communications and beyond. The advent of sophisticated design environments, such as Vivado from Xilinx, has opened new avenues for educators, engineers, and hobbyists alike to harness the power of these versatile devices. This book, "*Fundamentals of VHDL for FPGA Programming Using Vivado*," serves as a comprehensive introduction to the fundamental concepts and practical skills necessary for designing digital circuits with VHDL (VHSIC hardware description language) and implementing them on FPGA platforms. My goal is to equip readers with the knowledge required to navigate the complexities of VHDL programming and Vivado design flows, enabling them to bring their innovative ideas to life. Geared toward both beginners and those with some experience in digital design, this text begins with an overview of VHDL's syntax and semantics, emphasizing practical applications and real-world examples. Subsequent chapters delve deeper into the core principles of digital design, exploring topics such as simulation, synthesis, testbenches, and optimization techniques. The hands-on tutorials will guide readers step-by-step through the design process using Vivado, providing them with the confidence and competence to tackle their own projects. Throughout the book, I aim to balance theoretical knowledge with practical application. The provided examples, projects, and exercises are designed to reinforce learning, foster critical thinking, and encourage creativity in design. Furthermore, I address common challenges faced by designers and offer strategies and solutions drawn from my own experiences in the field. I believe that mastering VHDL and FPGA design is not just about understanding concepts and syntax, but also about cultivating the ability to solve problems and innovate. Whether you are a student, an aspiring engineer, or a seasoned professional looking to refresh your skills, I hope that this book will serve as a valuable resource on your journey into the exciting world of digital design. In conclusion, I would like to extend my gratitude to my family, colleagues, and fellow educators for their unwavering support and encouragement throughout the writing process. I also want to acknowledge the vibrant community of learners and practitioners who continue to inspire me with their passion for technology and creativity. Welcome to the world of VHDL and FPGA programming. Let the adventure begin!

Please note that all the figures in the book are printed in grayscale; however, they are reproduced in color on the book's website – please visit www.wiley.com/go/pakdelVHDL.

Acknowledgments

I would like to acknowledge the invaluable support and guidance I received from the Wiley team, especially Sandra Grayson, Becky Cowan, Nandhini Karuppiah, and others who played a crucial role in the publication process.

About the Companion Website

This book is accompanied by a companion website.

www.wiley.com/go/pakdelVHDL

The website includes:

- **MATLAB and Simulink Models:** This directory contains all the project files and models referenced in the book, particularly for chapters covering HDL Coder and Model-Based Design. These files allow readers to simulate systems in Simulink and generate the VHDL code discussed in the text.
- **Supplemental Chapters 19, 20, and 21:** These three advanced chapters are provided in their entirety as downloadable PDFs. They extend the book's coverage into critical areas such as deploying controller design on FPGA, combinational and sequential logic design, and the complete VHDL implementation of an 8-bit CPU, offering readers significant additional technical depth.
- **Vivado Codes:** This is a complete collection of all VHDL source files, test benches, and XDC constraint files for every project in the book. Readers can use these codes to run their own simulations, syntheses, and implement the designs on an FPGA board without needing to type the code manually.

1

Overview

1.1 Introduction

In this book, we are going to use the Vivado tool from Xilinx, which is their newest tool that we will be using to develop our projects on the A7 development board. What is this book about? Well, this book focuses on two main topics: VHDL and FPGA development. There is a lot to the VHDL language, so what we're going to do is go through and explain specific parts of the VHDL language, specifically how it relates to FPGA development. You can use VHDL for a variety of things, but in this book, we're going to focus on the part of the shell that's used to implement a design on an FPGA. This includes different types of digital logic or anything else you can think of that can be done digitally – we're going to put it on the FPGA. We're going to show you how to do just that. Thus, in this book, we're going to explain how to use Vivado, the software development kit for Xilinx, and how to develop on the A7 development board, which is a Xilinx development board. Now, the layout of this book is such that we are going to cover the basics of Vivado, and then work on a project. This means you will get some theory and an understanding of how different parts of the Vivado work. After that, we'll go ahead and do a project and then we'll cover more theory in another project. So, the idea is that once you come to a project, at that point, you should have learned everything you need to know to complete it successfully. As we go through the different theories and explain them, we will walk through and show how they relate specifically to the next project in the book as they step down. So, all of the different projects that we're working with include the following: we have a buttons and LEDs project, where we are going to press a button on our FPGA board, which will light up an LED. The whole idea behind this is to show you how to do the basics – how to take inputs and generate outputs. You're going to have to use the buttons as input and display the output on LEDs. Then we have a blinking LED project, where we're going to show you how to create a timer, specifically a counter, which control how LED blinks on and off at a specified interval. Then we have the brightness project, where we are going to vary the brightness of the LEDs. If you have the Arty A7 board, you will have both standard LEDs and an RGB (red–green–blue) LED. In this project, we are going to vary the brightness using pulse width modulation. We're going to explain how pulse width modulation works, how to implement it in VHDL, and then deploy it on the development board to vary the brightness of the LEDs on the board.

We then want to show you the UART demonstration project, which is a very basic introduction to using UART on your development board. Both boards have a UART port, and this is going to show you how we can take information from the actual developer board and display it on a computer. Well, you can use UART to take that information displayed on the computer. So, we're going to show you

Fundamentals of VHDL for FPGA Programming Using Vivado, First Edition. Majid Pakdel.

Companion Website: www.wiley.com/go/pakdelVHDL

how to do that. Then, we have the other UART IO project, where we will take information just like we did in UART demonstration project. But take it a step further, where we are going to take inputs in, so we'll type in commands on our computer to have certain actions occur on the development board, such as checking the input to tell us which buttons are pushed or which switches are high, and also sending a command to tell it to light up certain LEDs. We have an ADC processing project, where we are going to interface the board with a temperature sensor to demonstrate the analog-to-digital conversion. We're going to vary the temperature of the sensor to show how it responds – when the sensor gets colder, you'll see the temperature decrease, and when it gets hotter, the temperature will increase. We're going to take that temperature, convert it into both Celsius and Fahrenheit, and then display the values on the computer screen so that you can see the actual ambient temperature or the temperature that a sensor is indicating. We have an SPI interface project. This project is going to be working only with the Arty A7 development board, and what we'll do is we are going to take and store information on the onboard SPI flash. This is because these are volatile memories, which means that when you program the package and then power it off, the configurations are lost when it is powered back on. So, we're going to show you how to take information and store it on the external flash so that when you power it back on and reprogram it, you can pull that information from the external flash. So, we're going to work on it and show you how to do the interface between an FPGA and external flash using the SPI interface. Who's this course for? This is for anyone who has no previous FPGA or VHDL design experience and wants to understand how it's used as well as a design from scratch on FPGA. So, whether you're an engineer, student, or hobbyist, this course is specifically designed for you to show you how to do this type of project and use FPGA. Now that you know what to expect from this book, let's get started.

1.2 Introduction to VHDL

In this section, we're going to introduce and briefly explain what VHDL is. Well, it's a hardware descriptive language, and it's similar to Verilog. If you have ever heard of Verilog, they are not one and the same. They're two different languages, but they're doing the same thing. We're describing hardware. It's a hardware descriptive language, hence the name VHDL, and what that means is we're using the VHDL to describe a hardware circuit. We're not actually writing software, so you kind of have to wrap your mind around the idea. When you're working with the VHDL, you're not writing software. It's not like C, Java, or MATLAB, where everything you write happens sequentially. It's not that at all. What we're doing is describing how we want a digital hardware circuit to operate, so we can think of it as architecting a digital schematic. So, you'll say, OK, we want the specific AND gate or we want a comparator and we want to drive some sort of digital; whatever type of digital thing you're wanting to do, you implement it and describe it in detail, and that's the whole idea behind it. The big advantage of using VHDL is that it's very portable. So, if we have VHDL that we are running on Xilinx or Altera device, we can take that same VHDL out, as long as we're not using specific components on the actual device, such as memory, which means we have to change a little bit to use memory on the other device. We can take this one feature design and move it around a different logic device on FPGA, a couple or any other type of logic device you might be using. So, that's one of the big advantages of using the VHDL – you can build up a library so that if you're creating a project with an Altera device, you can use those same designs. When you create a project on a Xilinx device, you can use almost all of those same designs when designing a device. That's the whole idea – it's very portable. The whole reason that VHDL was developed is that designing digital hardware used to take a long time. If you wanted to design something digital, you had to place new chips on a board, which requires a new board layout, a new schematic, and long turnaround times. If you can just place one chip in the middle that's implementing all of this functionality and you want to change it, such as your algorithm changes or you find a small bug with an FPGA and VDL, you can then modify your VHDL circuitry and reload it on the FPGA and now your design change has taken place without an actual board layout change. So, that's one of the major advantages of using VHDL, or as you call it, configurable hardware, and the important thing we understand is it's not software. You're not writing software.

Some of the constructs, syntax, and structures are very similar, and they operate in the same manner. But you must understand that you're describing hardware and that's the overall concept. What you're doing when you're working with the VHDL is you're not actually writing a software program; you're describing a hardware circuit. Where is VHDL actually used? So, it's used in several different ways, and probably the most common is that it's used to configure – or we'll call it program, but you're not necessarily programming – you're configuring the circuit. CPLDs and FPGAs are types of configurable hardware devices, and CPLDs and FPGAs are similar in a lot of aspects. But typically, your FPGAs are larger devices, and they have SRAM-like base, whereas your CPLDs might be like EPROM or flash-based. So, the FPGA doesn't contain its configuration. When you power it off, you'll typically have to store that configuration on a memory chip offboard the FPGA. So, any power on it that sucks in that configuration or as in this book here, we're going to just program the JTAG. So, once your FPGA is powered up, you program and/or configure it. Then you'll notice that when you remove the power and power it back on, it loses its configuration. That's because we just loaded it directly onto the FPGA, which is just RAM memory. So, as soon as your power is down, it loses all the configurations. Another reason you can use VHDL is to simulate digital circuits. So, if you had a board, sort of a board layout you're working on, and you've got a section with several components where everything they're doing is digital, you can use VHDL to describe each one of those chips and how they interact with each other so that you can simulate it on your computer. Especially if you have an analog and digital interaction, you can say, "OK, we're going to stimulate my digital FPGA and we'll use PSpice to simulate my analog and verify that the interaction is occurring correctly." Another thing you can use VHDL for is to create an ASIC (application-specific integrated circuit). For example, if you're working with a microprocessor or an ASIC that handles Ethernet communications, a transceiver, or any similar type of ICs (integrated circuits), VHDL can be used to describe its hardware. If you work with circuit boards or develop any type of circuits in electrical engineering, you'll work with ASICs. When they're developing these, they'll use VHDL to describe the hardware that goes inside these ASICs, and we can use that to compile it and internally understand where they need to place all the gates for the actual chip to do the functionality you're describing. Finally, we're going to understand how we are going to use VHDL. Therefore, in this book, we are going to use the VHDL to configure programs on FPGA. Thus, we have these development boards and they've got different peripherals on them, including buttons, switches, LEDs, and communication ports, and we are going to use VHDL to interface these different ports to each other and just get an understanding of how VHDL works and how we can use it on these FPGAs. So, we'll be using these development boards. However, the overall idea is to use VHDL to describe a circuit inside the FPGA and we'll explain how these peripherals interact with each other. Therefore, this is just a brief introduction and explanation of the VHDL.

2

Creating a Project in Vivado

2.1 Introduction

Vivado is a design suite developed by Xilinx, primarily used for the synthesis and analysis of hardware description language (HDL) designs. It is a comprehensive toolset designed for the development of Xilinx's field-programmable gate array (FPGA) and system-on-chip (SoC) devices. Here are some key features and components of Vivado:

1. Design Entry: Supports VHDL, Verilog, and SystemVerilog for writing and managing design code. It also offers block diagram entry and high-level synthesis (HLS) from C, C++, and SystemC.
2. Synthesis: Converts high-level design descriptions into a gate-level representation optimized for Xilinx FPGAs.
3. Implementation: Includes processes like translation, mapping, placing, and routing of the design onto the FPGA.
4. Simulation: Provides integrated simulation capabilities for verifying the functional correctness of designs.
5. Analysis and Debugging: Offers various tools for analyzing and debugging the design, including timing analysis, power analysis, and on-chip debugging tools like the integrated logic analyzer.
6. Intellectual Property (IP) Integrator: Facilitates the integration and configuration of IP cores.
7. Design Environment: Provides a graphical user interface (GUI)-based environment as well as scripting capabilities for automation using Tcl.
8. Device Support: Continuously updated to support the latest Xilinx devices, including the latest UltraScale, UltraScale+, and Versal families.

Vivado replaces the older ISE Design Suite and offers enhanced performance, capabilities, and support for larger and more complex designs.

2.2 Creating a Project

We are going to create a project in Vivado in this section. We're going to walk through and show you how to create a project in Vivado using the Arty A7 development board. This is the board we are going to use to complete the buttons and LEDs project using Vivado. On the board, we will be using the pushbuttons, which are labeled from BTN3 to BTN0, as well as a switch for enable or

Fundamentals of VHDL for FPGA Programming Using Vivado, First Edition. Majid Pakdel.

Companion Website: www.wiley.com/go/pakdelVHDL

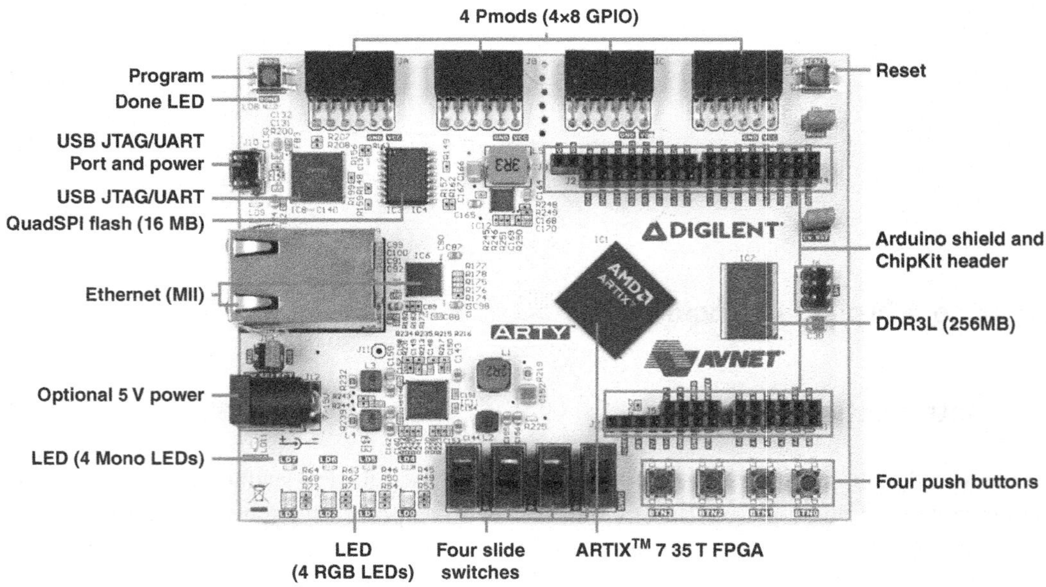

Figure 2.1 The Arty A7 development board. *Source*: Digilent.

disable operation, labeled from SW3 to SW0. Additionally, we will use four LEDs, labeled from LD7 to LD4, located on the Arty A7 development board, as shown in Figure 2.1.

What we want to do is after downloading Vivado HLS 2019.1, we'll go ahead and bring it up, and we're going to walk you through the steps required to create a project for the buttons and LEDs. Of course, once you bring up Vivado, if you'll notice, if you've opened projects previously, you'll see the recent projects – these are all the different projects that you have recently opened. It's just like a quick little shortcut to all the various projects you've worked on in the last month or so. If you go ahead, we want to, from the Quick Start menu, click on the Create Project, as depicted in Figure 2.2. This will bring up the New Project Wizard, which will guide us through everything we need to create our brand-new project, as shown in Figure 2.3.

So, we want to go ahead and click on the Next button, and we want to give the project a name. The name can be whatever you want to call it – we're going to call it Button_LED. You want to give it something that makes sense for the project you're working on. Then, you wanted to find the project location and make sure Create project subdirectory is checked. Right below that, you will see the specific location where your project will be created, as illustrated in Figure 2.4.

Then, you click on the Next button. Now, we want to select the project type. There are various types of projects. There is an RTL project, which is what we'll be using. You also have a post-synthesis project and an I/O planning project. You can import a project or create a new project from the Vivado example, as shown in Figure 2.5.

However, we're going to go with the RTL project for all the projects in this book. So, in this book, we'll be using RTL projects. These are the most common – you can import your IP, place IP that's already a part of Vivado, and create your own VHDL design sources, and we will click on the Next button. This brings up a window to add sources, as you can see in Figure 2.6.

This is where you will specify the VHDL or any other type of design source. If you want to import them, you would click on Add Files, and it'll bring up a navigation window where you can select any type of design source. If you look at the types of files it supports, there are VHDL, Verilog, as

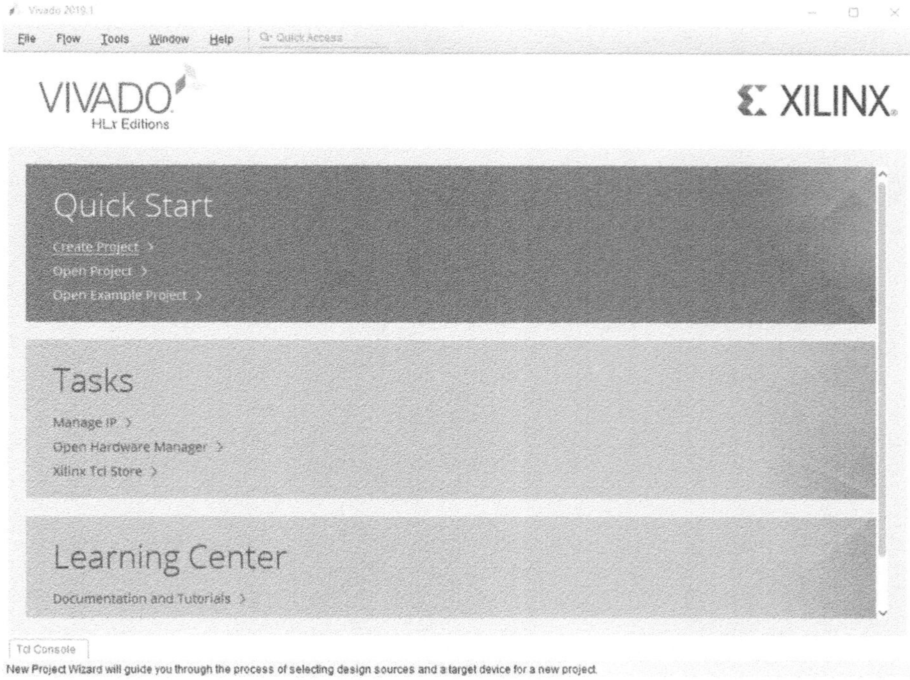

Figure 2.2 Clicking on the Create Project.

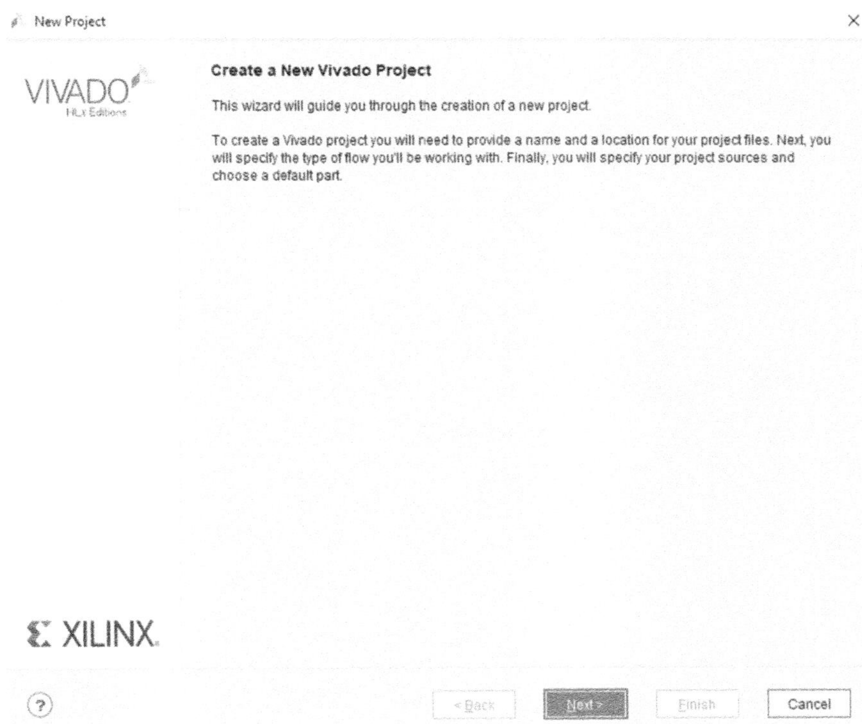

Figure 2.3 A New Project wizard.

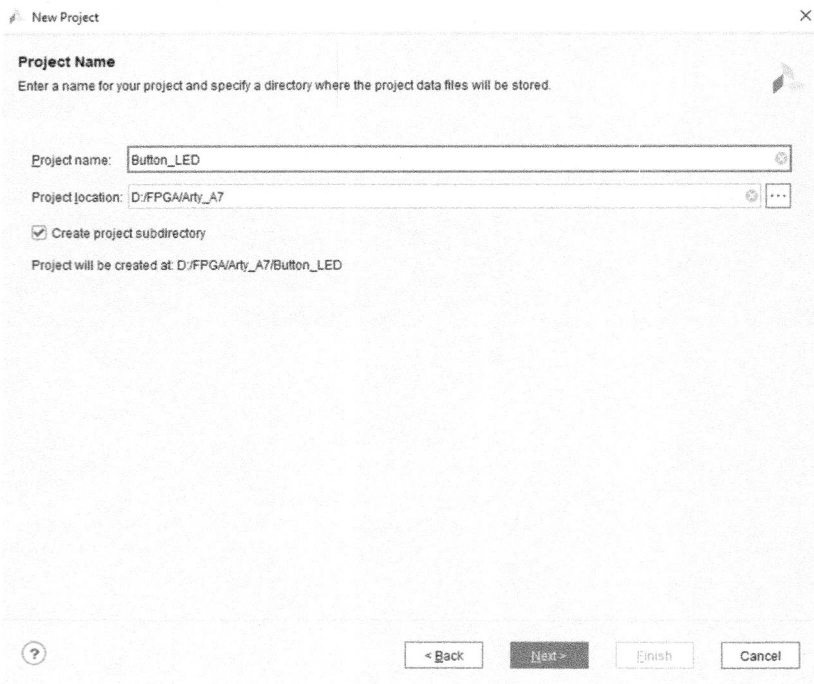

Figure 2.4 Giving the project a name.

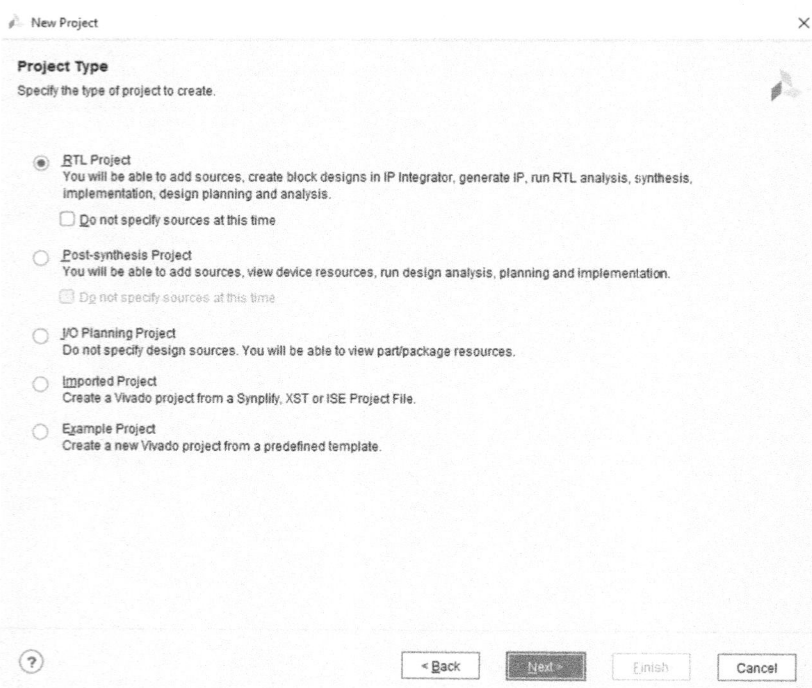

Figure 2.5 Selecting the project type.

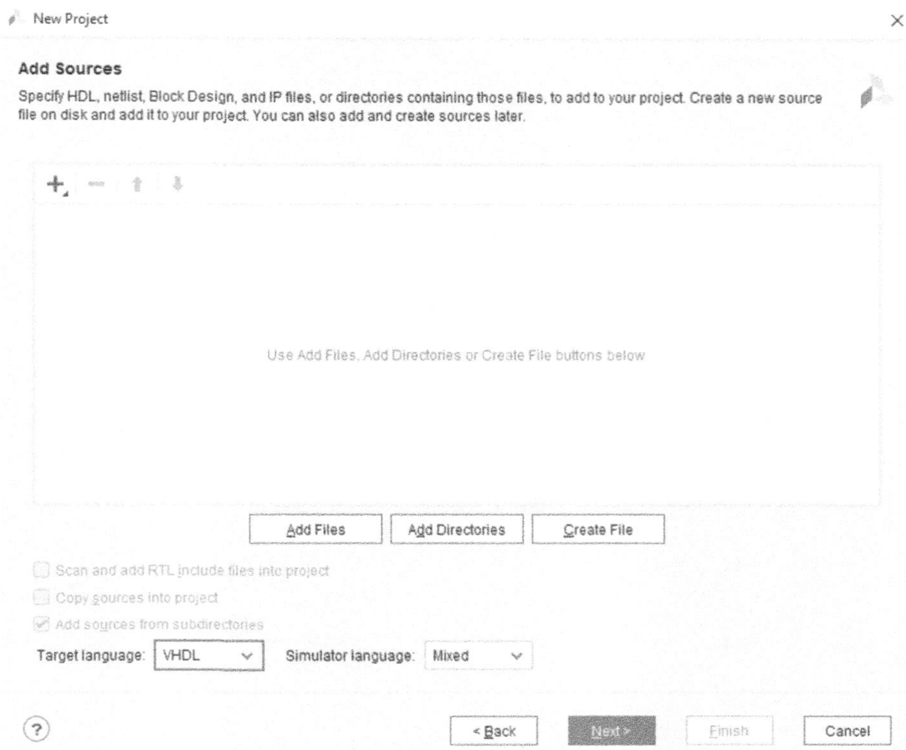

Figure 2.6 The adding sources window.

well as different scripts and all kinds of different files associated with projects in Vivado. However, we're going to go ahead and click Cancel because, after we create the project, we'll show you how to create a brand-new VHDL file, as well as how to import VHDL files. So, go ahead and click Next. This is where you can add a constraint file, which is where you define your inputs, outputs, and timings, as depicted in Figure 2.7.

However, we're going to add that later, and we'll show you how to do that after the project is created, and how to import the constraints file. If you already have one and want to import it, you would click on Add Files, which brings up the navigation window where you go out and select a constraints file, which would be the .xdc file, and we would cancel and click on the Next button. Now, we need to specify the actual part we're going to be using – this is the actual specific FPGA on the Arty A7 development board. If you click on it, you will see two different options: Parts or Boards. If we click on Boards, it brings up the specific boards that Vivado supports in this specific release. However, in this case, the Arty A7 is not supported, so we have to go to Parts and select the specific FPGA. For the Product Category, we will leave it as All; or the Family, we need to select the Artix-7, and for the Package, we select csg324. You can notice this by checking on the actual Artix board itself, you can see the csg324, or if you look in the manual, it tells you the specific part number that we're looking for, and the Speed Grade is −1L, which then filters the remaining options. On the Arty A7, we have an Artix-7 35T, which is the second option left in the Parts menu. So, if we select that FPGA, it displays all the different information for that FPGA, including the I/O Pin Count, Block RAMs, DSPs, Flip-Flops, gigabit (Gb) transceivers, available I/O blocks, etc., as illustrated in Figure 2.8.

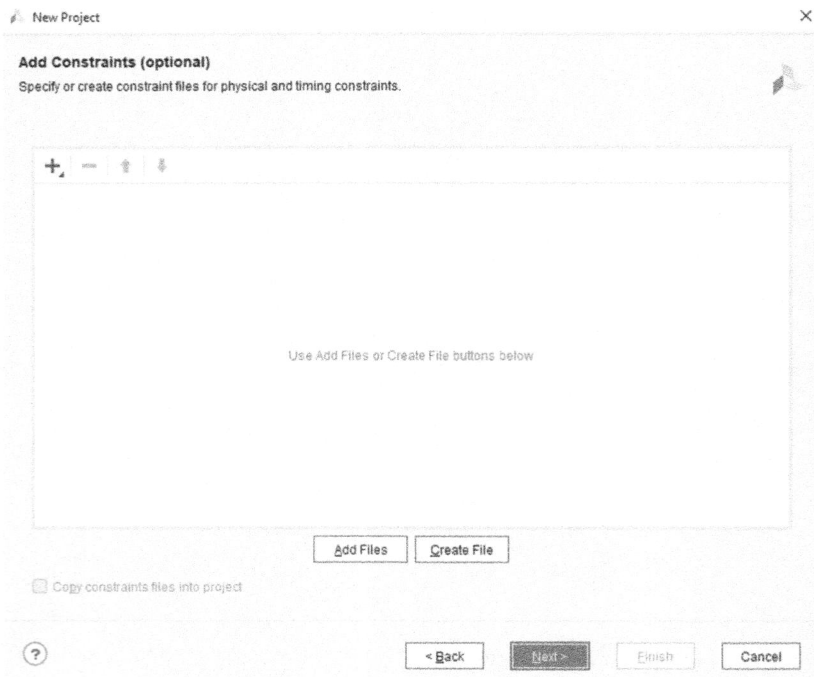

Figure 2.7 Adding a constraint file.

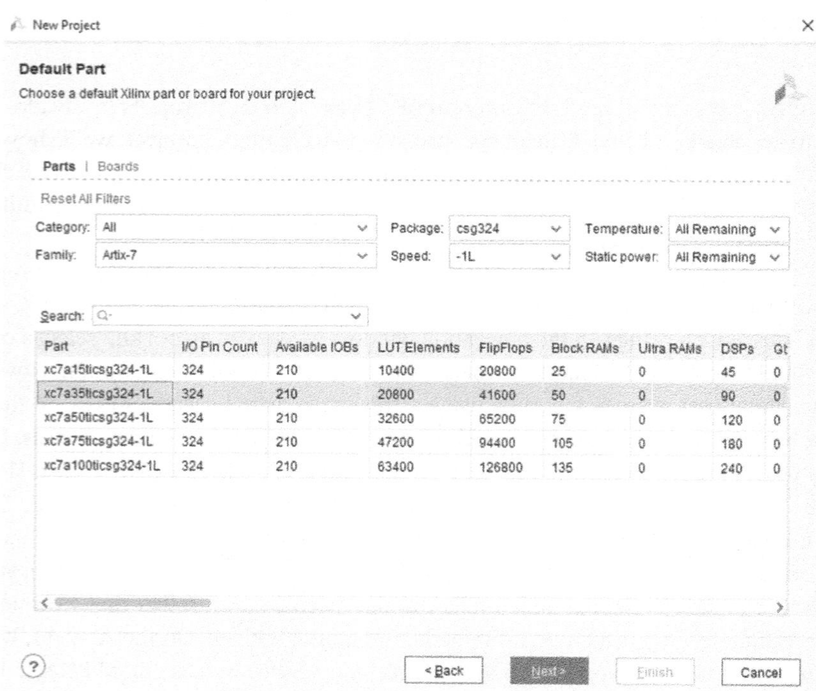

Figure 2.8 Choosing Arty A7 development board.

Look up table elements, and so on and so forth. The big difference between these parts is that you'll notice the xc7a series includes 15t, 35t, 50t, 75t, and 100t. The biggest difference is just the number of resources. The 35t has more than the 15t, but less than the 50t, and so on. For example, you can see that the DSP, which is a good indicator of how powerful your FPGA is, has 90 on the Artix-7 for 35t, but if you go with the Artix-7 for 100t, you have 240. So quite a bit more, however, it does cost more. This gives you a little overview of the number of resources you have on a specific FPGA for your project. Now, go ahead and click on the Next button, and this brings up the New Project Summary, which provides an overview of the project. We're having a new RTL project, the name, and it's giving us a little warning, saying we haven't defined any source files, and that's OK because we're going to go ahead and create the actual VHDL file for this project, as shown in Figure 2.9.

Then we'll click on the Finish button, and it's going to go ahead and create the project at the location we have defined, as shown in Figure 2.10. You'll notice that in our Design Sources folder, we have no files, so we'll need to add one, as depicted in Figure 2.11. However, as we progress through the chapters, we'll explore different libraries, keywords, and others. As we cover these topics, we'll create a VHDL file, walk you through the process, and use that file for the Button_LED project.

So, that is how you create a project using Vivado.

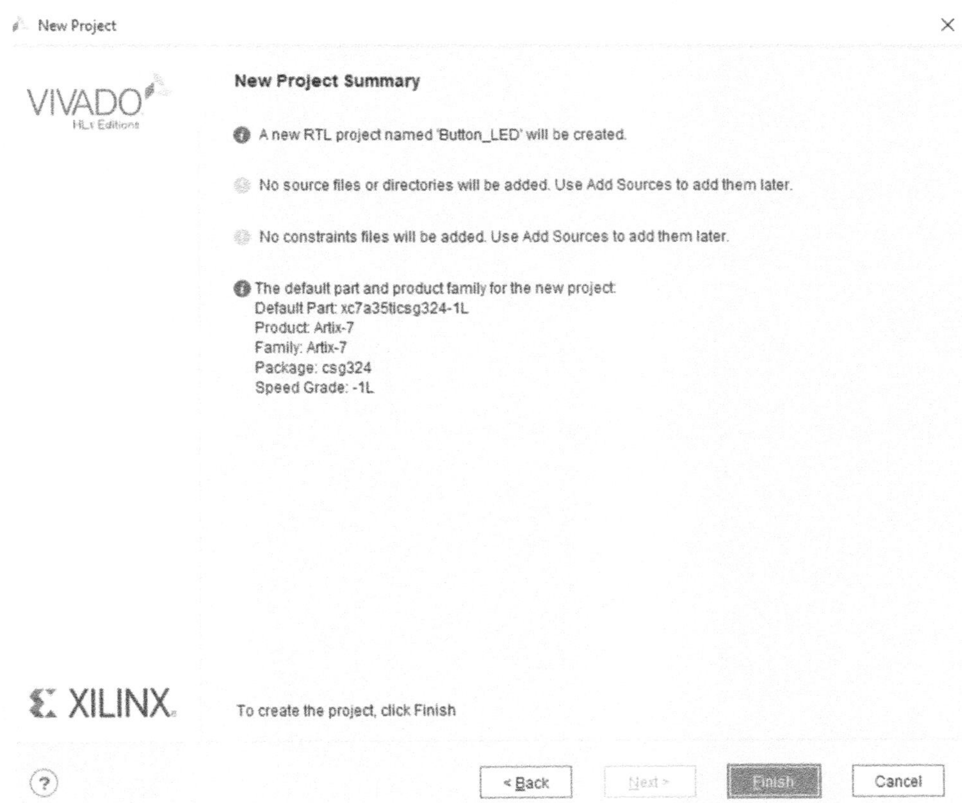

Figure 2.9 The New Project Summary window.

Figure 2.10 Creating the project.

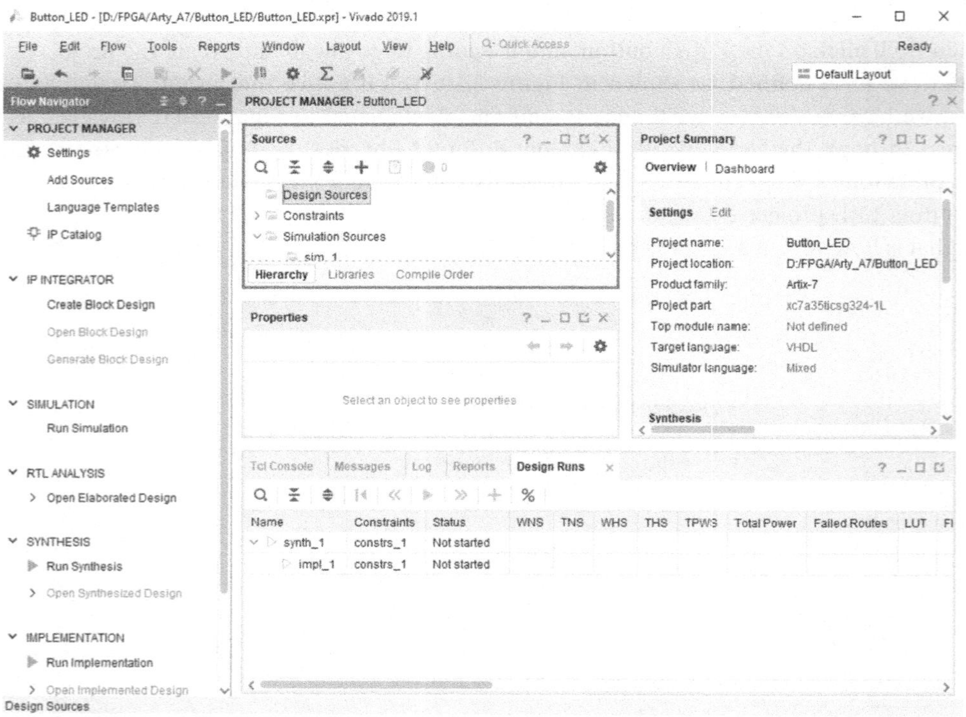

Figure 2.11 The project manager for the Button_LED project.

3

The VHDL Design Structure

3.1 Introduction

VHSIC Hardware Description Language (VHDL) is a hardware description language used in electronic design automation to describe the behavior and structure of electronic systems. The design structure in VHDL is hierarchical and modular, allowing designers to build complex systems by combining smaller, reusable components. The key elements of VHDL design structure are entities, architectures, configurations, and libraries/packages.

3.2 The VHDL Keywords

In this section, we will discuss the VHDL keywords and syntax. Therefore, we're going to explain the commonly used VHDL syntax and keywords. In the VHDL language, there are many different keywords and syntax that you need to perform certain actions. However, we're only going to cover the ones that you'll need to complete the projects in this book. So, the first keyword we're going to cover is "comment." A comment is similarly used as any other type of programming language. You'll put this in your actual code to kind of guide you to explain what's happening. Often, at the beginning of your VHDL code, you'll have a bunch of comments that explain who wrote the code, its version, or any other type of notes. A comment is denoted by a dash-dash (--) in the VHDL. The next keyword is "library," which is used to designate the specific library you're going to use in VHDL. Different libraries have different packages, and these packages provide different functionalities, such as functions, data types, and different operations, whether it's logical or math operations. If you're a C programmer or an embedded type of programmer, this is akin to a header file or a .h type of file. We also use a keyword, which is spelled out like it sounds "use," which makes a package available. So, we want to define a library and the packages we want to use in that library, for example, the following codes.

```
--This is a comment
library IEEE;
use IEEE.STD_LOGIC_1164.all;
```

The next keyword is "entity." In VHDL, an entity defines the inputs and outputs of your design. So, if you think of VHDL as a shell, the outside represents the entity, while the internal structure is defined by another keyword we are going to cover, "architecture." So, you have your entity that defines the inputs and outputs, whereas the architecture defines what's going on inside your design. The keyword is defined by spelling out "architecture," for example, consider the following codes.

Fundamentals of VHDL for FPGA Programming Using Vivado, First Edition. Majid Pakdel.

Companion Website: www.wiley.com/go/pakdelVHDL

```
entity example is
end example
architecture test of example
```

The next one is "generic," which allows you to pass certain parameters into your design without actually having to define them as an input or output – it's a generic parameter. For example, if you want to have an 8-bit, 16-bit, or 32-bit adder, you could have a generic parameter that you call the size of the adder and give it a number, and you don't have to define it as an input or output; for example, consider the following code.

```
entity example is
generic(
);
end example
```

The next keyword is "port," which defines the interface. Inside the entity, you have a port that defines whether it's an input (in), an output (out), or an input/output (inout), which means that it can be an input or an output, allowing data to be sent and received on that same line. The next keyword is "downto," which is used to define a range. If you have a specific data type, which we will cover later, you can use "downto" to say, OK, we have a data type that goes from 31 down to 0 and that just fills all the gaps in between. So, that means you have 31, 30, 29, 28, 27 all the way down to 0; for example, the following code.

```
entity example is
Port (
     My_input  : in std_logic;
     My_Output : out std_logic_vector(31 downto 0);
);
end example
```

Just as it sounds, if you want to go from 7 down to 0, use the "downto" keyword. The keywords "begin" and "end" are used as a begin–end pair, which are used kind of all over the VHDL language, when defining your architecture, you use a begin–end pair to specify the beginning and end of the architecture's logic. Similarly, if you're using an if statement, a loop, or any other type of statement, you'll say, OK, if this happens again and then it'll look for the end. So, if you have a begin–end pair as opposed to other languages, and if you're commonly working with C or any other type of language, you'll have curly braces ({}). Instead of that, we have a begin–end pair in VHDL. The next keyword is "end," and like we said before, you have "begin," and you have "end," and this just specifies the end of your "begin–end" pair, which are used in various ways. The next keyword here is "map," which is used to map parameters. So, if you have multiple designs inside one VHDL, you might have various files, and we'll do what we call a component instantiation. This is where you'll map certain parameters. You'll say, OK, the inputs from a top-level entity are mapped to different inputs on my components, which we have instantiated inside our design. We'll cover this much later in much greater detail, showing you how to do a component instantiation, how to map the certain inputs to the other inputs of your other design, and much more. Another important keyword is signal assignment, which is represented by the arrow equals, and this is assigned as a signal assignment. So, if we have signal A and signal B, and we're assigning about A equal to the value of B, then you can think of it as similar to working with a circuit, it's like taking a wire and going from A to B, and B is now equal to A. It's like connecting two points with a wire. Therefore, we're assigning the signal B, the value of A, as shown in the following codes.

```
architecture test of example
begin
-- Put your architecture code/design here
map -- this is used in component instantiations
My_input <= my_Output(1);
A <= B;
end test;
```

Now, we're going to look at our project. We have our Button_LED project, and we are going to go through and create a VHDL file using some keywords to start defining our project. So, if you open up your Button_LED project in Vivado, you'll have to go into Design Sources folder, right click, and select Add Sources... option, as shown in Figure 3.1.

So, this will bring up the Add Sources window, and we need to make sure that we have the second option selected, the "Add or create design source" option, and then we click Next, as shown in Figure 3.2.

Then, in Figure 3.3, we need to click on Create File because we're going to create a brand-new VHDL design file. We could hit Add Files and go out and pull a previously created VHDL file; however, we're going to create our own new VHDL file, so we'll select the Create File option, as depicted in Figure 3.3.

In the file type option, we want it to be VHDL. You can select other options, such as Verilog, but we're going to select VHDL. We want to give the file a name, and we're just going to call it BUTTON_LED. Typically, the file name doesn't have to be what your entity is, but you need to give it a name that kind of makes sense. Since this is our Button_LED project, we're going to call it

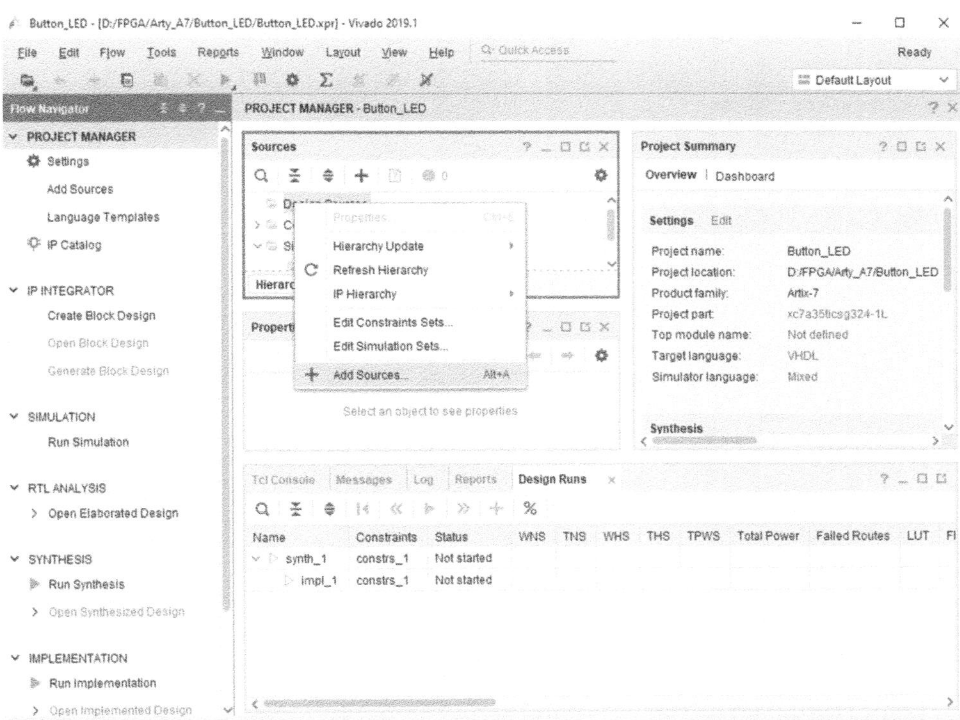

Figure 3.1 Adding sources in Design Sources folder.

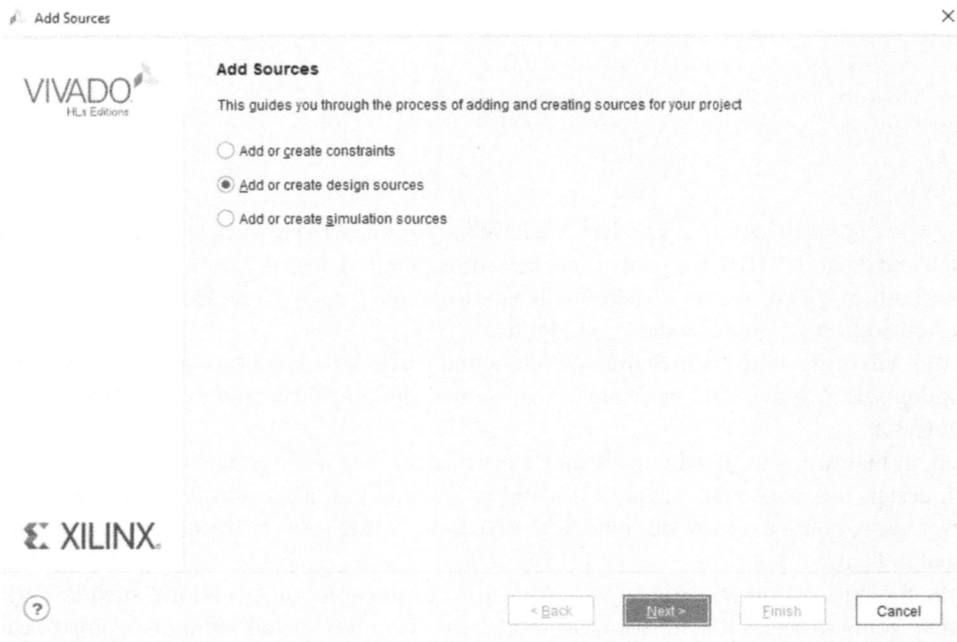

Figure 3.2 Selecting the Add or create design sources option.

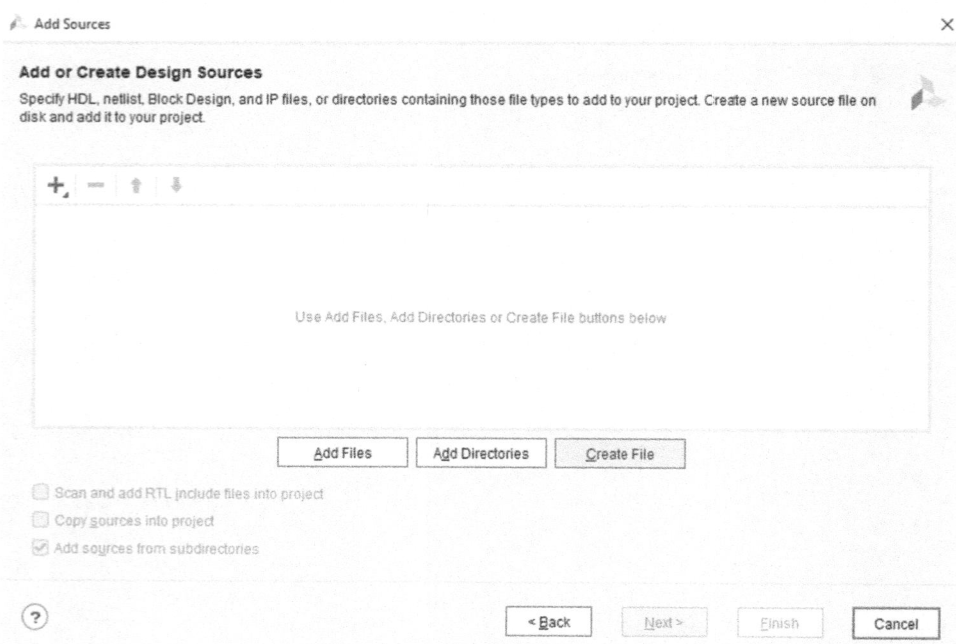

Figure 3.3 Selecting the Create File option.

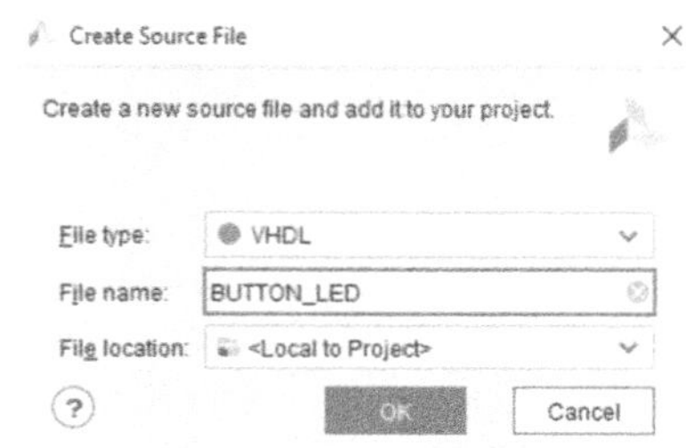

Figure 3.4 Calling the file name as BUTTON_LED.

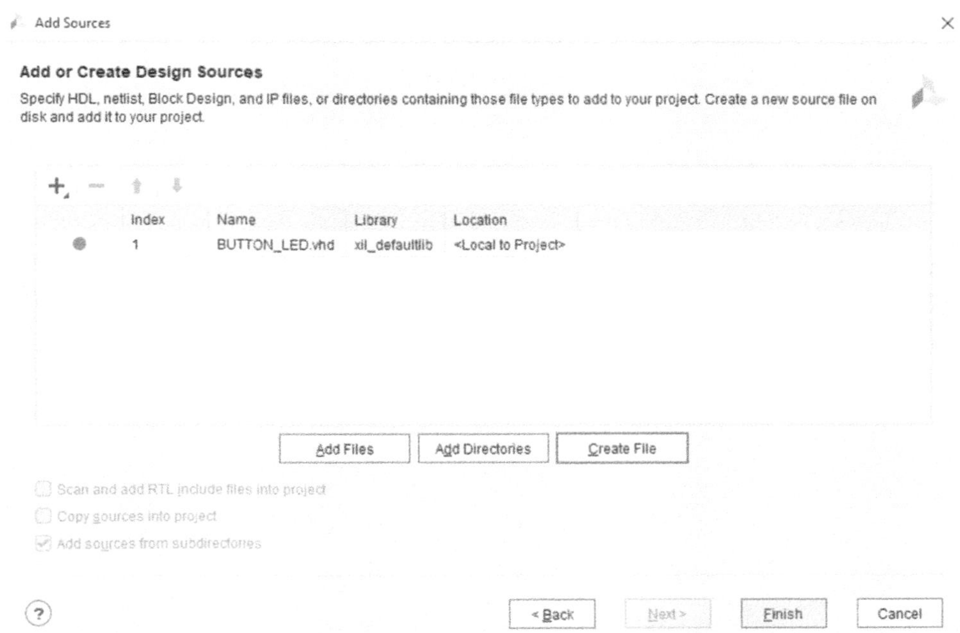

Figure 3.5 Clicking on the Finish button.

BUTTON_LED. For the file location, you can create a new location, choose an existing location, or navigate to where you want to save that file. However, we always like to have them local to project, as illustrated in Figure 3.4.

Therefore, when you go through and synthesize and do everything else with the design, it'll know where to find it. So, we'll click OK, and this is just telling us we're going to create the file BUTTON_LED, and we'll click on the Finish button, as shown in Figure 3.5. Once completed, a pop-up will appear asking, "Do you want to define the module?"

This is where we can define specific ports, specifying whether they should be inputs or outputs. However, in this case, we're going to manually enter them later. So, for now, we are going to click OK and leave it as it is, as shown in Figure 3.6.

By clicking on the OK button, the Define Module window appears, indicating that the module definition has not been changed. At this point, we click Yes, as depicted in Figure 3.7.

So, if we look at our BUTTON_LED file, it indicates behavioral because that's our architecture, which we'll cover later. But for now, we go ahead and double-click on our BUTTON_LED file. This will open the actual BUTTON_LED file, as shown in Figure 3.8. In this figure, you can see that

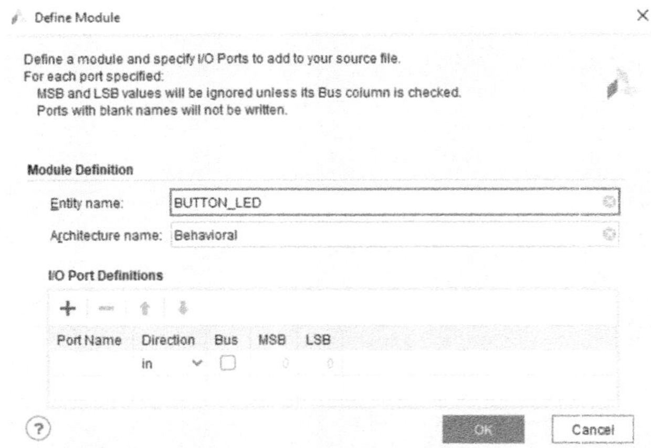

Figure 3.6 Clicking on the OK button.

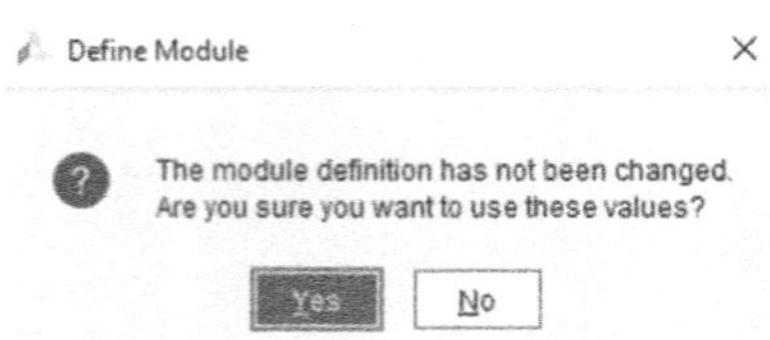

Figure 3.7 Clicking on the Yes button.

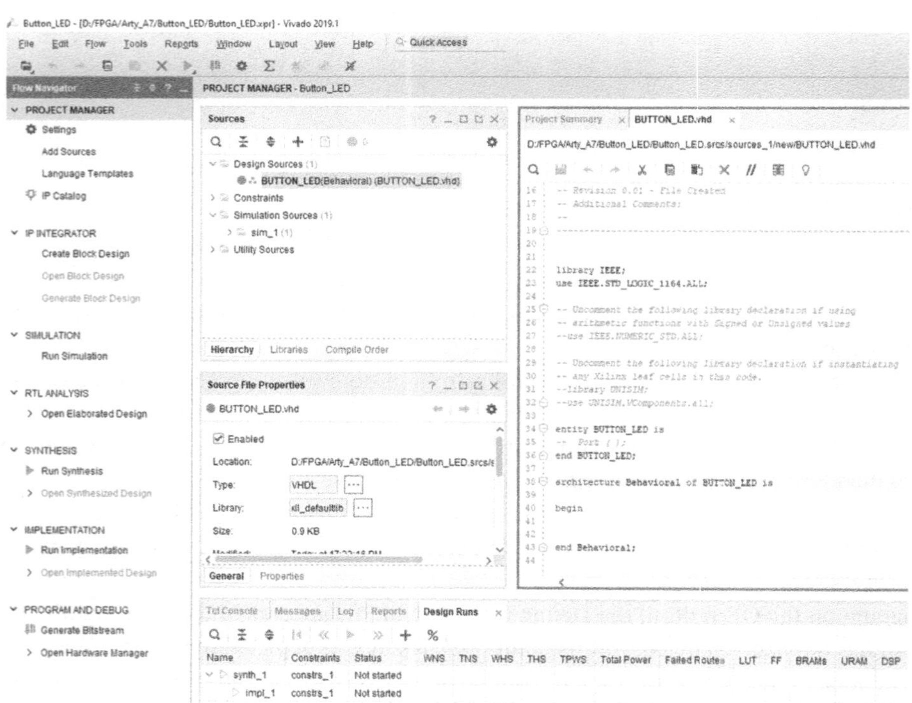

Figure 3.8 Bringing up the actual BUTTON_LED file.

there's already some syntax provided so that they define and give you this top-level type of description, and you can see the keywords we have are the comments.

We've got this library, which gives us the IEEE library, and we're using the IEEE standard logic 1164 package. This gives us an entity named BUTTON_LED. We have to define the port – what the inputs and outputs are. Then it also defines the architecture for us. So, it gives us the architecture, which we're calling behavioral for our BUTTON_LED and then it says begin and end behavioral. So, this is just the rough shell of our entity and architecture for our BUTTON_LED file, and so that is just a covering or summary of the keywords you'll see in VHDL. As you can see, once we create this file in Vivado, you can see a lot of these keywords.

3.3 The VHDL Libraries

This section is about the VHDL packages and libraries. In this section, we're going to tell you about certain packages or the most commonly used ones in the IEEE library. IEEE stands for the Institute of Electronics and Electrical Engineers, and they're the ones who manage the packages we'll be using in our project, individual language, or in the IEEE library. There are quite a few different packages, but we're going to cover the most common ones, specifically the ones that you'll need to complete all the projects in this book. These libraries or packages include the IEEE standard logic 1164, the IEEE math real, and the IEEE numeric standard packages. So, for the first one, the IEEE 1164, it has some key features. So, we'll review some of the key features for each library, and then we'll show you where to go to actually see all of the definitions and the actual code behind the library. The logical operations that are supported in the standard logic 1164 library include AND, OR, NAND, NOR, and all the basic logic gate operations. So, by using this package, it allows us to do all of these basic Boolean algebra and different logic operations. The data types it supports are standard logic and standard logic vector. These are the most common and basic data types used in VHDL, which we'll cover later in the data type section. The third most important feature of the standard logic 1164 library is the functions it supports, such as rising edge and falling edge. These functions work when you have a process. You'll use the rising edge function to trigger that process on the rising edge of a signal, and conversely, with the falling edge function, when you have a process and you call the falling edge, you'll trigger that process on the falling edge of that signal. The next library we're going to talk about a little bit is the IEEE math real library. Some key library features are that it basically supports some mathematical constants and operations. We won't use a whole lot of the math real library, but the biggest functions that we're going to use from here are the ceiling function and the floor function. So, with the ceiling function, you give it a possible number, and it takes you to the nearest integer that represents that number. The floor function does the opposite: you give it a number, and it'll give you the lowest integer used to represent that number. Now, in the math real library, the two different types of supports are integers and reals, where reals are floating-point numbers. However, if you're using this math real library, you can't use real data types when you want to synthesize your design on your FPGA. So, when we're using this, you can use it for simulation purposes. For example, if you want to simulate and have a value of π, it'll represent the value of π given a constant. But you can't actually synthesize that in your design. So, in all the designs we use, the only functions we will actually use from this library are the ceiling and floor. The next library or package we have is the numeric standard. Some key features of this library are that it supports logical operations such as greater than, less than, and greater than or equal to. The data types are unsigned and signed, which is key. The unsigned data type is very similar to a standard logic vector, except it gives you the ability to do different operations, such as addition, subtraction, multiplication, and division. You can still access the data at the bit level. The signed data type allows you to represent negative numbers. So, if you're doing something like a counter, you can count from zero and go negative or count upward. That's the key feature of the numeric standard library. Now, let's go ahead and show you where to go on your computer to find more libraries, and also take a closer look to understand what each one of these libraries provides at a more detailed level. First, we'll take a look at the standard logic 1164 library, and you can find this on the website https://standards.ieee.org/downloads/1076/1076.2-1996/std_logic_1164.vhdl. In there, you'll find a bunch of different libraries or packages in the IEEE library that you can use for your VHDL designs. However, as we mentioned, these three are the most common ones we'll be using. But if you visit the website, you'll find the required information. When we say we're going to use the package standard object 1164, it tells us the different data types that we have. Here, they give you a standard ulogic, which is the same as a standard logic except that the standard logic is a result of data type. So, if you have multiple drivers on your standard logic, you can resolve it to a single value, and this means, for example, in the standard ulogic, we have multiple values that it can take. It can be a U, which is uninitialized. You can have an X, which represents either an unknown value or a logic 0; a 1, which represents logic 1; and a Z, indicating high impedance. Additionally, W stands for a weak unknown, L for a weak 0, and H for a weak 1 – these values are similar to tri-state logic levels. So, you have a weak 1 or weak 0 and a dash is don't care, as illustrated in Figure 3.9.

So, if you have a standard ulogic, and multiple signals are driving an output, for example, if you have one signal driving a zero and the other driving it a weak high, a standard ulogic, if you're using Vivado or any sort of design tool, will give you an error saying you have multiple drivers for that signal. Now, if you use a standard logic, it will resolve that to a specific value. For example, if you're forcing it to a 1 and you have a weak 0, the 1 will most likely win, and the signal will be

std_logic_1164 - Notepad
File Edit Format View Help

```
-- -------------------------------------------------------------------

PACKAGE std_logic_1164 IS

    -------------------------------------------------------------------
    -- logic state system  (unresolved)
    -------------------------------------------------------------------
    TYPE std_ulogic IS ( 'U',  -- Uninitialized
                         'X',  -- Forcing  Unknown
                         '0',  -- Forcing  0
                         '1',  -- Forcing  1
                         'Z',  -- High Impedance
                         'W',  -- Weak     Unknown
                         'L',  -- Weak     0
                         'H',  -- Weak     1
                         '-'   -- Don't care
                       );
    -------------------------------------------------------------------
    -- unconstrained array of std_ulogic for use with the resolution function
    -------------------------------------------------------------------
    TYPE std_ulogic_vector IS ARRAY ( NATURAL RANGE <> ) OF std_ulogic;

    -------------------------------------------------------------------
    -- resolution function
    -------------------------------------------------------------------
    FUNCTION resolved ( s : std_ulogic_vector ) RETURN std_ulogic;


    -------------------------------------------------------------------
    -- *** industry standard logic type ***
    -------------------------------------------------------------------
    SUBTYPE std_logic IS resolved std_ulogic;

    -------------------------------------------------------------------
```

Ln 1, Col 1 100% Unix (LF) UTF-8

Figure 3.9 The standard ulogic data types.

forced to that value. If you continue down, it gives you some more details here. It goes through the functions, and you can follow this all the way through, but we're just going to skim through and point out the important parts of the file when you're looking at the package, trying to review it and determine if you'll need it for your design. They give you the subtypes, and this tells you what the value will result in. In this case, obviously, you can see we have the data type standard logic, as well as a standard logic vector, which is an array of standard logic, as shown in Figure 3.10.

Then, here we have the logical operations it supports: "and," "nand," and all the way down to "xor," "xnor," and "not" or inverter. In addition, they give us some conversion functions, so if you want to convert to a bit vector or to a standard ulogic, it gives you the functions that you need to call in order to do these types of conversions, as shown in Figure 3.11.

Also, they give you the functions at the very bottom, where we have our edge detection functions, including rising edge and falling edge, as shown in Figure 3.12.

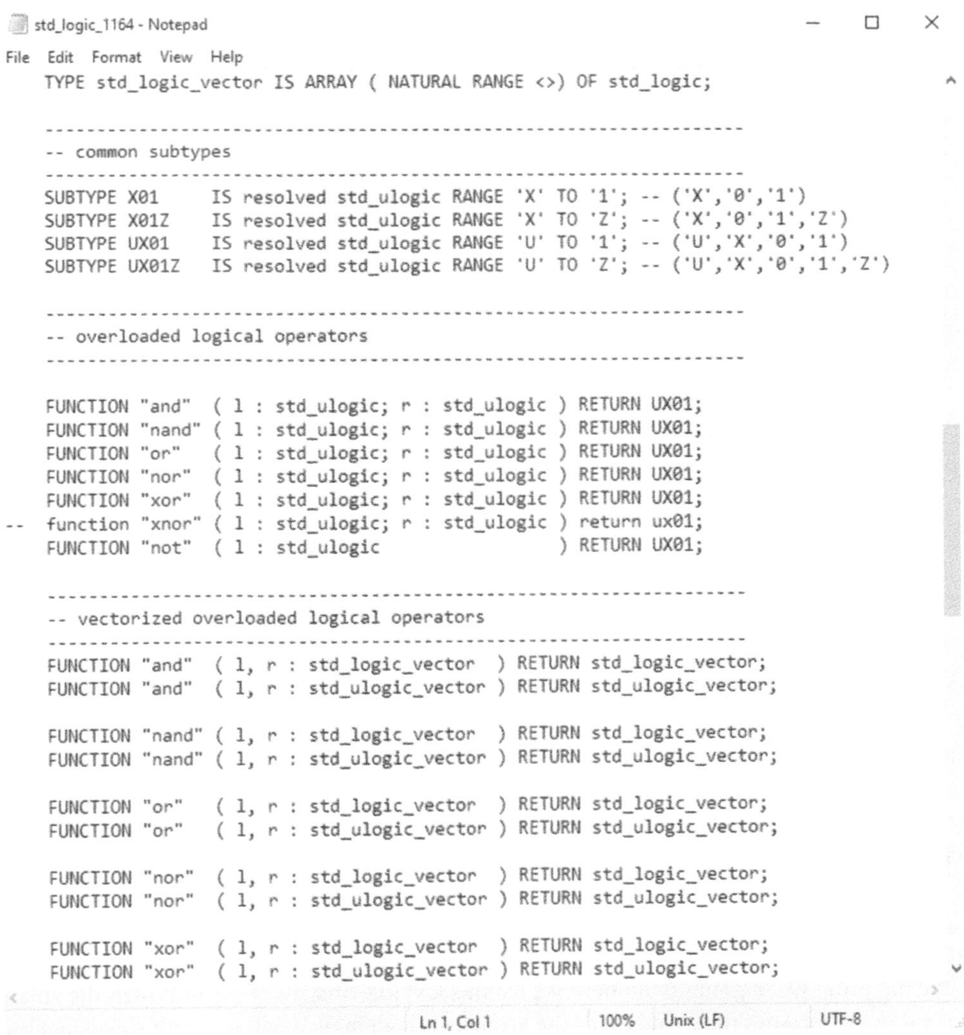


```
    TYPE std_logic_vector IS ARRAY ( NATURAL RANGE <>) OF std_logic;

    -------------------------------------------------------------------
    -- common subtypes
    -------------------------------------------------------------------
    SUBTYPE X01     IS resolved std_ulogic RANGE 'X' TO '1'; -- ('X','0','1')
    SUBTYPE X01Z    IS resolved std_ulogic RANGE 'X' TO 'Z'; -- ('X','0','1','Z')
    SUBTYPE UX01    IS resolved std_ulogic RANGE 'U' TO '1'; -- ('U','X','0','1')
    SUBTYPE UX01Z   IS resolved std_ulogic RANGE 'U' TO 'Z'; -- ('U','X','0','1','Z')

    -------------------------------------------------------------------
    -- overloaded logical operators
    -------------------------------------------------------------------

    FUNCTION "and"  ( l : std_ulogic; r : std_ulogic ) RETURN UX01;
    FUNCTION "nand" ( l : std_ulogic; r : std_ulogic ) RETURN UX01;
    FUNCTION "or"   ( l : std_ulogic; r : std_ulogic ) RETURN UX01;
    FUNCTION "nor"  ( l : std_ulogic; r : std_ulogic ) RETURN UX01;
    FUNCTION "xor"  ( l : std_ulogic; r : std_ulogic ) RETURN UX01;
--  function "xnor" ( l : std_ulogic; r : std_ulogic ) return ux01;
    FUNCTION "not"  ( l : std_ulogic                 ) RETURN UX01;

    -------------------------------------------------------------------
    -- vectorized overloaded logical operators
    -------------------------------------------------------------------
    FUNCTION "and"  ( l, r : std_logic_vector  ) RETURN std_logic_vector;
    FUNCTION "and"  ( l, r : std_ulogic_vector ) RETURN std_ulogic_vector;

    FUNCTION "nand" ( l, r : std_logic_vector  ) RETURN std_logic_vector;
    FUNCTION "nand" ( l, r : std_ulogic_vector ) RETURN std_ulogic_vector;

    FUNCTION "or"   ( l, r : std_logic_vector  ) RETURN std_logic_vector;
    FUNCTION "or"   ( l, r : std_ulogic_vector ) RETURN std_ulogic_vector;

    FUNCTION "nor"  ( l, r : std_logic_vector  ) RETURN std_logic_vector;
    FUNCTION "nor"  ( l, r : std_ulogic_vector ) RETURN std_ulogic_vector;

    FUNCTION "xor"  ( l, r : std_logic_vector  ) RETURN std_logic_vector;
    FUNCTION "xor"  ( l, r : std_ulogic_vector ) RETURN std_ulogic_vector;
```

Figure 3.10 The subtype and function data types.

std_logic_1164 - Notepad

File Edit Format View Help

```
-------------------------------------------------------------------
-- conversion functions
-------------------------------------------------------------------
FUNCTION To_bit       ( s : std_ulogic;        xmap : BIT := '0') RETURN BIT;
FUNCTION To_bitvector ( s : std_logic_vector ; xmap : BIT := '0') RETURN BIT_VECTOR;
FUNCTION To_bitvector ( s : std_ulogic_vector; xmap : BIT := '0') RETURN BIT_VECTOR;

FUNCTION To_StdULogic       ( b : BIT               ) RETURN std_ulogic;
FUNCTION To_StdLogicVector  ( b : BIT_VECTOR        ) RETURN std_logic_vector;
FUNCTION To_StdLogicVector  ( s : std_ulogic_vector ) RETURN std_logic_vector;
FUNCTION To_StdULogicVector ( b : BIT_VECTOR        ) RETURN std_ulogic_vector;
FUNCTION To_StdULogicVector ( s : std_logic_vector  ) RETURN std_ulogic_vector;

-------------------------------------------------------------------
-- strength strippers and type convertors
-------------------------------------------------------------------

FUNCTION To_X01  ( s : std_logic_vector  ) RETURN  std_logic_vector;
FUNCTION To_X01  ( s : std_ulogic_vector ) RETURN  std_ulogic_vector;
FUNCTION To_X01  ( s : std_ulogic        ) RETURN  X01;
FUNCTION To_X01  ( b : BIT_VECTOR        ) RETURN  std_logic_vector;
FUNCTION To_X01  ( b : BIT_VECTOR        ) RETURN  std_ulogic_vector;
FUNCTION To_X01  ( b : BIT               ) RETURN  X01;

FUNCTION To_X01Z ( s : std_logic_vector  ) RETURN  std_logic_vector;
FUNCTION To_X01Z ( s : std_ulogic_vector ) RETURN  std_ulogic_vector;
FUNCTION To_X01Z ( s : std_ulogic        ) RETURN  X01Z;
FUNCTION To_X01Z ( b : BIT_VECTOR        ) RETURN  std_logic_vector;
FUNCTION To_X01Z ( b : BIT_VECTOR        ) RETURN  std_ulogic_vector;
FUNCTION To_X01Z ( b : BIT               ) RETURN  X01Z;

FUNCTION To_UX01  ( s : std_logic_vector  ) RETURN  std_logic_vector;
FUNCTION To_UX01  ( s : std_ulogic_vector ) RETURN  std_ulogic_vector;
FUNCTION To_UX01  ( s : std_ulogic        ) RETURN  UX01;
FUNCTION To_UX01  ( b : BIT_VECTOR        ) RETURN  std_logic_vector;
FUNCTION To_UX01  ( b : BIT_VECTOR        ) RETURN  std_ulogic_vector;
FUNCTION To_UX01  ( b : BIT               ) RETURN  UX01;

-------------------------------------------------------------------
```

Ln 1, Col 1 | 100% | Unix (LF) | UTF-8

Figure 3.11 Some conversion functions.

These are commonly used functions in our designs. The next package we'll look at is the IEEE math_real file. So, it gives us some constant definitions, such as the math constant of MATH_E and math constant of MATH_PI. You can't use "REAL" for synthesis in FPGAs because it's not a synthesizable data type, as shown in Figure 3.13.

So, these are most commonly used when you're wanting to simulate a circuit. You'll be able to do that, and that's why a lot of these constants will come into play. The most common thing we're going to use this library for is specific functions like the ceiling and the floor function. So, the ceiling function returns the smallest integer value as a real number, not less than X, as shown in Figure 3.14.

So, the big thing we're going to do here is give this a value, and it's going to return the smallest integer value to represent that. And that's the key because an integer can be synthesized inside our FPGA, so we can use the function "ceil." Now, we consider the function "floor," which returns the largest integer value not greater than X. If you look at the function "round," it rounds the value to

```
-------------------------------------------------------------------
-- edge detection
-------------------------------------------------------------------
FUNCTION rising_edge  (SIGNAL s : std_ulogic) RETURN BOOLEAN;
FUNCTION falling_edge (SIGNAL s : std_ulogic) RETURN BOOLEAN;

-------------------------------------------------------------------
-- object contains an unknown
-------------------------------------------------------------------
FUNCTION Is_X ( s : std_ulogic_vector ) RETURN  BOOLEAN;
FUNCTION Is_X ( s : std_logic_vector  ) RETURN  BOOLEAN;
FUNCTION Is_X ( s : std_ulogic        ) RETURN  BOOLEAN;

END std_logic_1164;
```

Figure 3.12 Edge detection functions.

```
package MATH_REAL is
    constant CopyRightNotice: STRING
      := "Copyright 1996 IEEE. All rights reserved.";

    --
    -- Constant Definitions
    --
    constant  MATH_E : REAL := 2.71828_18284_59045_23536;
                                                -- Value of e
    constant  MATH_1_OVER_E : REAL := 0.36787_94411_71442_32160;
                                                -- Value of 1/e
    constant  MATH_PI : REAL := 3.14159_26535_89793_23846;
                                                -- Value of pi
    constant  MATH_2_PI : REAL := 6.28318_53071_79586_47693;
                                                -- Value of 2*pi
    constant  MATH_1_OVER_PI : REAL := 0.31830_98861_83790_67154;
                                                -- Value of 1/pi
    constant  MATH_PI_OVER_2 : REAL := 1.57079_63267_94896_61923;
                                                -- Value of pi/2
    constant  MATH_PI_OVER_3 : REAL := 1.04719_75511_96597_74615;
                                                -- Value of pi/3
    constant  MATH_PI_OVER_4 : REAL := 0.78539_81633_97448_30962;
                                                -- Value of pi/4
    constant  MATH_3_PI_OVER_2 : REAL := 4.71238_89803_84689_85769;
                                                -- Value 3*pi/2
    constant  MATH_LOG_OF_2 : REAL := 0.69314_71805_59945_30942;
                                                -- Natural log of 2
    constant  MATH_LOG_OF_10 : REAL := 2.30258_50929_94045_68402;
                                                -- Natural log of 10
    constant  MATH_LOG2_OF_E : REAL := 1.44269_50408_88963_4074;
                                                -- Log base 2 of e
    constant  MATH_LOG10_OF_E: REAL := 0.43429_44819_03251_82765;
                                                -- Log base 10 of e
    constant  MATH_SQRT_2: REAL := 1.41421_35623_73095_04880;
                                                -- square root of 2
    constant  MATH_1_OVER_SQRT_2: REAL := 0.70710_67811_86547_52440;
                                                -- square root of 1/2
    constant  MATH_SQRT_PI: REAL := 1.77245_38509_05516_02730;
                                                -- square root of pi
    constant  MATH_DEG_TO_RAD: REAL := 0.01745_32925_19943_29577;
                                  -- Conversion factor from degree to radian
    constant  MATH_RAD_TO_DEG: REAL := 57.29577_95130_82320_87680;
                                  -- Conversion factor from radian to degree
```

Figure 3.13 Some constant definitions.

```
function CEIL (X : in REAL ) return REAL;
    -- Purpose:
    --         Returns smallest INTEGER value (as REAL) not less than X
    -- Special values:
    --         None
    -- Domain:
    --         X in REAL
    -- Error conditions:
    --         None
    -- Range:
    --         CEIL(X) is mathematically unbounded
    -- Notes:
    --         a) Implementations have to support at least the domain
    --                 ABS(X) < REAL(INTEGER'HIGH)

function FLOOR (X : in REAL ) return REAL;
    -- Purpose:
    --         Returns largest INTEGER value (as REAL) not greater than X
    -- Special values:
    --         FLOOR(0.0) = 0.0
    -- Domain:
    --         X in REAL
    -- Error conditions:
    --         None
    -- Range:
    --         FLOOR(X) is mathematically unbounded
    -- Notes:
    --         a) Implementations have to support at least the domain
    --                 ABS(X) < REAL(INTEGER'HIGH)

function ROUND (X : in REAL ) return REAL;
    -- Purpose:
    --         Rounds X to the nearest integer value (as real). If X is
    --         halfway between two integers, rounding is away from 0.0
    -- Special values:
    --         ROUND(0.0) = 0.0
    -- Domain:
    --         X in REAL
    -- Error conditions:
    --         None
    -- Range:
    --         ROUND(X) is mathematically unbounded
    -- Notes:
    --         a) Implementations have to support at least the domain
    --                 ABS(X) < REAL(INTEGER'HIGH)
```

Figure 3.14 Using library for specific functions.

the nearest integer. As long as it returns an integer, we can use it, but if it returns a real number, we cannot. The next library we're going to look at in the package is the numeric standard, and the big thing about the numeric standard is they have the data types they define, such as unsigned and signed, and then also we have different operations that they support, such as absolute value, subtraction, addition, subtraction again, multiplication, division, as well as remainders, shifting, and modulus, as shown in Figure 3.15.

So, this gives you the ability to check if one signal is greater than the next, which is important depending on your design and what you're doing. If you only have standard logic, like if you're just using the standard logic 1164 library, you can't perform a greater than or an addition operation. So, if you want to do an addition or greater than type of operation, you need to use the numeric standard package so that you can define an unsigned data type and perform addition, multiplication, or any other type of arithmetic operation you'd like. Now, let's look at our project; for example, here in Vivado, notice that we have a keyword library, IEEE. This is the library we're using, and for the Button_LED project, we're only using the package by IEEE. STD_LOGIC_1164.ALL. This gives us access to create standard logic data types and standard logic vector data types.

numeric_std - Notepad

File Edit Format View Help

```
package NUMERIC_STD is
  constant CopyRightNotice: STRING
      := "Copyright © 1997 IEEE. All rights reserved.";

  --============================================================================
  -- Numeric Array Type Definitions
  --============================================================================

  type UNSIGNED is array (NATURAL range <>) of STD_LOGIC;
  type SIGNED is array (NATURAL range <>) of STD_LOGIC;

  --============================================================================
  -- Arithmetic Operators:
  --============================================================================

  -- Id: A.1
  function "abs" (ARG: SIGNED) return SIGNED;
  -- Result subtype: SIGNED(ARG'LENGTH-1 downto 0)
  -- Result: Returns the absolute value of a SIGNED vector ARG.

  -- Id: A.2
  function "-" (ARG: SIGNED) return SIGNED;
  -- Result subtype: SIGNED(ARG'LENGTH-1 downto 0)
  -- Result: Returns the value of the unary minus operation on a
  --         SIGNED vector ARG.
```

Ln 1, Col 1 100% Unix (LF) ANSI

Figure 3.15 The numeric standard data types.

3.4 The VHDL Entities and Architecture

This section is about the VHDL entities and architecture. Here, we're going to explain what VHDL entity and architecture are. Every design of VHDL requires one entity and one architecture. At a high level, the entity defines the overall structure of your design, including its inputs and outputs. So, when you have a design, you can think of what the inputs and outputs are?

The entity defines these inputs and outputs. Inside the entity, you define if it is an input, output, or bidirectional. These are the three types of modes. You can have either an input/output or an "inout," which is a bidirectional type of signal for your entity. When you define the entity, it's structured as follows: you have the name, the direction (which can be "in," "out," or "inout"), the data type, and optionally, you can define the initial value. For example, if you have an output, you would define its name as C, set it as an output, specify the data type (which we'll call standard logic), and give it an initial value, like zero (logical low or false). Now, let's go ahead and take a look at the entity example. We're going to start with a very basic example of an "AND" gate. So, if we have an "AND" gate, we have two inputs, A and B, and one output, which we will call C. This simple "AND" gate can be represented in VHDL, as shown in Figure 3.16.

Our entity is our inputs and outputs. So, we know that we've got our inputs, which are A and B, and our output, which is a C, defined in our entity. This is what we want to define inside our entity and how do we do that? Well, we would use the "entity" keyword as shown in the following codes.

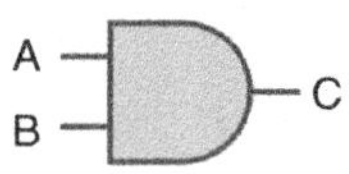

A	B	C
0	0	0
0	1	0
1	0	0
1	1	1

Figure 3.16 A very simple "AND" gate.

```
entity AND_GATE is
Port (
     A : in std_logic;
     B : in std_logic;
     C : out std_logic);
end AND_GATE;
```

So, this entity is defining the inputs and outputs of our design. For inputs, we know we have A and B; for the output, we have C. The data type tells us the bit, so we know we have two bits going into A and B and one bit coming out as C. The next part we want to talk about is the architecture of a design. In VHDL, architecture defines how your design operates. So, while the entity defines the inputs and outputs, the architecture is going to define and explain how this actually works. In the case of AND gate, which we used, we know we have two inputs A and B, and our architecture is going to tell us to AND those inputs logically to give us our output C. For example, in the Button_LED project, our architecture will define how pressing a button, which is our input, controls the LEDs, which is our output, where our architecture needs to define how we are going to process the inputs to determine the outputs. Because we also have a switch that either enables or disables the LEDs when the button is pressed, that's the type of stuff that will go on inside your architecture. So, you have a very complex VHDL design. Almost all of your VHDL is going to be inside that architecture. This is where the core logic of the design resides. Now, let's take a look back to the example of our AND gate and define its architecture. To do this, we use the keyword "architecture," as shown in the following codes.

```
Architecture behave of AND_GATE is
begin
    C <= A AND B;
end behave;
```

So, here we have effectively our entirety, our entity and the architecture of our AND gate design. We have our AND gate, which we define through the entity, telling us that A and B are inputs, and C is an output. We have our architecture, which defines the actual AND gate itself. So, our entity defines the inputs and outputs, and our architecture is inside the design, describing what's actually happening. So next, let's go ahead and open up the Button_LED project on your computer. We will take a look and see that we have the entity for our design, which is a defined entity, and we've named it BUTTON_LED. What we want to do is uncomment the port code, as we need to define the ports for our design. We are going to have what we call Led_Out, so we'll use the keyword "out" and specify that it is an output. Then, we'll have to give it a data type, and once we get to the data type section, we'll decide what data type makes the most sense. The next keyword or entity we want to describe is our input. We know we have a button, so Button_In will be the input, and this is going to be a bit of a type "in." This is going to be an input. We know we have one output, which is Led_Out, and we have two inputs: Button_In and Enable. This is what's going to enable or disable the output of the LEDs, and this is going to be input as well. Right now, we have some squiggly lines telling us that our entity is not properly defined, and that's because we have not defined the data type yet. We will get to that when

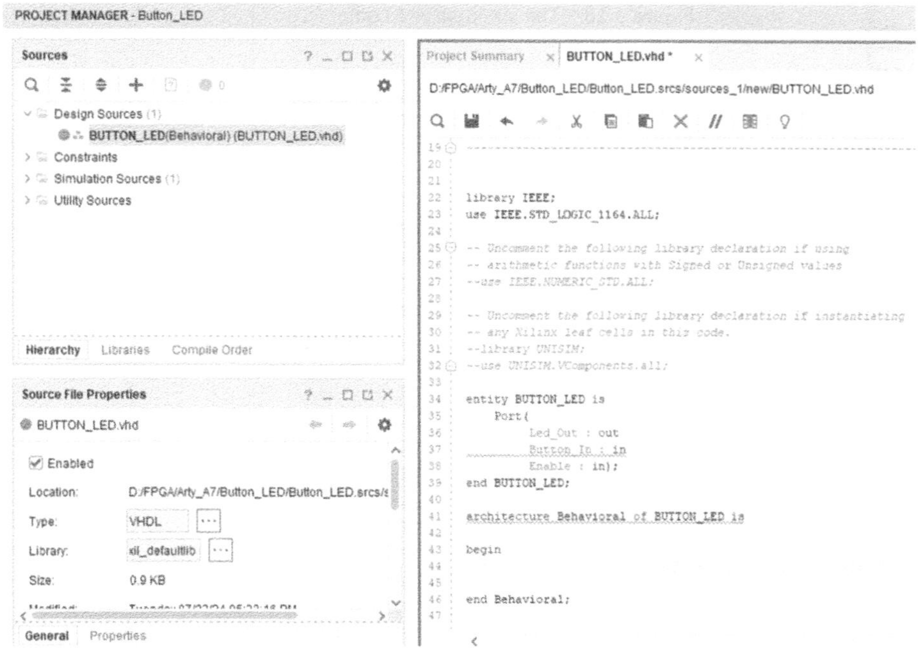

Figure 3.17 Adding Port to the entity "BUTTON_LED."

we cover the data types. We will define the data type for Led_Out, Button_In, and Enable. This defines our entity up to the point of what we need, as well as the direction of the signals. For the architecture, it gave us an architecture called "Behavioral" of BUTTON_LED, and then it gives us the "begin" and "end Behavioral." In the middle, we add the actual design, where we take Button_In and Enable and define what Led_Out should be based on these two inputs. When we cover that section, we'll put it there in the architecture of our design, as shown in Figure 3.17.

Therefore, that covers the entity and architecture section in this book.

3.5 The VHDL Generics

This section is about VHDL generics, and we're going to explain what a VHDL generic is and how we can use them in our designs. The VHDL generic provides a way of passing information into an entity externally. Within an entity, we have our inputs and outputs. However, with a generic, you don't have to define it as an input or an output; you can think of a generic as a constant. You can apply it externally, which is a huge benefit in that it allows you to customize your designs in a way that you can modify one line of code or essentially one value in your generic to change your design. We'll be using generics throughout our designs, and in any future designs that you work on, you'll most likely use them. They are a very powerful tool in a way, which lets you modify your design or defiant in a way that is very easy to customize. For example, let's take a look at what's taken and generate a generic. If we want a design that implements an AND gate, but with three inputs on each side, so a total of six inputs, we'll define two signals for A and B, each having three inputs. We then want to combine all of these into the AND gate, which will give us a single output, as illustrated in Figure 3.18.

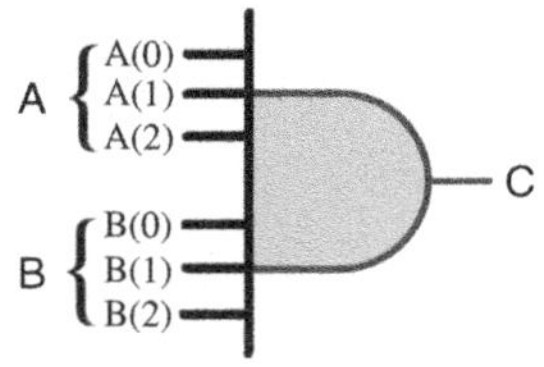

Figure 3.18 The six-input AND gate.

So, we'll take the top three as our A signal, and the bottom three as our B signal, with C being our output. So, in order to define this, we first need to make sure that we define our generic and the entity, as well as the architecture, as shown in the following codes.

```
entity Example is
Generic(
    Bits : integer := 3);
Port(
     A : in std_logic_vector(Bits-1 downto 0);
     B : in std_logic_vector(Bits-1 downto 0);
     C : out std_logic);
end Example;
architecture behave Example is
begin
   C <= A AND B;
end behave;
```

Now, this code defines the following AND gate, where we have a total of six inputs: A and B. A has three inputs and B has three inputs. Now, if we wanted to make this a four-input or a five-input AND gate, instead of having to modify any other piece of code, all we have to do is change this generic value. We can reference the Bits throughout our design – example here being in our port – to define the number of inputs we have, which is very powerful because, otherwise, if we wanted to modify this design, we would then have to go and come in the actual port and modify here. What's even more powerful is that when we do things like component instantiations, we can pass this generic into our design. The Bits is from a higher-level design and trickles down, so you don't even have to make any modifications to this specific design. This is an example of using a generic, and the key thing to note is that by changing just this one line, this generic value of 3, you can change the input width of your A and B signals. This makes it a lot easier. Now, in this case, it's a very simple design, but when you're working with complex designs that have lots of things going on, you might have more than just one generic – you could have two, three, four, or five. Of the available generics, you just change those three, four, or five values and it changes the way your design operates. A lot of times, the most common things you'll see in generics are, for example, like Bits. If you're working with different designs, such as the bus width (the number of data bits being transferred), the number of address bits, the clock rate, or the intended output clock rate. So, there are many things you can define as a generic, but it's a very powerful tool and something that you'll definitely be using a lot as you work through different designs in VHDL. So now, what we want to do is open up your Button_LED design, and here, we want to have a generic that we will be using. The generic we're going to have will define the number of buttons we have, because for every button, we want to have an LED to correspond to that button. So, as opposed to physically manually changing that, we're going to give it a generic. So, in order to do that, we have to define the generic before we define our port, as shown in Figure 3.19.

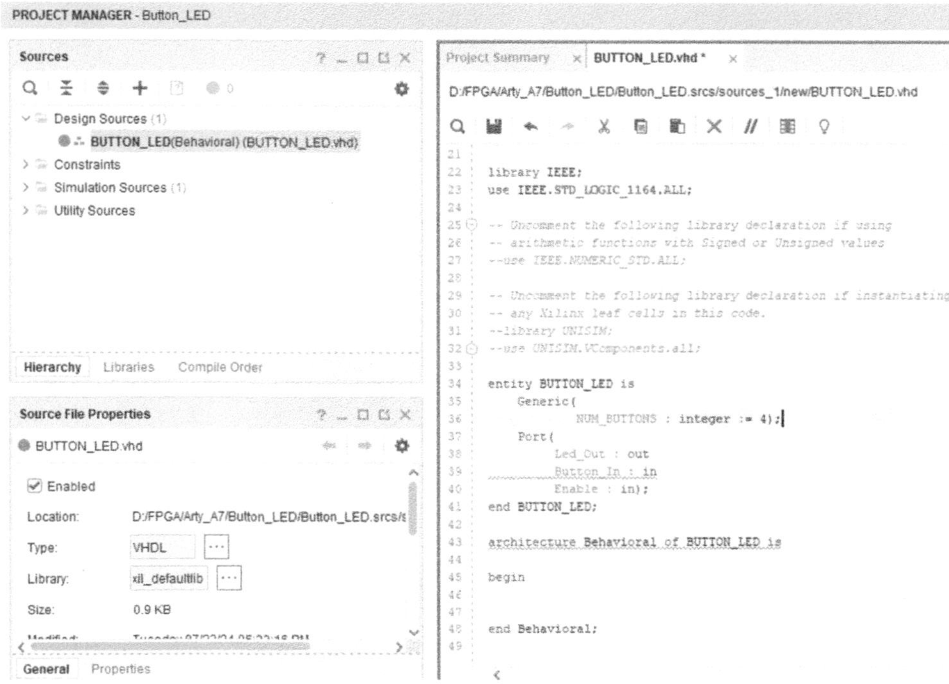

Figure 3.19 Definition of generic "NUM_BUTTONS."

We're working with the Arty A7 development board in this book, which has four buttons. So, we're going to give it a value of 4, so now the generic has been defined. We can now use this generic NUM_BUTTONS in our port as well as in our architecture for this design. Anywhere NUM_BUTTONS appears, it'll be replaced with an integer value of 4, and that is an explanation on generics in VHDL.

4

The VHDL Statements and Data Types

4.1 Introduction

VHSIC Hardware Description Language (VHDL) is used for describing digital and mixed-signal systems such as field-programmable gate arrays (FPGAs) and ASICs. Here's a brief overview of its statements and data types:

The VHDL statements include the following items.

1. Entity Declaration: Defines the name of the module and its interface (inputs and outputs).

```
entity ModuleName is
    port (
        input1: in std_logic;
        output1: out std_logic
    );
end ModuleName;
```

2. Architecture Declaration: Describes the internal implementation of the entity.

```
architecture Behavioral of ModuleName is
    begin
        -- Implementation
end Behavioral;
```

3. Process Statement: Describes sequential behavior. It can include signals, variables, and statements like if-else, case, loops, etc.

```
process (clk)
    begin
        if rising_edge(clk) then
          -- Sequential logic
        end if;
end process;
```

4. Concurrent Statements: These execute simultaneously and include component instantiations, signal assignments, and concurrent procedure calls.

```
output1 <= input1 and input2;
```

Fundamentals of VHDL for FPGA Programming Using Vivado, First Edition. Majid Pakdel.

Companion Website: www.wiley.com/go/pakdelVHDL

5. Component Instantiation: Includes instances of other modules within an architecture.

```
U1: ModuleName port map (
    input1 => signal1,
    output1 => signal2
);
```

The VHDL data types include the following items.

1. Scalar Types:
   - bit: Represents binary values ("0", "1").
   - boolean: Represents logical values ('true', 'false').
   - integer: Represents whole numbers.
   - real: Represents floating-point numbers.
2. Composite Types:
   - array: Represents collections of elements of the same type.
     type bit_vector is array (0 to 7) of bit;
   - record: Represents collections of elements of different types.

```
type Employee is record
     ID: integer;
     Name: string;
     Salary: real;
end record;
```

3. Access Types: Pointers to dynamically allocated objects.
4. File Types: Handles for file input and output operations.

   The commonly used standard data types include the following items.
   - std_logic: Represents a multi-valued logic type with values such as "0", "1", "Z" (high impedance), and others for unknown and uninitialized states.
   - std_logic_vector: Represents an array of "std_logic" values, used for buses and multi-bit signals.

These elements together form the foundation for designing and simulating digital systems in VHDL.

4.2 The VHDL Data Types

This section is about the VHDL data types. In this section, we're going to cover some of the commonly used types, specifically the ones we will be using for the projects in this book. These include the standard logic, standard logic vectors, unsigned versus signed, and integer data types. The first one to cover is the standard logic data type. The standard logic data type represents 1 bit of data. It can represent several values: X (unknown), 0 (zero or ground), 1 (high or logic one in digital systems), and Z (high impedance or tri-state). The Z state is typically used only in cases involving "inout" signals or shared buses where multiple drivers might access the bus, causing it to enter a high-impedance state. A standard logic is represented with the keyword "std_logic." When assigning values to a standard logic data type, you want to use a single quote. For example, if you have a standard logic of data type A and you want to assign the value of 1 (high), it would be a single quote, one, and a single quote ('1'). The next data type to represent is the standard logic vector, which is effectively an array of standard logic types. So, if you have 10 standard logic data types, you can create a standard logic vector, which is just an array of those. The same logic vector can be of any size. You specify it when you define the data type; For example, standard logic vector A is defined from 27 down to 0, giving you bits 27 to 0 – 28 individual standard logic data types. The keyword for a standard logic vector is std_logic_vector. Inside the parentheses, you define the size of the standard logic vector, from starting bit and then use the keyword "downto" with the ending bit to give you the size of that actual vector. When you're assigning values to a standard logic vector, you use double quotes. So, if you have a standard logic vector with a width of 3 and want to assign it all zeros, you would use a single assignment with a double quote, zero, zero, zero, and double quote, and that would define all three bits in that standard logic vector as all zeros. Just like the standard logic, the standard logic vector can represent the same values: X (unknown), 0 (low), 1 (high), and Z (high impedance or the tri-state buffer). So now that we understand the standard logic and standard logic vector data types, let's take a look at some of the following examples.

```
-- VHDL Data Type Examples
-- Standard Logic
signal sig_A : std_logic := '0';
signal sig_B : std_logic := '1';
```

So now we have signal A, which is a standard logic data type with a value of zero. We also have signal B, which is a standard logic data type with a value of 1. Now, we can start assigning values with the following codes.

```
sig_A <= '1';
sig_A <= '0';
sig_A <= sig_B;
```

So, these are the different ways in which you can represent your standard logic. You can assign it a 1 or 0, you can assign it the value of another signal, which has to be a standard logic, or you can also assign it an element from a standard logic vector. It is just the width has to match. For example, if you're working with a width of 1, it has to match the value you're actually assigning. The next data type we're going to look at is the standard logic vector. These, along with standard logic, are probably the most commonly used data types in VHDL. Almost every design you work with will have a standard logic or a standard logic vector. So, for a standard logic vector, we can write the following code.

```
signal sig_C : std_logic_vector(5 downto 0) := "000000";
```

We can also use another keyword called "others." So, we define another standard logic vector, and we'll make it from 5 down to 0, giving it the initial value. We want all the values to be one, as shown in the following code.

```
signal sig_D : std_logic_vector(5 downto 0) := "111111";
```

However, instead, what we can do is use the "others" keyword, as shown in the following code.

```
signal sig_D : std_logic_vector(5 downto 0) := (others => '1');
```

So, that will make every one of the six bits in sig_D to logic high, or equal to the value 1.

Next, we can go ahead and make sig_C equal to the value of sig_D because both sig_C and sig_D are 6 bits wide, so a width of 6. Thus, the following statement is valid.

```
sig_C <= sig_D;
```

We can also access individual bits. For example, bit 3 of sig_C can be equal to sig_B, as shown in the following code.

```
sig_C(3) <= sig_B;
```

Because sig_B has a width of 1 bit, or we can make it equal to bit 1 of sig_D, as shown in the following code.

```
sig_C(3) <= sig_D(1);
```

The next part we're going to look at here is the unsigned data types. An unsigned data type uses the syntax unsigned. They're very similar to a standard logic vector in that you can access the bits individually, as we've seen. When you initialize values or give them a value, you need to use double quotes, which is very similar to the standard logic vector. However, the difference with an unsigned data type is that you can perform addition/subtraction, so you can do an unsigned data type and say plus/minus one to two, or you can take two unsigned data types and add/subtract them together to get a third unsigned data type. With standard logic vectors, you don't have that freedom; you can't do simple additions like that, and it doesn't allow you to do that. So that's an advantage of using the unsigned data type. Additionally, these can only be used to represent positive numbers. If you want to represent negative numbers, you have to use the signed data type. The signed data type allows you to use or represent negative numbers. However, when representing negative numbers in the signed data type, you represent them in the two's complement form. So, it's not as simple as the bits directly representing the number; so, you'll take a number in binary and do the two's complement to get the negative number. We'll show you an example of how to do that. Just like the unsigned data type, the signed data type allows you the ability to use the addition/subtraction, as well as the greater than or equal to. For example, you can use these in an if statement for comparisons, which makes it very useful. So, the signed data type has advantages over the standard logic vector data type. Let's go ahead and take a look at some examples. First, let's look at the unsigned data type, as shown in the following codes.

```
-- Unsigned
signal sig_E : unsigned(3 downto 0) := (others => '0');
signal sig_F : unsigned(3 downto 0) := "0010";  -- 2
sig_E <= (others => '0');
sig_E <= sig_F + 1;  -- 3
```

The next data type will be the signed data type that uses the keyword "signed," as shown in the following codes.

```
-- Signed
signal sig_G : signed(3 downto 0) := (others => '0'); -- 0
signal sig_H : signed(3 downto 0) := "1011";  -- -5
sig_G <= (others => '0');
sig_H <= sig_G - 1;  -- -1 = "1111"
```

So that kind of shows you how to use this standard logic, as well as the unsigned and signed data types in VHDL. The next thing we're going to do is figure out how to represent these in our Vivado project. Now that we know what the standard logic, standard logic vector, and unsigned and signed data types are, let's see how to assign them in Vivado. Now that we have the project open, let's assign the data types to our project in Vivado. We know that we have the standard logic, standard logic vector, and unsigned and signed data types. So if we look at our project, we have the number of buttons, which is defined as an integer. We also have Led_Out, which we need to define and give it a data type. So, we're going to give it the standard logic vector data type, and we will use the generic NUM_BUTTONS, as shown in the following code.

```
Led_Out : out std_logic_vector(NUM_BUTTONS - 1 downto 0);
```

So, that's the size we define for our Led_Out. Now, for our Button_In, it's going to be the exact same as Led_Out, so we can copy from Led_Out and paste it into Button_In, as shown in the following code.

```
Button_In : in std_logic_vector(NUM_BUTTONS - 1 downto 0);
```

So, this makes our Led_Out and Button _In the exact same size, which is what we want, and that size is defined by our generic NUM_BUTTONS. Now, for Enable, we need to give it a data type, which we know is a single bit, represented as a standard logic, as shown in the following code.

```
Enable : in std_logic;
```

This defines our entity. We have our entity completely defined with our generic, ports, data types, and sizes for all of our inputs and outputs in the Button_LED project.

4.3 The VHDL Integers

This section is about the VHDL integer data types. In this section, we're going to explain the integer data type and how it's used in VHDL. In VHDL, the syntax for the integer is simply "integer," and it is 32 bits by default, if you don't define the range. We'll show you exactly how to define the range for your integer, and integers can represent both positive and negative numbers. By default, the integer is 32 bits, but when you define the range, you can have an 8-bit, 9-bit, or any other size. The number of bits required to represent that range is how you will define your integer. When you assign a value to an integer, no quotes are needed. However, you still need to use signal assignment, and then upon doing that, you assign a value without any quotes. With integers, you can perform operations like addition, subtraction, and comparisons (greater than, less than), similar to the signed and unsigned data types. Now, let's take a look and see how the integers actually work. So with the integer, we are going to add to our data type examples. Here's the code for the integer data type:

```
-- Integers
signal sig_I : integer := 0;
```

So, if we were to assign a value of zero, it would be a 32-bit integer with the value of zero. Now, we can also define the range. So we will say integer range and then we give it the low range; we're going to define the range from 0 to 255, which is equal to a value of zero.

```
signal sig_I : integer range 0 to 255 := 0;
```

Now, we have an integer that would be 8 bits. Why? Because the range from 0 to 255 requires 8 bits to represent that. Next, we define sig_J as follows.

```
signal sig_J : integer range 0 to 255 := 5;
```

This is also an integer, with the same range from 0 to 255, except we're going to give it an initial value of 5. So now we have two integers: sig_I, whose initial value is 0, and sig_J, whose initial value is 5. Now, we are going to assign them a value, as shown in the following codes.

```
sig_I <= 4;
sig_I <= sig_J + 5; -- 10
```

Therefore, this is an example showing how you can use the integer data type in VHDL.

4.4 The When-Else Statement

This section is about the VHDL when-else statement. In this section, we're going to explain how to use the when-else statement in VHDL. First off, what is the when-else statement? The when-else statement is used to form a conditional signal assignment. What this means is that we are going to form a signal assignment, but based on a specific condition. For example, in the Button_LED project, we are going to form a signal assignment for the LEDs based on a condition, and that condition is going to be the Enable switch. If the Enable switch is ON (enabled), we want the output of the LEDs to be the state of the buttons. If the Enable switch is OFF (disabled), we want the output to be a zero or no LEDs. So, the whole idea behind the when-else statement is to form a conditional signal assignment. Let's take a look at the actual syntax of the when-else statement, as shown in the following code.

```
signal_name <= expression1 when condition1 else expression2
```

For example, in the Button_LED project, our signal_name would be Led_Out, and this will be equal to expression1, which is equal to Button_In. When Enable is true (enabled), we want the output to be equal to the buttons; otherwise, we want it to be equal to a zero. Now, expanding on the when-else statement, you also can handle multiple conditions. In the first case, we only have a single condition, but we can also use multiple conditions. If we have two conditions, the syntax would be as follows.

```
signal_name <= expression1 when condition1 else
                expression2 when condition2 else
               expression3;
```

You have signal_name equal to expression1 when condition1 is true, or to expression2 when condition2 is true, or else to expression3. So, the signal_name can take one of the possibilities: expression1, expression2, or expression3, based on whether condition1 or condition2 is true. If condition1 or condition2 is true, you'll get expression1 or expression2; otherwise, expression3 will be assigned to signal_name.

Now, with the way you have this one statement, you could perform this type of operation. For example, in our Button_LED project, where we want to assign the output of the LEDs (Led_Out) based on a button (Button_In), you could use a when-else statement like we have here, or you could use an if statement or various other types of statements to do this. You could also do it just with logic using AND gates and OR gates. However, this is one of the ways you can do it in a single line. So, the next thing we want to do is open up our Button_LED project, and we're going to assign the when-else statement to define our LED output (Led_Out). We already have our library set up, our entity defined, and our inputs and outputs ready. The architecture is in place, and now we just need to begin the when-else statement for it. So, let's assign our LED output (Led_Out) as the following code.

```
architecture Behavioral of BUTTON_LED is
begin
    -- Set the output of LEDs
    Led_Out <= Button_In when Enable = '0' else (others => '0');
end Behavioral;
```

So, this statement is saying that when Enable is zero, our Led_Out will be equal to Button_In. If we press any of the four buttons, they will be high, and our LEDs will be high as well. Each button will be mapped to its corresponding LED. Now, if Enable is high (or one), the LEDs will automatically be reset to zero (turned off), regardless of which button is pressed. So, we were able to take the Button_In and Enable signals and map them to Led_Out in one single statement using the when-else statement. So, that's an example of how you would use the when-else statement in the Button_LED project.

4.5 Concurrent Versus Sequential Statements

This section is about the concurrent versus sequential statements. In this section, we're going to explain the difference between a concurrent statement and a sequential statement in VHDL. When we're working with VHDL, we know that we're describing a physical hardware circuit, and the advantage of using VHDL or something like an FPGA is that you can do multiple things in parallel, hence the concurrent statements. So, within concurrent statements, these are things that obviously happen at the same time. For example, you can do things such as signal assignments, and these can be inside a process, a case statement, or they can happen outside of a process. Now, a process statement allows you to rearrange process statements. However, inside a process, you can have various things that occur. You can put a sequential statement inside a process, or that these sequential statements have to go inside a process. These include an if statement, which is the infer priority, and a case statement, which is equal priority. These must go inside a process. With processes, you can rearrange them, and it doesn't matter because they all have the same signals in the sensitivity list. They're all going to happen at the same exact time, as shown in Figure 4.1.

So let's take a look and see some examples. You will see an example of a signal assignment and some processes, and just take a look at some VHDL code. Now, take note that when you're looking at this code, we're not expecting you to understand everything that's going on inside of it. We just want to show you some of the constructs so you can get an idea of what it actually looks like. So, if we take a look, let's see what a signal assignment looks like? Well, in the case of this design, we have a signal, and we have this start_proc signal, and we're giving it a signal assignment, as shown in the following code.

```
start_proc <= start_reg(1) and start_reg(0);
```

So, that is a signal assignment and that happens concurrently. If we look down and see data_out as follows, data_out is occurring at the exact same time as start_proc, even though start_proc is at line 103 and data_out is at line 175.

```
data_out <= data_reg(5 downto 1);
```

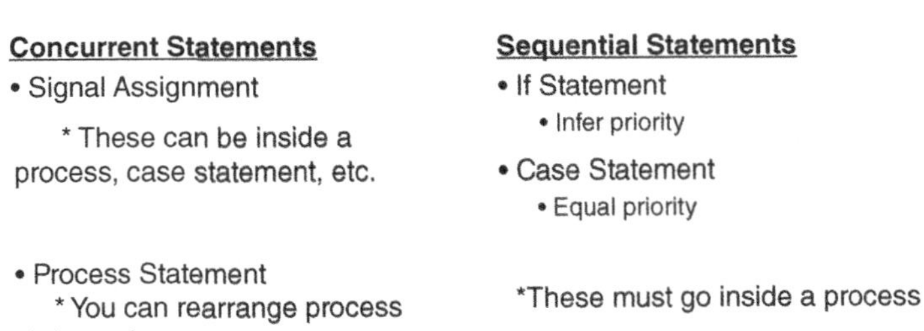

Figure 4.1 The concurrent versus sequential statements.

They're happening at precisely the exact same time because they're concurrent; they're happening in concurrency, so these are occurring at the same time. Now, this also applies to process statements. For example, we have number_bits_proc and done_reading_proc, both of which are processes occurring at the same time, as shown in the following codes.

```
number_bits_proc : process(clk)
begin
..........
end process number_bits_proc;
done_reading_proc : process(number_bits)
begin
..........
end process done_reading_proc;
```

However, since the process is sensitive to a number of bits (number_bits) rather than the clock (clk), swapping the order of the processes – starting done_reading_proc at line 134 and moving number_bits_proc to line 160 – would not affect the design; the functionality would remain the same. So, the way in which we order these process statements doesn't matter; the order makes no difference. We can move them around in our design, and it will remain exactly the same. Now, the time when it does make a difference is when you're inside your process. For example, if we look at this start_bit_proc process, this occurs every time we have a clock (clk) signal, and we check the if statement on the rising edge. Then we have an if, else-if, else statement. So, first, we're checking our statement, which is in first priority; if our reset is equal to zero or start_bit_counter is equal to the max_start_bit_count minus one, then start_bit_counter becomes zero and is assigned the value of zero; otherwise, if read_start is equal to 1, then that condition takes priority. The reason that reset has first priority is because, for example, if read_start is equal to one and reset is equal to zero, we're looking at a reset, and if that's the case, start_bit_counter is set to zero, and we exit the statement. So, as soon as one of the conditions in the statement evaluates to true, you right away jump out. That's how priority works: the item we have here at our top level is if the very first thing we have in the statement is the highest priority. So, when we say that for a priority, that's what we mean by that, meaning if you have a reset that takes the highest priority, you're going to set it to zero and just jump out of that if statement, as shown in the following codes.

```
start_bit_proc : process(clk)
begin
   if(rising_edge(clk)) then
      if((reset = '0') or (start_bit_counter = max_start_bit_counter-1)) then
           start_bit_counter <= 0;
      elseif(read_start = '1') then
           start_bit_counter <= start_bit_counter + 1;
      else
            start_bit_counter <= 0;
      end if;
   end if;
end process start_bit_proc;
```

If we look at, for example, a case statement, please note that it does not infer priority. Another reason for this is that when you have a case statement, you have to make sure that each case is unique and cannot overlap; you can't have the same condition happening twice. For example, we have this case statement where our state is init_state. We cannot be in any of these other states. So, if our state is read_start_bit_state, then we can't be any of these other states, as illustrated in the following codes.

```
nxt_state_proc : process (state, start_proc, read_bit, done_reading)
begin
      nxt_state <= state;         read_start <= '0';
      reading <= 0;         data_valid <= '0';
      case state is
            when init_state =>
                 if start_proc = '1' then
                       nxt_state <= read_start_bit_state;
                 else
                       nxt_state <= init_state:
                end if;
           when read_start_bit_state =>
                read_start <= '1';
                if (read_bit = '1') then
                       nxt_state <= read_bits_state;
                else
                       nxt_state <= read_start_bit_state:
               end if;
          when read_bits_state =>
               reading <= '1';
               if (done reading = '1') then
                     nxt_state <= done_state;
             else
                    nxt_state <= read_bits_state;
             end if;
         when done_state =>
             data_valid <= '1';
             if (start_proc = '1') then
                   nxt_state <= read_start_bit_state;
            else
                  nxt state <= init_state;
           end if;
         when others =>
                 nxt_state <= init_state;
     end case;
end process nxt_state_proc;
```

Therefore, the way in which you order these doesn't matter. We could have the read_start_bit_state at the top and place init_state at the very bottom. So, the order of your case statements doesn't matter; it's not going to change the way in which your design operates. However, with if statements, the order does matter and will make a difference. For example, if we would switch this line 120, this reset equals zero or start_bit_counter equals max_bit_counter minus one. Now, we would swap that with read_start equals one inside the start_bit_proc process. Well, if read_start is one, it doesn't matter whether we reset or not, because as soon as that condition evaluates to true, it will make our start_bit_counter equal to start_bit_counter plus one and won't look at that first term. So, when you're working with if statements, it does matter. When you have multiple conditions, you're checking against the order in which you set each of those signals. That's just a quick overview of concurrent versus sequential statements in VHDL. This explanation gives you an understanding of what the design looks like, as well as how you can move processes around or move parts of it around and still operate the same way because we're still describing the same hardware circuit. It's just a matter of whether you assign the signal on line 103 or line 142. It doesn't matter because it goes through and reads this and then generates the hardware based on that. So regardless of the order, you're going to assign that signal. So, whether you assign it on line 103 or 142, it doesn't matter – we're describing hardware, not writing software. That's just a quick overview of concurrent versus sequential statements.

5

The Simulations in Vivado

5.1 Introduction

This chapter is about simulation. In this section, we're going to introduce simulation in Vivado. So first off, what is this section even about? Well, we're going to explain why simulation is necessary and use industry-standard tools to simulate our VHSIC Hardware Description Language (VHDL) designs, specifically, the Vivado simulator tool from Xilinx, which is included within the Vivado package. With this action, we're going to help you understand how to run these simulations, which is going to help as we go through different projects, and how to verify whether your project is working correctly. So, when you use these, we're going to show you the steps required. You'll make your design, your test bench, and then you'll put them in a directory. We'll show you the exact buttons to click and where to go to actually get the waveforms showing your design being simulated. So, we'll go through all that within this section on VHDL simulation. Something that's very important – the reason it's very important – is because this is used to verify your designs are behaving properly with conventional code. If you're working with Visual Studios or any other type of tool, you can insert what they call breakpoints. This means that as your code runs, it stops, and you can see how the different variables are behaving internally because you can actually look at the values. Now, with VHDL, it's very hard to do that – or really, it's virtually impossible. You take the design, compile it, and throw it on your development board. Either it does what you expect, or it doesn't. And if it doesn't do what you're expecting, then it becomes a challenge to understand why. You'll look at your VHDL design and think everything looks fine – it synthesizes, it implements – but you still can't figure out what's going on. Well, that's where virtual simulations come into play, and they're very important because now, instead of having a breakpoint, you can see your entire design operate. You can see, OK, when these inputs are coming into my design, how my signals are behaving – what they are doing, and at what times they are doing it. You can see all this information within the simulation. So, it's a very powerful tool, especially for debugging, troubleshooting, or even generating documentation. If you're developing a design and you want to explain to somebody how that design is working, or if you need to create a test report or something of that nature, these simulation tools are great because you can see the waveforms that show your ones and zeros, or if you have an integer, it shows the actual integer value. You can take pictures of it and put it in your document, saying: when you give it this input, here's the output, here's the timing based on our simulation. At that point, it gives you high confidence. Once you simulate your design, you gain high confidence and can go ahead and throw it on your actual FPGA board. Now, when you're working with less complex designs – if you're just doing real simple stuff – you don't necessarily always have to simulate it. But when you're working with very complex or more

Fundamentals of VHDL for FPGA Programming Using Vivado, First Edition. Majid Pakdel.

Companion Website: www.wiley.com/go/pakdelVHDL

elaborate designs, these simulations become much more valuable and much more time-saving in the long run. You have to invest a little bit of time to set up the simulation, but it can save you a lot in the end. Instead of going in circles trying to figure out what's going on, you can just get to the point right away and say, OK, it's doing what we expect, make the change, check it, and put it on the board – it's complete. What will you need to do this? In order to perform a successful simulation, there are several things you'll need. First is the completed VHDL design – whatever your design may be. If you're working with the Button_LED and the blinking LED project, you'll need all associated files – whether it's one or multiple. You must have the completed design that you want to simulate. Second, you'll need to have a test bench. A test bench is a VHDL design that is not synthesizable. Its sole purpose is to test the actual design you want to run on the board. What you'll do in your test bench is called component instantiation – you'll instantiate your actual VHDL design inside your test bench so that you can then apply what's called stimulus. This is effectively just signals where you determine when they're on, off, or what values they're going to be. You then monitor the output from your test bench. Within your test bench, you can do this in several ways. You can calculate what the value should be and do a comparison to tell you if you passed, or you can just apply a stimulus and monitor the output via the waveforms you'll see in the Vivado simulator. So, there are various ways you can do that. For this section, we are going to use an example of a shift register. We'll generate a design – a shift register – and also design a test bench to test that shift register. Then we'll do the simulation demonstration in Vivado using the Xilinx Arty A7 development board. Now that you understand what this section is about and how we're going to do it, let's go ahead and get started with designing the shift register.

5.2 The Shift Register Design

This section is about the shift register design. In this section, we're going to explain how we want our basic shift register to operate, as well as how we're going to design and write the VHDL code for this shift register. Now, the shift register we are going to implement is a standard serial-in and parallel-out shift register, which means we'll have one input data line and four output data lines. We're then going to use the processor to develop a test bench, which will then be used to simulate this in Vivado. So, for our shift register design, we want to have a serial-in and parallel-out architecture. Let's draw a block diagram of what this register would approximately look like. Sometimes, having a block diagram helps us get a better understanding of what we're trying to do. So, we're going to have our serial-in defined by a value we'll call data_in. Then we'll have a clock and a reset. The way we want our shift register to work is on data_in, which is the single standard logic vector – it's going to be a single bit. We then have our clock, which is going to, on every clock cycle, shift the outputs. Since we have four outputs, we'll label them A, B, C, and D. This is a parallel-output shift register – A, B, C, D – so we want to assign the labels A, B, C, and D. On each clock cycle, we want to list out the things we want to happen – we want A to be equal to our data_in, since this is our input; B will be equal to the value of A, because we're going to take the value here and shift it every clock cycle; C is going to be equal to the value of B; and D is going to be equal to the value of C. Now, once we shift our value, it will essentially be discarded – it won't be going anywhere. So, on each clock cycle, the value of data_in will enter at A and shift over time to give it a clock shift. Next, we want to look at Figure 5.1. Given what we have described, we can now proceed with the VHDL design for this shift register.

To do that, we will begin designing our 4-bit shift register in VHDL using the behavioral architecture. Now, the key thing to note is that we have not yet covered all the topics needed to

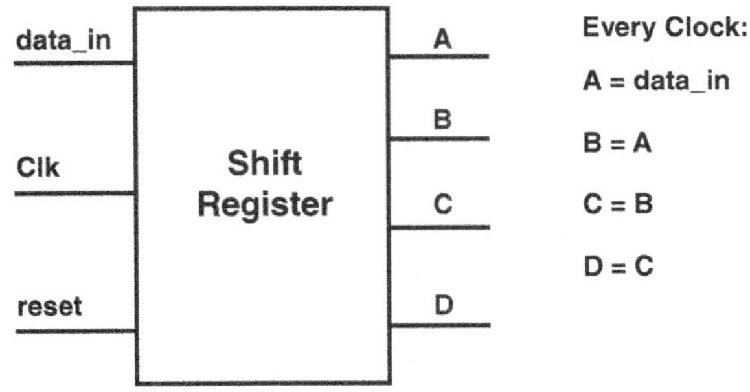

Figure 5.1 The 4-bit shift register block diagram.

complete the shift register. As a result, there will be parts of the design that you may not completely understand right now. However, we'll just explain them as we go, and once you encounter them later, you'll get a better understanding of how these different architectures in VHDL are being used. First, we'll start with our libraries. We're using the IEEE library and the following packages: IEEE.STD_LOGIC_1164.ALL and IEEE.numeric_std.all. Then, we'll define our entity as shown in the following codes.

```
-- Libraries
library IEEE;
use IEEE.STD_LOGIC_1164.ALL;
use IEEE.numeric_std.all;
-- Entity Declaration
entity Shift_Reg is
port(
    A : out std_logic;
    B : out std_logic;
    C : out std_logic;
    D : out std_logic;
    data_in : in std_logic;
    reset : in std_logic;
    clk : in std_logic);
end Shift_Reg;
```

So, now we define our architecture as the following codes.

```
-- Architecture
architecture behavior of Shift_Reg is
-- Defined Signals
signal A_reg, B_reg : std_logic := '0';
signal C_reg, D_reg : std_logic := '0';
-- Begin Architecture
begin
     -- Signal Assignments
     A <= A_reg;
     B <= B_reg;
     C <= C_reg;
     D <= D_reg;
     -- Process that is used to shift the values
     reg_process: process(clk)
     begin
         if(rising_edge(clk)) then
              if(reset = '1') then
```

```
                A_reg <= '0';
                B_reg <= '0';
                C_reg <= '0';
                D_reg <= '0';
            else
                A_reg <= data_in;
                B_reg <= A_reg;
                C_reg <= B_reg;
                D_reg <= C_reg;
            end if;
        end if;
    end process reg_process;
end behavior;
```

That completes the entire design of our actual shift register. We have our libraries called out, our entity where we define our outputs and inputs. We have our architecture for the behavior of our shift register, and we define the signals we're using. Then we use the begin architecture and our signal assignments, so we assign our signals. We take the registers and assign values to our A, B, C, and D registers, which are then connected to the output – the output of our design, the actual product of our shift register. At this point, we have our shift register completely designed. Now, we simply need to define and design our test bench so that we can then take the shift register and simulate it.

5.3 Shift Register Test Bench Design

This section is about the shift register test bench design. Here, we're going to walk through and explain what we're going to try and do with the shift register test bench, and then proceed to implement and design it. We'll be using Notepad++, though you can use any text editor that allows you to save a file with the .vhd extension, or you can also use the Vivado model, which allows you to actually write and edit VHDL designs. For our design, we have the shift register, which includes input_data clk, reset, and outputs A, B, C, and D. In the test bench, we don't actually have any inputs or outputs. All we're doing is instantiating the shift register inside the test bench design. From there, we'll analyze and look at the outputs – A, B, C, and D – while applying stimulus or data to the data_in, clk, and reset signals to see how the outputs respond. If we think about it visually, we can imagine our test bench design containing signals A, B, C, and D. We can enclose signals A, B, C, and D in a small box to indicate that they are our designated signals. The data_in signal will connect sequentially to each of the signals – starting with signal A, then to signal B, and so on. We also need to define our stimulus. For the stimulus, we'll create signals named data_in, clk, and reset, and we'll name them the same as our signals on the shift register so it's easier to follow. In the block diagram shown in Figure 5.2, the stimulus is represented. So, data_in is tied to the data_in signal, clk is tied to clk, and reset is tied directly to the reset signal, as shown in Figure 5.2.

So, we're going to apply values to the stimulus – our data_in, clk, and reset – and while we do that, we'll observe the output signals inside the simulation. We can see our signals A, B, C, and D and make sure they're doing exactly what we want. Now, if we think about our test bench, it encompasses the entire design. The design shown in Figure 5.2 is our Shift_Reg, and in this case, we're going to call it test_Shift_Reg. In our test bench, there are no inputs and no outputs – it is completely self-contained inside the architecture. So, there's nothing coming in or going out. Everything we're looking at is internal. All we're doing is applying stimulus internally to test the

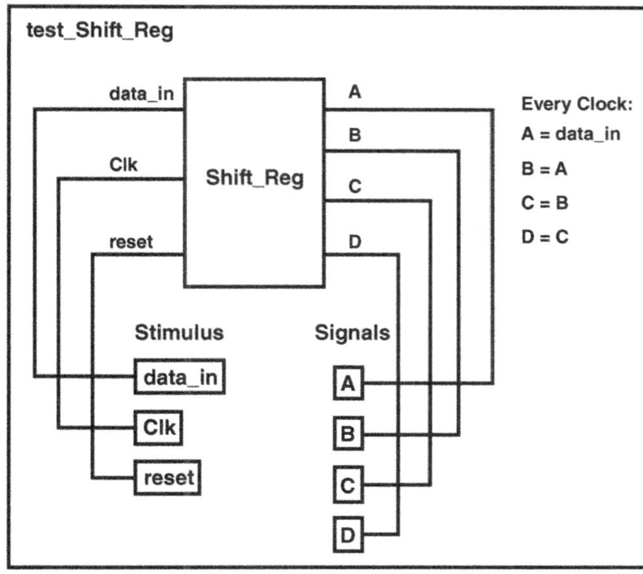

Figure 5.2 The shift register test bench.

design. That design – the Shift_Reg – is instantiated inside the test bench to apply these different values and see how they respond. Now that we understand the goal and have an idea of what we want to do with our shift register, let's open up the editor you're going to use to edit your VHDL design. We have already completed the shift register designed in the previous section, so now we'll focus on how to design the test bench. For our test bench, the first thing we want to do is to include a library. For example, an IEEE library, and we want to use IEEE.STD_LOGIC_1164.ALL and IEEE.numeric_std.all. Next, we have our entity. As discussed earlier, the entity is empty because we don't have any inputs or outputs. So, entity is as the following code.

```
-- Libraries
library IEEE;
use IEEE.STD_LOGIC_1164.ALL;
use IEEE.numeric_std.all;
-- Entity Declaration
entity test_Shift_Reg is
end test_Shift_Reg;
```

So, the entity is empty – there are no inputs or outputs in the design. We're only working with the architecture, and that's where we observe our signals to see what's going on. In our test bench, there are no inputs or outputs because a test bench is not synthesized. You cannot actually load it onto the FPGA; it's strictly used to simulate and test. So, the architecture code would be written as follows.

```
architecture test of test_Shift_Reg is
component Shift_Reg
port(A: out std_logic;
     B: out std_logic;
      C: out std_logic;
      D: out std_logic;
      data_in: in std_logic;
      reset: in std_logic;
      clk: in std_logic);
end component;
```

```
signal data_in: std_logic := '0';
signal reset: std_logic := '0';
signal clk: std_logic := '1';
signal A, B, C, D: std_logic;
begin
    dev_to_test: Shift_Reg
        port map(A, B, C, D, data_in, reset, clk);
    clk_stimulus: process
    begin
        wait for 10 ns;
        clk <= not clk;
    end process clk_stimulus;
    data_stimulus: process
    begin
        wait for 40 ns;
        data_in <= not data_in;
        wait for 150 ns;
    end process data_stimulus;
end test;
```

At that point, that is our entire design from start to finish. This is our top-level test bench design. We want to save it with the name, test_Shift_Reg. You don't have to follow the convention, but we like to use that naming convention so that when we put everything in the same directory and simulate using Vivado, it's clear. The design file is Shift_Reg, and by adding the keyword "test" in front of it, we know that that's the test bench for the shift register. So, you don't have to follow that naming convention, but that's just commonly what we like to use. So that is the entire test bench for the shift register design.

5.4 Vivado Simulator Tool Introduction

This section introduces the Vivado simulator tool. Here, we're going to briefly explain and introduce the Vivado simulator, which is a simulation tool integrated into Vivado. No external tool needs to be downloaded; when you download Vivado with the free license or the web version, the simulator is already included. All you have to do is bring in your test bench and run the simulation; it's all included in one big package. So, it's nice that you don't have to download a separate tool or anything. It's already included in the same project, so you don't have to have two separate projects. One of the nice things about Vivado is that everything is integrated into a single tool. With the Vivado simulator, you can view your waveforms and analyze how signals behave in relation to each other. For example, when one signal goes high, you can see when another signal goes high or low. You can also measure the timing differences between those signal transitions. Additionally, you can save waveform images, zoom into specific portions, and set the simulation runtime. One feature we really like is the ability to simulate at different design stages. First, there's behavioral simulation, which simulates your VHDL design. Then, after synthesis, you can simulate your synthesized design as well as your implement design. Your implemented design is going to be as close as a simulation can get to your actual FPGA device, because with your implemented design, it's going to know all of the specific delays – from when a certain event occurs to when the next event occurs – because it's already routed on your FPGA. They've already got the routing mapped out, and they'll have a really good estimate on the certain timings and delays.

So, you can go ahead and run a simulation after that to verify that you'll meet your timing. So, the first step is to do a behavioral-type simulation, and that's just to verify that you're logically doing exactly what you want your design to do. For example, you're pressing a button – you want to make sure that lights up an LED. In this case, you want to say, "OK, we've got input A; we want to make sure when input A goes high, output B goes high." Behaviorally, we're just going to make sure that happens. Now, when you implement and run that simulation, that's going to make sure when we press A and B goes high, what's that time delay between them – does it meet the timing specifications? If you specify those timings, the tool will make sure they are met. But you can then simulate and verify that there are no issues with things like that. When you're working with the Vivado simulator, it can only be used on Xilinx designs. You have to create a Vivado project. You can pick just some sort of generic device, make a generic design, simulate it, and verify it works. Then, you can take that design out. If you're working with Xilinx-specific features, such as internal memory, you want to use Vivado because it can simulate things like that. It understands the initialization files you can place to say that we want our memory initialized with these values. So, for simulation, we can account for those, knowing they're in there and how we are actually going to use them. A lot of time, we'll be using the simulator tool. For example, we're going to use it to simulate our shift register design. So, we'll be needing two files: the shift register design file – the actual physical design we want to test – and the test bench. We will load the test bench and tell it to run the simulation. It will simulate the actual design we're going to be implementing. We want to use this to verify that our designs are behaving as expected, which is the main overall purpose of using VHDL simulations. In the future, you can use this whether you're working at your job or on other projects at home. If you're working with Xilinx devices and using their tools, you can use this understanding to apply these simulations. The shift register simulation is a very simple and basic simulation, but it's just here to give you the first steps in understanding the simulation process and how to run a simulation and design a test bench. You can then apply this to future projects or more complex things you may be working on. So now you have an understanding of what the simulator tool is and how we're going to be using it. Let's get started – the next step is running a simulation in Vivado.

5.5 Running a Simulation in Vivado

This section is about running a simulation in Vivado. We're going to show you how to use Vivado to simulate a design – in this case, a shift register that we have designed. We'll be using a test bench specifically written to test this shift register. Once you have your project set up – in this case, a basic project with the shift register already created or implemented – you'll begin by locating the design source file. Under the Design Source section in Vivado, you'll see a file named Shift_Reg.vhd. Double-clicking on the file opens it and shows the shift register design you have designed. Next, under the Simulation Sources, you'll also see the shift register (Shift_Reg). As shown in Figure 5.3, Vivado recognizes the shift register as the top-level module for simulation. This is because it is currently the only module defined for simulation within the project.

So, what we want to do is import the test bench. To do that, we'll select the Simulation Sources, right-click, and click on Add Sources. Make sure you select the third option: Add or create simulation sources, as depicted in Figure 5.4.

Then, we'll click Next, and leave the specified simulation set as the default, sim_1. This allows you to run different simulations, such as sim_1, sim_2, etc. But for our cases and everything we're

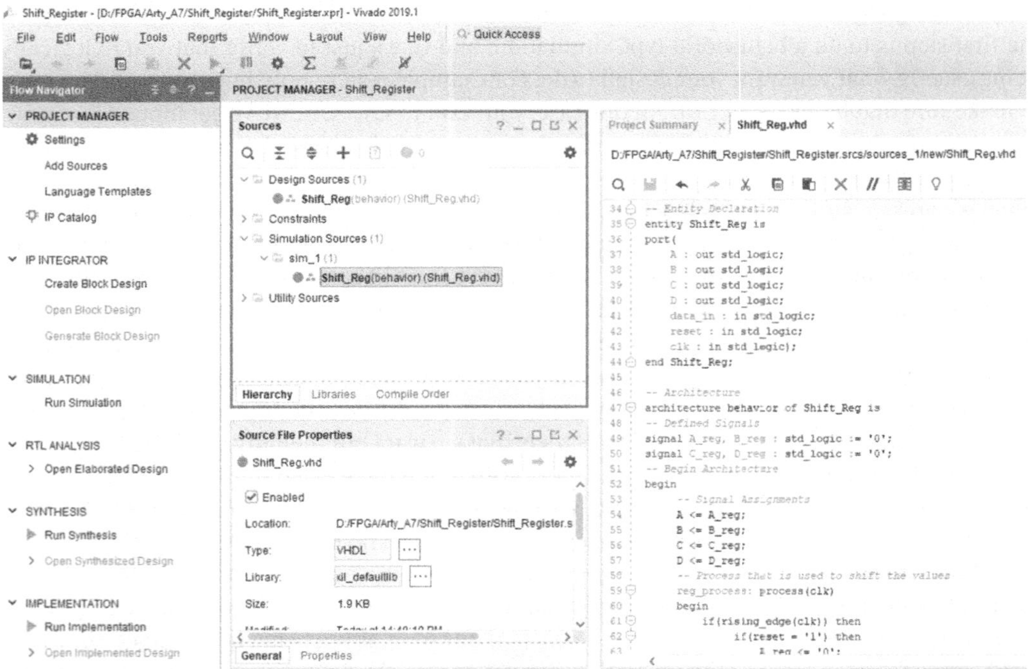

Figure 5.3 The design and simulation sources of shift register.

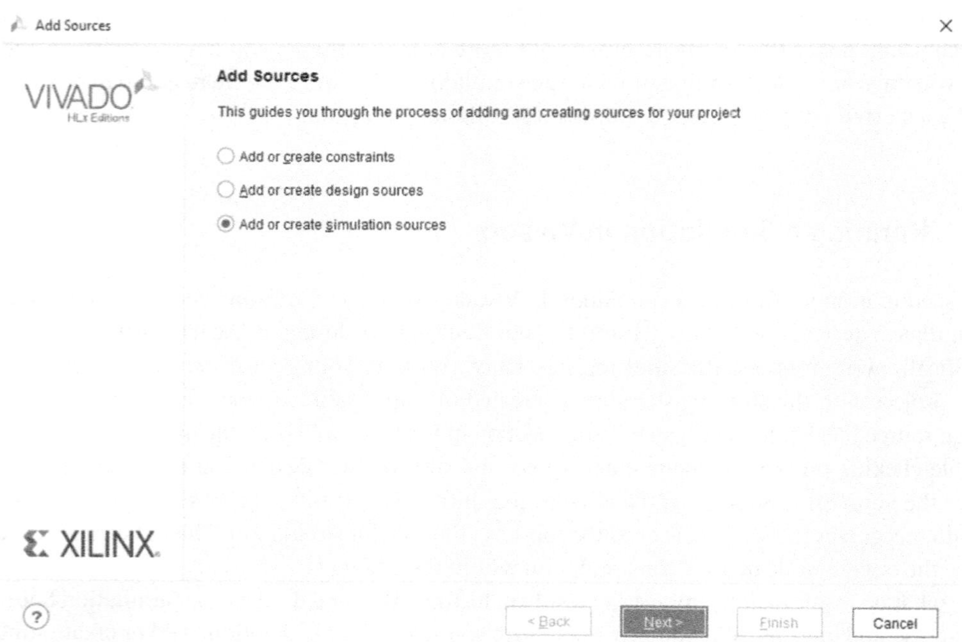

Figure 5.4 Selecting the Add or create simulation sources.

going to be using in this book, we can get by with just having a single simulation set. Now, we'll create a VHDL file (test_Shift_Reg.vhd), as illustrated in Figure 5.5.

Next, we click OK, and in the Define Module window, we set the architecture name as "test," and then click OK again, as shown in Figure 5.6.

Next, we can copy and paste our designed test bench into test_Shift_Reg.vhd. It'll automatically update, and now it will recognize test_Shift_Reg as the top-level design for our shift register, as shown in Figure 5.7.

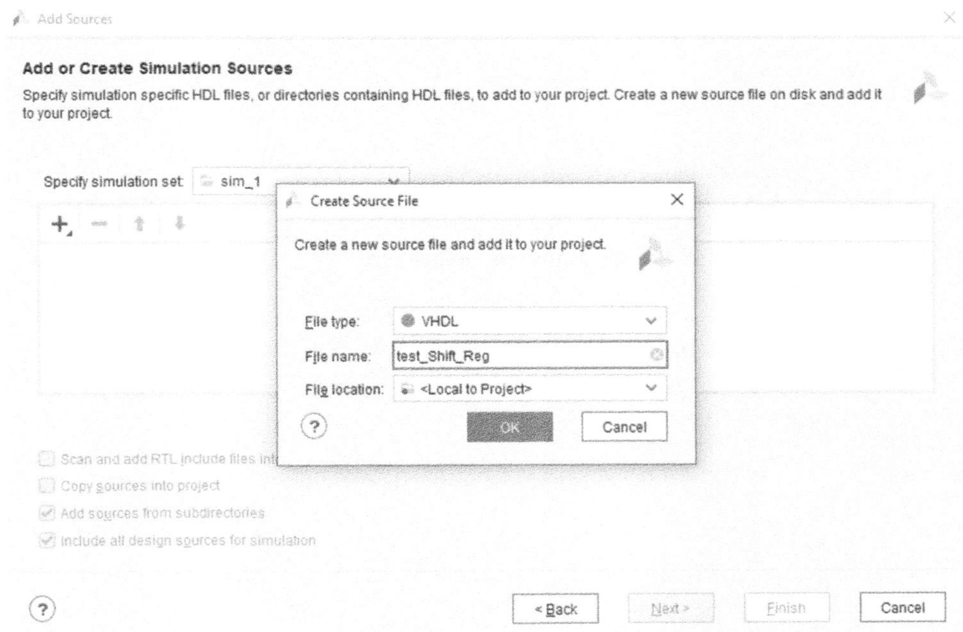

Figure 5.5 Creating the new file.

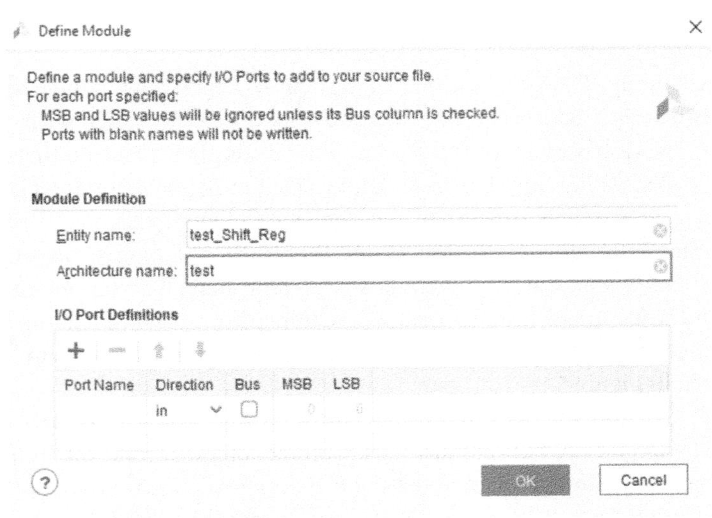

Figure 5.6 Setting the architecture name as "test".

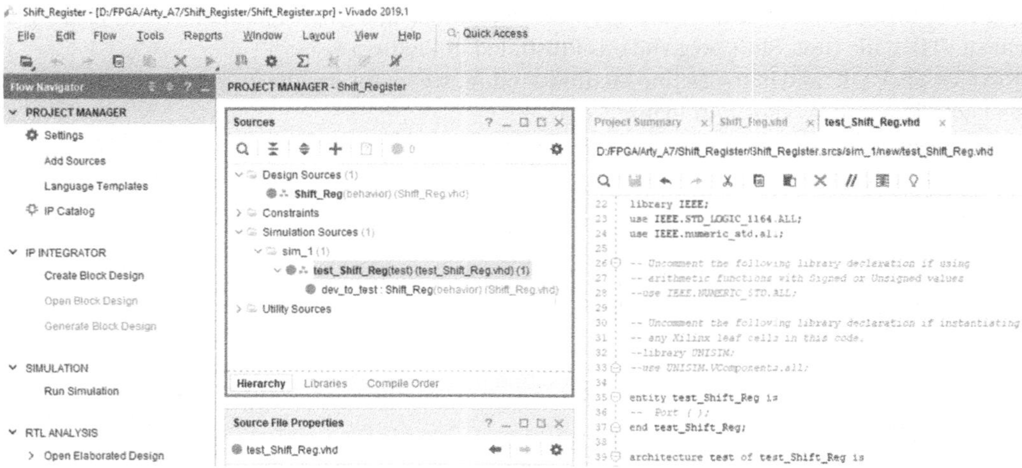

Figure 5.7 Recognizing test_Shift_Reg.vhd as the top-level design.

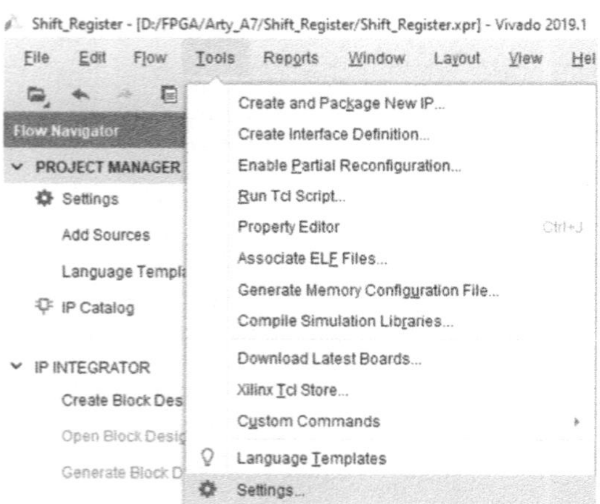

Figure 5.8 Selecting Settings option.

By expanding test_Shift_Reg, you'll see dev_to_test: Shift_Reg. If we highlight and double-click it, it brings up dev_to_test under test (Shift_Reg.vhd). So, this is the design we're actually testing in our test bench. The next step is we run the simulation. But before we do that, we'll go to the Simulation Settings. This brings up some basic settings for simulation. Our target simulator setting can be assessed by going to Tools → Settings..., as depicted in Figure 5.8.

We'll leave the simulator language as Mixed and keep the simulation set as sim_1. If you had different sets like sim_2 and sim_3, you could select them here. The top module name is this test_Shift_Reg, and it just picks it up by default, we'll just leave it all in default. The only other value we will change is that under the Simulation tab. We will set the runtime to 350 ns and click Apply, as illustrated in Figure 5.9.

Now we're ready to simulate. You can always change the simulation time later within the simulator, so this isn't a huge deal, but that just tells you how long you want the initial run to simulate. When you click Run Simulation, you'll see a few options. Since we have not synthesized or implemented our design, we can only run a Behavioral Simulation, which is what we'll start with. This will make sure that our design behaves as expected before synthesis. Later, you can then run

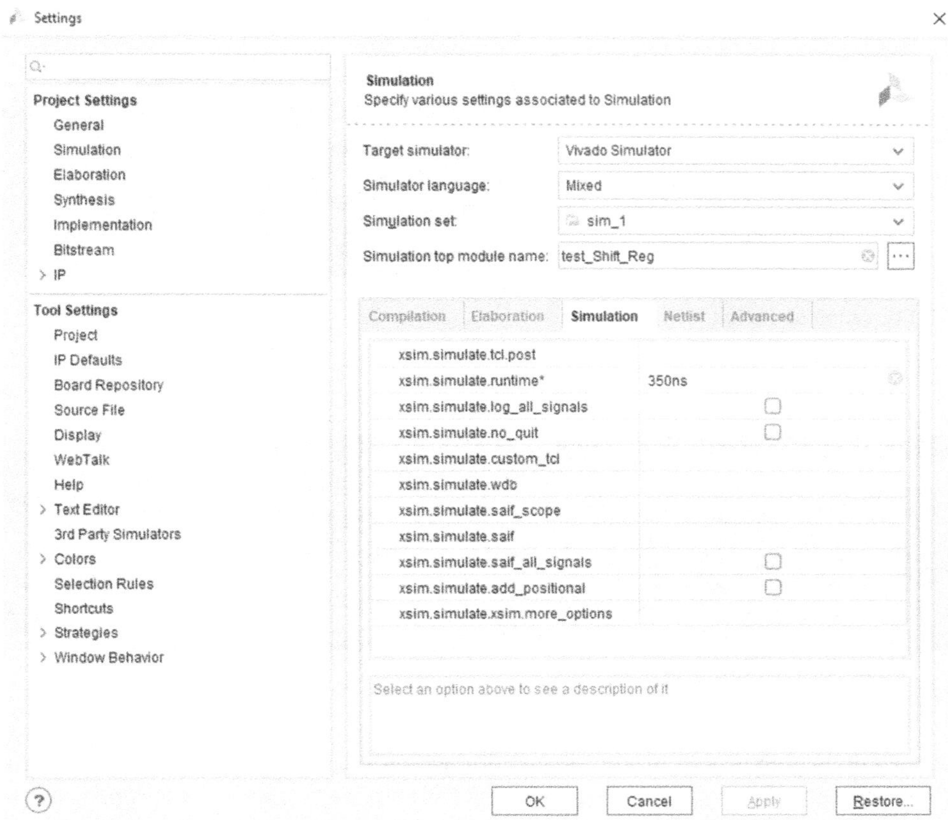

Figure 5.9 Setting the simulation runtime to 350 ns.

Figure 5.10 Running the behavioral simulation.

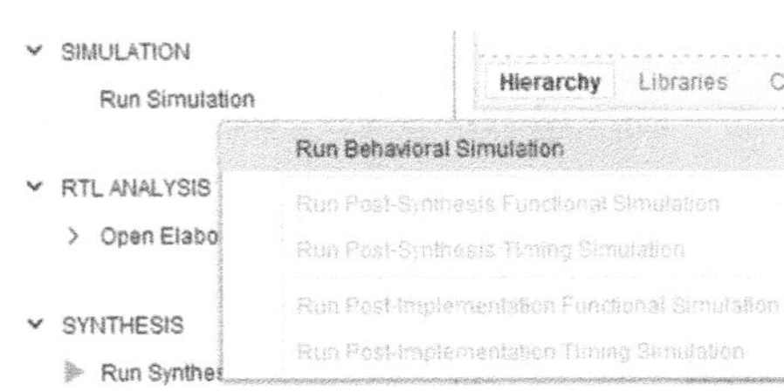

the post-synthesis simulation after it synthesizes your design and you can run to make sure it's still operating as expected. Finally, once you implement your design, you can make sure you still meet your timing. So, let's go ahead and run the behavioral simulation, as shown in Figure 5.10.

The behavioral simulation opens the simulator inside Vivado. So we're still in Vivado, but now within the simulation tool. In Figure 5.11, under Name, you'll see our design: test_Shift_Reg. We're simulating dev_to_test, and it shows all of the objects: data_in, reset, clk, A, B, C, and D. On the right side, you can see the waveforms. Right-click and choose Full View, as shown in Figure 5.11.

In Figure 5.12, you can see the waveforms for data_in, reset, clock (clk), and the output signals A, B, C, and D.

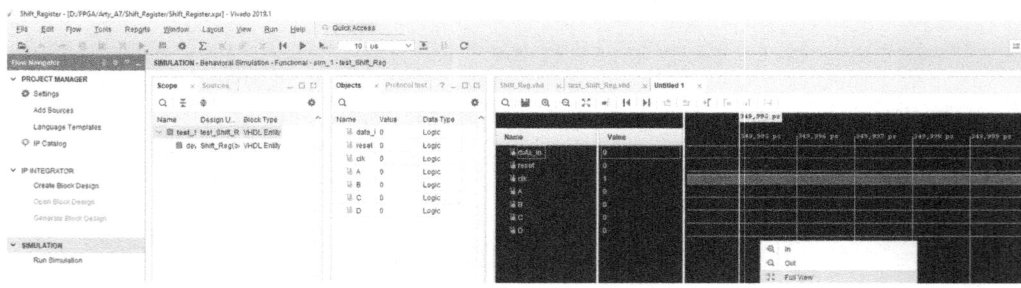

Figure 5.11 Going to the full view.

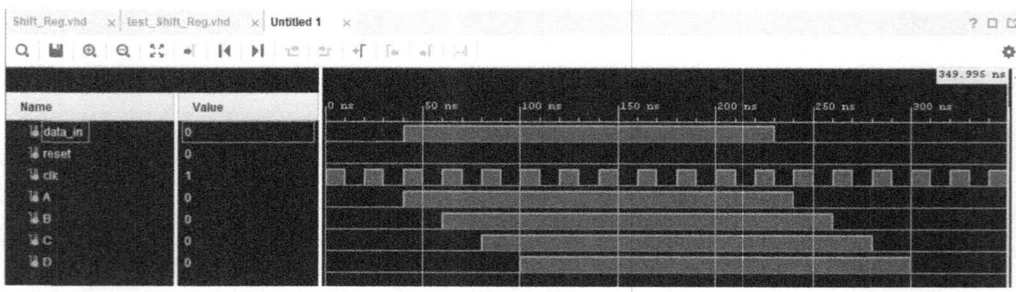

Figure 5.12 The waveforms of inputs and outputs.

That is all the required steps to run a simulation within Vivado. Whether you're doing this for the shift register, the Button_LED project, the blink LED project, or any future design, the steps are the same. You have your main design, then you add a test bench to your Vivado project to run the simulation. Note that when you go through synthesis, implementation, and bitstream generation, the test bench file is not included in the final output. So you can continue developing your main design and still use the test bench to verify everything within a single project.

5.6 Navigating Vivado Simulations

This section explains how to use the Vivado simulation tool. So, the tool in Vivado that we use to simulate is all part of the Vivado tool, which is referred to as the Vivado simulator. Within this tool, there are many different options, capabilities, and features. We're not going to go through every single one of those, as that would take quite a bit of time. There's a lot you can do with the Vivado simulator. We're going to show you, at a high level, how to simulate, so that throughout this project you can use this tool to verify that your designs are working as expected – before you go ahead and load them onto the FPGA. Once you're there, it's not like working with regular software where you can pause, debug, or set breakpoints. You don't quite have those same capabilities, and unfortunately, once it's on the FPGA, you also do not have any indicators like LEDs or UART output. So, this is a great way to test your design before deploying it to the FPGA. To start, we already know how to run our simulation in Vivado. You have created a project, written a test bench, and launched the simulation. Now, you have your simulation running, as shown in Figure 5.12. You can save the simulation window and re-open it later, but for now, we'll just leave it running. The first thing we're looking at is how you can arrange windows or expand your design view. You can click and drag the window edges to make the display area larger. It's helpful to see everything clearly. You don't want to scrunch everything together, as it becomes difficult to read the signals and quite annoying to work with. However, at the same time, when you have a lot going on and limited screen space, it can be challenging. One nice feature is that you can float the waveform window. If you click the icon shown in Figure 5.13, it'll take your actual waveform and break it off into its own window so that you can click on Maximize.

Now you've got this thing taking up your entire screen, which is really nice, especially when you're working on a laptop where you don't have a lot of screen space or screen real state to work with. So, in here, we're going to move some things around. You can move these tables, such as your value column, and since these are all standard logic, they're only going to be a one or zero. You can make that small because it doesn't really matter. Then, we right-click and select Full View, as depicted in Figure 5.14.

Now the waveforms window takes up your whole screen, which is a lot nicer and easier to look at. Let's say we're done looking at it and want to put it back and look at everything else again. We just go back to this dock icon (the reverse float icon), and it puts the window right back in the spot we pulled it out from. That's a nice little feature when you want to get a screenshot or take a snippet – whether it's for a test report or if you have a question or something like that. It's always great to simulate the design, capture the waveforms, and include them in your documentation. The next thing we want to look at is how we zoom in. We know that if we right-click in the area, we can go to Full View, and that takes your entire simulation time – from 0 to 350 ns in this case – and fills up the wave window or the screen space with it. Now, if you want to zoom in, you can right-click and choose Zoom In (+). To zoom out, just right-click and choose zoom out (–). If you keep zooming out, it eventually gets to the point where it doesn't show much, as shown in Figure 5.14. So, go back

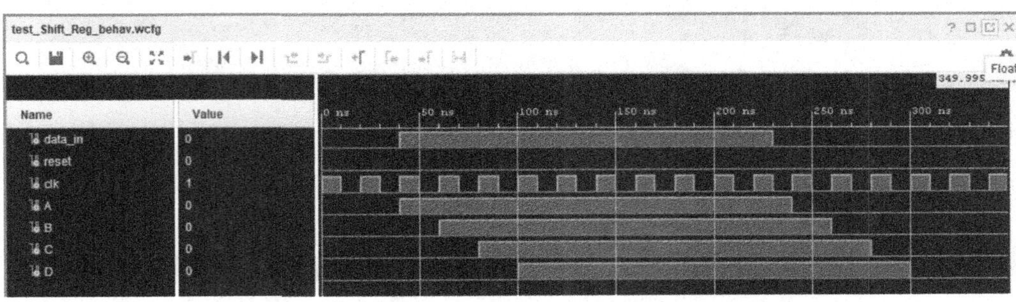

Figure 5.13 Making float the waveforms window.

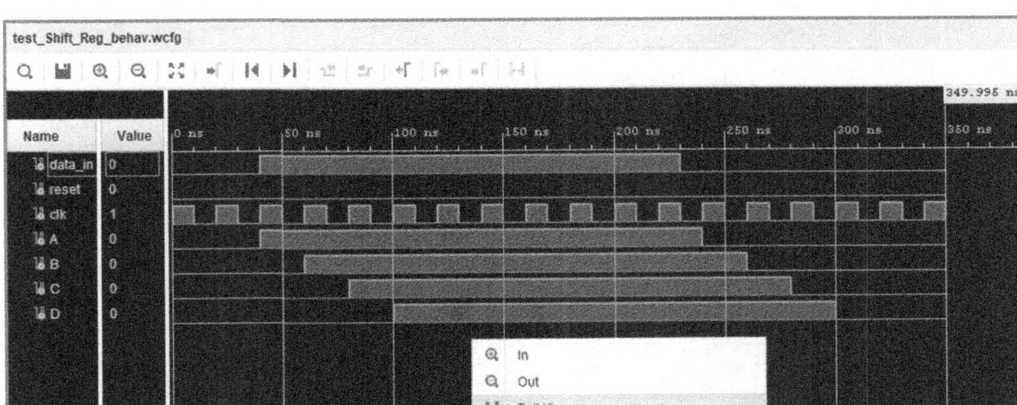

Figure 5.14 Right-clicking and selecting Full View.

to Full View. If you want to zoom into a specific range, you can do that by clicking, holding, and dragging the mouse. Now, when you drag up, it zooms out. If you drag down or back, it zooms in. So, drag up to zoom out, and drag back to zoom in – just by doing a left mouse click and dragging. And again, if things get messy, you can always right-click and go to Full View. It's always nice to start from there, as shown in Figures 5.15 and 5.16, respectively.

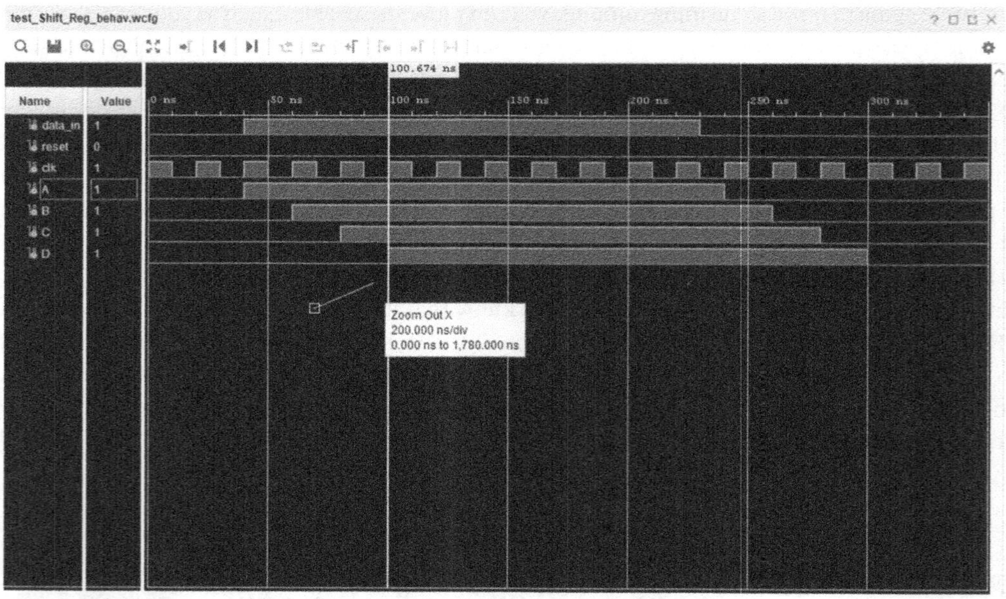

Figure 5.15 The zooming out.

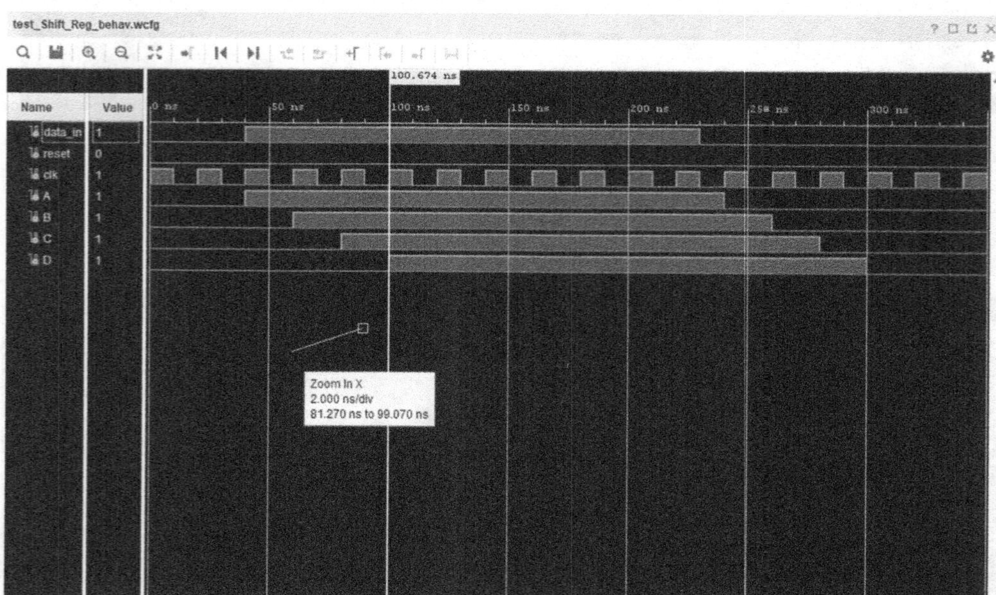

Figure 5.16 The zooming in.

And if you want to zoom in to a specific section – let's say we want to zoom in from 50 ns to 100 ns – we take our mouse and left-click at 50 ns. When we drag, we want to drag directly horizontal from the point we have selected. If we move up or down, we're going to miss it. But as soon as we go back, it'll go ahead and zoom in. However, if we go right in the middle, you'll notice it by seeing the red window, as illustrated in Figure 5.17.

So, when we do that, it zooms the range from where we clicked, and we'll get a little icon on the window. In this example, we're zooming more like from 50 ns to 100 ns. When we do this, we can see just the range between 50 ns and a little after 100 ns – it zooms us into that specific range. Now, let's say we want to look at the range from 60 ns to 70 ns. We can click and drag again to define that red window, and that zooms us into 60 ns to 70 ns. All of this depends on how long you run your design, how fast your clock (clk) is running, and what exactly is happening during that time. If everything you need to see is within 100 ns, there's no point in running the design for 500 ns. Just run it for 100 ns and focus on what's happening in that specific range. Now, if we're trying to find the exact point where something is happening, but we're not sure when it occurs, we can go back to our Full View. From there, we can look at specific edges. For example, if we select a signal – like the clock (clk) – as soon as we select the clock (clk), you'll see the outline edge of that signal highlighted, as shown in Figure 5.18.

If you'll notice, when the signal is high, it's shaded underneath; when it's low, there's no shade. This makes it much easier to see when a signal is high versus when it's low. You can pick out those transitions pretty easily. However, they are also tools to help you identify transitions. At the top, there's a little toolbar. We know we can zoom in by right-clicking, or we can click on the zoom in or zoom out buttons. You can also use Zoom Fit, Go to Cursor, Go to Time Zero, or Go to Last Time. These are just shortcut buttons that let you move around the simulation timeline. Now go back to Full View and bring the cursor right here. We have transitions, and we're looking at the next transition. Say we're on the clock (clk). If we go to the next transition, it'll take us from a low

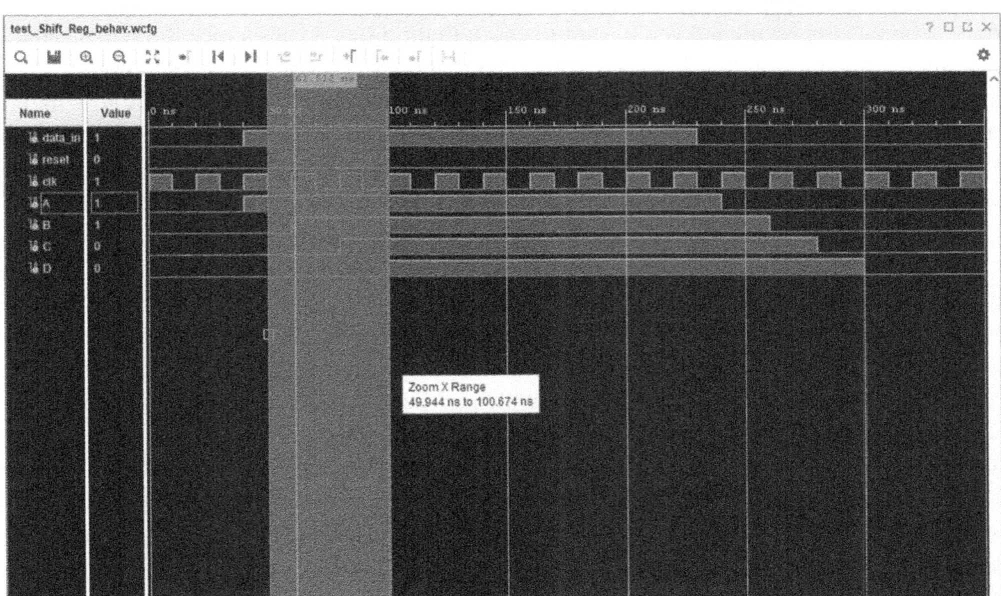

Figure 5.17 Getting the red window when zooming the range. The color rendition of this image is available on the book's landing page on wiley.com.

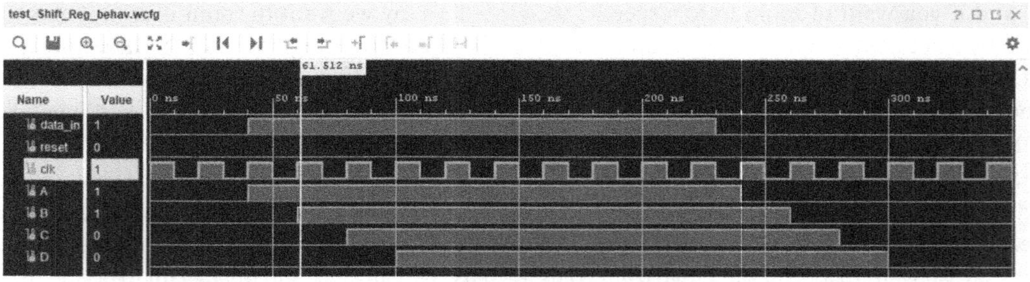

Figure 5.18 Highlighting the outline edge of clock signal.

to a high. A transition is defined as when your signal changes from one state to the next, and for each of these signals, they only have the state high or low. So, a transition is going to be when you go from a low to a high or from a high to a low. You can keep clicking through the transitions. On the clock, it transitions every 10 ns: up, down, up, down. Now, take another signal, say A. We can see that our next transition is from high to low. If we go to our next transition, it moves to the next one. We can also do a previous transition to go back. If we click it again, it will take us back to the very beginning. Once you reach the end, it won't take you any further. So that's kind of a nice way if you have a signal that you're looking to see what happens only when that signal is making a transition, where you can use that tool to kind of figure out where you're at. So, it's a nice little way to see the different transitions. Next, we will look at how to measure time between two events. For example, let's say we want to know the time difference between when output A goes high and when output C goes high. We can look at the scale and estimate that A goes high around 40 ns and C goes high around 80 ns, so we can calculate it out. But when you're working with different signals – especially in time-critical applications –something might be running at two different clock rates, or a signal might not be occurring exactly on a clock rate. You might be looking at an asynchronous signal coming in, and you'll want to check that timing. So, it's a nice thing to be able to do. In order to do that, you'll want to take your marker and select the first edge. For example, we're going to select our A signal. We'll go ahead and move to that transition – so now A goes high at 40 ns – and we're going to click the Add Marker button. That's going to add a marker at our transition A, which gives us a point at 40 ns. Now, we're going to select our marker, make sure we have our cursor selected, and we want to select C. So now we're at our signal C and we want to know the time at which C goes high. We'll go ahead and select that transition. That gives us our two transitions. At this point, we know the first marker is at 40 ns, and the second is at 80 ns, as depicted in Figure 5.19.

So, the difference between them is 40 ns. We can go ahead and add another marker and continue exploring. For instance, let's say we want to add a marker here – these markers simply indicate specific time points. Our start time is 40 ns, which is the first marker, and 80 ns is our second marker. This is a nice way to measure the time between certain events, as illustrated in Figure 5.20.

It requires you to do a little math, but you can see that it's 40 ns and 80 ns, and that tells you the difference. So, that's a little indicator to say, OK, this occurred at exactly this time right now with this design. It works out so that these things happen exactly at nice times, like 40 ns versus 80 ns. In some designs, you'll have things where, for example, it may have started at 42.87 ns and the next one started at 82.46 ns, so when that's the case, it's a lot nicer to be able to see the marker on your map, do the math right away, and understand if that's the time you'd expect to see.

The next thing is, what if you have a design that runs for a really long time? In this case, it's only 350 ns, so it's pretty short. But, for example, let's say all we're interested in – if there's a lot going

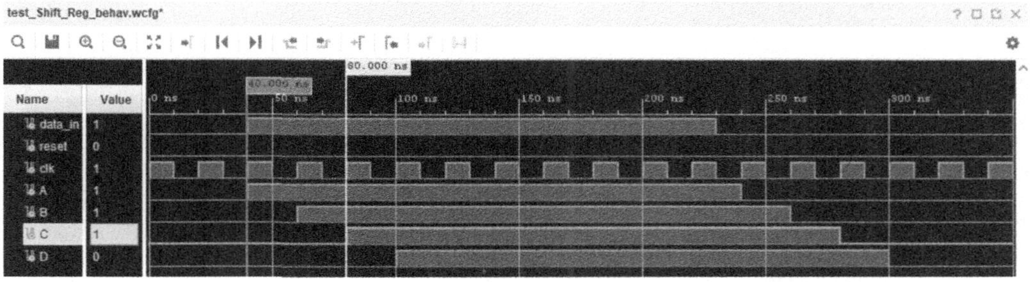

Figure 5.19 Selecting the first transition by Add Marker button.

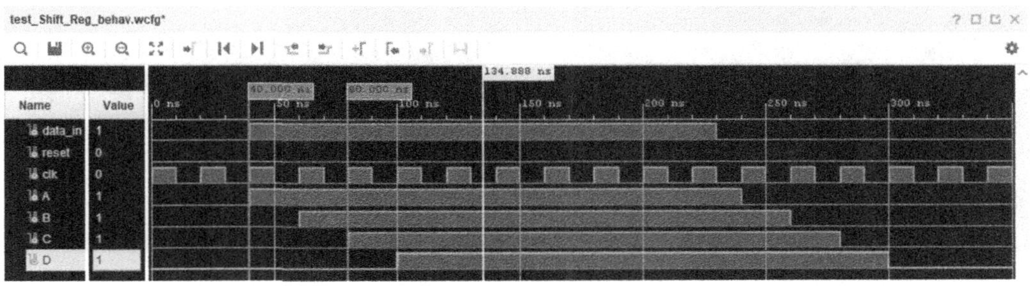

Figure 5.20 Measuring time between certain events.

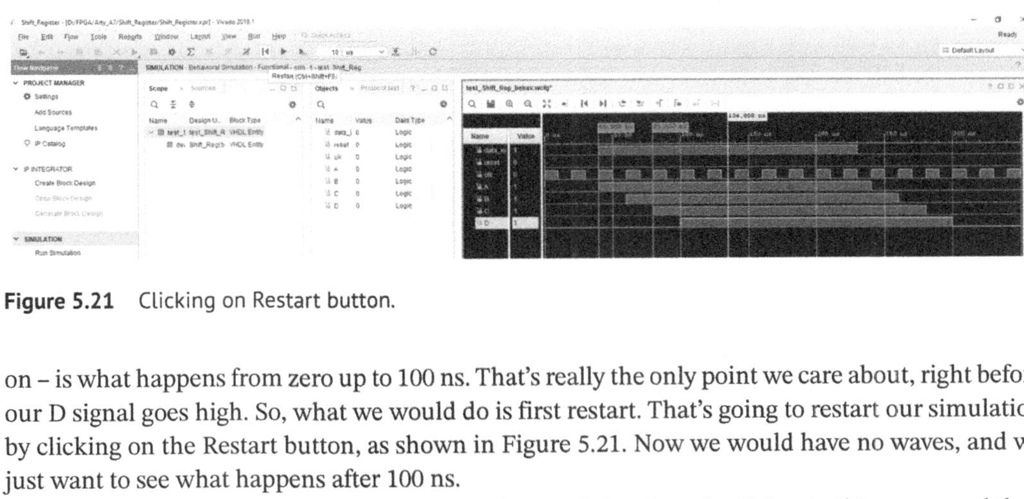

Figure 5.21 Clicking on Restart button.

on – is what happens from zero up to 100 ns. That's really the only point we care about, right before our D signal goes high. So, what we would do is first restart. That's going to restart our simulation by clicking on the Restart button, as shown in Figure 5.21. Now we would have no waves, and we just want to see what happens after 100 ns.

We type that in more time to make sure our time scale is selected – 100 ns in this case – and then we click Run for 100 ns, as shown in Figure 5.22. This is going to run for 100 ns.

So, for 100 ns, that's what it looks like, as shown in Figure 5.22. Now we can switch to the full view to get a better look at what's going on, if needed, as depicted in Figure 5.23.

In this case, we don't really need to, because we can clearly see what's going on. There aren't too many signals, but this becomes useful when you're dealing with a long design. You'll notice that some simulations can take a long time to run. So, if you have a simulation that might take –like 20 minutes to run – it makes sense to just say, OK, let's just run for 100 ns and see if we can observe what we need. So, it's nice to be able to run for just a short time and figure out exactly what we need to see. Then, you can design your test bench to test that specific case. When you run into that

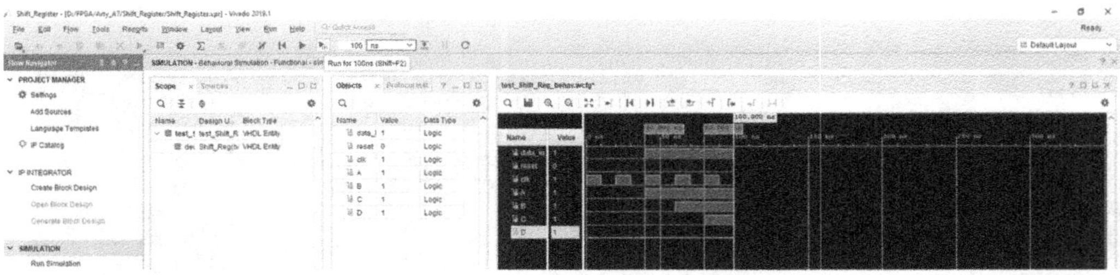

Figure 5.22 Clicking Run for 100 ns.

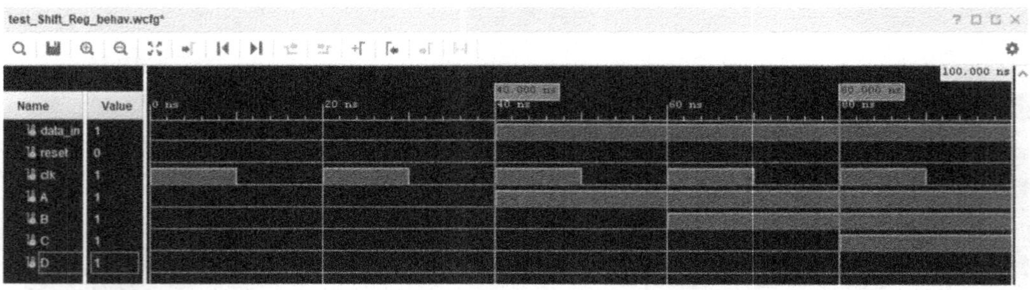

Figure 5.23 The full view for 100 ns run.

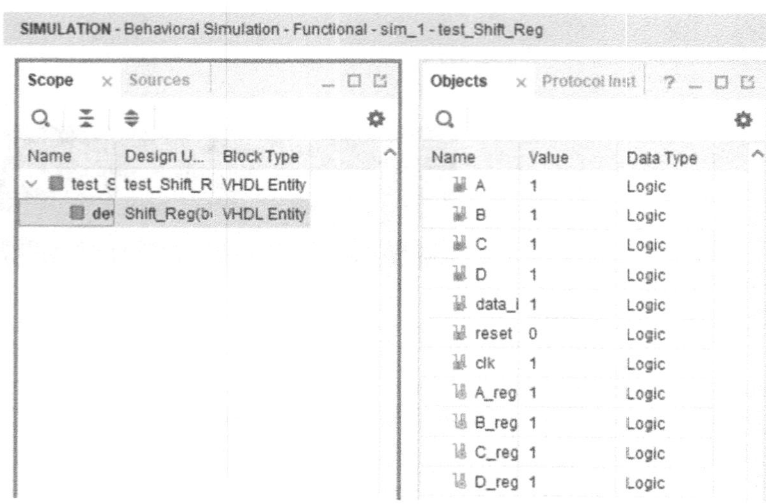

Figure 5.24 All of the signals in dev_to_test.

little bit, you can catch these kinds of details. Now, in this case, we're looking at every single signal in our design. So, if we look under Scope, we've got our test_Shift_Reg, and within that, we have our dev_to_test. When we select dev_to_test, it shows us all of the signals we have, as illustrated in Figure 5.24.

If, for example, A_reg doesn't currently exist in the waveforms window, then we only have data_in, reset, clock (clk), and the outputs (A, B, C, and D). So, if we want to see A_reg and what that register is doing, we right-click and select Add to Wave Window, as shown in Figure 5.25.

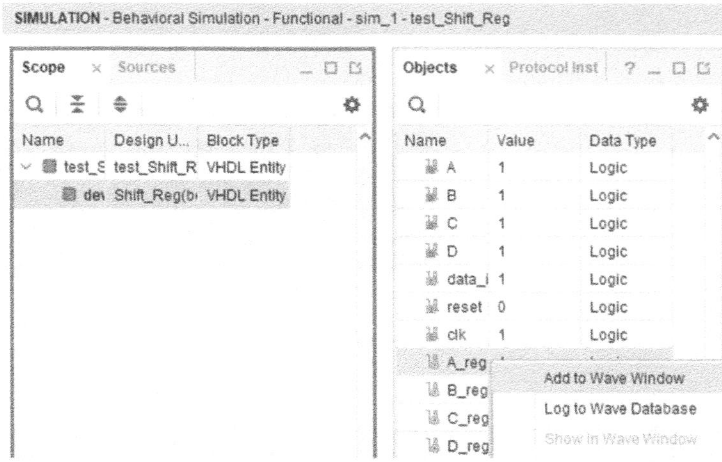

Figure 5.25 Adding A_reg to the wave window.

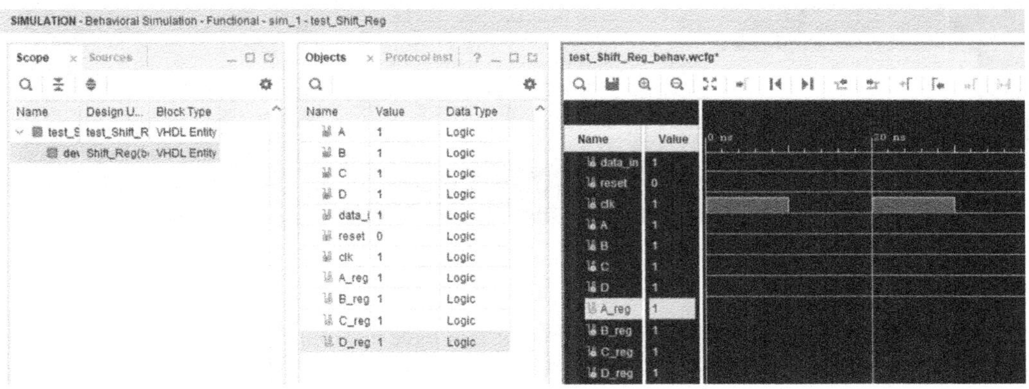

Figure 5.26 Added registers to the wave window.

Now that we have got A_reg, we want to add B_reg to the wave window, then add C_reg, and finally D_reg. So now we've got A_reg, B_reg, C_reg, and D_reg all to this wave window, as shown in Figure 5.26. This lets us see what these guys are doing.

However, if you notice, as soon as we added them, they do not show any values because we have not simulated what those values should be yet. So, we'll go ahead and restart the simulation. You'll notice we still have A_reg, B_reg, C_reg, and D_reg in the same window. Now if we click Run for 100 ns, the simulation calculates the values. You'll notice that A_reg, B_reg, C_reg, and D_reg are calculated with these values, as depicted in Figure 5.27.

So that's pretty cool. Go ahead and run the simulation for another 100 ns, and another, and so on –we have now run for a total that covers every 100 ns step. From our data_in, we can see the value shifting out through A, B, C, and D. And now, we also see A_reg, B_reg, C_reg, and D_reg. This is useful when you have a design with lots of signals – you might not want to have all of them in your wave window. So, you can right-click and use the Cut action to remove signals from the wave window, as illustrated in Figure 5.28.

So let's go ahead and do that. Just select the signal, right-click, and choose Cut to remove it. That's how you can basically take signals that are in your design, especially the ones that are internal.

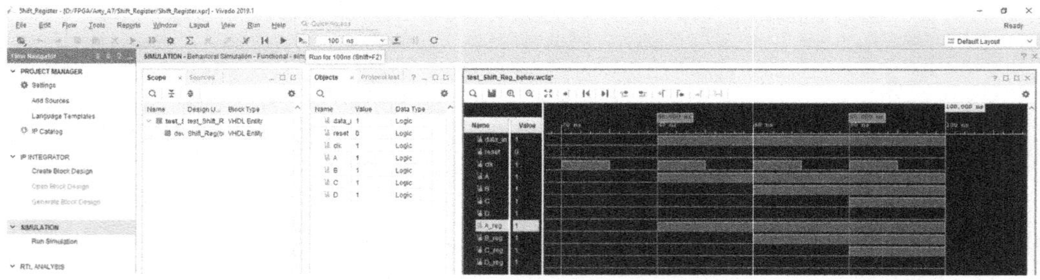

Figure 5.27 The calculated values for A_reg, B_reg, C_reg, and D_reg.

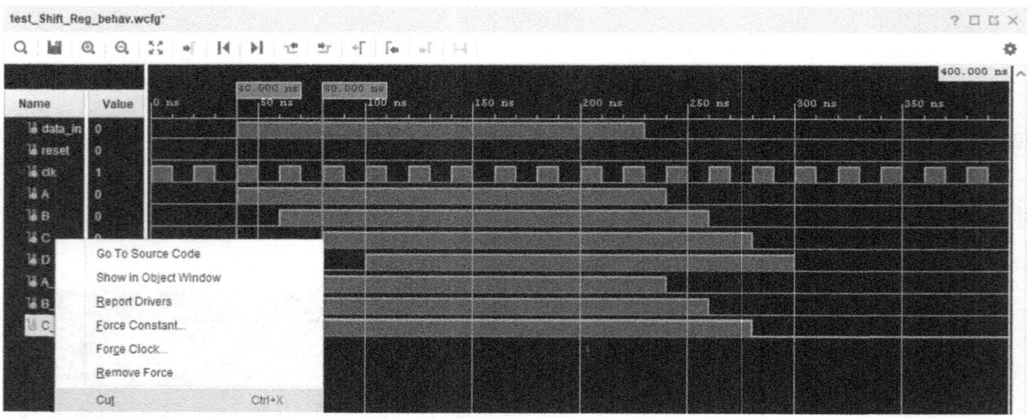

Figure 5.28 Removing D_reg and C_reg from the wave window.

You already have your inputs and outputs, but you also have internal signals you want to look at. It's helpful to add those to the wave window. For example, if you have a big design with 50 signals, you might only be interested in five. Just add those five to the wave window to see what they're doing. Based on those values, you may add more and rerun the simulation to figure out what kind of help you debug and determine that your design is working as expected.

6

The Buttons and LEDs Project

6.1 Introduction

This chapter is about the introduction of buttons and LEDs on the Arty A7 boards. In this chapter, we're going to explain what we're trying to do with the project buttons and LEDs. For this project, we're going to look at the Arty A7 development board. Our goal is to incorporate the four push buttons on the Arty A7 development board, a single slide switch, and the four regular LEDs on the board. What we want to do is use an enable signal, which is going to be used by one of the slide switches. When enabled, pressing a button should cause the corresponding LED to light up. For example, if we press button zero (BTN0), which is the button on the farthest right, we want LD4, the regular LED closest to the right, to light up. Similarly, for button one (BTN1), we would like the second LED (LD5) to light up. If we press all four buttons at once, all four LEDs light up at the same time. That is the goal for this chapter. This project incorporates two different types of inputs: four push buttons, a single slide switch, and then four LED outputs. It serves as a basic introduction to the workflow of using tools for Vivado, developing a basic VHDL design, including the entity, the architecture, and the basic ins and outs of standard VHDL design, as well as the process to implement it using the tools, program it, and run it on this Arty A7 development board. That is the overview of the introduction to the development board for the buttons and LEDs project.

6.2 Buttons and LEDs Complete Design Walk-through

This section is about the complete design walk-through. In this section, we're going to show you how and what we do to have all these things in the Button_LED project, what everything is doing, and why it's there. So, in the Button_LED project, we know that we have our specific inputs – our buttons and an enable – and our outputs, which are the LEDs. So, at the very beginning, let's just walk through the design and explain everything we have in the design and why we have it, as shown in Figure 6.1.

So, at the very top, we have some comments that you can fill out to help explain the design. This way, when you look at it later, you get an understanding. For example, there are fields for the company name, the engineer's name, the design name, and the project name. The important thing you include here is a description explaining what this file actually does. That way, if you open it up, you can just read the description and get an understanding of what's going on. Additionally, there is a revision, which is useful for keeping track of all the changes that

Fundamentals of VHDL for FPGA Programming Using Vivado, First Edition. Majid Pakdel.

Companion Website: www.wiley.com/go/pakdelVHDL

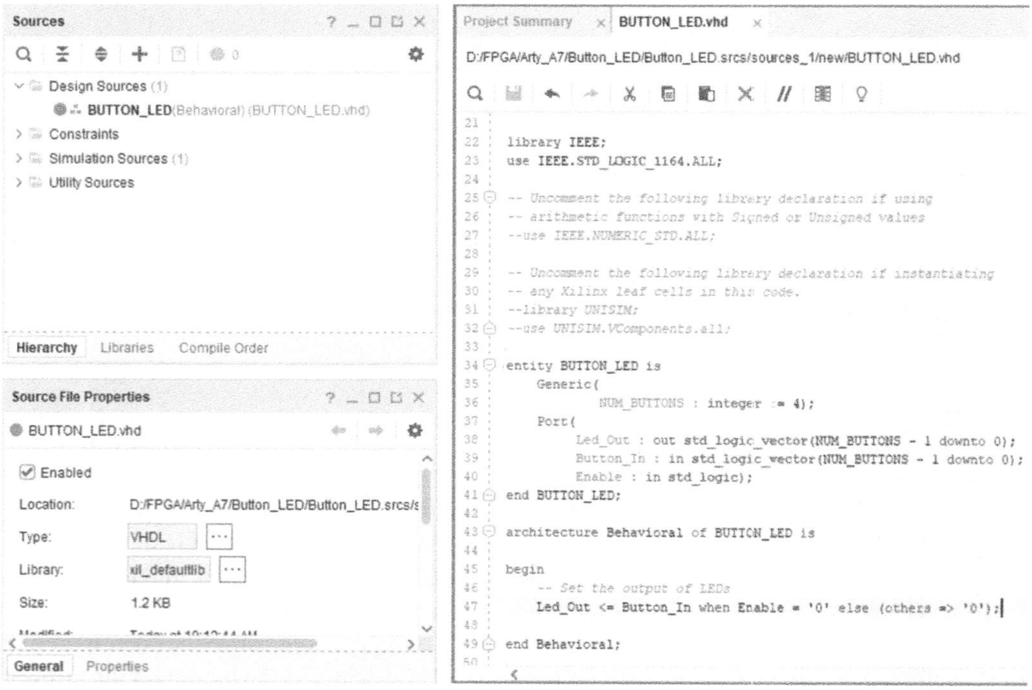

Figure 6.1 The Button_LED project.

have been made. Next, we have the library and we're using the IEEE library, which is very common, and we know that all of our designs will be using this library. From the IEEE library, we're using the standard logic 1164 package, which gives us access to standard logic as well as the standard logic vector data types. It also allows us to perform different Boolean expressions, such as a logical AND, OR, NOR, and other logical operations. Now, we have our entity, which we just call BUTTON_LED. We define our entity with a generic, which is NUM_BUTTONS, and this value is going to change based on the development board being used. For the Arty A7 development board, NUM_BUTTONS is set to 4.

So, when we say we have four buttons, that means we have four output LEDs and four input buttons. We also have Enable, which is just one. The Enable is just a standard logic data type, while Led_Out and Button_In are standard logic vectors. This defines the entity for our Button_Led project design. Next, we have the architecture, which we have named Behavioral. The Behavioral architecture is describing the BUTTON_LED entity, and then we have the begin-end pair to just describe the actual architecture of what's going on inside our design. For this design, we use a when-else statement to control the output of the LEDs. We set the output of the LED (Led_Out) to match the button input (Button_In) when Enable is equal to zero; otherwise, the output is equal to zero. If the buttons have a high pull-up resistor, they are high by default. So, you just have to do an inversion on the buttons input, and that's just done by putting a not in front of Button_In. So, all this statement is doing is basically deciding the output of the LEDs, if it's a Button_In or if it's a zero, based on Enable, which is tied to a slide switch. That provides an overview of our design and a complete walk-through from top to bottom, what we're trying to do with this Button_LED project, that is the entire VHDL code needed to get the project to run successfully.

6.3 The Button_LED Arty A7 IO Placement

The section is about the Vivado FPGA IO placement. Here, we're going to show you how to place the IO that we'll be using for our design on the Arty A7 development board within Vivado. So, at this point, our project is completed, where we have the VHDL design completed. Now, we need to specify where the actual items we have defined in our entity, specifically Led_Out, Button_In, and Enable, are located on the actual FPGA device. Specifically, we need to determine which IO pins are tied to the LEDs that we're going to use. In order to do that, we need to add what we call a constraint file, which has the extension ".xdc." So, if we go into Constraints, we can see that once we expand it, we have no constraints on our project. So, we need to right-click and then click on Add Sources, as shown in Figure 6.2, and make sure you have the first selection, Add or create constraints, and we'll click Next, as depicted in Figure 6.3.

Then, you click on "Add Files," as illustrated in Figure 6.4. At this point, you just need to navigate to the constraints file that we have provided, titled Arty_A7_Master.xdc. This is the master constraints file for the Arty A7 development board, so you will need to navigate through this constraint file, as shown in Figure 6.5.

So, we go ahead and click the Finish button, as depicted in Figure 6.6. After it updates, we can now see if we can expand our constraints. If we double-click it, it'll open up the .xdc file, which contains all the IOs on the Arty A7 board, which are defined, and you just have to uncomment the ones you want to use and rename them accordingly.

So, for example, if we go to our Button_LED, we know we have this Led_Out, so we copy it, and it's going to be from 3 down to 0. If we go to our Arty A7, we have our standard LEDs. So, we just have to replace led with Led_Out, and that will then map out our Led_Out to its location. If you look at the actual syntax of what's going on here, the only thing we're changing is the actual name. Previously, it was led, and we just changed it to Led_Out. So, we have Led_Out from 0 to 3, and we can go ahead and select all those lines and click Uncomment. Now, we can see that the set_property and get_ports keywords have been highlighted, and we'll go ahead and click Save. So, at this point, we've got Led_Out mapped to the actual outputs. Now, Button_In needs to be quickly mapped. To do this quickly, we copy Button_In from 3 to 0 in Button_LED. Just like we do with the latest, we want to replace the btn with the IO we have specified in our entity. So, we change it to Button_In, and we'll do that for all four of these here. Then, once again, uncomment – you can

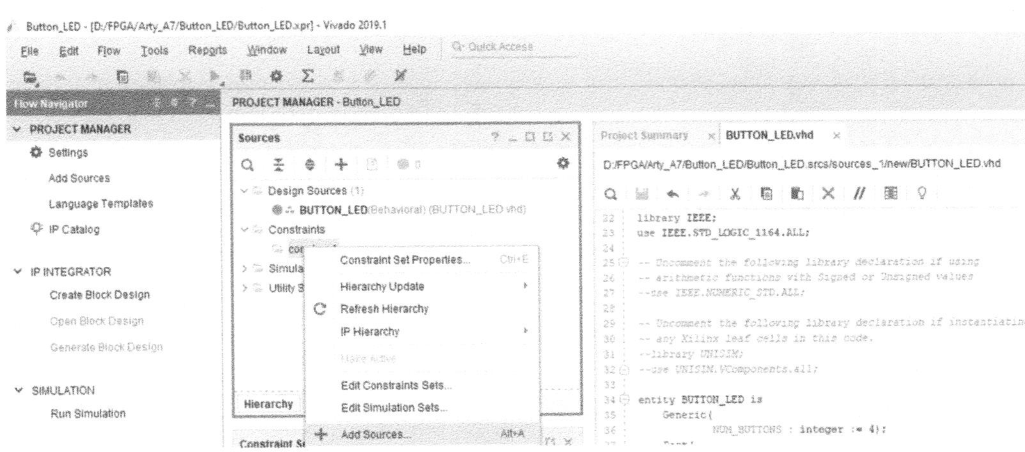

Figure 6.2 Selecting the Add Sources... option.

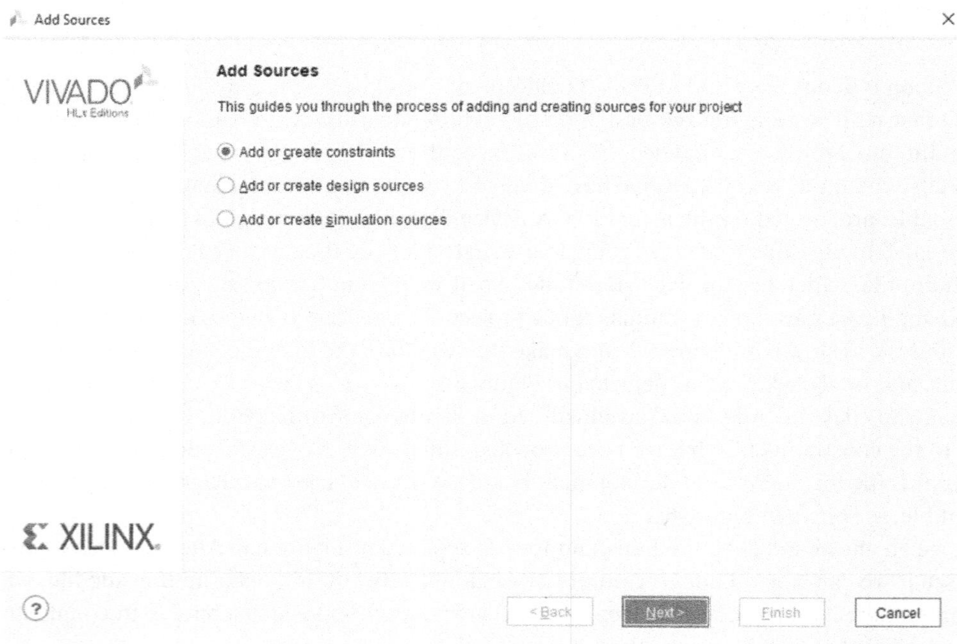

Figure 6.3 Selecting the Add or Create Constraints.

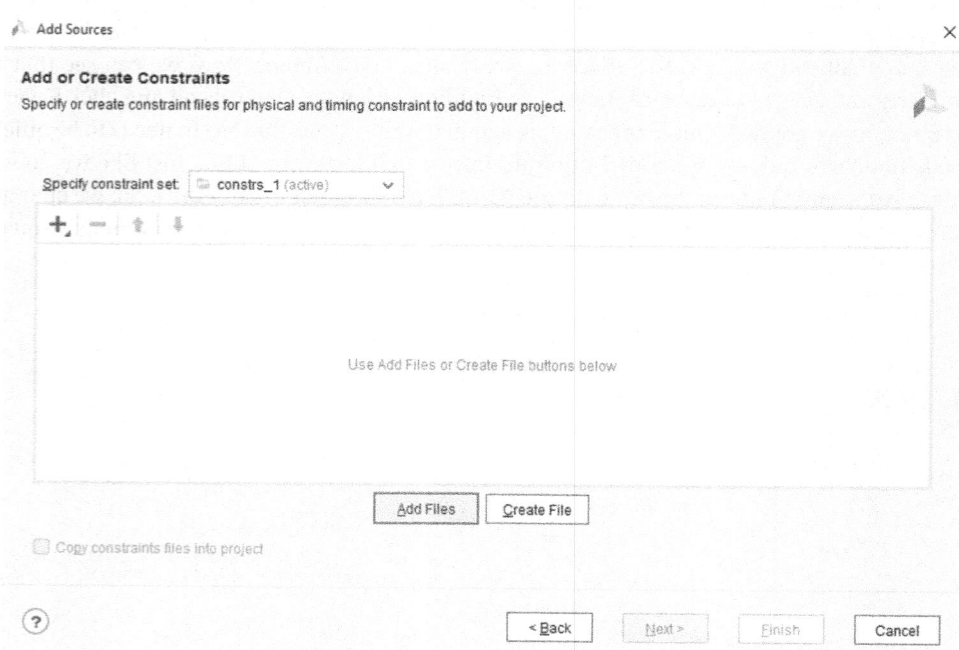

Figure 6.4 Clicking on the Add Files button.

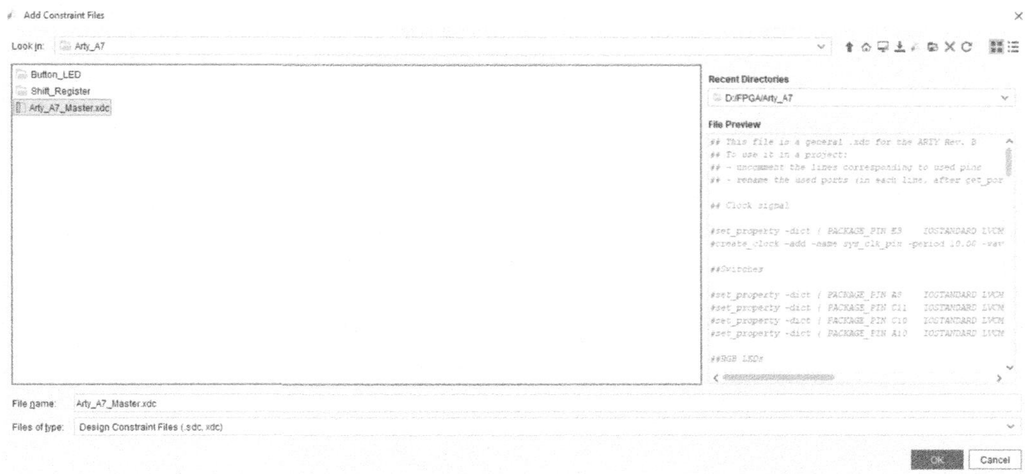

Figure 6.5 Selecting the master constraints file for the Arty A7.

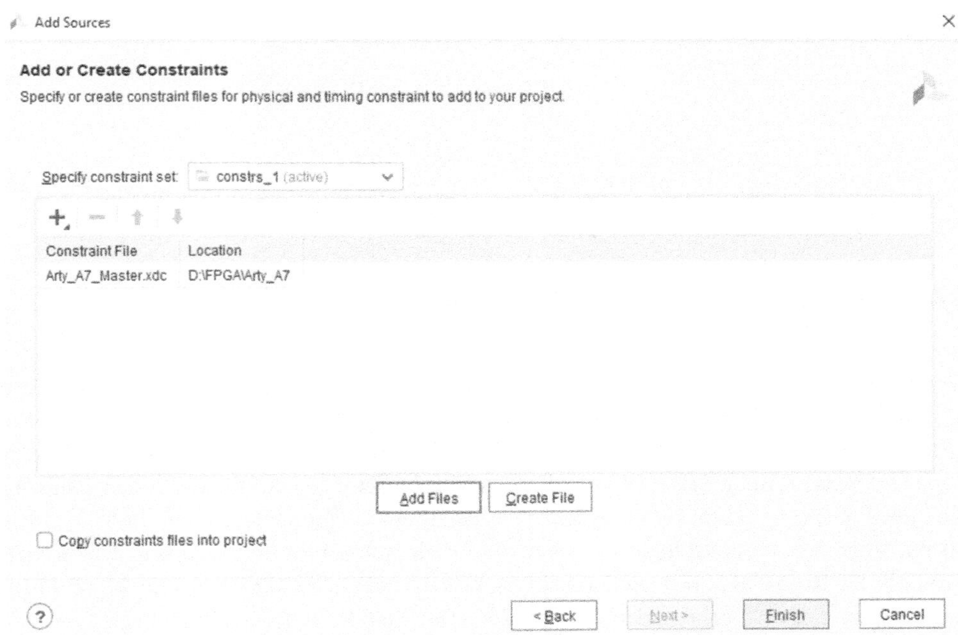

Figure 6.6 Clicking on the Finish button.

click the comment line and press backspace. The hashtag is what generates a comment here. So, go ahead and click Save. If you look through the rest of the file, you can see we've got the RGBs, the switches, the input clock, all the different Pmod headers – JA, JB, JC, and JD – and your USB to UART interface, as well as the ChipKit digital I/O. So, all of the IOs that are available on that board can be accessed through here. They have all the specific IO pins that tie to the interface for Ethernet, quad SPI flash, power measurements, and so on. So, if we go back to our design, we now have Led_Out and Button_In, so we just need to tie Enable. If we right-click and copy to pick, what do we want Enable to be tied to? We know we went through one of the slide switches. If you look

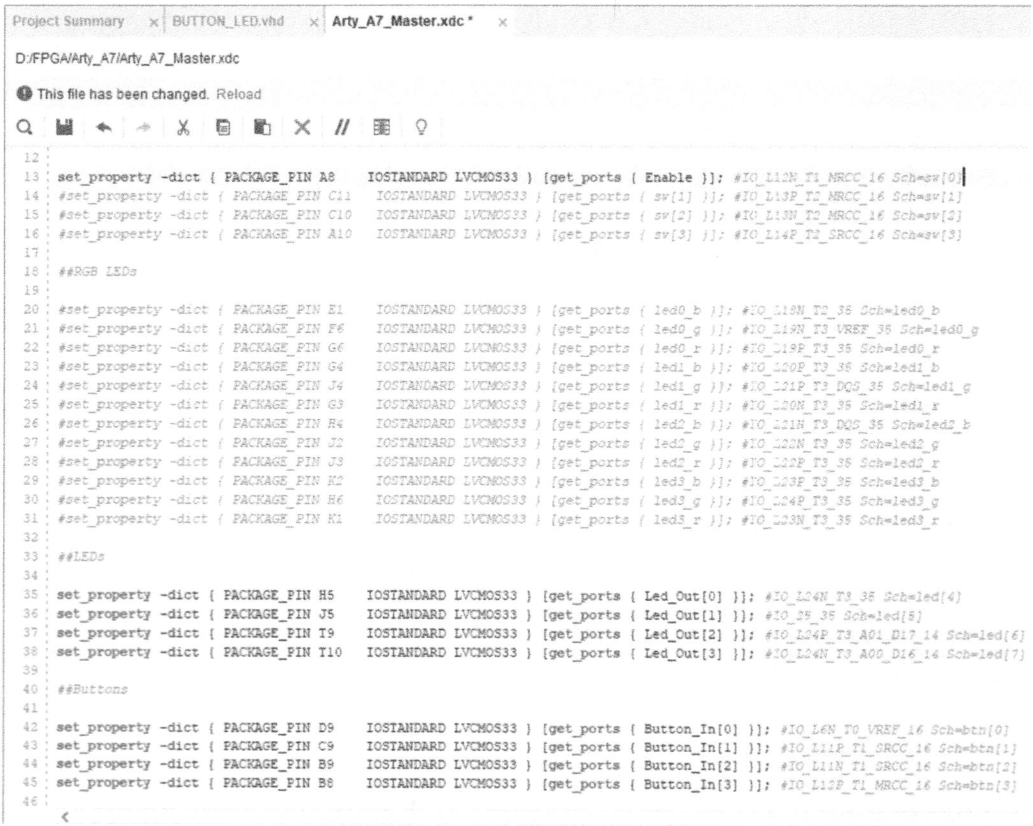

Figure 6.7 The uncommented ports in Arty_A7_Master.xdc.

at the schematic, it tells you that switch zero is named sw[0]. So, there are switches 0, 1, 2, and 3, and switch 0 (sw[0]) is the one furthest to the right on the Arty A7 development board. We're going to pick that to cover the entire interface. Now, a key thing to note: Enable is a standard logic data type and Button_In and Led_Out are standard logic vectors. Since Enable is just a single bit, we don't need to specify a bracket to say zero because there's just one. However, the Led_Out and Button_In, which are standard logic vectors with four elements, require brackets to access each of the elements. So, from 0 to 3, you need to use brackets in there to tie to that specific bit of the standard logic vector. So, we'll go ahead and uncommon our Enable, as shown in Figure 6.7.

At this point, our design is ready to generate a programming file. We have our VHDL design completed, and all of our inputs and outputs are defined in our ".xdc" file. The generic does not need to be defined in our file because it's not an input or an output, and it's not tied to any specific pin on the actual FPGA. That's everything you need to do in order to map or place the IO for FPGA in Vivado.

6.4 Generate a Xilinx Programming File

This section is about the generation of a Xilinx programming file. Here, we'll walk you through the steps required to generate a programming file for a Xilinx development board. A Xilinx programming file is a .bit file or a bit, and this file holds all the configuration data derived from our VHDL

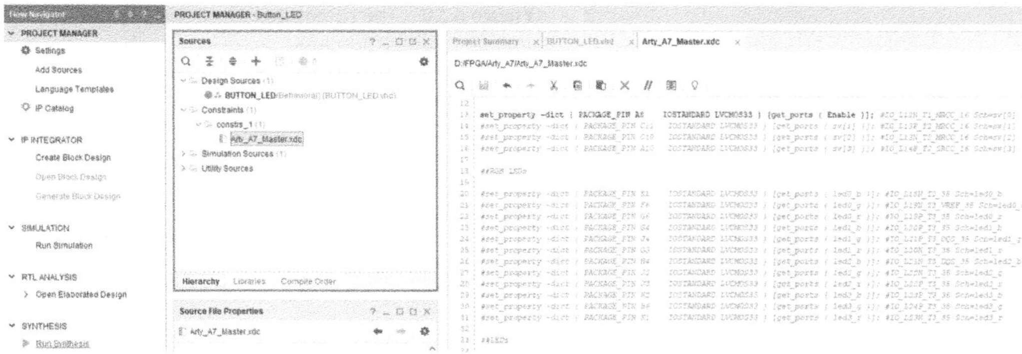

Figure 6.8 The constraints file.

Figure 6.9 The Launch Runs window.

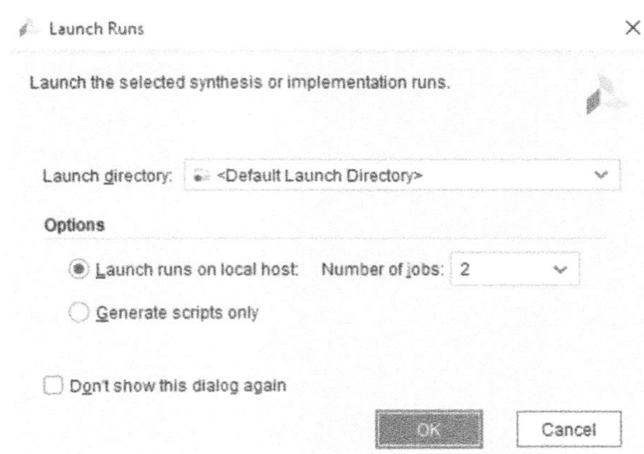

design, our constraints for the pin-out location, timing constraints, and more. It uses all this information to configure the FPGA to perform exactly as specified. The steps to generate this file begin with ensuring that a top-level design file is defined. In this case, the top-level design file is a VHDL file. However, it could also be a block design, IP core, or whatever you have in your top-level design. You'll notice it by the three squares – two blue and one green on top. If you select and right-click, you could do a set as top-level file. However, it already recognizes that Vivado is smart enough to know the top-level file. In addition, you'll also need to have a constraints file before generating a programming or a .bit file. We already have one there, as shown in Figure 6.8.

A constraints file contains the pin-out locations, timings constraints, and any other types of constraints you want to apply to your project. These are defined in your .xdc file. So, the first step is to go ahead and click Run Synthesis, as shown in Figure 6.8. This'll bring up a small window, as depicted in Figure 6.9. Simply click OK, because we will go ahead and run the synthesis.

Synthesis is the process of analyzing your VHDL design to verify that you have the correct syntax and that the design makes logical sense. There are several steps involved, and regardless of the type of tool you're using, you'll need to synthesize your design and then do a fitting or implementation. In the next step, it is going to ask us if we want to implement our design, as illustrated in Figure 6.10.

Now that the synthesis has been completed successfully, and there are no errors or warnings, this means our VHDL is syntactically correct, and the compiler – or the synthesizer for

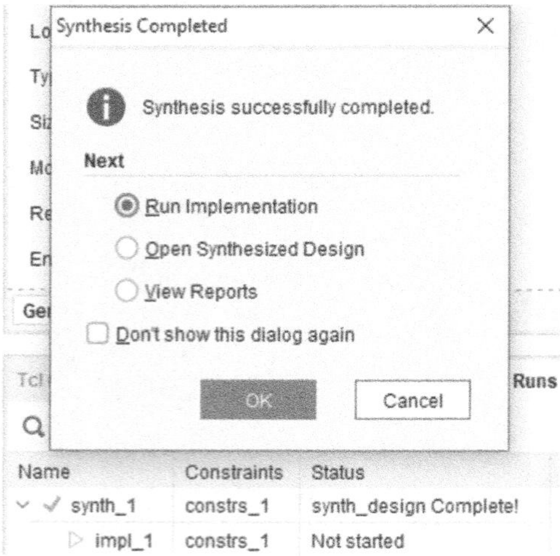

Figure 6.10 The completed synthesis and Run Implementation window.

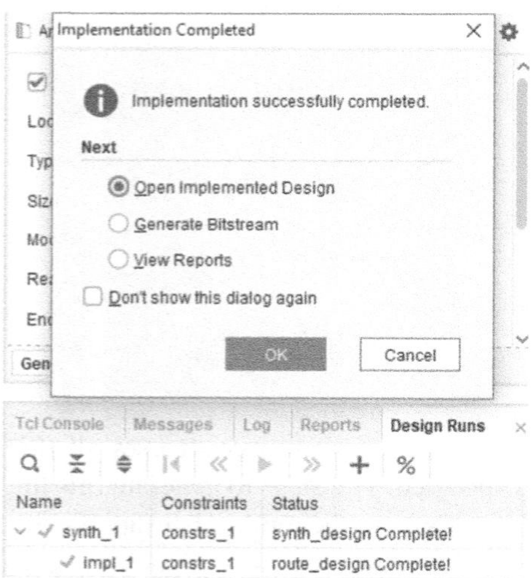

Figure 6.11 Successful completion of implementation.

Vivado – understands it and can implement the VHDL design we have described. So now, we want to run our implementation. It'll bring up a little window, as depicted in Figure 6.9, and we're going to click OK because we're going to go ahead and run the implementation. This is going to take our synthesized design and attempt to fit it on the actual device. You could have a VHDL design that passes synthesis but takes up so many resources that you can't actually fit it on your FPGA development board. If that happens, you'll get a failure or error thrown, saying, "Hey, the implementation failed because we can't fit it on that design." For example, it might take over 100% of the lookup table or it could define more IOs than are available on that FPGA, which will result in a failure there. If your .xdc file does not define any of the IOs, it's going to throw a lot of warnings. It will indicate that while the design can be implemented, it doesn't know where to place the inputs or outputs on FPGA. So, in this case, our implementation has completed successfully, as shown in Figure 6.11.

So, as you go through the different steps, you can see that we've passed synthesis with no errors or warnings and that our implementation is successfully completed. The final step is to generate the bitstream.

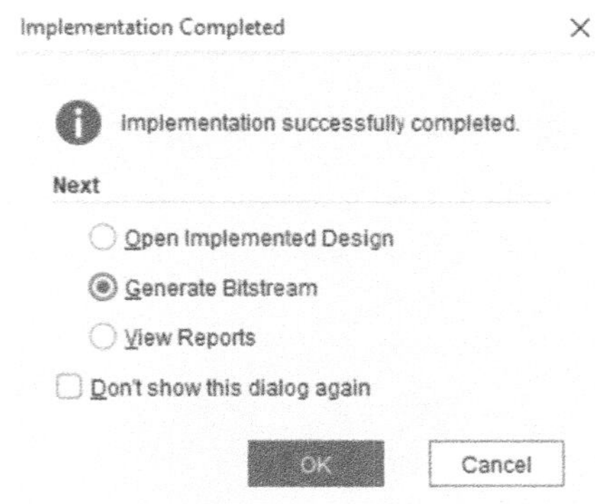

Figure 6.12 Selecting the Generate Bitstream.

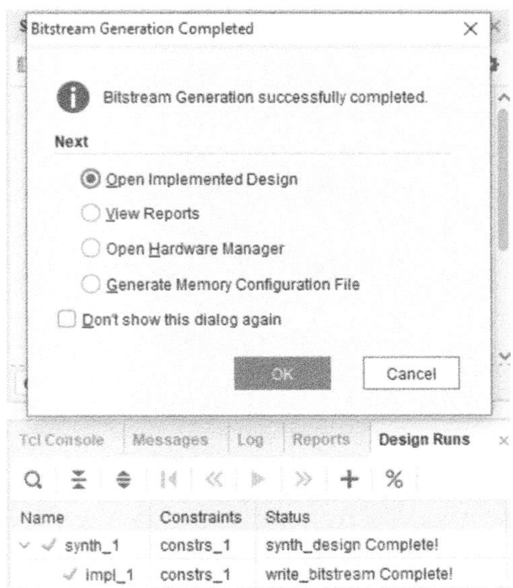

Figure 6.13 The completed bitstream generation.

Click on Generate Bitstream and select OK, as shown in Figure 6.12. This process takes our synthesized and implemented design and generates the ".bit" or the bitstream file, which we'll use to program the actual Xilinx board or device.

A lot of times, when you're working with Code C/C++, people tend to focus on the number of lines of code or how much memory it takes up. However, with FPGA, it just comes down to how many resources the design uses, which makes a big difference. You can have a lot of lines of code or just a few, but that doesn't necessarily make a big difference. Now, our bitstream generation has completed, as depicted in Figure 6.13.

At this point, we'll want to click on Open Hardware Manager and then select OK, as illustrated in Figure 6.14.

This brings up the Hardware Manager, which is what at this point you will use to actually program the physical device as shown in Figure 6.15.

So far, we've run synthesis, completed implementation, and then generated the bitstream. If you look at the left side of the Vivado interface, you'll see the Flow Navigator, which outlines the flow of your project. At the very beginning, you can see project settings. This is where you can define parameters depending on the type of project or a specific device. If you have specific constraints

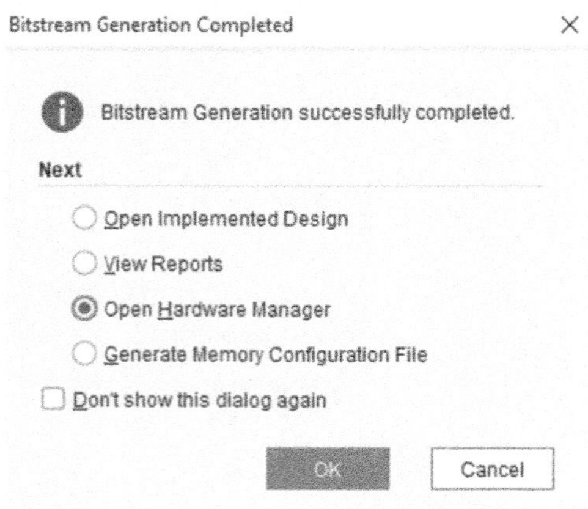

Figure 6.14 Selecting Open Hardware Manager.

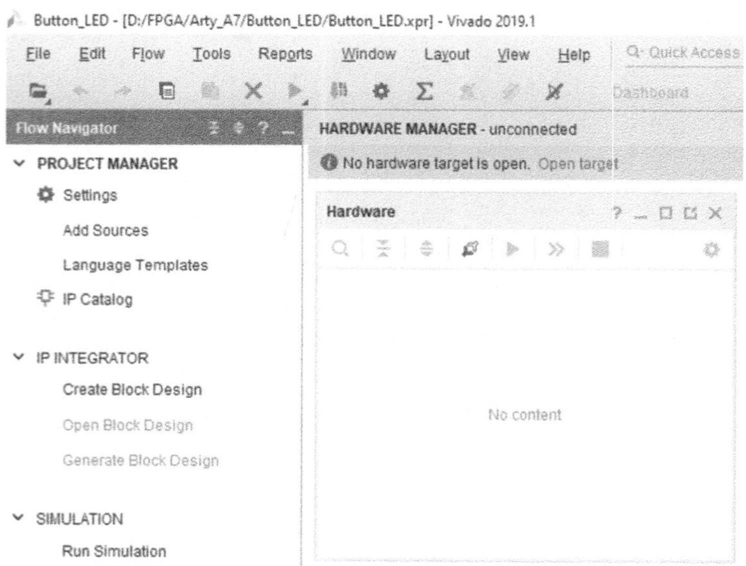

Figure 6.15 Bringing up the Hardware Manager.

such as a clock or anything that's absolutely specific and necessary, you go through the project settings to get that set. This also selects the chip you're targeting – if it's a Zynq, Artix, or any other device in the Xilinx family. Next, you have Add Sources, where you can add your VHDL or Verilog files. It also kind of goes with the IP catalog, where you can add an IP as well, which is basically a block diagram representation of VHDL. So, we can create a block design, and you'll drop these IPs in there from the IP catalog. Then, from that, you'll generate a VHDL file that becomes your top-level source file. You can then hook in your own VHDL, or you can take your own VHDL and generate an IP. After that, you'll have simulation. If you want to simulate your design, that would be the next step. Then, there's RTL analysis. If you want to look at and analyze your design, it'll give you maybe some suggestions on things you could do next. The key thing you need to do to actually program is run synthesis, run implementation, and then generate the bitstream. Those are the three key steps that you need to do. But if you look at the Flow Navigator, they're listed right there, and it's very easy to say, "OK, run synthesis, run implementation, and generate bitstream." After each step, they bring up a dialog box that helps you and asks, "What do you want to do next?" Do you want to look at the design or do you want to run implementation? OK, now implementation is complete – do you want to look at your warnings or do you want to generate your bitstream? So, we're going to go ahead and close out the hardware manager and go back to look at the messages we got. If we click on the messages, it shows that we have three warnings. So, we will uncheck the info messages, and it says we have three warnings. In the Route Design, it says there are no user-specific timing constraints. Timing constraints are needed for proper timing analysis. The next warning says that no user-defined clocks were found in the design. That's because, for this specific project, we don't have any clocks, and therefore we don't have any timing constraints. And since I'm pressing a button and looking to see how fast it outputs on the LED, we're not worried, so we didn't put timing constraints in the file. However, if we wanted to remove these warnings, we would put timing constraints in the file to remove them. These are just warnings, so just because you have warnings doesn't mean your design won't work. These are just things Vivado wants you to be aware of. So, these are the steps required to generate a programming or a .bit file for your Xilinx development board.

6.5 Simulation of Button_LED Project

This section is about running the simulation in Vivado. Here, we're going to show you how to use Vivado to simulate the Button_LED project. In order to do that, we'll select the simulation sources, right-click and click on Add Sources. Make sure you select the third option "Add or create simulation sources," as depicted in Figure 5.4. Then we'll click Next, and it's a specified simulation set, which we'll leave as the default sim_1. After that, we'll create a VHDL file (test_Button_LED.vhd), as illustrated in Figure 6.16.

Next, we click on OK and Finish buttons, and in the Define Module window, we set the architecture name as "test" and then click OK, as shown in Figure 6.17.

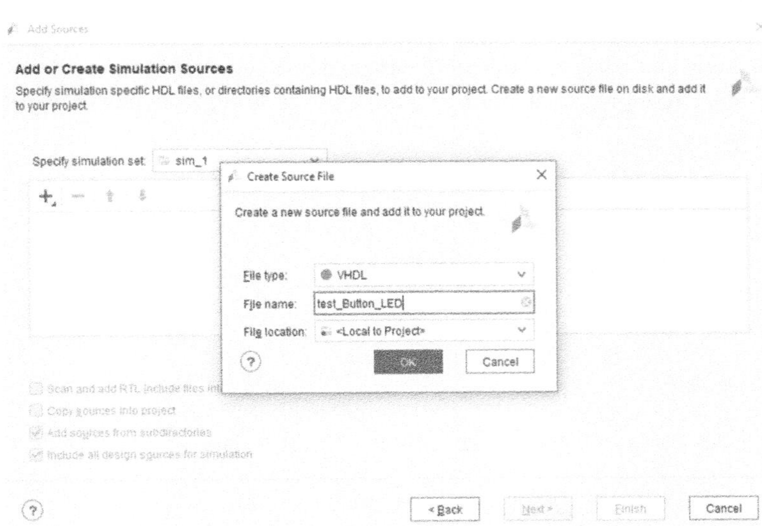

Figure 6.16 Creating a new file.

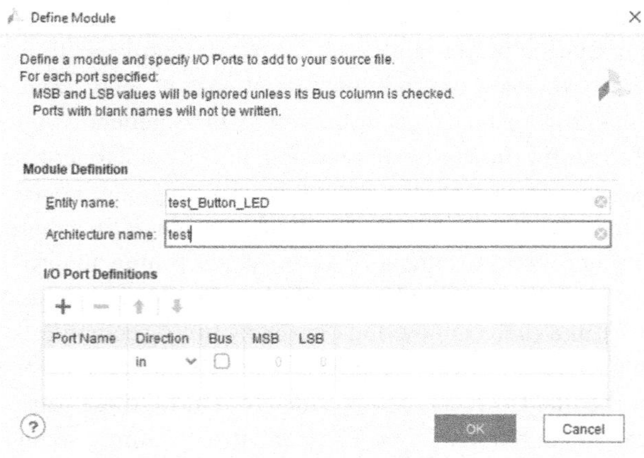

Figure 6.17 Setting the architecture name as "test."

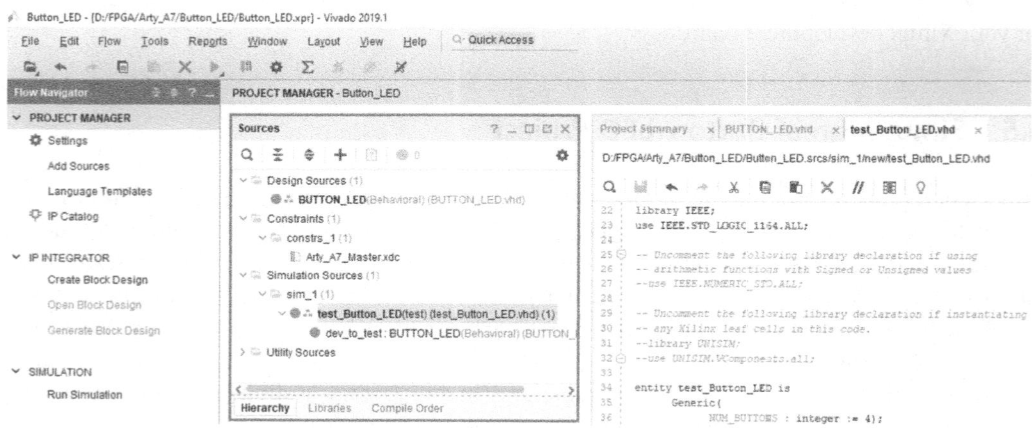

Figure 6.18 The test_Button_LED.vhd as a top-level design.

Then, we can copy and paste the following designed test bench code inside test_Button_LED.vhd. Once updated, it recognizes test_Shift_Reg as a top-level design for our shift register, as shown in Figure 6.18.

```
library IEEE;
use IEEE.STD_LOGIC_1164.ALL;

entity test_Button_LED is
      Generic(
            NUM_BUTTONS : integer := 4);
--  Port ( );
end test_Button_LED;

architecture test of test_Button_LED is
component Button_LED
port(Led_Out: out std_logic_vector(NUM_BUTTONS - 1 downto 0);
     Button_In : in std_logic_vector(NUM_BUTTONS - 1 downto 0);
     Enable : in std_logic);
end component;
```

```
signal Led_Out: std_logic_vector(NUM_BUTTONS - 1 downto 0);
signal Button_In: std_logic_vector(NUM_BUTTONS - 1 downto 0) := (others => '0');
signal Enable: std_logic := '0';

begin
    dev_to_test: Button_LED
        port map(Led_Out, Button_In, Enable);
    data_stimulus: process
    begin
        wait for 10 ns;
        Button_In(0) <= not Button_In(0);
        wait for 10 ns;
        Button_In(1) <= not Button_In(1);
        wait for 10 ns;
        Button_In(2) <= not Button_In(2);
        wait for 10 ns;
        Button_In(3) <= not Button_In(3);
        wait for 50 ns;
        wait for 10 ns;
        Button_In(0) <= not Button_In(0);
        wait for 10 ns;
        Button_In(1) <= not Button_In(1);
        wait for 10 ns;
        Button_In(2) <= not Button_In(2);
        wait for 10 ns;
        Button_In(3) <= not Button_In(3);
        wait for 50 ns;
        wait for 10 ns;
        Button_In(0) <= not Button_In(0);
        wait for 10 ns;
        Button_In(1) <= not Button_In(1);
        wait for 10 ns;
        Button_In(2) <= not Button_In(2);
        wait for 10 ns;
        Button_In(3) <= not Button_In(3);
        wait for 10 ns;
        Enable <= '1';
    end process data_stimulus;
end test;
```

So, the next step is to run the simulation. However, before we do that, we need to configure the simulation settings, which brings up some of the basic settings for simulation. We'll open Tools → Settings..., with the simulation settings, as depicted in Figure 5.8. The only other value we will change is that under the Simulation tab. We will set the runtime to 300 ns, and then click Apply, as illustrated in Figure 6.19.

Now, we'll select Run Behavioral Simulation, as shown in Figure 5.10. Once inside the simulation tool, as seen in Figure 6.20, we can observe that we have test_Button_LED running, simulating our dev_to_test. It displays all relevant signals: Led_Out, Button_In, and Enable. If we look over to the right, we can see the waveforms. Right-click on the waveform and select Full View, as shown in Figure 6.20.

That completes all the steps required to run the simulation of the Button_LED project within Vivado.

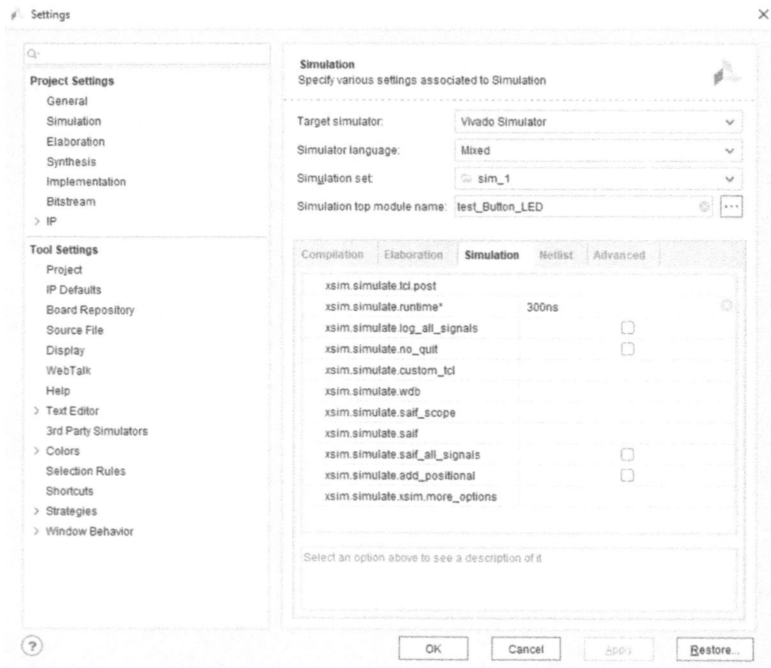

Figure 6.19 Setting the simulation runtime to 300 ns.

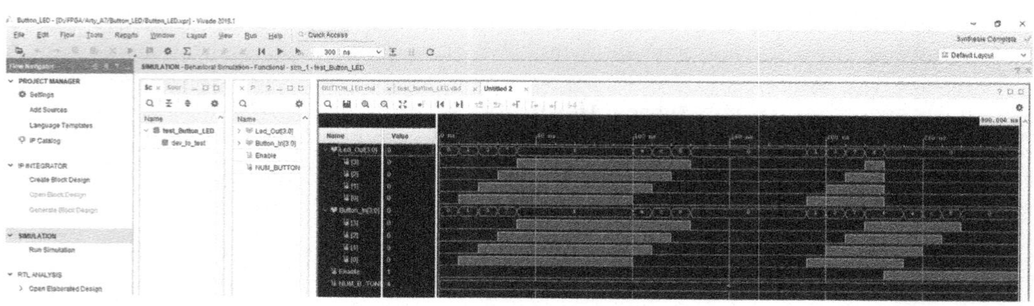

Figure 6.20 The Behavioral Simulation window.

7

The Blinky LEDs Project

7.1 Introduction

In this chapter, we are going to review some VHDL statements, including constants and signals, VHDL processes, VHDL if statements, as well as an example of if statement priority. Finally, the blinky LEDs project is implemented with the Arty A7 development board using Vivado.

7.2 Constants and Signals

This section is about signals and constants. In this section, we're going to explain how signals as well as constants are used in VHDL. Now, a signal or constant is not a data type. You can have a constant that is an integer, a standard logic, or a standard logic vector. Similarly, you can have a signal that is an integer, a standard logic, or a standard logic vector. Thus, a signal and a constant are types of object in VHDL that we use to define our designs. First, we're going to start with constants. What constants do in the VHDL language is give you the ability to have a better documented code. For example, if you're using a baud rate throughout your design, you could just type in that value. So, if you're running at 9600, or 9600 bits per second baud rate, you could type that in everywhere you want to have baud rate. But if you use a constant, it just makes it much easier to read, where in one line you calculate what the baud rate is or you define a constant, which would be like an integer to a specific baud rate, and then everywhere else in your code where you reference that baud rate, you just use that constant name. So, the big thing about a constant – the key thing to know – is it allows you to document your code much better, and it looks much easier. When you come back later and look at the code, you'll be able to understand what's going on more clearly. The syntax for a constant is just "constant" spelled out. To define it, you would write "constant", then the constant name, a colon, the constant's data type, an equals sign, and finally, a semicolon. We'll show some examples of how to define constants. The next thing we're talking about is a signal. A signal is different from a constant. It's an object that you define in your design. Signals are used to connect different parts of a design – you can think of them as internal wires. In a physical design on a breadboard, you use wires; however, in VHDL, you use signals. For example, the entity of one design can be connected to the entity of another design using signals. It's a way to connect dynamic data between designs. So, if you have two AND gates, and the output of one AND gate goes to the input of another AND gate, that would be a signal because that's dynamic data. That value is constantly changing based on the inputs. The keyword for defining a signal is just signal. You give it a signal name, followed by a colon, and then the signal type, which would be

Fundamentals of VHDL for FPGA Programming Using Vivado, First Edition. Majid Pakdel.

Companion Website: www.wiley.com/go/pakdelVHDL

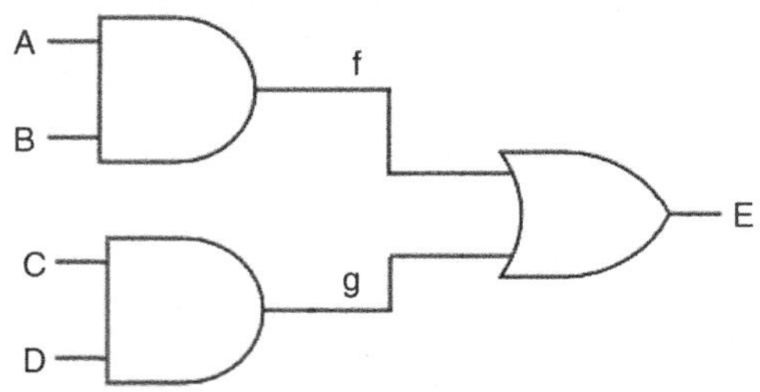

Figure 7.1 Digital design with two AND gates and one OR gate.

an integer (unsigned or signed), a standard logic vector, or a standard logic. You can then optionally assign it an initial value using colon equals. If you don't give it an initial value, the compiler will resolve it to a default value or to whatever value you specify during reset, which becomes the signal's initial value. Now, let's take an example and go through a signals and constants example. The first thing we'll look at is a signal example. Before we do that, we are going to sketch a rough digital design. It will include two AND gates and one OR gate, as shown in Figure 7.1. This is just a very basic VHDL design.

So, we have three logic gates, and we know we have our entity, which defines our inputs and outputs. We have four inputs, which we'll call A, B, C, and D, and we're going to have one output, which we will call E. So, our entity will define these four inputs – A, B, C, and D – as well as the output, E. Now, we have two signals that come from the outputs of the AND gates, and they go to the inputs of the OR gate. These are what we would call signals, because the inputs and outputs have already been defined, but these internal signals are communicating dynamic data. We'll just name these signals f and g, and if we know that, those are going to be our signals. From a VHDL perspective, we have our entity, which defines the inputs in this case, a single output in architecture, which would define the two AND gates and the single OR gate, and the inside of our architecture, which defines two signals, f and g. This is an example of how signals are used. You can think of them as internal wires inside your actual logic design. Next thing we look at is a constant – how or why you would use a constant? There are many different ways you could use a constant. In this example, we're going to use a comparator. Let's say we have a two-bit comparator. So, the two-bit comparator has to have four inputs. We'll have our A and B inputs, and since it's a two-bit comparator, A and B are both two bits, and we need to make sure we specify that they are two bits. The way the comparator works is it's going to compare our A and B inputs and output a greater than, an equal to, or less than. So, we are going to label them as G, E, and L, as shown in Figure 7.2.

So, if our A is greater than B, which we will define here with two bits, then we have a one on our G output. If our A is equal to B, then we'll have a one on our E output, and if our A is less than B, we'll have a one on the L output. Now, in this case, we're going to say our entity is just A and B, and the G, E, and L. So, in this case, we have this B, which is an internal value. Now, for this specific case, if we make B a signal, we would have to tie it off to something else. So, you can give it an initial value and it never changes, which, in that case, is by definition a constant. So, for this case, if we wanted to make B a constant, we can write the following code.

```
constant B: std_logic_vector(1 downto 0) := "11";
```

So, in this case, we know that we have our B, which is a "11." We've defined the constant B as a standard logic vector, so anywhere we use B in our design, it's going to be a two-bit value of "11," or 3 in this case. On the other hand, A is changing because it's part of our entity. So, this value (A) is changing, and it's always going to be compared to our constant B, which is defined as "11." Based on that comparison, our output will be greater than, equal to, or less than – so that's an example

Figure 7.2 The two-bit comparator.

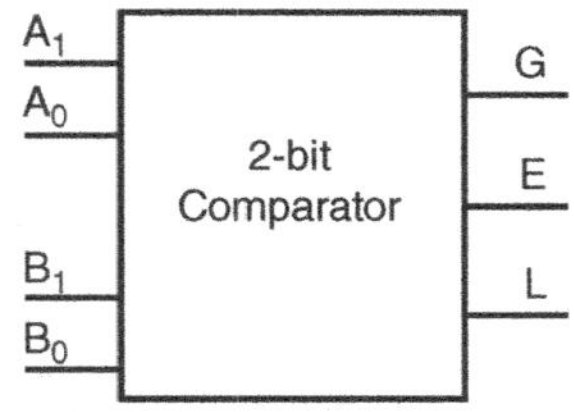

of a basic way you can use a constant in your VHDL design. The next thing we'll take a look at is actually choosing and determining what constants and signals we want to use inside our design for the blinky LEDs project. So, we open up the blinky LEDs project and the BLINKY_LED actual VHDL design, as depicted in the following codes.

```
-- Libraries
library IEEE;
use IEEE.STD_LOGIC_1164.ALL;
use IEEE.numeric_std.all;
use IEEE.math_real.all;
-- Entity
entity BLINKY_LED is
    Generic(
        NUM_LEDS: integer := 4;  -- 8 LEDs default
        CLK_RATE: integer := 100000000;  -- 50MHz default
        BLINK_RATE: integer := 2);  -- 2Hz default
    Port(
        Led_Out: out std_logic_vector(NUM_LEDS - 1 downto 0);
        Clk: in std_logic;
        Reset: in std_logic);  -- Active Low
end BLINKY_LED;
-- Architecture
architecture behavior of BLINKY_LED is
begin

end behavior;
```

We want to go ahead and set up our entity, and our libraries are already defined in the packages. We have our architecture defined, except we don't actually have anything inside it. In order to use constants and signals inside our architecture, we need to define them before the begin statement. So, before the begin and after the architecture keyword, we want to actually put our signals and constants. The first thing we want to do is define our constants. We'll start by defining a constant called maximum value. This will be used to calculate the count value required to achieve the blink rate, which we have defined in our entity. We're going to calculate this count value to determine the blink rate from the generic by taking our clock rate and dividing it by the blink rate, which gives us the maximum value we need to calculate since we need to count in order to achieve the specified blink rate. The next step is to create a constant called bit depth, which will determine the number of bits required to count to whatever value because our count value is going to be changing, and that's based on the clock rate and blink rate, the number of bits that are required to count that will change as well as the following codes.

```
-- Architecture
architecture behavior of BLINKY_LED is
constant MAX_VAL: integer := CLK_RATE / BLINK_RATE;
```

```
constant BIT_DEPTH: integer := integer(ceil(log2(real(MAX_VAL))));
signal count_reg: unsigned(BIT_DEPTH - 1 downto 0) := (others => '0');
signal led_reg: std_logic_vector(NUM_LEDS - 1 downto 0) := "0000";
begin

end behavior;
```

So, that bit depth is going to tell us how many bits are required to count to the maximum value. The next thing we're going to need is a signal, which we'll call "count_reg". This will be used to hold the actual count value. As we increment our count, the value will be stored inside this signal. So, we want to make it a signal because it's going to be constantly changing. If we made it a constant, we can never change it. So, this one has to be a signal and we're going to make it an unsigned type. The reason we use unsigned is that it allows us to perform addition and subtraction operations on it. With a standard logic vector, you can't take a standard logic vector and add one to it, since you have to physically make it do that. So, an unsigned type is a lot easier for doing counting type of operations, and we want to make it a bit depth minus one down to zero, and we will give it an initial value of zero. Since we don't know what the depth is, we can't manually type out all the zeros. Therefore, we will have to use the others keyword, and this will select all the bits that we haven't defined, which in this case are all of them, and we'll set them all to the value we specify, which we are going to tell it to set all the values to zero. Now, we need one more signal, and this signal is going to be what we would call led_reg or LED register, which is going to hold the value that's going to be output on our LEDs. So, this will make a standard logic vector, and we're going to make it number of LEDs (NUM_LEDS) minus one down to zero, and we'll give it an initial value in this case, because we know that all of our LEDs we have defined in this case is four. If you have eight, you have to put eight zeros here. However, in this case, we have four LEDs. Now, we could use the others keyword as well. But the reason we're going to go with zeros is that if we want to have it where we toggle and say instead of having it all zeros and ones, we want it to be 1010, and we can do that by just changing the definition we have of our signal "led_reg." Now, at this point, we have our four blinky LED design, and we have the constants and signals that we need for our project defined in our VHDL design file. At this point, all we're going to do then is just update the comments to include those. But we'll leave that up to you because you can add whatever makes the most sense in terms of helping you come back later and understand what these actual constants and signals are doing.

7.3 The VHDL Processes

This section is about the VHDL processes. In this section, we're going to explain what a process is in VHDL. In VHDL, we can use logical statements such as AND gates, OR gates, or any type of operation, and these are evaluated instantaneously. For example, if we have an AND gate with two inputs and one output, any time one of the inputs changes, the output changes almost immediately. However, if we want something to occur at a specific time or in response to a specific event, such as a signal changes, we would use a VHDL process. So, what these processes do is define a portion of your design that you want to be evaluated whenever a certain signal changes state. There are several parts to a VHDL process. First, there's the process label, which is used to name your process. When you have multiple processes in your

Figure 7.3 The VHDL process statement syntax.

[process_label:] **process** (sensitivity_list)
process_declarations
begin
sequential_statements
end process [process_label];

design, giving them labels helps you identify each one easily. Then, there's the sensitivity list. Inside the sensitivity list, you include all the signals that you want to be monitoring. Whenever they change, you evaluate your process. Most commonly, the clock signal is included in the sensitivity list. So, you want the process to be evaluated any time the clock changes. When the clock transitions from a low to a high state, the process is evaluated; likewise, if the clock goes from a high to a low, you're going to evaluate the process again, and on the same side, you can put any type of signal in there. For example, if you have a standard logic vector with eight elements, you might include that in the sensitivity list. When any one of the elements in that standard logic vector goes from a low to a high or from a high to a low, you will then evaluate that process, and so the only time a process block is evaluated is whenever any one of the signals in the sensitivity list is evaluated. The syntax for a VHDL process statement is illustrated in Figure 7.3.

We have the process label, which is optional. You don't have to give it a label – you can just call it a process. But a lot of times, we think it's best to actually give it a label. It helps make your code more readable, and when you come back, you can give it a comment. Plus, the label helps explain what's going on. So, if your process is counting, it makes sense to just say something like count and then process, so that when you come back and look through something, you can say, "OK, this process is implementing a counter." Next, you have to have the colon and then the keyword process, and then inside of the brackets you want to insert a sensitivity list, where you'll include all the signals that you want to be looking at. Whenever any of the signals inside that sensitivity list change from high to low, you'll then evaluate everything inside your process. Underneath that, you have your process declarations. This is where you can declare variables, signals, or constants. If you declare them within the list, they'll only be seen inside your actual process and not throughout the entire design. As we know from our signals and constants, once we define those before the begin statement of our architecture, we now then evaluate and use those inside our process statement. Then we have the begin, which is referred to as the sequential statements section. After the begin is where we can include constructs such as if statements or any other type of statements that evaluate what the output of our process should be. We then use the end process, which completes the begin–end pair of the VHDL process. After that, you include the process label again to indicate that the process defined earlier is officially ending. Once you open the blinky LED (BLINKY_LED) project, we are going to add our processes into the project. We already have the libraries and entities defined, as well as the architecture. We've got defined the constants and signals we'll be using. Now, we need to define the processes. So underneath the begin, the first thing we'll do is assign our output LED values. Therefore, we'll create a concurrent statement. We'll do this by assigning Led_Out the value of led_reg. This means that at all points in time, Led_Out, which we have defined in our port as an output, will reflect the value of led_reg, which we have defined as a signal. This setup allows us to assign a value to led_reg within our process, and that value will automatically be passed to Led_Out, While we could assign a value to Led_Out directly, using led_reg gives us the advantage of being able to read its value. In contrast, we can't actually read the value of Led_Out directly, since it's an output port. We don't know what that is, but we can read the value of led_reg, and by reading that, we know what is on our output. So, for the first process, we're going to create a count process. We'll add a little note that says process increments the counter on every rising clock edge. To start, we'll give it a label – let's call it count_proc – followed by a colon and then space and we use the process keyword. After that, we'll define the sensitivity list. In this case, we only care about changes on the rising clock edge of the input clock (Clk), which is defined in our inputs. So, Clk will be the only signal in the sensitivity list. With the sensitivity list complete, we won't be defining any specific signals within this process, so we move on to the begin and end process. We'll label the end of the process with count_proc, completing the begin–end pair for the first process. Later, inside the structure, we will define what we actually want to happen each time the clock toggles. So, this is going to be a process that counts, and what we want to do is increment our count register to determine when we reach our max value. The next process we want to define will basically toggle the LED output every time our count reaches the calculated maximum value. In the first process, we're simply going to increment the count register. Then, we'll check whether count_reg is equal to our max value. When it is, we'll toggle the led_reg. So that's going to toggle between all zeros and ones and therefore present the blinky LED output we're desiring. So, we'll add a comment that says "process that will toggle the output every time the counter reaches the calculated MAX_VAL." It's a good practice to include these kinds of comments to make your code look more readable and understandable. Now we want to define our actual process. We'll give it a name – output_proc – because it defines the LED output. So, we'll write output_proc, and use the process keyword. Then, we need to give a signal for sensitivity list. In this case, we want to evaluate the process on every clock cycle, so we'll include the clock signal (Clk). After that, we'll use the begin and end process pair, ending the process with the label output_proc. Inside this process, we'll check if count_reg is equal to MAX_VAL. We're going to want to toggle the signal led_reg, to give us that blinking effect. These are the processes described in the video, and they represent the processes we'll use in our blinky LED project, as shown in the following codes.

```
-- Architecture
architecture behavior of BLINKY_LED is
constant MAX_VAL: integer := CLK_RATE / BLINK_RATE;
constant BIT_DEPTH: integer := integer(ceil(log2(real(MAX_VALUE))));
signal count_reg: unsigned(BIT_DEPTH - 1 downto 0) := (others => '0');
signal led_reg: std_logic_vector(NUM_LEDS - 1 downto 0) := "0000";
begin
    -- Assign output LED values
    Led_Out <= led_reg;
    -- Process that increments the counter every rising clock edge
    count_proc: process(Clk)
    begin
    end process count_proc;
    -- Process that will toggle led output every time the counter
    -- reaches the calculated 'MAX_VAL'
    output_proc: process(Clk)
    begin
    end process output_proc;
end behavior;
```

7.4 The VHDL If Statements

This section covers VHDL if statements and how to use them within the VHDL language. The VHDL if statements are used to conditionally execute sequential statements based on whether a given condition evaluates to true or false. For example, if a condition is true, a certain operation will be executed. If it's false, a different operation will be executed. Just like if you're working in C/C++ or almost any type of programming language, the if statement in VHDL works the same. Now, when you're working with the if statement, the caveat is you need to have it in a process. You can't have this statement just flowing in an architecture, because inherently, an if statement does form a sequential type of operation. So, in your architecture, you need to have all these consecutive assignments where they happen simultaneously. If you have a bunch of AND gates or logic gates, they all happen at the same time. However, with an if statement, you have your if else and end if, and it has to happen sequentially. So you need to put these in a process – that's an important part of the if statement. When you're working with them, the keywords are if, then, elsif, and else. You know, with else if in the if statement, it is spelled "elsif" – there's no "e," which is slightly confusing. But when we show you the syntax of how it actually works in the VHDL language, you'll see it, and it'll make more sense what we're saying. You can test with your statements; you can have multiple "elsif." So, you can have an if ... end if where, if it evaluates to true, you'll operate on that condition, and if not, then you'll just ignore it. You can have an if ... else, or you can have an if ... elsif ... else ... end if statement. You can have a bunch of elsifs – there's no limit on that amount. You can also do nested if statements – you can have an if statement and then inside that statement have another if statement. When you're working with if statements, they do infer priority, and what that means is if the first condition evaluates to true, then that condition gets executed and exits out of that statement. So, if you have three conditions that you're checking against and if they're overlapping where the first and third conditions evaluate to true, it doesn't matter because it's going to execute on that first condition and then ignore the next two. So, when designing your logic, place the highest-priority condition at the top of the if statement – that's an important thing to know. Figure 7.4 illustrates the syntax for if-then-end if, and if-then-elsif-end if constructs as used in VHDL.

So, if we look at the first one, we have our if statement, and then a condition – something that must evaluate to either true or false. For example, if we have an unsigned data type, we can say if this unsigned data type is greater than or equal to a specific number, then a sequential statement where we set an output or write an input, which is something that we'll evaluate to a true or false. Next, we have the if-then-else structure. In this case, we're still checking one condition, but if that condition evaluates to false, the second sequential statement inside the else block

Figure 7.4 The VHDL if statements syntax.

```
if condition then
  sequential_statements
end if;
```

```
if condition then
  sequential_statements
else
  sequential_statements
end if;
```

```
if condition then
  sequential_statements
elseif condition then
    sequential_statements
else
    sequential_statements
end if;
```

will execute instead. If the condition evaluates to true, then the first sequential statement will be executed. Now, if you have multiple conditions that you would like to check, you'll need to use the if-elsif-else-end if structure. In this case, the first condition is checked, and if that evaluates to true, the first sequential statement is executed, and the program exits the conditional block – no further conditions are evaluated. However, if the first condition evaluates to false, it will check the second condition. If that condition is true, its sequential statement will be executed. If both the first two conditions evaluate to false, then the statement in the else block will be executed. This structure is very similar to if statements used in most programming languages, which is a common tool for controlling flow based on logic. Next, go ahead and open your blinky LED project. Within that project, we'll insert the if statements. As mentioned earlier, these statements can only be placed inside a process block. So, inside our count process, the first thing we're going to do is ensure that our operation is performed on the rising edge of the clock, as shown in the following codes.

```
count_proc: process(Clk)
begin
   if rising_edge(Clk) then
   else
   end if;
end process count_proc;
```

So, this is our first if statement block. If we have a rising edge clock, then we want to do an operation; otherwise, we want to do another operation. Now, if we look at those codes on falling edge, we're not actually doing anything, so we're going to get rid of that else condition. We're going to just have a straight if and end if condition as the following codes.

```
count_proc: process(Clk)
begin
   if rising_edge(Clk) then
   end if;
end process count_proc;
```

So, we have our if rising edge clock – what do we want to do? Well, we're going to include a second if statement. The first thing we will check is the Reset, which is an input to our design. If Reset is equal to zero – or in this case, if it's equal to a logical low – then we're going to use a logical "or" such that reset is equal to zero or count_reg is equal to MAX_VAL, which we have defined as a constant. So, if count register is equal to the max value or reset is equal to zero, then we want to perform the following operation, as shown in the following codes.

```
count_proc: process(Clk)
begin
   if rising_edge(Clk) then
      if((Reset = '0') or (count_reg = MAX_VAL)) then
         count_reg <= (others => '0');
      else
         count_reg <= count_reg + 1;
      end if;
   end if;
end process count_proc;
```

So, we have a nested if – we check if it's the rising edge of the clock, then we want to see if Reset equals zero or count_reg equals MAX_VAL. If that's the case, we want to perform a specific operation – which, in this case, is updating our count_reg. Since count_reg is an unsigned data type, we can perform mathematical operations on it. If the condition is true, we want it to be equal to zero. To do this, we'll use the others assignment, because we don't know what size it is. The size of count_reg is defined by BIT_DEPTH, which is derived from MAX_VAL, which itself depends on CLK_RATE and BLINK_RATE, all of which are passed in through generics. So, since we don't know the size of it, that forces us to use "others" statement, which is just going to set count_reg equal to zero. Now, if we detect a clock signal, and we're not in a reset or enable state, and if count_reg does not equal MAX_VAL, then we want to increment count_reg. That's done by assigning count_reg to itself plus one. Since count_reg is an unsigned data type, we can do an addition of one. So, this is effectively just creating a counter where this process is just going to be constantly, continually evaluated every time our clock goes off and we're just going to be counting up count_reg until it equals MAX_VAL, at which point we reset it. So, count_reg just goes from zero to MAX_VAL over and over, unless reset is active, in which case we stay in a zero or reset state. Next, we have our second process, which is the output process (output_proc). We want to do the same thing here –the first thing we do is check if it's the rising edge of the clock, because we want to do everything on either the rising or falling edge (in this case, we're using the rising edge). If we're on the falling edge, we don't want to do anything. So, we're not going to use an if-else condition – just an if condition for the rising edge. Inside that, we'll have a second if statement, checking if count_reg equals MAX_VAL. This condition could also occur during reset, which is why we want to make sure that this is done in a separate process. Although we already check count_reg is equal MAX_VAL in the first process (count_proc), putting LED toggle logic there would cause the LEDs to toggle every time we reset, which we don't want. That's exactly what we're going to do here. So, if count_reg equals MAX_VAL, we'll go ahead and perform the toggle. We don't include an else statement here, because if count_reg does not equal MAX_VAL, we simply don't want to do anything. We'll then tab over and assign the LED output. We say that led_reg, which is the value that gets sent to the actual LEDs, is equal to not led_reg, as shown in the following codes.

```
output_proc: process(Clk)
begin
   if rising_edge(Clk) then
      if((count_reg = MAX_VAL)) then
         led_reg <= NOT led_reg;
      end if;
   end if;
end process output_proc;
```

So, all that's going to do is take whatever value is on our led_reg and invert it. Since we start with all zeros in led_reg, inverting that will result in all ones. The reason we don't want to specify something like "set all to zero" or "set all to one" is because using a single inversion statement is a lot simpler and easier. The reason we have a second process is exactly what we discussed earlier – if we were to include led_reg in the first process, count_proc, where we're checking if the reset is zero or if count_reg equals MAX_VAL, then every time we reset, we would end up toggling the LEDs, which we don't want to happen. So, that is how you use the if statement in VHDL and how we're using it to conditionally check and blink the LEDs for our blinky LED project. We can now

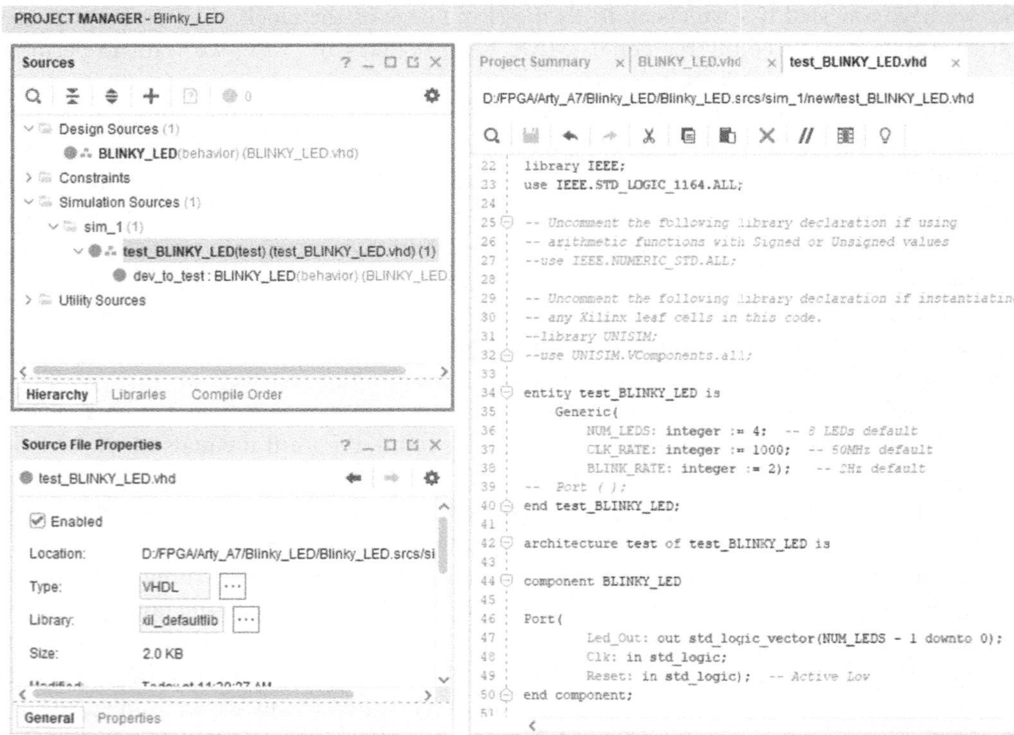

Figure 7.5 The test_BLINKY_LED.vhd as a top-level design.

copy and paste the following designed test bench code into test_BLINKY_LED.vhd. It's going to update accordingly, and the tool will recognize that test bench as test_BLINKY_LED, which is the top-level design, as shown in Figure 7.5.

```
library IEEE;
use IEEE.STD_LOGIC_1164.ALL;
entity test_BLINKY_LED is
    Generic(
        NUM_LEDS: integer := 4;  -- 8 LEDs default
        CLK_RATE: integer := 1000;  -- 50MHz default
        BLINK_RATE: integer := 2);   -- 2Hz default
--  Port ( );
end test_BLINKY_LED;

architecture test of test_BLINKY_LED is

component BLINKY_LED

Port(
        Led_Out: out std_logic_vector(NUM_LEDS - 1 downto 0);
        Clk: in std_logic;
        Reset: in std_logic);  -- Active Low
end component;
```

```
signal Led_Out: std_logic_vector(NUM_LEDS - 1 downto 0);
signal Clk: std_logic := '1';
signal Reset: std_logic := '1';
signal count_reset: integer := 0;
begin
   dev_to_test: BLINKY_LED
        port map(Led_Out, Clk, Reset);
    data_stimulus: process
    begin
        wait for 10 ns;
        Clk <= not Clk;
        count_reset <= count_reset + 1;
        if (count_reset = 100) then
           Reset <= not Reset;
           wait for 10 ns;
           Clk <= not Clk;
           wait for 10 ns;
           Clk <= not Clk;
           wait for 10 ns;
           Clk <= not Clk;
           wait for 10 ns;
           Clk <= not Clk;
           wait for 10 ns;
           Clk <= not Clk;
           Reset <= not Reset;
        end if;
    end process data_stimulus;
end test;
```

7.5 The If Statement Priority Example

This section is about the if statement priority example. Here, we're going to demonstrate how an if statement can infer priority. We're going to simulate a priority encoder. The way a priority encoder works is that you have an input, and the output tells you which is the highest bit that is set to '1'. For example, if our input is 8 bits wide, ranging from bit 0 to bit 7, and bit 7 is high, then the output will be a 3-bit value (from 2 down to 0) that indicates bit 7 is the highest active bit that is set. In this case, the output would be 111 to indicate that bit 7 is high. Because if we look at this, we have an encode process where we're processing a_in, and when signal changes, we define our output. If our seventh bit is high, our output goes to seven (111). Now, if our seventh bit and sixth bit are both high, the encode output just says 111, or tells you seven. It tells you the highest bit or the most significant bit that is actually high. So, the idea behind this is that it shows you all of these conditions can evaluate to true. But if our seventh bit is high, that's the one we want to know. So, the way this operates is by using "if". It's going to check if bit seven is equal to one, then our encode output is equal to 111. Then we're going to end this if statement – we do not even check the rest of the statements, as shown in the following codes.

```
library IEEE;
use IEEE.STD_LOGIC_1164.ALL;

entity Priority_Encoder is
```

```
Port(
     encode: out std_logic_vector(2 downto 0);
     a_in: in std_logic_vector(7 downto 0));
end Priority_Encoder;

architecture behavior of Priority_Encoder is

begin
   enc_process: process(a_in)
   begin
       if(a_in(7) = '1') then
           encode <= "111";
       elsif(a_in(6) = '1') then
           encode <= "110";
       elsif(a_in(5) = '1') then
           encode <= "101";
       elsif(a_in(4) = '1') then
           encode <= "100";
       elsif(a_in(3) = '1') then
           encode <= "011";
       elsif(a_in(2) = '1') then
           encode <= "010";
       elsif(a_in(1) = '1') then
           encode <= "001";
       elsif(a_in(0) = '1') then
           encode <= "000";
       else
           encode <= (others => 'X');
       end if;
    end process enc_process;

end behavior;
```

So, if our seventh bit is zero, then we check the next one. If our sixth bit is high, we disregard the fifth, fourth, third, second, first, and zeroth bits, setting them to zero. We disregard all of those and just say, "OK, our sixth bit is high," and move on, disregarding the rest. Then we check again: if the sixth bit goes low, we say, "OK, that fifth bit is high," and we just keep going down the line. We're going to simulate this to show how it works. Otherwise, if you just looked at this, you might think that if all seven bits are high, every one of these conditions in the statement would evaluate to true, which would mean if you go through it, you'd say the encode is equal to 111. But now, our encode is equal to 110 – you just keep going down the line. If all your bits are high, your encode would end up being zero (000), because the last "elsif" statement would be the one that gets evaluated. That's how it works: you evaluate the conditions one by one, and as soon as you find one that's true in an if-elsif-else type of statement, the rest are skipped. You only execute the code associated with the first true condition. In this case, we're just setting the encode output to some value, and that ends our process. So, we'll go ahead and run the simulation to show what this actually looks like when simulated.

We can copy and paste the following test bench code into test_Priority_Encoder.vhd, and it'll update accordingly. Now, it recognizes the test bench as test_Priority_Encoder, which is a top-level design, as shown in Figure 7.6.

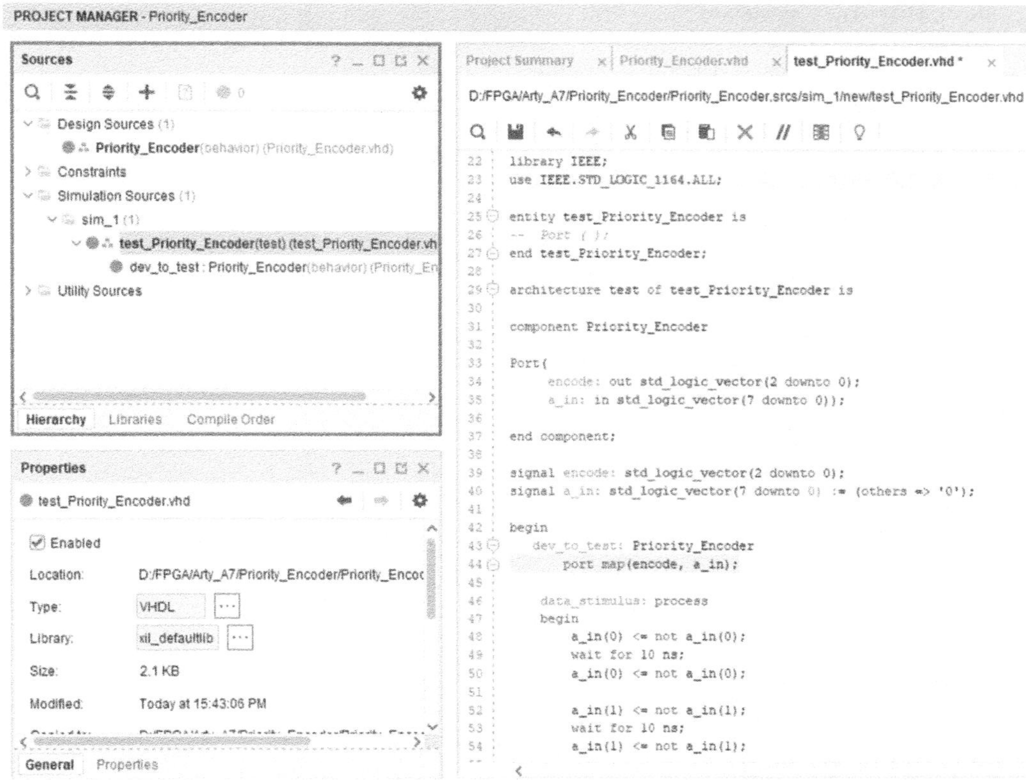

Figure 7.6 The test_Priority_Encoder.vhd as a top-level design.

```
library IEEE;
use IEEE.STD_LOGIC_1164.ALL;

entity test_Priority_Encoder is
--  Port ( );
end test_Priority_Encoder;

architecture test of test_Priority_Encoder is

component Priority_Encoder

Port(
     encode: out std_logic_vector(2 downto 0);
     a_in: in std_logic_vector(7 downto 0));

end component;

signal encode: std_logic_vector(2 downto 0);
signal a_in: std_logic_vector(7 downto 0) := (others => '0');

begin
   dev_to_test: Priority_Encoder
       port map(encode, a_in);
```

```
    data_stimulus: process
    begin
        a_in(0) <= not a_in(0);
        wait for 10 ns;
        a_in(0) <= not a_in(0);

        a_in(1) <= not a_in(1);
        wait for 10 ns;
        a_in(1) <= not a_in(1);

        a_in(2) <= not a_in(2);
        wait for 10 ns;
        a_in(2) <= not a_in(2);

        a_in(3) <= not a_in(3);
        wait for 10 ns;
        a_in(3) <= not a_in(3);

        a_in(4) <= not a_in(4);
        wait for 10 ns;
        a_in(4) <= not a_in(4);

        a_in(5) <= not a_in(5);
        wait for 10 ns;
        a_in(5) <= not a_in(5);

        a_in(6) <= not a_in(6);
        wait for 10 ns;
        a_in(6) <= not a_in(6);

        a_in(7) <= not a_in(7);
        wait for 10 ns;
        a_in(7) <= not a_in(7);
    end process data_stimulus;

end test;
```

We'll go ahead and right-click on the Full View, then zoom in on a specific area to get a clearer look at what's going on. If we take a closer look, we can see our data_in, and we'll expand that to display the actual bits and encode_out. The encode_out shows the value of the highest active bit, as depicted in Figure 7.7.

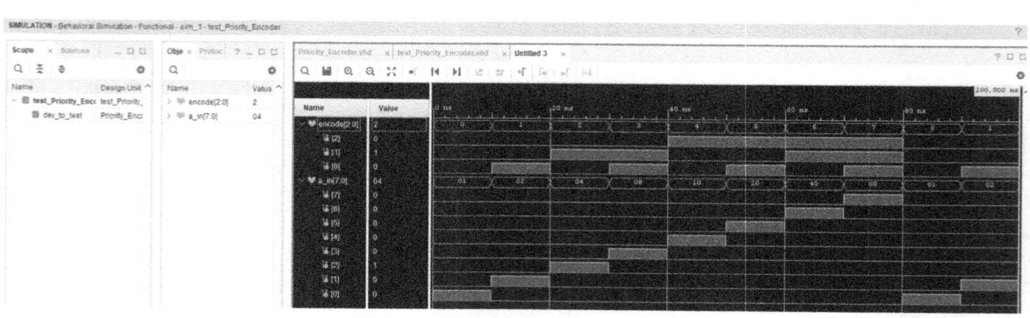

Figure 7.7 The Behavioral Simulation window (part 1).

Now, what we're going to do is restart the simulation (Ctrl+Shift+F5). We know we only need to run it from zero to approximately 100 ns. Next, we'll set bit seven to high by forcing a constant with a binary value of 1. We'll cancel that force after about 20 ns, as shown in Figures 7.8 and 7.9, respectively.

We're just going to force it to be high for a short period and then say, "OK, now let's run it for 100 ns." After switching to the full view, we can see that encode out is set to 7, as illustrated in Figure 7.10. The reason it's 7 is because both bit seven and bit zero are high now, and are one.

If we look at our priority encoder, we can see that both bit seven and bit zero are set to one. So, both conditions evaluate to true. However, the condition for bit seven comes first; for example, it has higher priority. So, the higher the priority you want a bit to have, the earlier you place its conditions in the if statement. So, bit seven has the highest priority, but bit six has the second highest priority. Since bit seven is a one, the output is set to seven, and it's going to disregard the rest of the bits. Bit seven has the highest priority, which we can see is exactly what's happening right there. So, our bit seven has priority because it's high, and then as soon as we drop it low, whichever bit is high – in this case, bit two – now our encode output is two. So that just shows you this is a very good example explaining how these if statements infer priority. So, in this exact example, when you're designing a priority encoder, it makes the most sense. An if statement will make the most sense as a way of implementing that. But when you work with different designs, if you have specific situations where you say, "OK, we want

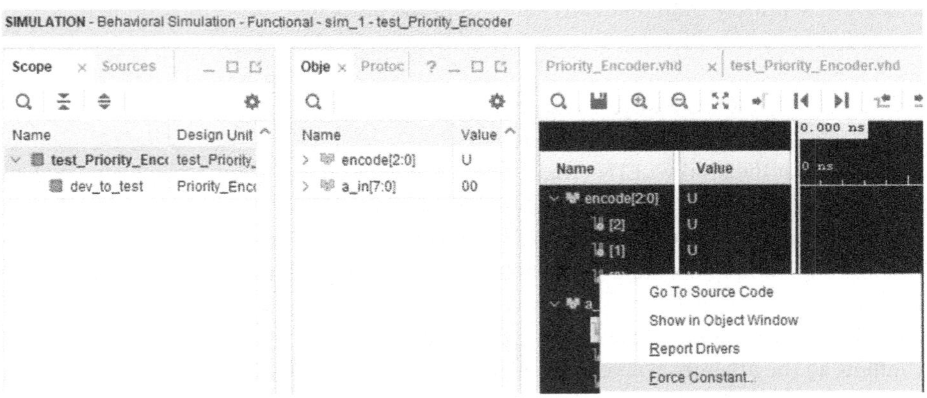

Figure 7.8 Selecting the force constant... for a_in(7).

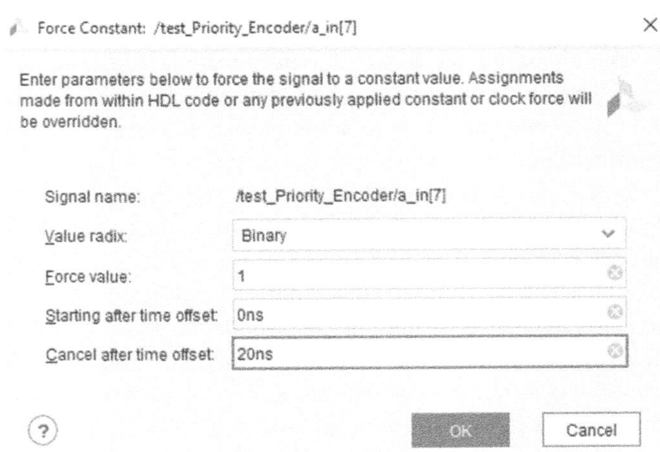

Figure 7.9 The Force Constant window.

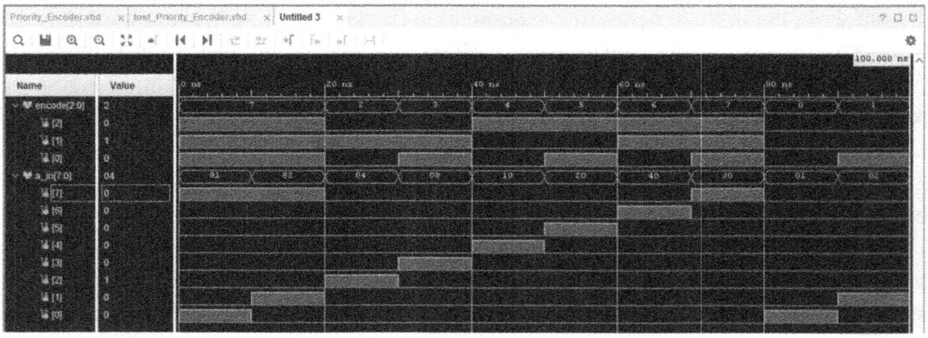

Figure 7.10 The Behavioral Simulation window (part 2).

to make sure that there're multiple conditions that could be true. But if this one is true, then we want this to be evaluated," then the next statement, which has lower priority, should go below. So, the first condition you test, the big thing is when you're working with these statements, the first condition you check is your highest priority, in this case, bit seven. The last condition you check, in this case bit zero, is the lowest priority, and it just goes in descending order: highest priority, next highest, and all the way down to the lowest priority. And that's an example of how the if statements infer priority.

7.6 The Blinky LEDs Project IO Placement

This section is about the blinky LEDs Project Arty A7 IO placement. Here, we're going to show you where and how to place the specific IOs required for the blinky LEDs project to run on the Arty A7 board. So, we open up the Arty A7 blinky LEDs project. We've already gone through the complete design walkthrough and understand the entire VHDL code and what's going on. The next step is to assign the IOs. Then we'll run the synthesis, run the implementation, and generate the bitstream. Let's go ahead and assign the different IOs. If you open up the attached Arty_A7_Master.xdc file, you'll see that it contains all the different IOs you can work with on the Arty A7 development board. Go ahead and double-click to open it. Inside, you'll notice they are already called out – just comment it out. To use any of them, we simply need to uncomment the lines. At the top of this master.xdc file, it gives you a reference of what to do in order to use it in the project, uncommented lines corresponding to these pins. In our blinky LEDs project, if we look at the entity, we see it's generic – we do not need to map any generics. We only have three signals that need mapping: LED output (Led_Out), clock (Clk), and reset (Reset). The LED output is an output defined as a standard logic vector from the number of LEDs, which we have defined as four minus one down to zero. So, if we take our Led_Out, it will decrease down from three, two, one, to zero. If we open a master.xdc file and scroll down to LEDs, we'll see three, two, one, and zero, or four different LEDs. By replacing each instance of led with Led_Out on all four of these lines, we effectively assign the signals from our design to the actual physical LEDs on the Arty A7 board. Next, select all four lines and click the Toggle Line Comments button (Ctrl+/). If we hit it again, it'll comment or lineout. Once uncommented, you'll notice that set property and get ports are highlighted because in the .xdc file, these are specific keywords that are recognized. Now, let's go back to our BLINKY_LED. vhdfile and examine the next input, which is the clock (Clk). Copy the clock signal, return to the .xdc file, and scroll to the very top. We can see we have that clock signal, so we have our set property with package pin E3, our IOSTANDARD as LVCMOS33 and it's connected to the clock port at 100 MHz. So, we are going to go ahead and paste it there. This is going to set our property and tell us we're using the clk port. We want to go ahead and uncomment that

Project Summary × | **Arty_A7_Master.xdc** ×

D:/FPGA/Arty_A7/Blinky_LED/Arty_A7_Master.xdc

```
## This file is a general .xdc for the ARTY Rev. B
## To use it in a project:
## - uncomment the lines corresponding to used pins
## - rename the used ports (in each line, after get_ports) according to the top level signal names in the project

## Clock signal

set_property -dict { PACKAGE_PIN E3    IOSTANDARD LVCMOS33 } [get_ports { Clk }]; #IO_L12P_T1_MRCC_35 Sch=gclk[100]
create_clock -add -name sys_clk_pin -period 10.00 -waveform {0 5} [get_ports { Clk }];

##Switches

set_property -dict { PACKAGE_PIN A8    IOSTANDARD LVCMOS33 } [get_ports { Reset }]; #IO_L12N_T1_MRCC_16 Sch=sw[0]
#set_property -dict { PACKAGE_PIN C11   IOSTANDARD LVCMOS33 } [get_ports { sw[1] }]; #IO_L13P_T2_MRCC_16 Sch=sw[1]
#set_property -dict { PACKAGE_PIN C10   IOSTANDARD LVCMOS33 } [get_ports { sw[2] }]; #IO_L13N_T2_MRCC_16 Sch=sw[2]
#set_property -dict { PACKAGE_PIN A10   IOSTANDARD LVCMOS33 } [get_ports { sw[3] }]; #IO_L14P_T2_SRCC_16 Sch=sw[3]

##RGB LEDs

#set_property -dict { PACKAGE_PIN E1    IOSTANDARD LVCMOS33 } [get_ports { led0_b }]; #IO_L18N_T2_35 Sch=led0_b
#set_property -dict { PACKAGE_PIN F6    IOSTANDARD LVCMOS33 } [get_ports { led0_g }]; #IO_L19N_T3_VREF_35 Sch=led0_g
#set_property -dict { PACKAGE_PIN G6    IOSTANDARD LVCMOS33 } [get_ports { led0_r }]; #IO_L19P_T3_35 Sch=led0_r
#set_property -dict { PACKAGE_PIN G4    IOSTANDARD LVCMOS33 } [get_ports { led1_b }]; #IO_L20P_T3_35 Sch=led1_b
#set_property -dict { PACKAGE_PIN J4    IOSTANDARD LVCMOS33 } [get_ports { led1_g }]; #IO_L21P_T3_DQS_35 Sch=led1_g
#set_property -dict { PACKAGE_PIN G3    IOSTANDARD LVCMOS33 } [get_ports { led1_r }]; #IO_L20N_T3_35 Sch=led1_r
#set_property -dict { PACKAGE_PIN H4    IOSTANDARD LVCMOS33 } [get_ports { led2_b }]; #IO_L21N_T3_DQS_35 Sch=led2_b
#set_property -dict { PACKAGE_PIN J2    IOSTANDARD LVCMOS33 } [get_ports { led2_g }]; #IO_L22N_T3_35 Sch=led2_g
#set_property -dict { PACKAGE_PIN J3    IOSTANDARD LVCMOS33 } [get_ports { led2_r }]; #IO_L22P_T3_35 Sch=led2_r
#set_property -dict { PACKAGE_PIN K2    IOSTANDARD LVCMOS33 } [get_ports { led3_b }]; #IO_L23P_T3_35 Sch=led3_b
#set_property -dict { PACKAGE_PIN H6    IOSTANDARD LVCMOS33 } [get_ports { led3_g }]; #IO_L24P_T3_35 Sch=led3_g
#set_property -dict { PACKAGE_PIN K1    IOSTANDARD LVCMOS33 } [get_ports { led3_r }]; #IO_L23N_T3_35 Sch=led3_r

##LEDs

set_property -dict { PACKAGE_PIN H5    IOSTANDARD LVCMOS33 } [get_ports { Led_Out[0] }]; #IO_L24N_T3_35 Sch=led[4]
set_property -dict { PACKAGE_PIN J5    IOSTANDARD LVCMOS33 } [get_ports { Led_Out[1] }]; #IO_25_35 Sch=led[5]
set_property -dict { PACKAGE_PIN T9    IOSTANDARD LVCMOS33 } [get_ports { Led_Out[2] }]; #IO_L24P_T3_A01_D17_14 Sch=led[6]
set_property -dict { PACKAGE_PIN T10   IOSTANDARD LVCMOS33 } [get_ports { Led_Out[3] }]; #IO_L24N_T3_A00_D16_14 Sch=led[7]
```

Figure 7.11 The uncommented ports in Arty_A7_Master.xdc.

line and click Save. This sets our input clock, which is 100 MHz, just based on what's written there. But also, if you look in the manual, it will confirm that the input clock for this design comes on pin E3 and is 100 MHz. If we go back to our blinky LED project, we also have the reset (Reset). So, we want to go ahead and copy that, and for this, we're going to just tie it to a slide switch. So, we go back to the .xdc file. We have our four slide switches, so we can set from switch zero to switch three. We're just going to use switch zero, and since we only have one, we no longer need the brackets. So, we'll highlight the entire sw[0] and paste it. Now, that will set our switch zero to the reset, and we'll go ahead and remove that comment. So now we have our clock, reset, and LED outputs all mapped. Now, the other thing to note is that we already have the set property for the clock (Clk). We also want to uncomment the line create_clock, and this is going to say, create clock add and then dash name (-name), so, we will have at name clock period 10 waveform [0 5]. What we're doing here is defining our input clock so that when we run our design, it can verify that we meet the specified timing. For this line, we have our get_ports keyword. We want to copy that and make sure it's using clock (Clk). So, we're going to tell it that we have a clock, and we'll go ahead and keep this as our add name. Effectively, we're telling it, "OK, we have a clock that is coming in on this pin (E3)," and that's where it is physically connected. Now, when we run our synthesis and implementation, it's going to tell us whether we can meet the specific requirements, where we want our period to be 10 ns and our waveform to have a 50 percent duty cycle ([0 5]). If you comment this out, when you run your synthesis, you'll probably just end up with a warning. But if you get a warning and it is incorrect, you can go ahead and uncomment it out. So that is everything that is required to set up IO. We've mapped the LED output, clock input, and the reset signal, as shown in Figure 7.11. At this point, we're ready to go ahead and run synthesis, run implementation, and generate the bitstream.

8

The LED Brightness Project

8.1 Introduction

In this chapter, we will review the pulse width modulated (PWM) signals and explain what a PWM signal is, because we are going to use PWM signals for the LED brightness project. We're going to use VHDL to generate an actual PWM signal, and the first thing we're going to explain is what PWM stands for. The PWM signal, as mentioned earlier, stands for pulse width modulated signal. All this means is that you have a standard digital signal with a value of one or zero, and what we're doing is varying the duty cycle of that specific signal. This technique can be used for various applications. The application we're currently focusing on is using it to vary the brightness of an LED. On the Arty A7 board, we have standard LEDs as well as an RGB (red, green, blue) LED. We will send these PWM signals to the RGB LED to vary its brightness, allowing us to generate a wide range of colors. If you've ever worked with colors on a computer, you know that you can mix different parts of red with different parts of green and different parts of blue to create various colors. In order to get different intensities of red, blue, or green, you vary the amount of time the signal is high versus the time it is low, or vary the duty cycle of each signal being sent to each corresponding part of that LED. Another common application of PWM is in servo motors. In robotics or RC airplanes, you'll have a servo motor where you send a certain duty cycle of a signal to that motor, and that duty cycle corresponds to a specific position. For example, if you have a motor that can rotate 180 degrees, you might define zero degrees as a duty cycle of 50% and 180 degrees as a duty cycle of 90%. Therefore, by varying the duty cycle of 50% and 90%, you can vary between zero and 180 degrees. Now, we'll take a closer look at what an actual PWM signal is. Let us make a few notes about PWM. To generate a PWM signal, we know that this is simply where we are varying the duty cycle. So, this is a key thing to know – by modulating, we're just varying the duty cycle. The next logical question would be: what is duty cycle? The formula for duty cycle is equal to time on (t_{on}) divided by time on (t_{on}) plus time off (t_{off}), and this is for just one period of your signal. So, let's take an arbitrary digital signal that goes high for a certain amount of time, then low, then high again, and so on, as shown in Figure 8.1.

So, we know that we have one period, which would be from our first rising edge to our second rising edge, and so between this period right here, we have a time on, t_{on}, which is effectively the time that your signal is high. So, this is your time on and your time off, as you can see in Figure 8.1. So, for example, if we take the signal we have shown right here, and our time on is 50 ms and our time off is 50 ms, we can solve for our duty cycle by taking our time on. So, we know that's 50 ms, and we divide that by our time on plus our time off. So, the duty cycle would be equal to 0.5, and duty cycle is almost always represented as a percentage. So, that would be equal to a 50% duty cycle

Fundamentals of VHDL for FPGA Programming Using Vivado, First Edition. Majid Pakdel.

Companion Website: www.wiley.com/go/pakdelVHDL

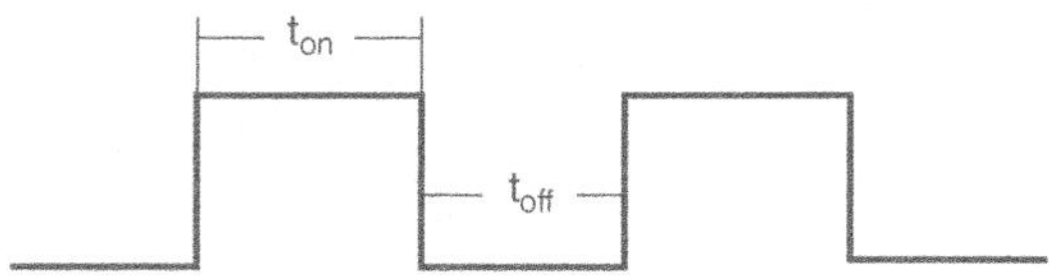

Figure 8.1 The PWM signal.

of that signal, and that's commonly seen in clocks, where you have a signal that's high just as much as it's low, or 50% duty cycle, you just have a nice square waveform that aligns. Now, if you have a signal where the duty cycle varies – for example, let's say your time on is 70 ms and your time off is 50 ms – then we know that our time on is greater than the time off. So, we're going to have a higher duty cycle. Using the same formula, the duty cycle is equal to 0.583. Since we represent it as a percentage, that's equal to 58.3% duty cycle. So, the wider this width is, the brighter the LEDs will be. In the case of a standard RGB LED, this controls brightness. Now, if you're controlling a robot, the width of this is going to determine the position of the servo or the motor you're trying to control. And that is an introduction and explanation of PWM.

8.2 The PWM VHDL Design

This section is about the PWM VHDL design. Here, we are going to determine how to design a PWM controller in VHDL. We already know that PWM stands for pulse width modulation, and we understand that it involves varying the duty cycle of a standard digital signal, where we have a zero and one, and that is going to vary the amount of time the signal stays high. So, in order to do this, we understand there are two components: the signal frequency and the duty cycle. We'll be outputting a signal at a specific frequency, and that's an important thing to remember and know. We must define the frequency and the duty cycle, which is the key factor. While we can fix the frequency, we want the design to allow the duty cycle to be adjustable. This will make the design more flexible and reusable for other projects beyond just controlling LED brightness, allowing it to be used in a variety of applications. So, in order to do that, we want to make sure we define the frequency as a generic and the duty cycle as a port. So, what this allows us to do is, when we create our design via the generic, we can, on the fly, specify the duty cycle through the port. This allows us to set the frequency to 50 Hz, but the duty cycle can be changed just on the fly throughout the design. Another important thing to note is that we have to define the entity. When we define the entity, we need to focus on two key parts: generics and the port. For the generics, we want to have a bit, and this will determine how many bits we use to represent the duty cycle. For example, if we had one bit to represent the duty cycle, we would only be able to represent two cycles: 100% or 0%, because we would only have 0 and 1. The more bits we use, the higher the resolution and the finer the granularity we can get for both the output duty cycle and the input clock. This is important because it will define the input frequency of our clock. For example, on the Arty A7, the input clock would be 100 MHz. This clock is used to derive the output frequency that we generate in our design. The frequency is going to specify what our output signal should be. Our entity will also define the actual duty cycle of that signal, but this is the actual frequency of that signal. For the entity, we need one output: the PWM output, which is our pulse width modulated output signal. This is a standard logic signal where the duty cycle determines the width. The width of this signal is going to be determined by bit depth. We'll also need a clock input, which is the input clock signal defined by the generic input clock frequency. So whatever frequency we specify here will be the frequency of a clock that feeds into our design. Additionally, we need an enable input. It's always good to have the ability to enable

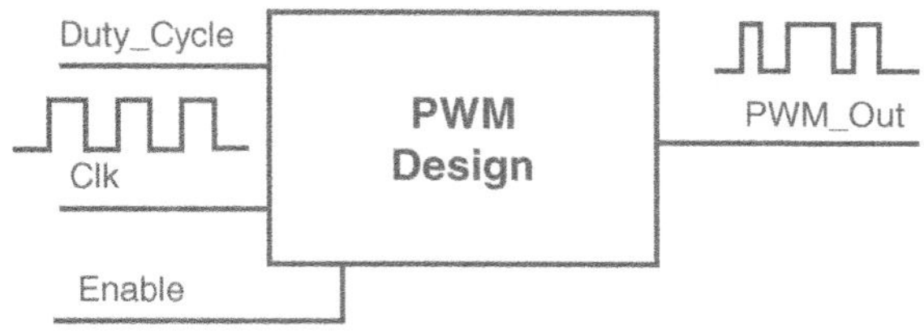

Figure 8.2 The PWM design block.

or disable your design. So, we're going to make it an active low, meaning that when we pull it high, the output will stay low and will no longer be active. So, in order to use this pulse width modulated design, we'll want to have to put a zero on the enable to allow it to actually run. So now let's take a look at how exactly we're going to implement this. We'll call these our PWM design notes, which is just to outline our ideas so that we know what we actually need to do to generate this PWM design. The first thing we want to do is generate a block that explains what's going on at a high level. We will create a block for our PWM design because, initially, we care about the inputs and outputs, and later we'll work on making the inputs produce the desired output. This is our PWM design, as shown in Figure 8.2.

We know that one of the key inputs we want to have is the duty cycle. So, this is going to be one of our inputs. We're going to have the duty cycle, and we're going to have a clock input – the clock that comes into our development board. Then, we want to have an enable signal, which allows us to enable or disable our design. The clock is going to be a signal that toggles at a very high rate. So, it's going to look something like this – likely with a 50% duty cycle or more. But that doesn't necessarily matter. We're just going to be looking at the rising or falling edges of that clock signal. The important part of our design is the output, which will be defined as PWM_Out. That's our output, and it's going to have the varying duty cycle. So, that's where the duty cycle will change on the fly, based on the duty cycle input to the design. The amount of time the signal stays high is determined by the duty cycle value we input. So, what does that actually mean? How are we going to make this happen from our duty cycle right here? Well, let's first look at a standard waveform. If we have a waveform, it will be high for a certain amount of time, then low for a certain amount of time, repeating periodically. If we look at the signal, we know it has a specific frequency. Because in our generic redefined, we want to be able to set the output frequency regardless of what the duty cycle is. So, we need to know the timing between rising edges. So, from one rising edge to the next, we'll have a time interval, and we'll call this max_freq_count. This is what we'll use to determine the frequency of our PWM design. It's constant and will be generated based on our input clock and the frequency defined in the generic. So, based on the speed of our input clock and the desired frequency, which we're going to set in our generic, we calculate this max frequency count. The next key thing is the actual PWM, or more specifically how long the signal stays high. So, we're going to have a second counter that determines how long our signal is high. We'll call this PWM_count and that's going to determine this width right here, and that'll be the same as right there, as depicted in Figure 8.3.

Now, pwm_count is going to be a dynamic value that can change because it is defined by our duty cycle. What this means is: if we think about it, we're going to have a signal that is going to reset or disabled below. Then, as soon as we enable it, we are going to pull our signal high. We're going to have two processes at the same time: we need to track both pwm_count and max_freq_count. The first thing we'll look at is the frequency count (max_freq_count). This is derived by pulling the signal high, which we'll call pwm_out, and now, we'll pull that high at the beginning of our max frequency count (max_freq_count). Once we reach our max frequency count again, we'll

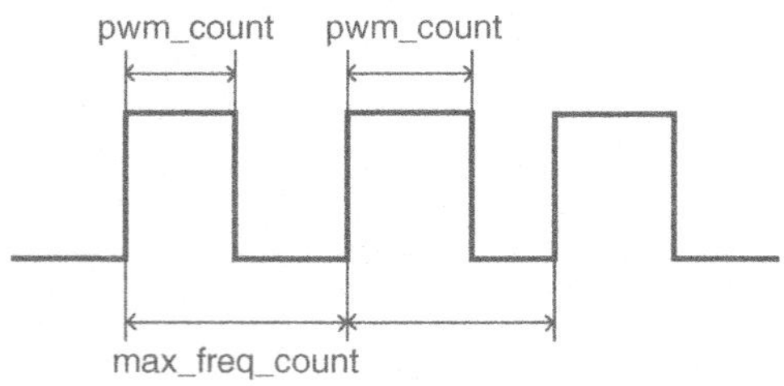

Figure 8.3 The PWM signal parameters.

pull that signal high again. If we have 100% duty cycle, this would be high the entire time, and even though it's already high, we're just going to pull it high once again with our max frequency count. So, this count simply tracks up to the max_frequency_count and triggers the pwm_out to go high again. Simultaneously, a second process is tracking pwm_count. This counter counts up until our pwm_count value is reached, then it's going to pull our signal low. So, we have two opposite things going on: pwm_count, once that's reached, will pull low, and then our signal will stay low until our max frequency count (max_freq_count) is reached. That's when it's going to pull up our pwm_out. It'll pull it higher, then our pwm_count is going to count again and it's going to pull low. So, the max_freq_count and the pwm_count are going to start at the same time. They're both starting on this rising edge, and so we have two opposite things going on. You have your max_freq_count, which is always pulling high, and pwm_count, which is always pulling our output signal low. So, based on the duty cycle, the higher the number, the higher your duty cycle is, and the larger this pwm_count value will be. As this pwm_count value reaches all the way to the max_freq_count, that's going to be 100% duty cycle. If you have a 0% duty cycle, your count just won't occur, and you'll have a signal that stays low all the time. So, the next thing you want to think about is: how are we going to handle doing all of this in VHDL? This is a very important thing to think about. You're going to have all kinds of stuff going on. So, we have to worry about our inputs, our outputs, and then we know roughly what we want to do inside our architecture. So, the first thing let's do is open up whatever tool you're using – you're working with Vivado – so open it up. We'll want to define our libraries. We're using the IEEE library, just like all of our other designs and projects. And we want to figure out what packages we want to use. We know that our output signal is a single bit and uses standard logic, so we know we want to use IEEE.STD_LOGIC_1164.ALL. We want to use this library to allow us to use the standard logic and standard logic vector data types. We also want to use IEEE.numeric_std.All, and this will allow us to use the unsigned and signed data types. In this case, we want to use unsigned, as we're doing our counting and want to perform addition, which is a key part. If you're using just standard logic, it's hard to do a straight addition and comparisons. So now we want to define our entity. We know we're going to give it a name, and we'll call it PWM. That seems to make the most sense. We know we want to define a generic, and for a generic, we know we want to have BIT_DEPTH. This will be an integer data type, and by default, we're just going to give it "8." That gives us eight bits of resolution, which is a pretty standard and common number. That allows you to go from 0% to 100% duty cycle across 255 values. Next is our input clock (INPUT_CLK), and this is used to determine the clock frequency. This is also an integer, and we're going to give it 50 MHz by default. This value will depend on the clock – whatever one you're working with. For the final generic, we're going to define FREQ, which is the actual output frequency you want to have. This will also be an integer value, and we'll give it a default of 50 Hz to start with. Then we want to define the port, so our port should have the following: a PWM output, a duty cycle, a clock, and an enabler. So, we have a PWM output, which is a standard logic. We have a duty cycle input, which is of width BIT_DEPTH minus one down to zero. We have our clock, which is an input and a standard logic. We also have an enable, which is what we're going to use to enable or disable our design. Finally, we need to close off our entity by using the keyword as the following codes.

```
-- Libraries
library IEEE;
use IEEE.STD_LOGIC_1164.ALL;
use IEEE.numeric_std.all;
-- Entity
entity PWM is
    Generic(
        BIT_DEPTH: integer := 8;
        INPUT_CLK: integer := 50000000;  -- 50MHz default
        FREQ: integer := 50);   -- 50Hz default
    Port(
        Pwm_Out: out std_logic;
        Duty_Cycle: in std_logic_vector(BIT_DEPTH - 1 downto 0);
        Clk: in std_logic;
        Enable: in std_logic);  -- Active Low
end PWM;
```

So, there we have the beginning of our PWM design. We know what we want to do – we have our libraries and packages defined. Also, our entity is defined, and so the final part of this is to actually implement and design the architecture of our PWM design. We are going to finish the PWM design we started in this section. We have discussed the design and clarified what we're trying to achieve. At this point, we have the libraries and packages we're going to use. We have all our comments, and we have our entity defined. So, what we need to do is define our architecture. We already know what we're trying to do with our architecture, we just need to define it. So, we make a note to know where the architecture begins and use the architecture keyword. We'll call it behavior, which is a common name for the architecture. Next, we define the constants we'll be using. The first constant is max_freq_count, which is the maximum frequency count. Make this an integer and that's going to be equal to the input clock divided by the frequency. So, this is going to tell us what we need to count to generate our specific frequency. We know we are working with two things here: frequency and PWM or duty cycle. So, the next constant we want to create is pwm_step, which will also be an integer. It will be equal to maximum frequency count divided by two to the power of BIT_DEPTH. This tells us the size of each step for the PWM, based on how we're taking our maximum frequency count and dividing that by 2 to the power of BIT_DEPTH, or how many different individual steps we have inside for our PWM. So, for one-hundredth of a duty cycle, this PWM step is like the maximum it can be. So, we now want to define our signals: in the first signal, we have a signal PWM_value, and this is a standard logic. We will initialize it to a value of zero. In our second signal, we have what we're calling our frequency_count, and this is an integer. It's going to be ranged from zero to our maximum frequency count. We initialize it to a value of zero. The next thing we want to have is our signal PWM_count. We'll make that an integer as well, and it's going to range from zero to two to the power of BIT_DEPTH, and we'll initialize it to the value of zero as well. We also want to have another signal called max_PWM_count, and that would be in your range from zero to two to the power of BIT_DEPTH as well, and in the next signal, we want to create one more signal, and this is our PWM_step_count. This is an integer, ranging from zero to max_frequency_count, and we initialize it to a value of zero. So now we have our constants and signals – all have been defined. We will use the begin statement, and the first thing we want to do is have a conversion where we're going to convert the duty cycle to our maximum_PWM_count.

So, in order to do that, we'll say max_PWM_count to be assigned the value of using the function to_integer and then unsigned and inside the unsigned we want to put the Duty_Cycle, so it's going to take the value of our Duty_Cycle and convert it to an integer for our next max_PWM_count, which is an integer. Now, our Duty_Cycle is an unsigned because it's a standard logic vector because it's our input there. Pwm_Out is equal to our PWM_value. So, our PWM_value is a signal that we're going to change. In order to write to the PWM output, we're going to write to our signal PWM_value. The next thing we want to do is define a process so this process will run the signal out at the correct frequency. So, we need a process that runs our signal out at the correct frequency. Now we're going to do this entire process as a single process because, ultimately, we care about the Pwm_Out, and we don't want to have multiple processes driving the same signal. So, instead of having separate freq_counter and PWM processes, we'll do everything in one process. However, we want to have it in the same process because we're both trying to drive the Pwm_Out signal. So, we will then end our process freq_counter, and we'll do it in process freq_counter. Now, within this process, we first want to do a check, on the rising or falling edge. So, we are just going to pick rising_edge, and you could do this on the falling edge, too. There's no reason for that; we just prefer to do on rising edge, and then if we're not on the rising edge, we don't want to do anything. So, we'll end that if statement, and the next we want to do is we need to check if our Enable is zero. If our Enable is not equal to zero; otherwise, we're disabled then our PWM value is going to equal to a zero, and then we'll do our end if, so first thing we have going on here is we are checking if we're at the rising edge and at the rising edge of "Clk," we're checking to make sure our design is enabled, and if our design is not enabled, then we're going to set our PWM value out to zero. So, if we disable it, our PWM value signal is just going to ride low and stay that way. So, the next thing we want to do is check if the frequency count is less than our max frequency count. So then if it is, we'll have to do multiple things; if it's not, then we're going to set our frequency count to zero and our PWM count to zero. So, we need to check if our frequency count is less than our max frequency count; if it's not, then we're greater than or equal to the max frequency count. So, we need to set our frequency count and set that to zero and we want to set the PWM count to zero as well. So, if we're greater than that, it means we need to do something else. So we need to set them back to zero. Now, inside of it, we want to check if our frequency count is left there, max frequency count. The next thing we want to do is – within our next if statement, if our frequency count is less than our max frequency count, we want to set our frequency count equal to our frequency count plus one – we want to increment that frequency count. Now we want to do another check to see if our PWM account is less than our max PWM count. So if our PWM count is not less than our PWM max count, then we want to set our PWM value to zero. Now, we need to have the parentheses there because we know that this is a standard logic data type. So, for PWM, if count is less than our max PWM count, then we want to set our PWM value to a one, and that makes sense because if our PWM count is less than the max PWM count, we're high. So, we're going to count for how long we stay high; otherwise we set it down to low. The next thing we want to check is if our PWM step count is less than our PWM step, then we want to perform the following and take our PWM step count and increment it. So, we'll say our PWM step count is now equal to our PWM step count and add one or increment it. Otherwise, we want to set our PWM step count to a zero and then set PWM count to a zero as well. That gives you the entire design process of PWM design, so we end our behavior to finalize our design. At this point, this is our PWM design. We have our entity and our architecture and in here we're just checking multiple if statements, because the ultimate goal of this is to output the PWM out values. So, it's going to evaluate this entire if statement logical processing and determine what the value of this PWM is, which we care about. So, we're checking if our frequency count is less than max frequency count, then we increment our frequency count.

If our PWM count is less than our max PWM count, then the PWM value stays high. However, if it's not, then the signal goes low. We also check if the Enable is disabled or if the PWM value is zero. In addition, the PWM value goes to zero; if our PWM count is less than the max PWM count, then our PWM value goes low; it shows this right there. The way this is setting up our PWM value is going to be high and low based on the duty cycle, so the input or duty cycle is going to determine the amount that our PWM value output signal is high versus how much it's low. Therefore, the codes for our PWM design are as follows.

```
-- Architecture
architecture behavior of PWM is
-- Constants
constant max_freq_count: integer := INPUT_CLK / FREQ;
constant pwm_step: integer := max_freq_count / (2**BIT_DEPTH);
-- Signals
signal pwm_value: std_logic := '0';
signal freq_count: integer range 0 to max_freq_count := 0;
signal pwm_count: integer range 0 to 2**BIT_DEPTH := 0;
signal max_pwm_count: integer range 0 to 2**BIT_DEPTH := 0;
signal pwm_step_count: integer range 0 to max_freq_count := 0;
begin
    -- Convert Duty_Cycle to max_pwm_count
    max_pwm_count <= to_integer(unsigned(Duty_Cycle));
    Pwm_Out <= pwm_value;
    -- Process that runs signal out at the correct frequency
    freq_counter: process(Clk)
    begin
       if rising_edge(Clk) then
          if (Enable = '0') then
              if (freq_count < max_freq_count) then
                  freq_count <= freq_count + 1;
                  if (pwm_count < max_pwm_count) then
                     pwm_value <= '1';
                     pwm_count <= pwm_count + 1;
                     if (pwm_step_count < pwm_step) then
                         pwm_step_count <= pwm_step_count + 1;
                     else
                        pwm_step_count <= 0;
                        pwm_count <= 0;
                    end if;
                else
                    pwm_value <= '0';
                end if;
             else
                freq_count <= 0;
                pwm_count <= 0;
             end if;
          else
             pwm_value <= '0';
          end if;
       end if;
    end process freq_counter;
end behavior;
```

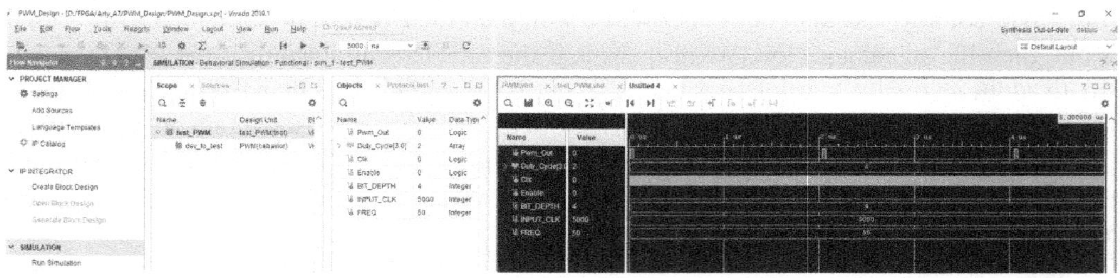

Figure 8.4 The Behavioral Simulation window.

If we take a look at the simulation, we have set the bit depth (BIT_DEPTH) to 4 and the Duty_Cycle to 2. So, if we look, we can see our resolution with regard to that of two. If we look, we can see that we're only high for a very little bit. If we go and try to zoom in to see what's going on right there, we can see that as we go through, we start and we're counting, we're incrementing our count until we reach max frequency count – it's 12, 13 right there – then our PWM value goes low. So, this just shows that if we go back to zoom full, you can just see that this PWM output is directly related to the input, which is the duty cycle. So, for our design, we'll be using this to dim an LED, so as we change this duty cycle, input value, the PWM output of our signal with that duty cycle will change, and that very thing – the duty cycle – will change the brightness of the LED. Or, if you're using it to control a servo motor, it will change the position of that specific servo for the motor. That is a PWM design step showing an example of how to do the entire code and simulation of our design, and we will be using this design inside the LED brightness project. We can copy and paste the following designed test bench codes inside the test_PWM.vhd, and it's going to go ahead and update, and now it recognizes that test bench as our test_PWM, which is a top-level design, as shown in Figure 8.4.

```
library IEEE;
use IEEE.STD_LOGIC_1164.ALL;
use IEEE.numeric_std.all;
entity test_PWM is
Generic(
        BIT_DEPTH: integer := 4;
        INPUT_CLK: integer := 5000;   -- 50MHz default
        FREQ: integer := 50);    -- 50Hz default
--  Port ( );
end test_PWM;

architecture test of test_PWM is

component PWM
 Port(
        Pwm_Out: out std_logic;
        Duty_Cycle: in std_logic_vector(BIT_DEPTH - 1 downto 0);
        Clk: in std_logic;
        Enable: in std_logic);

end component;

signal Pwm_Out: std_logic;
signal Duty_Cycle: std_logic_vector(BIT_DEPTH - 1 downto 0) := "0010";
```

```
signal Clk: std_logic := '1';
signal Enable: std_logic := '0';

begin
   dev_to_test: PWM
       port map(Pwm_Out, Duty_Cycle, Clk, Enable);

   data_stimulus: process
    begin
        Clk <= not Clk;
        wait for 10 ns;
    end process data_stimulus;
end test;
```

8.3 The VHDL Generate Statement

This section is about the VHDL generate statement; in this lecture, we're going to explain how the generate statement is used in the VHDL language. In this case, we're going to be using the generate statement and a design for a counter, which is going to be used in our LED brightness project, and so what is a generate statement used for VHDL? Well, the definition of it is a mechanism for iterative or conditional elaboration of a portion of a description. Now, that sounds kind of confusing, but basically what it's doing is it allows you to iteratively place an object so you can use it in two different types of schemes in a for loop as well as an if statement, and generate statements consist of three main parts. These include a generation scheme, and that is either an if statement or a for loop, and then you have a declarative part and then concur concurrent statements. So, for example, if you have a design you want to replicate multiple times in a row, you can use a generate statement to do that. If you have a circuit that takes a lot of multiple components and just copy and paste, a for loop would be a way to do that. If you want to have some sort of condition where you're checking a value such as and your generic, you have a value that's telling you whether you do a synchronous or asynchronous type of operation; you would use it to generate an if statement. So, we'll go through the examples of both of these. The next thing we will look at is the structure of a generate statement. So, the structure of a generate statement is the first part, where you have a label, and this is just any name you want to give it. That explains what this generate statement is doing. First, you use a colon and then you'll have a for keyword for the generate statement and then you specify the parameter. There's a keyword in, followed by the range. After that, you use the keyword generate. Then, you can include declarations, which are optional. That might be signals or constants that you plan to use. Then you have the begin statement, the begin keyword, and then the concurrent statements underneath the begin. After that, you have the end generate keyword, which is part of begin, and end pair, and you put the label that you've used to describe your generate statement. Now, the other way, if you're using the generate an if type of condition, you still have your same label, your colon, then you have the if keyword, and you put the condition and the keyword generate, and then underneath that you have your declarations, which are your signals or constants. If you're using those, those are optional and then you need the begin keyword, and underneath the begin keyword, you put your concurrent statements and then at the end generate and the label of whatever label you use to label that specific generate statement. Let's take a look at some examples of what these look like. So, the first example we're going to look at is the for loop. So, if we have a for loop, this will say a for loop generate statement because we know we have a for

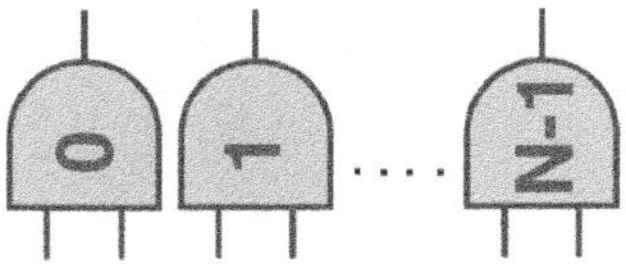

Figure 8.5 The N times of AND gates.

loop, and if so, we'll go with the for loop first, for loop example, and for this example, we are going to say something basic like, if we want to have multiple AND gates. So, for example, if we say we have or draw three of them, but we know we want multiple of these, so we'll say up to n – 1, we're going to use N to describe the number of AND gates we want to have 0, 1, ..., N – 1. We have our inputs and we have our outputs. So, in order to place all of these AND gates, we could go in VHDL and specifically have the logical AND operation for N times, or what we can do is use a generate statement, which is a lot cleaner and easier in VHDL, because we could do this in just a couple of lines and within those several lines will be able to define and describe any number of AND gates in place. Now, this example is for just an AND gate for simplicity, but it could be any design. It could be a completed design, like when we generate our counter, we could use a for loop to place AND counters, or how many counters we want. In order to do this, we would do the generate label, so we're going to call it our G_label. Then you do the colon and we're going to use the for keyword and we'll say I in 0 to N generate, and then we'll put our AND gate. So, we'll say this is a component instantiation of an AND_GATE, and we will say this is AND a number of the port mapping, so we'll just port map, we're not going to go into the details of how we do port mapping, but we understand how port map works. So, and then we use our end generate statement and semicolon as the following codes and also shown in Figure 8.5.

```
G_label: for I in 0 to N generate
    AND_GATE: AND(port map)
end generate;
```

At a high level, we're looping through from 0 to N times, generating AND gates, and for the first time, we have our zero AND gate, one AND gate, all the way up to the N-1 AND gate. So, this whole example here is just showing that you can place a lot of different AND gates by using a simple three lines of code, whereas if you would have to, if this N number is, say, 10 or 12, it could possibly take 10 or 12 lines of code and your VHDL design versus just doing this in three lines. If you want to adjust the amount of AND gates replacing or whatever design it is, you just have to change this N value, and its N value could be a constant in your design. It could be a generic or whatever way you decide to pass this into your VHDL design. This is a number that you can change to change the number of AND gates you're going to be creating in your design. So, the next example we want to take a look at is the if example. So, within our design, if we want to use an if example for the generate, we'll call this our if example; let's say we have a design that could use a JK flip-flop or a D flip-flop. For example, if we have a design, this could be any type of design, we have our J input, clock, and K input, which is just a JK flip-flop. We have our Q output and Q-not, and we also have a D-full flop. So, we have two different designs and this was our D input and our clock input. Then we have our Q and Q-not outputs. It's a D flip-flop, so we could do this in several different ways. We could have two different designs: one that utilizes the JK flip-flop and the other that utilizes the D flip-flop and save them off as two VHDL designs, or we can have a single VHDL design that contains both the JK-flip flop option and the D flip-flop option. The way we would do that is by using a generate example and using the if configuration. So, if we wanted to do this, we would have, for example, our design, where we'd have our entity and it could be whatever we decide, or inputs and outputs, or whatever our design is, then we'd have our architecture. Inside our architecture, we are

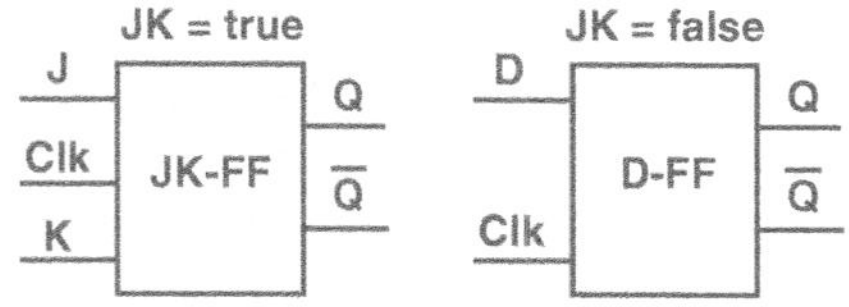

Figure 8.6 Using if example in generate statement.

going to have two generate statements that are going to check whether or not we decide to define our JK-flip flop or D-flip flop. So, for example, we would have a generic that we're going to call inside our entity, a generic value, which we're going to call it JK and it's a Boolean, as it's going to be equal or true or false. What we want to do if JK is equal to true is use the JK-flip flop option. If JK is equal to false, we want to use the D-flip flop option. In order to do that, we would use our label. We'll call it JK_gen for generate use our colon and then we use that if JK referring to a generic; if JK is equal to true, then we want to say generate, where we would generate, so that's where you put your JK-flip flop design and then we'll do the end generate semicolon. Now within the same architecture, we're going to have another label that we will call D_gen for the D-flip flop generate and we're going to check if JK is equal to false, then we're going to say generate the following, and underneath there we would have our D-flip flop design. We'll put the end generate, and both of these designs are in the same architecture, the same VHDL design file. Then we'll have our just end of architecture here. That's just to show you that this is all in one design. So, they're sharing the same entity, we have the same architecture or architecture within here. We're going to have this JK_gen generate statement and we're going to have that D_gen generate statement as the following codes and depicted in Figure 8.6.

```
entity ….
     generic (JK: boolean);
end entity …;
architecture ….
    JK_gen: if JK = true generate
        -- JK flip flop Design here
    end generate …;
    D_gen: if JK = false generate
        -- D flip flop Design here
    end generate …;
end architecture …;
```

Then we'll also have different processes or different concurrent statements to determine how these Qs are output or if we have the JK_gen option, how we're connecting our J and K versus if we go with the D_gen option, how we are connecting our D, and how our clocks are going to be connected the same. So this allows you to have flexibility within a single design. In this way, we're going to use this within a counter to have the flexibility to determine whether or not we want to have the counter do a synchronous or asynchronous type of reset. So that is an example of showing how to use the generate statement in two different configuration examples.

8.4 The Counter Design

This section is about the counter design. Here, we are going to cover how we can create a counter in the VHDL. So, the first thing we need to think about is we've done counter before. So, there's nothing crazy happening with a counter. You're just effectively having a clock set on coming in and you're counting that clock signal. Now, what we want to do is do a little bit more than that. So, we have our standard counter, but we want to add a few more features to it. So, there are a couple of parameters or specifications we'd like to have. The first one being that we want to be able to have a counter be synchronous or asynchronous with regard to the reset. We're going to count on the clock edge, but for the reset, we want to have an asynchronous or synchronous type of reset. A synchronous reset means that if you're counting up, it won't go to zero until the rising edge of the clock. An asynchronous reset, however, will go to zero as soon as the button or your reset signal is pressed, and so we'll talk about how to design a counter to count for both of these different options. We want to have a configurable maximum count value. When you're working with counters, one of the key things you need to do is understand what your maximum count value is. When you're working with a counter, the goal is to create some sort of delay, and that delay is going to be defined by how high you count to. So, you have your input clock rate and you're going to count to a certain value, which will equate to a certain time delay, and we also obviously want to be able to reset the counter if maximum value is, say, 100; we want to be able to, at any point in time, get to 50, be able to hit reset, and have our counter go back down to zero again. So, it can also be used like a watchdog timer where you have this counter that counts. Every time you perform a certain operation to check on your watchdog timer, you'll just have to reset and know that, if we ever get at our time or ever reach the maximum value, we know we have an issue. So, we need to shut things down or generate a warning like a watchdog timer fault. Obviously, the most important thing is to have an output to indicate that our maximum count is reached. So, we'll have an output that is zero, except more just count based on the input clock we have coming in. Once we reach the maximum count value, then our output will flag high and toggle to say, "Hey, I reached the maximum count now and then go ahead and start cycle over and bring our count to a zero and continue again." So effectively, the output is going to be toggling high one clock cycle, every time it reaches the maximum count. Now, the important thing to think is in order to do this, we're going to need to figure out how we are going to work with the generic and the entity. So, for the generic, we're going to have a maximum value, which will be our maximum count value. We have a synchronous reset, and this is going to determine whether or not if we reset synchronously or asynchronously and this will be a Boolean value. So, it's going to be a true or false to determine what type of reset we're going to do, and that's key for our design. Then, our entity is going to contain three different inputs and outputs. So, we have one output, which is our maximum clock count and our max count, and this will toggle high when we achieve our maximum value, which we define in our generic, and then we have our clock and reset. Our clock is whatever input clock you're bringing in, which will most likely be the system clock on your board. So, it'll be the 100 MHz for Arty A7 development board. The reset determines when we reset the counter, and this is going to be correlated with our synchronous reset. If we set our synchronous reset to true, we want to have our reset take effect on the rising clock edge. If our synchronous reset is set to false, we want to reset to take effect as soon as we hit that reset value.

So, let's take a look and go through and understand our counter design. For our counter design to get an idea of what we want to do, let's just call this our counter design. The key thing for this is to understand the synchronous versus asynchronous reset. As far as indicating a maximum value in account, we know that we can have a process that evaluates on the clock and every rising edge increment. So, what we need to understand is: how does a synchronous versus asynchronous reset occur? First, we're going to create a signal that is going to be our clock and we'll try to make it a 50% duty cycle and just a clock signal. So, we'll just make several pulses here to show what that clock signal will look like. We're going to have this be a steady signal that doesn't ever change. It's coming in from the board, from an external oscillator, so we won't really ever

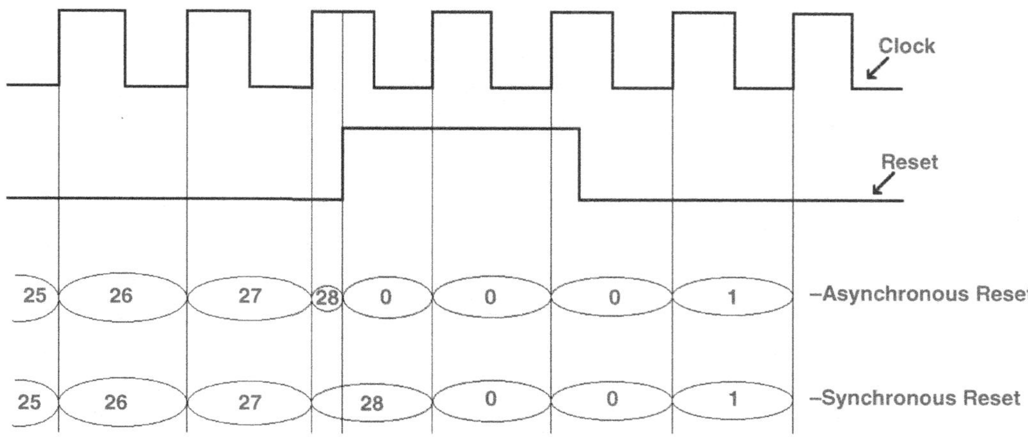

Figure 8.7 The counter design.

see that change occur. We are going to have another signal, which we'll call reset. So it's going to stay low, and we will say at a certain point it's going to hit reset. Then we'll stay in reset for a while, until it decides to go back down and have a reset. Now what we need to do is define the edges. So, if we are going to just assume we're going to work off the rising clock edges, we could also work off the falling clock edges. However, we're just going to go with the rising edges. So, make those dashes just to say that these are the times at which these rising edges have occurred. The key thing to note right here is that our reset has occurred after the rising edge of the clock. So, there is a bit of a delay there, not quite halfway through the rising edge, but slightly after, the reset has occurred, so it looks like if you were to simulate your design, this is roughly what it will look like, as illustrated in Figure 8.7.

We'll give you a nice straight line showing how the synchronous versus asynchronous design looks, but this is just to give you an understanding of what we want the design to do. So, we're going to start with our asynchronous clock value. So, we start with our clock value at 25, and as the signal counts, since our account value is not like a standard logic or the logic vector, it doesn't have a higher or lower or just put the actual count value, so it's going to be counting, which becomes 26 and it counts again. Since this is an asynchronous reset, we'll have a reset check later. We will also have our synchronous counter, as shown in Figure 8.7. The main difference is that for an asynchronous reset, you monitor both the reset and the clock, whereas for a synchronous reset, you need to have a process that looks at just the clock. On the high end, on the rising edge of the clock, you can then evaluate your reset signal; however, in your asynchronous reset, you'll need to have a process that looks at the clock and the reset. If the reset changes state, you need to check that and determine if it's the rising edge of the reset, and you're going to have to go low or reset your count. So that's the two biggest differences; we're going to have two separate processes: one for the asynchronous and one for the synchronous. Now, in order to implement this in VHDL, the key thing we're going to use is the generate statement, because we're going to have a generate, which checks if asynchronous or synchronous is true. This goes back to our generic value synchronous reset. If we set that to true, we're going to generate a synchronous counter or a process that looks only at the clock and evaluates the reset inside that process. If our synchronous reset is false, then we need to have two processes where we look at the two signals that are our stimulus: the clock and the reset. If the reset changes state, we have to reset right away. So that is kind of the difference between asynchronous and synchronous clock cycles, and that's exactly what we'll be implementing in our counter design. Now, we are going to write the code for our counter that can be either an asynchronous or synchronous reset with the generic maximum value we're passing in, and then we're going to go ahead and simulate a counter just to see what exactly it should be doing or what it will be doing. So, for the first part of the design, if you're working with the Xilinx board, you can do this in Vivado. So, the first thing we want to do is just put some sort of comment back up front to kind of explain what's going on, and the next thing we're going to put is libraries, and we'll use library IEEE, and then use IEEE.STD_LOGIC_1164.ALL. This is going to allow us to use the standard logic and standard logic vector data type for our outputs. We have a maximum count, so we need to be able to have access and use the standard logic data type. So, we need this standard logic 1164 package. We also include the IEEE.math_real.All, and this is going to allow us to do specific things; for example, when we're determining our counter, we need to know the number of bits we're going to use. The math_real will allow us to determine and use the specific number of bits. So, we want to say use the IEEE.numeric_std.all, and the next thing we want to do is go ahead and define our entity, so we're going to say entity and we're going to call it COUNTER keyword. Then we need to define our generic and port, so we know we're having a generic. So, for the generic, we want to have the open parenthesis, and we're tab down to the next line and say MAX_VAL. However, we like to do this because we think it just makes it easier if you have an all-caps variable or signal that, you know, is either constant or generic, so it's not going to be dynamic and change, and will say integer, and we will make it equal to 2 to the power of 30, so this is just a generic MAX_VAL is going to change, and so we're just going to put some sort of value in there. So, it is effectively a 30-bit counter, which in this case when we go through our brightness project will change it anyway. So, the values you put in a generic don't necessarily matter if you're going to use them later. They will matter if you don't fill them in or if you're using them. But if you're going to use these like a component instantiation sort of situation, your rewrite over those generics anyway, and so we're going to come up with our next generic, which is the SYNCH_Reset for synchronous reset, and this is going to determine if we have a synchronous reset or not, and this will be in Boolean data type and we will have a default to a true, and so we are going to close off a bracket there, so that completes our generic. We now want to have the port, so go ahead and use the port keyword and we will go to the next line. In the port part, the first thing we're going to have, is our Max_Count, which is our output. It's a keyword now that tells us when we reached our maximum count. So, this will flag high when we have our maximum count has been reached. So, a clock (Clk) this is an input and it is a standard logic, and our third is a reset (Reset), and this is an input and it's a standard logic as well. So that gives us for our generic we have maximum value and synchronous reset and for our port we have our maximum count, our clock, and our reset. So, we'll go ahead and do that end COUNTER. So, our counter entity has been officially defined. Now we want to define the architecture, so architecture and we would give it a name behavior of the word of COUNTER is, now we need to define any constants or signals we're going to use so we have a constant. We want to use that here constant, and we are going to call that bit_depth and we will make it an integer and for this constant, we are going to calculate and determine the number of bits required based on our MAX_VAL. So right now, it's going to be 30 because we're taking you to 2 to the power of 30. But this max value can change and so we don't know how it's going to be 30. So, we want to make sure we don't want to have more bits that are required just due to resources. So, we'll say go to integer(ceil(log2(real(MAX_VAL)))). So that's going to give us the number of bits required to count to the MAX_VAL, and we'll go ahead and do a signal, which we will call Count_reg, for our count-register, and this will be an unsigned data type, and that's because this will give us the flexibility of being able to do addition and subtraction to this count-register. So, we're going to give that bit_depth minus one down to zero, and we are going to initialize that to a value of zero, so we will use the other's keyword, and that's going to say no matter how big it is, because we don't know what size count is going to be until it's based off of whatever the MAX_VAL is, that's where all these other values are derived from. So, the next part, we'll go ahead and use our begin statement to begin our architecture and we'll go ahead, and before we put everything else in there, let's just do our end statement so we don't forget it. Since, we think it makes it easier that way, so you don't forget put your end statement there. So, first thing we want to do is check if we want to do a synchronous reset. So, if we're doing a synchronous reset, we want to generate a synchronous sort of resetting operation. So, let's give it a label, and we'll call it synch_rst and colon, and then we'll do our if statement. So, if SYNCH_Reset is equal to true, then we're going to generate the following, so before we even put everything on it, let's do our end generate. So that's our first generate process, so inside of there, we want to create a process, call it Count_proc for a process, which is a process that is going to look at the clock (Clk). So, beings as a synchronous reset, we only care about the clock. So put that in the sensitivity to us and we want to go ahead and do our end process. Inside of our count process (Count_proc), what do we want to do? So first, before we do that, we don't forget to write begin and end process, inside our process, we want to check that it's the rising edge of the clock. So, if rising edge, and we're looking at our clock. So, if the rising edge of clock, then we'll do the following if it's a falling edge of a clock, we're not going to do anything. So, make an end if statement right there. If rising edge of the clock, then what do we want to do? We want to check if our reset is zero or if our count register is going to the max value, then we'll set our count register to zero. So we'll say there are Reset and this will give us that active low reset, equal to zero, or our count register is equal to the max value, then we want to do the following, and in this case, we're going to have an else, and end if. So we'll say about Count_reg, and we will use the others keyword to set it all to zero, so this is going to set our count register zero, otherwise we want to take our count register and add one to it. So, this is why we're using the unsigned data type that allows us to add a one to our count register. So now we have our first synch_rst, our first generic statement defined. We have our synchronous reset, now we need to define our second one where we're going to have an asynchronous reset. So, we'll give it the asynchronous reset definition, the label, and inside of there, we're going to replicate very similar to what we did inside our synchronous reset, however, except we want to have a check if a Reset is zero. So, go ahead and put our label, which is we call it asynch_rst and this is we're checking the SYNCH_Reset is false, so if SYNCH_Reset is false, only one of these will be evaluated, and then we'll do our end generate. So inside of here, we want to have a process and we're going to call it Count_proc again, except we are going to have two signals that for our sensitivity we have our clock and a reset, so if our reset changes state, we're going to evaluate this as well. So, we are using begin and we'll go ahead to end process so we don't forget our any of our begin–end pair inside of our count process for asynchronous reset. The very first thing we want to check is if our reset is equal to zero, so we're not even worried about that, which edge of the clock or anything, we don't even care where the clock is at all, we care about it, if our reset is equal to zero, then we need to reset, that's the case, we need to set our count register back to zero, else if in the rising edge of the clock, then we want to do the following, then we basically want to follow the same methodology we have there for synchronous reset, we want to check if our count registers equal to our max value, if it is we will set our count register to zero else we will add one to our count-register, so, if inside this else, if we're going to have a second another if statement, if our count-register is equal to the max value, then our count register going to be set to zero, meaning we've reached our maximum count else, our count register is going to be equal to count register plus one, we're going to increment it now we have to think about our first if statement. So, we're only going to be care about if a reset is zero or if the rising edge of the clock, then we're going to either increment our count register or reset it to zero. So now we have our count process. We want a second process, so the generic statement is going to pick either the synchronous reset count process or the asynchronous reset count process. So only one process will be within our architecture. Now we need an output process to determine what is the maximum count, what is the output of our design. So, we'll call it output_proc, go and say process and we're going to put our Count_reg, so every time our count register is updated, we are going to evaluate this process. So, within here we'll say it begin and we'll do our end process. That's important and we want to do that. So, within this process, what we're going to do is we're going to check and we're going to set our max count, Max_Count is equal to zero. However, then we're going to check if the count register, Count_reg, is equal to the max value (MAX_VAL). If it is, we're going to do the following – and we will end that with an end if. So, if our count register equals the max value, what are we going to do? Well, we're going to set our Max_Count to 1. After this process evaluates, one of two things will occur, a max count is going to be equal to a one or a zero. So initially we set our max count to zero, and since we're inside the process, it's a sequential statement. If our max count, Max_Count, is equal to zero, if our count register is equal to maximum value, then we're going to set our max count to a 1. So, after this process is evaluated, our max count is going to be either a zero or a one, and so that is the entire design of the counter where we have and it's being set there synchronous or asynchronous through the use of the generate statements, go ahead and save our design. The codes for this counter design are as follows.

```
library IEEE;
use IEEE.STD_LOGIC_1164.All;
use IEEE.math_real.all;
use IEEE.numeric_std.all;
entity COUNTER is
    Generic(
         MAX_VAL: integer := 2**30;
         SYNCH_Reset: boolean := true);
    Port(
         Max_Count: out std_logic;
         Clk: in std_logic;
         Reset: in std_logic);
end COUNTER;
architecture behavior of COUNTER is
    constant bit_depth: integer := integer(ceil(log2(real(MAX_VAL))));
    signal Count_reg: unsigned(bit_depth - 1 downto 0) := (others => '0');
    begin
    synch_rst: if SYNCH_Reset = true generate
        Count_proc: process(Clk)
        begin
           if rising_edge(Clk) then
              if((Reset = '0') or (Count_reg = MAX_VAL)) then
                 Count_reg <= (others => '0');
              else
                  Count_reg <= Count_reg + 1;
              end if;
           end if;
        end process;
    end generate;
    asynch_rst: if SYNCH_Reset = false generate
        Count_proc: process(Clk, Reset)
        begin
```

```
            if (Reset = '0') then
               Count_reg <= (others => '0');
            elsif rising_edge(Clk) then
                if(Count_reg = MAX_VAL) then
                   Count_reg <= (others => '0');
                else
                    Count_reg <= Count_reg + 1;
                end if;
            end if;
         end process;
      end generate;
      output_proc: process(Count_reg)
      begin
          Max_Count <= '0';
          if(Count_reg = MAX_VAL - 1) then
              Max_Count <= '1';
          end if;
       end process;
end behavior;
```

We can copy and paste the following designed test bench codes into test_COUNTER.vhd, and it's going to go update, and now it recognizes that test bench as test_COUNTER, which is a top-level design, as shown in Figure 8.8.

```
library IEEE;
use IEEE.STD_LOGIC_1164.ALL;
use IEEE.math_real.all;
use IEEE.numeric_std.all;
entity test_COUNTER is
   Generic(
         MAX_VAL: integer := 2**3;
         SYNCH_Reset: boolean := true);
--  Port ( );
end test_COUNTER;

architecture test of test_COUNTER is

component COUNTER
 Port(
         Max_Count: out std_logic;
         Clk: in std_logic;
         Reset: in std_logic);
```

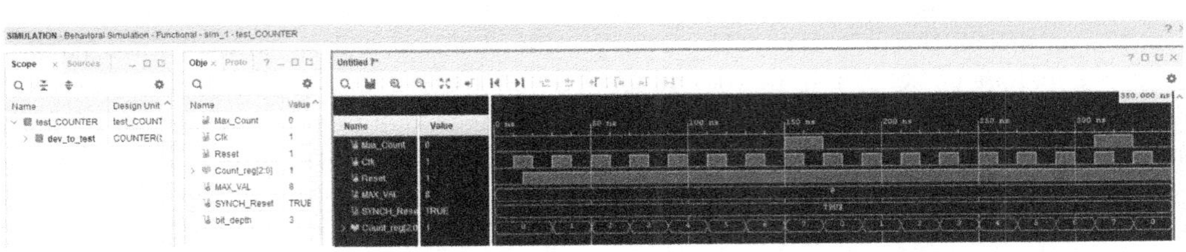

Figure 8.8 The Behavioral Simulation window.

```
end component;

signal Max_Count: std_logic;
signal Clk: std_logic := '0';
signal Reset: std_logic := '0';

begin
   dev_to_test: COUNTER
       port map(Max_Count, Clk, Reset);

   data_stimulus: process
    begin
        wait for 10 ns;
        Clk <= not Clk;
        wait for 5 ns;
        Reset <= '1';
        wait for 5 ns;
        Clk <= not Clk;
    end process data_stimulus;
end test;
```

We created a quick simulation just to show you what this counter locks like. So, we have our count signal, and it doesn't start until our reset, which we can see here is a zero, so our count still stays at zero. Once the reset goes high, you'll notice our count stays zero until the rising end of the clock, and that's because we're looking at the rising edge of the clock. At that specific point, we get a count of one. You will notice that each count occurs at the rising edge of a clock. So, we can tell that everything occurs at the rising edge of the clock and our reset. Since it occurred right there, the counter doesn't count until we get there. So that is just a basic example of just a simulation showing what this counter is going to do. Now that you have this counter, this same counter will be used in our LED brightness project.

8.5 The VHDL Component Instantiations

This section is about VHDL component instantiations. Here, we're going to explain what a component instantiation is, give an example of how it can be used, and show how to apply it in our LED brightness project. At this point, we have a PWM design as well as a counter design, and we want to take both of those individual VHDL designs and use them in our LED brightness project. First, let's start off with what actually is a component instantiation; these allow your VHDL designs to take designs from other files. For example, if you have a project, whether it's in Vivado or any other type of development tool, and you have multiple designs, you'll need a top-level design. This top-level design is where you map your individual IOs. From that entity, you're going to map to all the different IOs on the FPGA you're using. However, you may have other designs written in separate VHDL files that are embedded inside your design. This approach follows a hierarchical structure: you have a top-level design, and within it, you can include or 'pull in' other sub-designs. These sub-designs don't actually instantiate the entities, and those entities don't actually get mapped to physical IO pins from the .xdc file or whatever type of the pin mapping file. Those can actually be mapped to signals inside your design, and this allows you to instantiate the design multiple times. For example, if we have created a counter module, we can use it multiple times in our design – say, 20 counters – while only needing one VHDL file to define that counter; we simply instantiate it

20 times. Now, when you instantiate it multiple times, you will use more resources. So, it's going to be like you're just taking, copying, and pasting that design several times. You just no longer write it out 20 different times. So, it saves you in that aspect and makes your designs a little nicer, and a lot of times when you're working with more complex and bigger designs, this will become more and more relevant, especially if you're working on a big project and you have multiple people designing. You have somebody working on this portion. They'll say, OK, make this design, then we'll instantiate that component here. You'll instantiate that component here, and you'll work on the top-level or the wrapper, if you will, and take all those designs in time together into one VHDL design. When you define when you're working with this component instantiations, there are several parts to it. First, you need to define the component, hence the name "component instantiation." This is done before the architecture or inside the architecture, specifically before the begin and the declarative statement section. The component definition outlines its inputs and outputs. So then inside your architecture, if you want to use that component, you then just go ahead and do a port map and a generic map to map signals to go into that actual component itself. Next, we're going to take a look at an example of this with some notes. The component instantiation is the term we call to represent a second design inside our design. For example, if we look at a block diagram, like the one shown in Figure 8.9, we can say this represents our top-level VHDL design. Let's say one of them is called Example.vhd. So, this is a VHDL design that is defined somewhere in our project, but maybe somebody else wrote it, and we have a couple of inputs to it. It has two inputs and one output. The output gets mapped and comes outside of our top-level VHDL design. For the inputs, let's assume we're working with something like a JK flip-flop. This is just an example to illustrate the point. So, we might have signals like K, Clk, and J as inputs, and outputs like Q or Q-not. These inputs would be mapped to the external inputs, so we have these three inputs. Then we have another one, which we'll call test.vhd. This is a second design that also uses the clock signal and has one input. It produces two outputs that we then we do AND together to form our final output. So, in our top-level VHDL design, we are instantiating two components: the first one is Example.vhd and the second is test.vhd. These are two separate VHDL files. Including the top-level file itself, we now have three VHDL files in total. The top-level file can be named something like Top.vhd. This Top.vhd file serves as the highest level in our hierarchy. It defines all the inputs and outputs that get mapped to the physical IO pins on the actual FPGA device. For example, these four inputs we see on the left and the two outputs we see on the right are shown in Figure 8.9.

The way this work is if we're doing this within our VHDL design, we're not going to write the entire design at once, but what we're going to do is define the main parts. First, we have our entity, and within our entity, we define both generics and ports. Then, we have our architecture, and within our architecture, we would define the components. In this case, we would define the

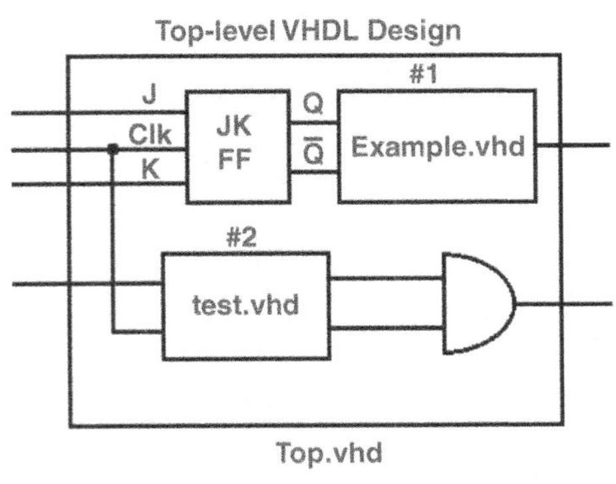

Figure 8.9 The component instantiation.

Example.vhd and test.vhd components. By defining them, we're going to do is to define what the inputs and outputs are. The actual design itself will be in separate files that contain the architecture definition. Here, we are also specifying the inputs and outputs, along with any constants or signals that will be used within the architecture. Then, within the architecture, we use the begin statement, followed by instantiating the components, along with the rest of our design. Finally, we close with the end statement, forming the begin–end pair. What you have is a top-level design that allows you to reuse components easily. For example, if we want to use Example.vhd a second time, all we need to do is instantiate it a second time inside the architecture, we don't have to define it a second time. You only need to instantiate it again, or you can instantiate it a third time, fourth time, or even more. This is where you can use generate statement to define the number of times you instantiate a component, or both components. This is just an overview of how this component instantiation works, as shown in the following codes.

```
entity…
   Generic();
   Port();
end;
architecture
-- Define component
-- Constants
-- Signals
begin
  -- Instantiate components
end;
```

The VHDL design is quite useful. As mentioned earlier, when you're working with more complex designs and utilizing multiple designs, it's valuable to have a structured approach. This is especially true when you have components that you frequently use. For example, we have a counter and a PWM design that we use in our LED brightness lab. These two components can be reused in many other labs or whenever you need PWM. Once the design is completed, you can simply instantiate it in any of your other projects, saving time and effort. When designing components or working on your own designs, it's always a good idea to think about how you can make them more flexible, so they might be reusable in the future. Now, let's open up the LED brightness project. Within the project, you'll see that we already have the initial setup: descriptions, comments, and everything is kind of explained out. We have also defined our libraries and packages, and our entity is defined. We have also defined the generics and ports. In the architecture section, the architecture name is already provided, along with the components to instantiate. Signals and constants have also been defined. The first step is to define the components. The first component we're going to define is the PWM design. The key here is that defining a component is very similar to an entity, except we use the keyword component. Then, we refer to the entity of our design. For example, in PWM.vhd, we can see that the entity name is PWM. So that is the name we want to use for our component. To proceed, open the LED brightness component PWM, and we'll now define the generic. If we look at our PWM design, we can see that our generics are BIT_DEPTH, INPUT_CLK, and FREQ. We can simply copy these generics from our PWM design because it's the exact same for our component instantiations in the LED brightness project. Now we need to define the ports. So that is done by using the keyword port, just like it would in an entity, and if we go to our PWM design again, we can see that the PWM has the port Pwm_Out, Duty_Cycle, Clk, and Enable. So, we can take and copy all that over just the same, because we have to define the data type as well as it's inputs or outputs, which we're already doing in the entity description. So, we go back to LED brightness and tab over and we'll go ahead and paste our port into there. So now we have our component defined, last part we need to do is the key part of what we say, the end component, and then we'll give it the name PWM. So now our component is defined and we can instantiate our pulse width modulation (PWM) design throughout our project. The next part we want to do is we want to define or instantiate our counter. So, we'll just say counter so we know what we're instantiating here, and for this part, that follows the same structure as our PWM, we say component and then COUNTER, and just like we did with the PWM, we can do our copy and paste just to kind of that way. So we go to our COUNTER, we know our generic has two of these MAX_VAL and SYNCH_Reset, so use a copy and we will paste these into our generic. Now we need to do our port, just like we did on our PWM design, the port here in just the same, we can copy that from our COUNTER because it's going to be the same ports, same signals, the same names, and that gives us our Max_Count, Clk and Reset, now we just need to do our end component and the name COUNTER. So now we have our PWM and COUNTER both defined, what we want to do is inside our architecture, we want to instantiate them. So, we already have our signals and constants, they are all defined and set up. We have our begin of architecture, we're assigning the outputs and now we want to add before over processes and after we do this invert enable signal for counter, we want to start instantiating components. So, let's do a quick component instantiation, so the first one we're going to do is this LED counter. So, we're going to have a counter that's counting for the LEDs. So, we have to give it a label and we're going to call it LED_COUNTER, and then we go with the colon and we have to give it the components name, so it's the COUNTER, and now we need to do a generic map and the port map. So, for generic map, and when you do the generic map, you have to give it the value. So, if we look at our COUNTER up there, we had the first values are MAX_VAL and SYNCH_Reset. So, we need to give it the MAX_VAL, which in our case, our MAX_VAL is LED_MAX_COUNT, and this is defined by our constant LED_MAX_COUNT and then our SYNCH_Reset, we already have this defined as a constant SYN_RESET says that is true. So, we just want to go ahead and put that in there. So now our generic map has been completed and we do not want to put a semicolon at the end of that, we want to go ahead and do our port map and for port map, so, in this case will say port map, and we want to have Max_Count, which is the value we see on our COUNTER. So, we're going to say, OK, Max_Count, because that's the value that's our entity, that is the actual what we call our output. Now we do the equals arrow and we need to assign it a value. So, we're going to assign it the signal led_max_cnt for LED maximum count and we'll do a comma. Now we want to do a second one. So, our Clk (clock), we are going to map to our clock (Clk), which is a clock in our design, and then for the third one, we want to do a reset. So, our Reset is mapped to our led_counter_rst, and then at that point we go ahead and do close our parentheses off, and if we save, we can see that we have underneath our LED brightness behavior, we have this LED_COUNTER-COUNTER behavior because we just defined this right there. Since, the LED_COUNTER is defined within our LED brightness design. So, we just instantiate the COUNTER. That was part of our LED brightness project is we have this COUNTER that's instantiated, which is defined as a separate design. Now you notice it recognizes we have this PWM design, but it's not underneath this LED brightness because we have not instantiated it, as depicted in Figure 8.10 as well as the codes below, and so that is the example showing you how a component instantiation works as well as how to use them in your designs, and as soon as you instantiate them in your VHDL designs, your tool will recognize and understand that it's being defined or it's in the hierarchical structure and it's underneath your top-level design.

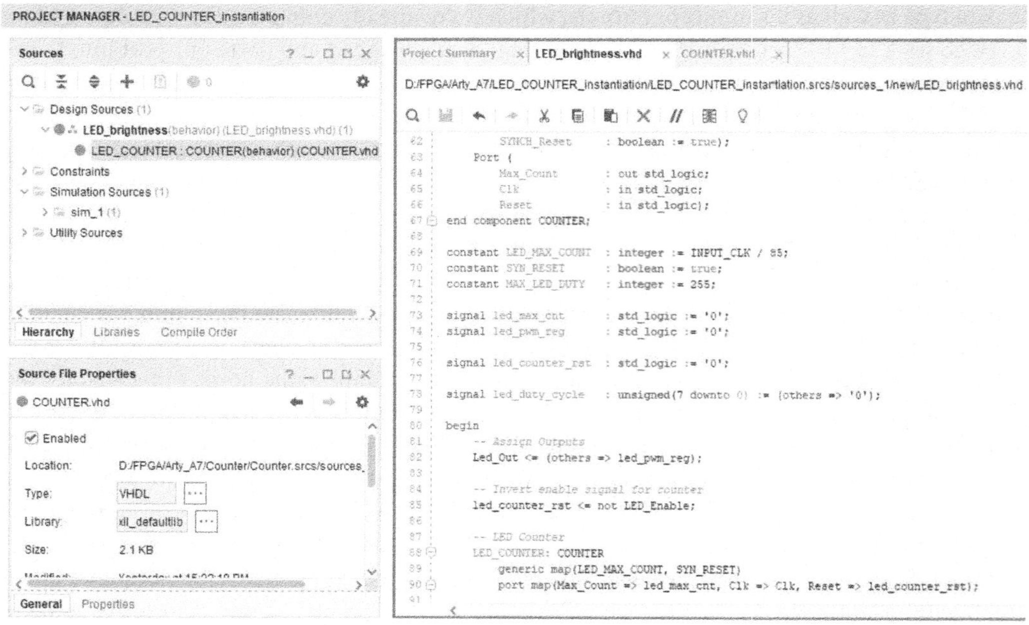

Figure 8.10 The instantiation of component COUNTER.

```
-- Libraries
library IEEE;
use IEEE.STD_LOGIC_1164.ALL;
use IEEE.numeric_std.ALL;
use IEEE.math_real.ALL;
-- Entity
entity LED_brightness is
  Generic (
      INPUT_CLK: integer := 100000000; -- 50 MHz default
      NUM_RGB_LEDS: integer := 4;
      NUM_LEDS: integer := 4);
  Port (
      Led_Out: out std_logic_vector(NUM_LEDS - 1 downto 0);
      Clk: in std_logic;
      LED_Enable: in std_logic);  -- Active Low
end LED_brightness;
-- Architecture
architecture behavior of LED_brightness is
-- Components
-- Pulse Width Modulation Design
component PWM
   Generic (
      BIT_DEPTH: integer := 8;
      INPUT_CLK: integer := 50000000; -- 50MHz
      FREQ: integer := 50); -- 50Hz
   Port (
      Pwm_Out: out std_logic;
      Duty_Cycle: in std_logic_vector(BIT_DEPTH - 1 downto 0);
      Clk: in std_logic;
```

```
      Enable: in std_logic);
end component PWM;
-- COUNTER
component COUNTER
   Generic (
     MAX_VAL: integer := 2**30;
     SYNCH_Reset: boolean := true);
   Port (
     Max_Count: out std_logic;
     Clk: in std_logic;
     Reset: in std_logic);
end component COUNTER;
constant LED_MAX_COUNT: integer := INPUT_CLK / 85;
constant SYN_RESET: boolean := true;
constant MAX_LED_DUTY: integer := 255;
signal led_max_cnt: std_logic := '0';
signal led_pwm_reg: std_logic := '0';
signal led_counter_rst: std_logic := '0';
signal led_duty_cycle: unsigned(7 downto 0) := (others => '0');
begin
-- Assign Outputs
   Led_Out <= (others => led_pwm_reg);
-- Invert enable signal for counter
   led_counter_rst <= not LED_Enable;
-- LED Counter
   LED_COUNTER: COUNTER
   generic map(LED_MAX_COUNT, SYN_RESET)
   port map(Max_Count => led_max_cnt, Clk => Clk, Reset => led_counter_rst);
-- LED PWM Count update Process
   Led_Count_Proc: process(Clk)
   begin
     if rising_edge(Clk) then
        if (led_duty_cycle = MAX_LED_DUTY) then
           led_duty_cycle <= (others => '0');
        elsif (led_max_cnt = '1') then
           led_duty_cycle <= led_duty_cycle + 1;
        end if;
     end if;
   end process Led_Count_Proc;
end behavior;
```

At this point, we have defined the LED counter instantiation, where we instantiated the counter design. We already have our counter, as well as our PWM designs defined. We just need to instantiate the components so we have an LED counter. We want to have an RGB counter that counts for – if you have an RGB LED (if you're using the Arty A7 board), you will need this. So, it is called RGB_COUNTER and colon, and then we have to define the component we want to use, which we call COUNTER. Just like our LED counter, we have a generic map, and we will give it the RGB_MAX_COUNT, and we're going to give it the SYN_RESET, and then we'll do our port map. Within our port map, we need to define what we want to use. So, we have our MAX_Count, and that is going to take on our rgb_max_cnt to tie it to that, and our clock (Clk) is going to be tied to clock (Clk), just like on our LED counter. Then we have our Reset, and our Reset is tied to our rgb_counter_rst signal. We're basically instantiating the same thing as our

LED counter, except that we have a different count value (rgb_max_cnt) because our RGB LEDs are counting to a higher value. We've got more of them, so that's the reason we have two separate counters. Now we want to define our actual PWMs to be instantiated like these as well, and now we need to do our PWM instantiations. So, the first one we're going to work with is our LED PWM signal generator, and in our next one, we're going to put comments for all these so that we know which ones we have to do. So, we have the LED PWM signal generator – we only need one for our LED because we're going to take the same value, and whether we have four LEDs or eight LEDs, we're going to feed all the LEDs the same PWM value. So, we can get by just having a single signal value. For RGB, we're going to have a red, a green, and a blue. So, we're going to say our RGB LED PWM signal generator, and we'll give out so we know it's for the red, and this is a four-bit and 50 Hz signal. So, the four bits is used to define our duty cycle, and we have a 50 Hz output frequency for our signal. So, if we take and copy that, we know we're going to want to have the green as well as the blue. It's always nice to put comments in there so when you go back through, you can understand what's going on. So, for this first one – our LED PWM signal generator – where we want to instantiate, this is very much the same as our counter. We just have to – instead of counting – put them in our maps. We map to the actual PWM and not the counter. So, for example, this LED – we're going to call it LED_PWM, and we use our colon and then the PWM, and we have to go with our generic map. For our generics, we look at our PWM – we know that we have a BIT_DEPTH, INPUT_CLK, and FREQ. So, in this case, we're going to put our constant eight for our BIT_DEPTH; our input clock frequency is defined by INPUT_CLK, which we pull from our generic on top – so the generic we have at our top-level design (i.e., this LED brightness). We're going to take that generic and feed it into this generic, and then we add the 50 Hz. Since we have that there, we're going to a little bit more description in our comments, just so when we look at this later on, we can quickly understand and make it a little easier to read later in the future. The port map is where you want to map your outputs – in this case, we have our output Pwm_Out, and its value is being mapped to led_pwm_reg. The next thing we have is the duty cycle. The duty cycle is mapped to a specific value, and in this case, our Duty_Cycle is going to be mapped to the following. However, our led_duty_cycle is defined as an unsigned type, so we need to perform a conversion. This means that the Duty_Cycle will be converted to a standard logic vector of led_duty_cycle. So, our Duty_Cycle is being mapped to our led_duty_cycle, which we converted to a standard logic vector. Otherwise, this operation would not be possible, because of way led_duty_cycle is an unsigned data type, and that our clock (Clk) is being mapped to our clock (Clk), and that's pretty common when you have a clock. A lot of times, you're just going to map that same signal to all these because you want all of your designs running off the same clock (Clk). So, when you have a rising edge, it occurs at the same time for all your different designs, and finally, we have, as we look at our design, a generic map and a port map, the Pwm_Out, a Duty_Cycle, and the clock (Clk). We now need to define the Enable. So, we have the Enable, and we are going to map to our signal, we call it LED_Enable. So, that is our LED PWM design (LED_PWM), and we've now defined that. So, if we go ahead and say it's going to update the sources, and if we look at our LED brightness, the top-level design, it tells us we're using the counter twice. We only have one COUNTER, but we're using the PWM, once because it gives you the name. So, we got the name LED_PWM, which was the label. We've given our component instantiation and we're using the PWM, so like in your dash and after dash, we have the PWM behavior, and that behavior comes from within the PWM design. We've defined the architecture of PWM behavior, so that's where it's pulling that from. So now we want to take and replicate this, except we want to do it with our RGB red, RGB green, and RGB blue. So, another way in which we can instantiate is if we do not want to use the arrows, the equal arrows, to assign it, you can do it without that. For example, we'll show you how to do that on the Red LED, so we'll give it a name and that is RGB_RED_PWM, colon, and it's going to be the PWM, still the same PWM, now we need to use, or just like before we do our genetic map, and we're going to give it the value of four bits and its input clock (INPUT_CLK) because we're still the pull from that, and then we're going to make it 50 Hz. So the port map stays the same, because it is going to change. So, instead of defining this amount, Pwm_Out => red_pwm_reg, we're just going to sign that so that this becomes red_pwm_reg, which is the first value. This is equivalent to doing this right there; however, if we do it this way, you'll see once we're finished, in Vivado, they will yell at you a little bit and just say, hey, if you do the arrow equals zero, then you know specifically which port maps to which signal. So, we're just going to finish it off this way just to show you that it can be done this way. So that's red PWM register. Now we want to put our led_duty_cycle, which is unsigned. So, we're going to copy our conversion, by converting it to a standard logic vector so that it matches on the PWM. If we look at our duty cycle in PWM.vhd, it is the standard logic vector. So, we need to do the conversion in sort of led_duty_cycle. We have the signal we're calling our RGB duty cycle and we are taking it from three down to zero, because we have a signal with RGB duty cycle; if we look, this is a signal that goes from 11 to 0, so it's of 12 bits. That's why we have three different signals: red, green, and blue. So, we start by dividing the signal into four different parts. We get four bits for each of our four different signals: red, green, and blue. So, we're just pulling the bottom three bits of that signal. Next, we want to have the clock (Clk), so we just say Clk, comma, and the final we have our RGB_Enable to close it off. If we have a semicolon, we'll save and notice right there a highlight saying this positional port connectivity increases the risk of design errors there. It's more or less like a warning saying, hey, if you do it this way, it's harder to understand which signals are getting mapped to which port. If you do it the way you can see in the LED_PWM, it doesn't give you any error because it tells you this is a lot clearer. So, it just comes down to whatever way you prefer; there's not a specific issue, this will still work. This synthesis simply gives you a warning saying that you recommended it, and in Vivado, you probably disabled that. So, we're going to do since we have this right here, we know we want to do it the same for the blue, the green, and the blue, we're just going to copy this and edit so much as way we're doing a lot of component instantiations instead of typing every single one out. If you already did, just go ahead and copy it, paste it, and do not modify. So instead of RGB Red, it would be green (GRN) is going to shorten and the generic map stays the same, in port map is now instead of red_pwm_reg it's grn_pwm_reg, and then we have standard logic vector, and duty cycle instead of three down to zero it's seven down to four, and our clock and RGB_Enable to stay the same, so we can take and copy that guy, and we'll play it again for our blue, it looks like you're so set up green, what's called RGB_BLU_PWM and our genetic map once again stays the same, that hasn't changed your generic map. In port map, instead of green, it becomes blu_pwm_reg, and this RGB duty cycle signal instead of seven down to four, we're going to take 11 down to eight, and then we have our clock, and RGB_Enable stays the same, so we go ahead and save this. We'll see that we don't have any errors. It's just giving you a little warning to your saying they recommend you do this equal arrow, where you specifically tell it which item on the port maps to which signal. But if you look in our design sources under our LED brightness, it has picked up all of these instantiations. So, we have two counter instantiations, and four different PWM instantiations. However, in our project, we only have three files, we have these LED_BRIGHTNESS.vhd, COUNTER.vhd and PWM.vhd. But since we did component instantiations, it's the same as if we typed out this PWM and COUNTER for different times, except we only had to type it out once, and we have multiple component instantiations where we get to reuse effectively copy and paste that PWM design as well as the COUNTER design. So, when you're building these once again, we're going to reiterate that when you make a design such as the COUNTER and PWM, you should consider making it generic such that you can use them in other places and other design projects. The codes for LED brightness are as follows.

```
-- Libraries
library IEEE;
use IEEE.STD_LOGIC_1164.ALL;
use IEEE.numeric_std.ALL;
use IEEE.math_real.ALL;

-- Entity
entity LED_BRIGHTNESS is
  Generic (
     INPUT_CLK: integer := 100000000; -- 50 MHz default
     NUM_RGB_LEDS: integer := 4;
     NUM_LEDS: integer := 4);
   Port (
     Led_Out: out std_logic_vector(NUM_LEDS - 1 downto 0);
     RGB_R_Out: out std_logic_vector(NUM_RGB_LEDS - 1 downto 0);
     RGB_G_Out: out std_logic_vector(NUM_RGB_LEDS - 1 downto 0);
     RGB_B_Out: out std_logic_vector(NUM_RGB_LEDS - 1 downto 0);
     Clk: in std_logic;
     LED_Enable: in std_logic;
     RGB_Enable: in std_logic);        -- Active Low
end LED_BRIGHTNESS;

-- Architecture
architecture behavior of LED_BRIGHTNESS is

-- Components
-- Pulse Width Modulation Design
component PWM
  Generic (
     BIT_DEPTH: integer := 8;
     INPUT_CLK: integer := 50000000; -- 50MHz
     FREQ: integer := 50); -- 50Hz
  Port (
     Pwm_Out: out std_logic;
     Duty_Cycle: in std_logic_vector(BIT_DEPTH - 1 downto 0);
     Clk: in std_logic;
     Enable: in std_logic);
end component PWM;

-- COUNTER
component COUNTER
  Generic (
     MAX_VAL: integer := 2**30;
     SYNCH_Reset: boolean := true);
   Port (
     Max_Count: out std_logic;
     Clk: in std_logic;
```

```
     Reset: in std_logic);
end component COUNTER;

-- Signals and Constants
-- 85 is from taking 255, the max PWM value, and dividing by 3 seconds
constant LED_MAX_COUNT: integer := INPUT_CLK / 85;
constant RGB_MAX_COUNT: integer := INPUT_CLK / 1365;
constant SYN_RESET: boolean := true;
constant MAX_LED_DUTY: integer := 255;
constant MAX_RGB_DUTY: integer := 4095;

signal led_max_cnt: std_logic := '0';
signal rgb_max_cnt: std_logic := '0';
signal led_pwm_reg: std_logic := '0';
signal red_pwm_reg: std_logic := '0';
signal grn_pwm_reg: std_logic := '0';
signal blu_pwm_reg: std_logic := '0';

signal rgb_counter_rst: std_logic := '0';
signal led_counter_rst: std_logic := '0';

signal led_duty_cycle: unsigned(7 downto 0) := (others => '0');
signal rgb_duty_cycle: unsigned(11 downto 0) := (others => '0');

begin
-- Assign Outputs
   Led_Out <= (others => led_pwm_reg);
   RGB_R_Out <= (others => red_pwm_reg);
   RGB_G_Out <= (others => grn_pwm_reg);
   RGB_B_Out <= (others => blu_pwm_reg);

  -- Invert enable signal for counter
  rgb_counter_rst <= not RGB_Enable;
  led_counter_rst <= not LED_Enable;

  -- LED Counter
  LED_COUNTER: COUNTER
  generic map(LED_MAX_COUNT, SYN_RESET)
  port map(Max_Count => led_max_cnt, Clk => Clk, Reset => led_counter_rst);

  -- RGB Counter
  RGB_COUNTER: COUNTER
  generic map(RGB_MAX_COUNT, SYN_RESET)
  port map(Max_Count => rgb_max_cnt, Clk => Clk, Reset => rgb_counter_rst);

 -- LED PWM Signal Generator (8 bit, 50Hz)
  LED_PWM: PWM
  generic map(8, INPUT_CLK, 50)
  port map(led_pwm_reg, std_logic_vector(led_duty_cycle), Clk, LED_Enable);

  -- RGB LED PWM Signal Generator [RED] (4 bit, 50Hz)
  RGB_RED_PWM: PWM
  generic map(4, INPUT_CLK, 50)
```

```
  port map(red_pwm_reg, std_logic_vector(rgb_duty_cycle(3 downto 0)), Clk, RGB_
Enable);

-- RGB LED PWM Signal Generator [GREEN] (4 bit, 50Hz)
  RGB_GRN_PWM: PWM
  generic map(4, INPUT_CLK, 50)
  port map(grn_pwm_reg, std_logic_vector(rgb_duty_cycle(7 downto 4)), Clk, RGB_
Enable);

 -- RGB LED PWM Signal Generator [BLUE] (4 bit, 50Hz)
  RGB_BLU_PWM: PWM
  generic map(4, INPUT_CLK, 50)
  port map(blu_pwm_reg, std_logic_vector(rgb_duty_cycle(11 downto 8)), Clk, RGB_
Enable);

 -- LED PWM Count update Process
  Led_Count_Proc: process(Clk)
  begin
  if rising_edge(Clk) then
    if (led_duty_cycle = MAX_LED_DUTY) then
      led_duty_cycle <= (others => '0');
    elsif (led_max_cnt = '1') then
      led_duty_cycle <= led_duty_cycle + 1;
    end if;
  end if;
end process Led_Count_Proc;

-- RGB PWM Count Update Process
Rgb_Count_Proc: process(Clk)
begin
  if rising_edge(Clk) then
     if (rgb_duty_cycle = MAX_RGB_DUTY) then
        rgb_duty_cycle <= (others => '0');
     elsif (rgb_max_cnt = '1') then
        rgb_duty_cycle <= rgb_duty_cycle + 1;
     end if;
   end if;
end process Rgb_Count_Proc;

end behavior;
```

So, when you start a new project, you already have bits and pieces you can work with, and you're not starting from scratch. That is the component instantiation in VHDL and the LED brightness that gives you the completed project for your LED brightness. At this point, you're ready to assign your IO and start your synthesis for your implementation and generating your bitstream or programming file to run this on your development board.

8.6 The LED Brightness Arty A7 IO Placement

This section is about the LED brightness and Arty A7 IO placement. In this section, we're going to show you how to place the IO for the Arty A7 development board. So, for this design, if we look at the entity, we can see that we have Led_Out, which is a standard logic vector from the number of LEDs minus one down to zero. Now, on the Arty A7, we know that we have four regular LEDs and four RGB LEDs. We also have an input clock (INPUT_CLK) and we're going to have an LED_Enable and an RGB_Enable. So first, let's start from the bottom with our RGB_Enable. We copy it, and we're going to go into the Arty_A7_Master.xdc file. Under switches, we will make switch 3 the RGB_Enable. Paste it and then uncommented that line. So there, we're assigning package pin A10 to the IO standard LVCMOS33 and setting that to RGB_Enable. We go back to our design, and RGB_Enable is an input. Now, copy LED_Enable and we're going to make that switch zero, just so we have them on opposite sides. You can put them on any slide switch you want. Just note that whichever switch you tie RGB_Enable and LED_Enable to will determine which switches control the RGB LEDs or the standard LEDs. Go ahead and do that. So now we have the clock, which we will go ahead and do next. Copy the clock. If we go to the .xdc file, we have the clock signal up top, and they define it as clock 100 MHz. We will go ahead and place that and just change the clock. On the second line, we're going to create a clock, give it a name sys_clk_pin, and leave the period and waveform as default. Because by default, it describes the 100 MHz clock that is already coming down the Arty A7 board, so we don't want to change that. We'll highlight both lines and click on the toggle comments. So now we have our clock. We've set the property for the specific pin, the clock is coming in on, and we've also created the syntax so that when it goes through the timing of the design, it'll make sure that the clock can come in at 100 MHz with no issues. Now, in our design, we have RGB_B_OUT, which is the blue output of our RGB LED. Since we have four different RGB LEDs, the RGB_B_OUT has to be associated with each one of them. So, if we go to the master .xdc file for the RGB LEDs, we have LED zero, one, two, and three. Since we're taking the RGB_B_OUT, which will be the blue output, we want to copy and paste that in place of the corresponding blue lines. So we want to take that and paste it into our blue, and we'll set that to zero. Then we'll go down to the next blue line for led1_b, paste it, and assign that to one. Then we move to the next blue line for led2_b, paste it, and assign that to RGB_B_OUT[2]. Finally, we paste again and again it in place of led3_b using RGB_B_OUT[3]. That gives us our blue RGB outputs: RGB blue output zero, one, two, and three. Now, for each of those lines, we'll go ahead and uncomment them. So, with those four lines uncommented, we now have the blue LED output defined for the RGB LED. Next, we want to configure the green outputs. So, we copy RGB_G_OUT, and we'll repeat the exact same steps we did for blue, just now for green. So, everywhere we have the green output set to zero, we'll paste and make that zero because it's a standard logic vector. Then we set the next one to one, then paste and set the next to two, and paste again to set the third element. After that, we go ahead and remove the comments from those lines. Now, we move on to defining the red output (RGB_R_OUT). So, copy that, and for red we want to: set our LED zero, make it the zero, then LED one, making that one, our LED two, make that two, and our LED three, make that value three. Then we go ahead and remove the hashtag for the comment at the beginning of all these, so that we have all of the RGB LEDs defined. We now only have to define the Led_Out, which we know – if we copy our number of LEDs – we only have four of them. So, we know that we will go from three down to zero. If we go to our LEDs, we have three to zero. So, in the first part of that, we just want to replace "led" with Led_Out, and we'll do that for all four of them there. We can then just highlight the whole chunk and hit the toggle comment, and click on the Save button. So that gives us all of the IO that have now been placed for the Arty A7 on the LED brightness project, as shown in Figure 8.11.

Also, for simulation, we can use the following codes.

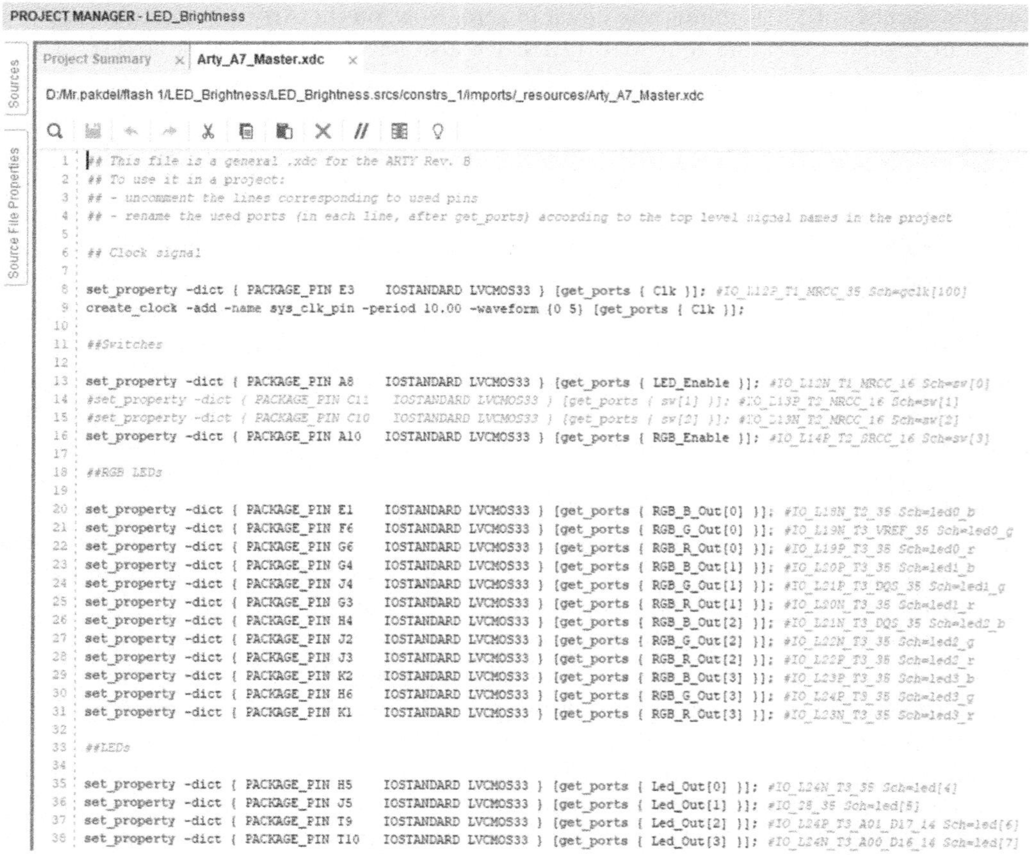

Figure 8.11 The uncommented ports in Arty_A7_Master.xdc.

```
library IEEE;
use IEEE.STD_LOGIC_1164.ALL;
entity test_LED_BRIGHTNESS is
  Generic (
     NUM_RGB_LEDS: integer := 4;
     NUM_LEDS: integer := 4);
--  Port ( );
end test_LED_BRIGHTNESS;

architecture test of test_LED_BRIGHTNESS is

component LED_BRIGHTNESS
 Port(
     Led_Out: out std_logic_vector(NUM_LEDS - 1 downto 0);
     RGB_R_Out: out std_logic_vector(NUM_RGB_LEDS - 1 downto 0);
     RGB_G_Out: out std_logic_vector(NUM_RGB_LEDS - 1 downto 0);
     RGB_B_Out: out std_logic_vector(NUM_RGB_LEDS - 1 downto 0);
     Clk: in std_logic;
     LED_Enable: in std_logic;
```

```
     RGB_Enable: in std_logic);          -- Active Low
end component;

signal Led_Out: std_logic_vector(NUM_LEDS - 1 downto 0);
signal RGB_R_Out: std_logic_vector(NUM_RGB_LEDS - 1 downto 0);
signal RGB_G_Out: std_logic_vector(NUM_RGB_LEDS - 1 downto 0);
signal RGB_B_Out: std_logic_vector(NUM_RGB_LEDS - 1 downto 0);
signal Clk: std_logic := '0';
signal LED_Enable: std_logic := '0';
signal RGB_Enable: std_logic := '0';     -- Active Low
begin
  dev_to_test: LED_BRIGHTNESS
     port map(Led_Out, RGB_R_OUT, RGB_G_OUT, RGB_B_OUT, Clk, LED_Enable, RGB_
Enable);

  data_stimulus: process
    begin
        wait for 10 ns;
        Clk <= not Clk;
    end process data_stimulus;
end test;
```

9

The UART Demonstration Project

9.1 Introduction

In this chapter, the UARTs are explained, and we're going to explain what UART is and how it's used at a detailed level. In this chapter, we have the UART demonstration where we are going to just try and use one of these development boards. They have a UART port, or they support UART to USB conversion, so they'll take that, and you plug it into your computer. When we have a serial port emulator, we can send data back and forth from our computer to the FPGA or actual design, and when we do that, we want to explain and understand what you can see at a detailed level – what's going on, what bits are being transmitted to send certain pieces of data. So first, let's ask: what does UART stand for? If you work with microcontrollers or if you work in embedded systems or with computers, you've probably heard of UART before doing any sort of embedded systems type of application. And what that stands for is universal asynchronous receiver/transmitter. What that means is that it's asynchronous, and by universal, it means you can use this for various transmission layers. So, if you want to go with RS-232, RS-485, 10BUS, or any of the other physical layers, you can do that with UART – you just need a transceiver. So, it's universal in that it's most commonly used. You send it into a different hardware chip; the mindset says it has a UART port; you go ahead and connect using this interface, and connect to it – the chip will transfer the data and send it out through RS-485 or whatever type of transceiver it is. Asynchronous means you can transmit and receive at the same time. There's no clock source that is synchronized to a lot of communication protocols. When you're working with them, you'll have a specific clock source that you're running at, and you need to have your data aligned with that clock source or you'll send a clock and your data is correlated directly to that clock source. With UART, there's none of that. There are only two signals: Rx and Tx. Asynchronous means Rx and Tx can be happening at the same time, and they're not correlated by a clock at all. Obviously, you have the receiver and transmitter, meaning you have the Rx, or receiver, and Tx, or transmitter, and what do we actually use this for? So, if you use microcontrollers or development boards, you'll hear this term UART a lot. What do you actually use that for? It is just a common way to communicate between devices. In this case, we're going to use it to communicate between the FPGA development board and your computer. You can also use it to communicate between individual devices. If you have like two development boards and you want them to talk back and forth, you can use UART to do that. It's a predefined language to talk back and forth – just like we're talking to you in English, UART is predefined in these devices to both understand and communicate that way. These are most commonly found inside the microcontrollers or in integrated circuit chips. You'll see them when you're searching

Fundamentals of VHDL for FPGA Programming Using Vivado, First Edition. Majid Pakdel.

Companion Website: www.wiley.com/go/pakdelVHDL

on DigiKey or any kind of computer distributor. If you're building your circuit board and looking for different interfaces, UART is commonly listed as one of those interfaces. You can do this in various ways. You can implement UART using VHDL by physically describing when to set your bits high or low and do all of that the way we're going to use it. Because on the Arty A7 board, the UART circuitry is connected to a transceiver chip, so it takes that UART signal and converts it to USB and sends it to a computer. So, you just plug that USB in, and those pins are already physically tied. So, we can't just map it to whatever we want. So, inside the actual FPGA, we are going to use a softcore processor or a hardcore processor to generate this data. That means we'll be writing C code to process and transmit data. So, we'll use a softcore processor, which, if you're using Xilinx, is the MicroBlaze, and we'll use the different tools in there to write specific code that tells it to send out data. We're going to explain here what that data looks like, and if you put a scope on it, you would see what those waveforms look like and what that actually means. So, there are some common things that let you know what UART is. There are two signals: Tx for transmitting, and Rx for receiving. As mentioned earlier, you can transmit and receive at the same time. It's a full duplex system and there're no clock signals to synchronize. You just have these two signals. We'll show you what that signal looks like. When you're sending data, it's transmitted in a predefined sequence. This allows UART communication to work without needing to interpret each condition individually. This means you have a predefined sequence, meaning you have a start bit, a stop bit, a parity bit, and data bits. You can also define whether you're transmitting little endian or big endian. Now, let's take a look at how this actually looks like and what this will look like if you did have a scope or some sort of equipment for that UART data. We're going to take some UART notes. Just so when we go back, it gives us something we can refer to. So, if we take a signal and we are just going to generate an arbitrary signal, that's going to toggle between 3.3 V and 0 V. So, when we're idle, we're going to say we're high, then we'll toggle it low for a specific amount of time, and then it'll go back high and stay for a while. We'll define this specific duration as the bit time, which is determined by the baud rate. Once we draw this entire signal, we'll explain how to calculate this bit time or baud rate when you're working with UART. That's a common term you'll hear: bit rate. So, we'll get in to explain that. This first bit you'll see is what we call our start bit, and afterward, we'll explain how this all ties in. So, you have your start bit, which is defined when the line is idle or there's no data being transmitted. You're at your high level, which we'll say is 3.3 V. It could be 5 V or whatever logic will be running at near low 0 V, so we'll stay high here, and we'll say that we come down, that is the same. So, every time, we have the space between all of these dash lines, we're not measuring exactly, but we're going to say that's all the same exact amount of time. So, the start bit is equal to this line is high for the equivalent of two start bits. So, here's another time division that we'll call when we have another time division, and the third time division comes to high. This is our start bit. Now, since this is high during this time division, we're going to give that a value of one, the logic value of one. Now, that is low during that time division, so this gets a logic value zero, as well as two ones and three zeros. Now we're going to say we're high for one bit time. So, that gets a logic value of one and then we're going to pull low again. Now, our next bit time has a logic value of zero, and the next one, because our next time division is the logic value of zero, and now at this point, we can see we have our start bit, and if we're transmitting little endian, that just means we're transmitting with our least significant bit first. So, we're going to assume that this is transmitting little endian, and you can have little endian or big endian, and a little endian just means you're the first bit you transmit at least a little bit. So, we're going to write in top-level design what these are. So, that would be a bit zero, bit one, and so on until the bit seven, so we know from zero to seven, which gives us eight bits and that's another parameter that's defined in UART protocols for us as little endian. Then it's eight data bits, and that tells us that, OK, the first bit we see is that zero, because we're a little endian. For a big endian, that would be seven, six, and all the way down to zero. So, we have a little endian, eight data bits, and we're going to say we have even parity, and what parity is, there is a basic form of error correction. So, since we have even parity, that means in our transmission of data, we have eight data bits; we need to give a parity bit that's going to determine the number of ones we have, and we want to make them even. In this case, we have one to three ones, and since we're even parity, we need to add a one to make it even, so we have four ones. So, we'll take and add one, and this is our parity bit. So, it's not part of our data, but it's used as an error checking. For example, if our parity bit is set to zero, we would throw these data bits away, because we're like, we're even parity, but we only have three ones, including a parity of it. If this was a zero, we would then just discard this message saying, OK, something got corrupted along the lines, please retransmit, we don't trust that, so that becomes a one, and then we have a stop that would just stop, and it is defined by a high. So, that's just our bit time. Then we have another bit time, and this tells us this is our stop bit. So then after that, after our stop bit, our line is going to stay high and then wait to see if we're receiving, and then we wait to see until we drop low again till we get another start. But once we get another start, we'll start this whole process over again, and we send our data, our parity, and our stop bit. You have different options where you might define there's no parity bit, so you just have two stop bits. If that's the case where you have this parity bit, you'll always see a one. Now, if we were odd parity, if that was what was going on, we would make the parity bit to a zero because we have three ones, which is an odd number of ones. So, we'd make that a zero, but since we're even parity, we want to make this one. Now our actual data, if we look at our data we sent, we set down and if we look at it in binary, you would think your one is your most significant bit, but it's not. So, it's kind of a backward order. But if we look at it, it would look to us something like the following: 00100011. So, if we take that, we could say, you know, that's equal to a hex number of 0x23. If we got this UART transmission, the actual value you'd process or the data you would see is this hex number 23 (00100011); that's the actual data you would work with. Now, in our project design, we are working with ASCII because we would look up to see where this ASCII code of 23 is and that be the character that you received and you would send most of these down. So, if you're sending down like hello or my name is whatever, and whatever you want to send down, you would send down the first character for hello. So, you look up the ASCII character for h and send down that because it's an 8-bit value, then the ASCII character for e, the ASCII character for l, and so on, and you just keep going down that line. So, you'd have multiple of these packets being set – one of these packets for each individual character. Now we talked about these time divisions here, and this is defined by your baud rate. So, baud rate is similar to a bit rate. So that's the number of bits you can send per second. For example, if we have a 19200 baud rate, we can send out 19200 bits a second. So, in order to determine this time, you would say, OK, that's our bit rate, and our time is determined by one divided by our baud rate, or 19200, which is equal to 52.08 μs. When you're working with it, there's always a bit of a tolerance level, so if you're not exactly at 52.08 μs, it'll still work as long as you're within a certain tolerance. For example, if you made this like 100 μs, then you would get all sorts of errors because nothing would match up, and you get all your data checking would not work out, or you would just end up seeing weird characters on your terminal. So that being said, each one of these individual time stamps means the bit time is 52.08 μs, and the reason this is important is because you have two devices: one device that's transmitting this and another device that's receiving this. So, the guy that's receiving this is going to look and see that while his line is high or at 3.3 V, he is going to disregard it, and you're doing nothing until he sees from a high to a low transition. When he sees that, he knows that this is the first line (start bit) and we know that we have 52.08 μs until this is over, so we're going to divide that by 2, and so approximately at 26 μs right here is where we're going to sample. So, we're going to sample there and see that, OK, since we are low, we have started because you may get noise, it goes down for a little bit and comes back up. So, when you check here (at 26 μs) and you see your start bit, a one there, just disregard and start over again and then you're going to say, OK, wait, 52.08 μs, and we're going to sample there. So, what you're going to try and do is sample in the middle of each one of these bit times. So that gives the transmitter; he doesn't have to be exactly at 52.08 μs; he might be a little off, but as long as he gets there before the halfway point, like 26 μs, the receiver should see it as a low or a high, whichever the transmitter is trying to send. Now, if something gets corrupted, you've got various ways to correct that. The very first and most rudimentary is this parity bit. Now, you could send the wrong data at your parity, but might still say it's valid. For example, if we have even parity and we sent three ones. Well, what if we accidentally sent five ones (and also parity of 1)? We would still send this bit; even parity bit would be fine. So, it's a very first line of defense, and a lot of times what you'll do is if you're working with a different protocol, they'll send a CRC checksum or something afterward to verify that it is actually the right data. Now, with our project, we don't need to have that so much because we'll see it clearly on our terminal that if we're trying to say Hello World and we see Hello World, we know the data was transmitted correctly, and we don't have to worry about the error checking. Now, to show you the lines, for example, if we have our FPGA board and a computer, it really doesn't matter; the idea is just that you have two separate devices, so we can connect them via UART; you would have two lines that are important here. Obviously, you have a ground, so they share a common ground. For the computer, this would be our Rx line or our receive line, and the transmit line, Tx line, goes from my computer to the FPGA. Now, our FPGA board is going to be our Rx line, and that is going to be the transmit line (Tx). This is important because if you try to hook the transmit line from your FPGA board or from one device to the transmit line of a second device, that's not going to work at all. You can see once again your transmitter (Tx) has to go to your Rx, and your Rx has to go to Tx. So, it has to make sense because it's a very common mistake to take and connect both the Rx lines together and both the Tx lines together. But you don't want to do that, because if you

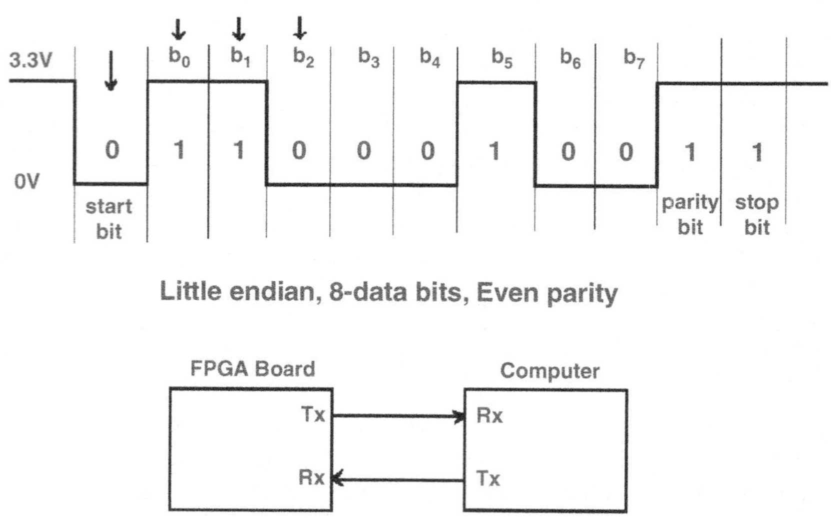

Figure 9.1 The UART protocol.

do that, you'll end up getting errors and nothing will communicate. You'll think there's something wrong with your code, but the fact of the matter is that your cables are not connected correctly. Now, for our project, you don't have to worry about that because it's in the USB. So, you're just plugging a USB cable in from the FPGA to the computer, and it handles all of that. So, we don't have to worry about physically matching these different ports. But if you do this anywhere else or you're going to work with this later, that becomes important, and it's just a thing to take note of. So that is the UART explained here and how we're going to use it in our project, as shown in Figure 9.1.

9.2 The MicroBlaze Softcore Processor

This section is about the Xilinx MicroBlaze softcore processer. In this section, we're going to explain what a MicroBlaze softcore processor is and how you can instantiate to use one for your Xilinx development board, inside of Vivado. If you're using the Arty A7 development board, to do the UART demonstration project, you're need to use a MicroBlaze softcore processor. So, the first step is understanding what a softcore processor is. If you've heard of it or familiar with processors and microprocessors, you probably would have heard of Arduino or Microchip, or Texas Instruments, which are hard processors essentially baked into an actual integrated circuit. You can't change them, and they're not definable. But a softcore processor is a definable processor that is implemented from FPGA fabric, which, when you think of FPGA, you can think of as basically digital fabric. You can do whatever you want with them, and it's just wide open. It's gives the raw ingredients to create a digital design. One of those things you could create is a processor, and so Xilinx includes a drop in block that represents one of these processors. So, if you have an FPGA, you could drop multiple of these processors on there, which would then turn your FPGA effectively into a multicore processor. So, it's pretty cool, and what's really cool is that we can define all of the resources. So, if we want to have more memory, we either attach external memory or use more of the internal block memory of FPGA. If we want more SPI ports, UART ports, or IO, we can do so by just defining that softcore processor and adding more. Now, there obviously are limits, but a lot of times when you get a development board that has a microcontroller on there or a processor, you might only have three or four timers you can use. But if you develop an FPGA, you can configure the MicroBlaze processor to suit your needs. You can specify if you want more timers or less timers and the number of bits for each of those timers. On top of that, you have all that FPGA fabric around your processor. So, if you want to add functionality, you can create the design of the VHDL and a little interface such that the microprocessor can communicate directly to your ports in the VHDL chip. So instead of having a separate chip on the board, it's all in one chip, and our communication is also very fast. So, there are a lot of advantages and benefits of using software processors. Some of the peripherals you can add to the MicroBlaze softcore processors have various flavors, the high-end ones that use more resources, but they give you more features. They have a low end and are called MCS, which just has a few small subsets of features. For example, what we're doing here, we're just sending data out on UART port so we don't need the high end, and all these super fancy features, we just need a UART port and that's it. But you can add features such as, obviously UART, you have SPI or I^2C or GPIO, which is general-purpose inputs, outputs, or timers, as well as additional features you can add, and another cool thing is we can create, as we talk about custom peripherals so they can create an interface in AXI is what they call this, and so if you generate AXI interface on your VHDL designs, you can now have, for example, the PWM we generated a PWM on block, and if we put an AXI interface on that, we could then have our

MicroBlaze software processor start talking to this actual PWM input or design and actually talk to it so then we can interface it, just writing the C code, which is pretty cool. However, you can do it in much easier ways, and you would just put a general-purpose input output. We want to have a general-purpose output where we're writing out eight bits, and those eight bits correlate to the duty cycle of my pulse width modulated (PWM) signal, or you can make it nine bits where one of the bits ties to the enable so you can enable or disable your PWM design. So, there's a lot of different things you can do with them. It's very customizable, and the coolest thing about these softcore processors is that you're not constrained to certain resources, you're constrained to what you have available in your FPGA. But if you want to use up more resources, if you have the resources there, you can add those features and functionality. So, it's a pretty cool thing, and just some things that are keynote to these processors, they're going to use the following resources. On your FPGA, they'll use the internal BRAM, which is Block RAM, and they'll use a multiplier. So, in your processor, if you want to do a multiplication, that's going to require a multiplier inside the arithmetic logic unit of that processor, as well as logic, which takes digital logic just to implement it. You have to have your memory tables and there's a whole slew of information on it. But effectively, you can't drop it out of processors and it is not free. It does take resources, so when you say you want to create a multicore processor on FPGA, you will be limited by the resources that are on that FPGA. If you drop in, say, five MicroBlaze softcore processors, you might be OK, but you might not be able to get six. You might have to upgrade to a larger package part, which just costs more money, and it takes more power. So, there are considerations to it, and you can't just drop them in freely. Then, once you synthesize the design, the idea behind it is you'll take MicroBlaze processor, define it, implement it in your design, and then once you synthesize that design, you can then export it into another tool, Xilinx software development kit (SDK) tool, where you can then write C code that then gets programmed onto that MicoBlaze processor and then we can take that C code and generate what's called a.elf file, and that gets associated with your softcore processor so that when you program it on the board, you don't have to program it and then program the C code, if you can just program one file, one bit file onto your FPGA board that will then do everything you described in your C code, and we'll show you how to do that later. So, the next thing we want to take a look at is see open up your UART demo project, and within UART project, you'll see that it is blank, and there're no design sources. So, in the design sources, we want to create a MicroBlaze softcore processor. Now, if you want to do that directly with VHDL code, it's going to be a challenge, and you have to use special scripts and special functions to get that. What we're going to do is create a block diagram. So, we're going to click on the Create Block Design and give our design a name, UART_Demo, and for directory, we want to leave that local project and specify source set as design sources, and then will say, OK, so this is going to create a block design for us, and this will allow us to drop in a MicroBlaze softcore processor, from which we then can generate a VHDL file to program, as depicted in Figure 9.2.

So, we want to click on Add IP, and in the search bar, type MicroBlaze. You should see three options: MicroBlaze, MicroBlaze Debug Module (MDM), and MicroBlaze MCS. For this project, we'll use the MicroBlaze MCS, as illustrated in Figures 9.3 and 9.4, respectively.

The MicroBlaze is the one that has a lot of features and functionality, but it requires more resources and power to implement. While it's fine to use and can be handled by the board, for this project, we'll go with the MicroBlaze MCS instead. It's easier to instantiate, less complex, and it can do everything we need. So, we can double-click on the MicroBlaze MCS. A small window will appear notifying that it's adding that IP to the block diagram. Once it drops it in there, we want to go ahead and double-click on MicroBlaze MCS object to open a window "Re-customize IP." Within this window, we are going to set and define all the parameters associated with MicroBlaze

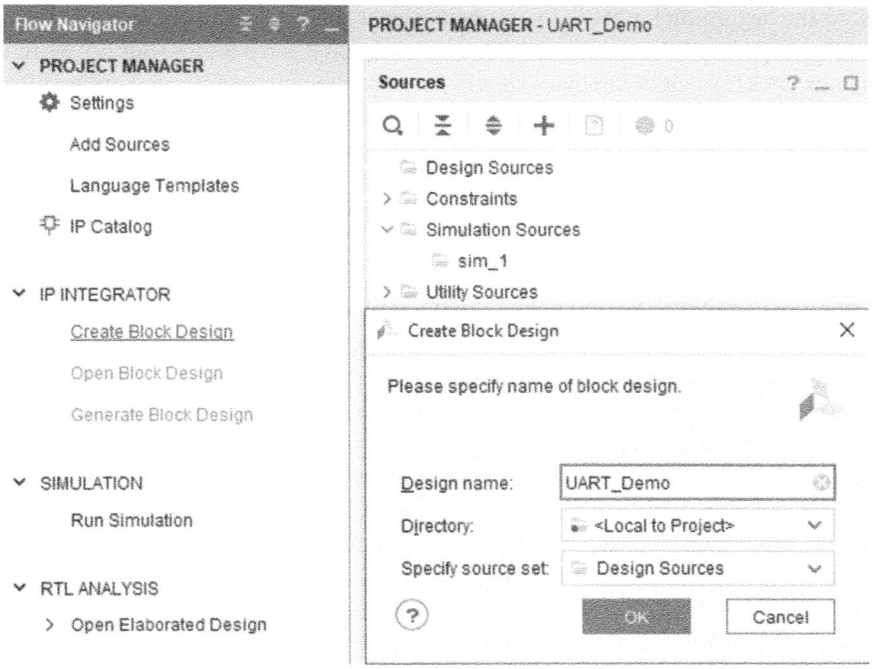

Figure 9.2 Creating the block design (UART_Demo).

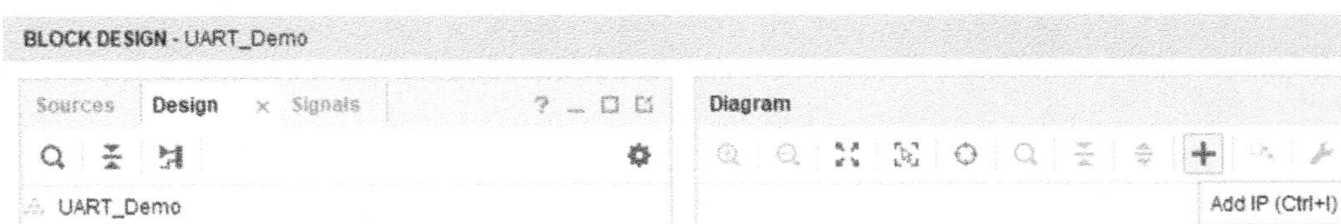

Figure 9.3 Clicking on the Add IP (Ctrl+I) button.

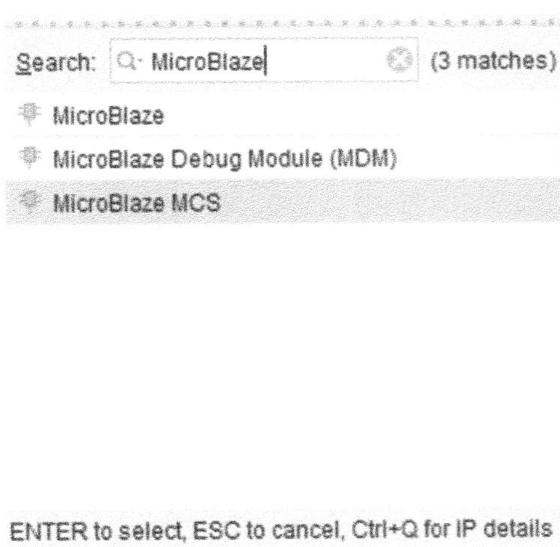

Figure 9.4 Selecting the MicroBlaze MCS.

processor. At the beginning for MCS, the memory size is important – it determines whether your code will run properly. If there isn't enough memory, you'll have issues. But all we're doing is outputting our UART, so eight kilobytes should be more than enough, for the optimization, we'll select Area. We don't really have a huge performance issue, we're just writing out your data on the UART port. We're not doing some sort of signal processing. If we were doing signal processing or a highly intensive task, we want to go for Performance where we don't care if it costs more area to do so, and we want the performance in this case. We want to optimize in our area, and for buses, we will leave both options blank, and for debug, we'll select None, as shown in Figure 9.5.

Next, we'll move to the UART tab. Here, we just want to enable the transmitter and define our baud rate to 115200. For Programmable Baud Rate, we're just going to leave that unchecked. We're going to have this as a very basic MicroBlaze processor, and we're not going to have too many parameters in this one, so the rest of these will remain unchecked, as depicted in Figure 9.6.

The following tabs allow you to use a fixed interval timer (FIT) and the programmable interval timer (PIT). However, since we're not using these in our project, we'll leave them unchecked. Similarly, for the GPO (general-purpose output) and the GPI (general-purpose input), we're not using any of those either, so we'll leave all of those unchecked as well. Same goes for the Interrupts, so that option will remain unchecked as well. At this point, we can go ahead and click OK. This is going to take our configurations we have set, which in this case is a very basic setup. All we're doing is just enable UART transmission. We're not even using the receive line (rx) because for this demonstration, the only task is to transmit data from the FPGA board to the computer. Since we don't need to have an Rx line, this is just a single line of data transmission. If you click on it, you see it out of this, UART plus, and if we pull that down, you can see UART_txd line. That's our UART transmit line, as illustrated in Figure 9.7.

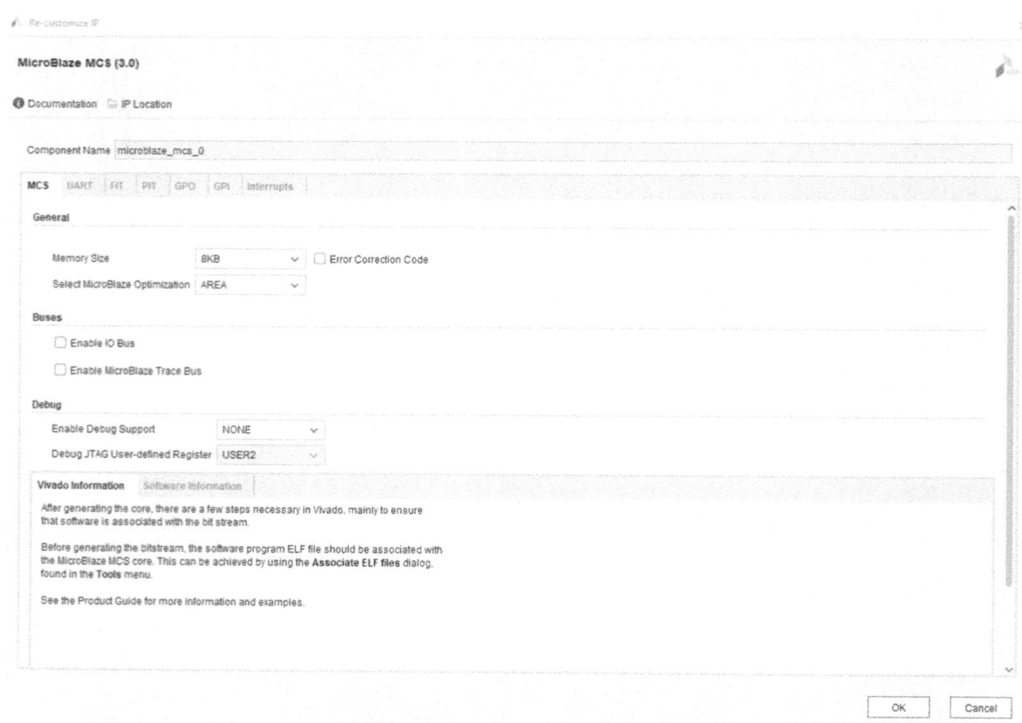

Figure 9.5 The MCS tab of Re-customize IP window.

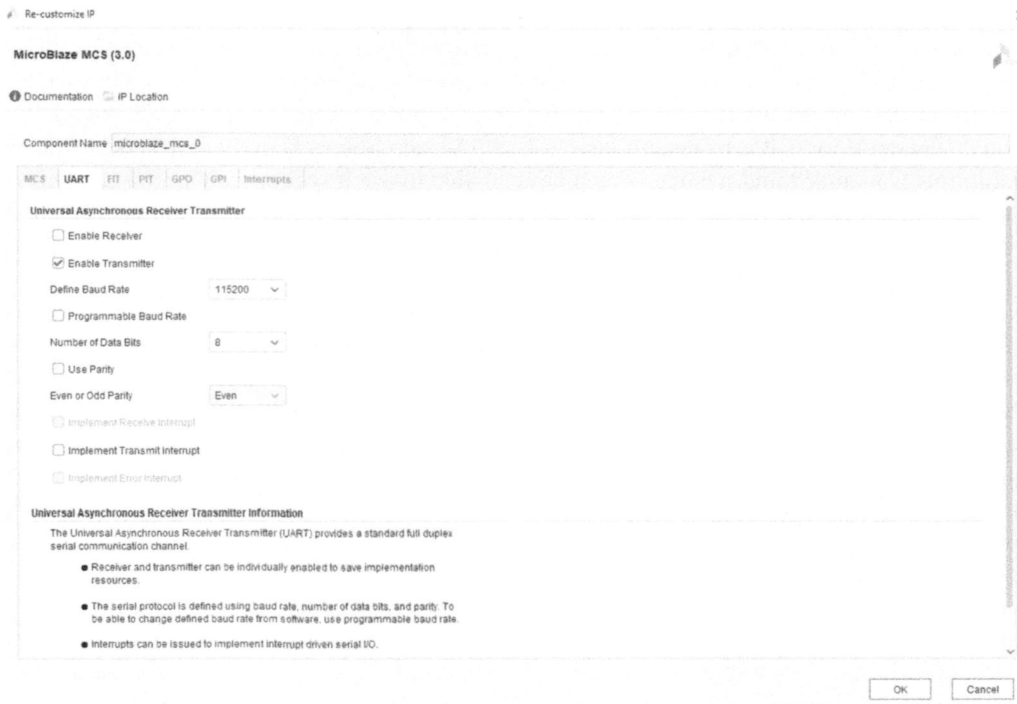

Figure 9.6 The UART tab of Re-customize IP window.

Figure 9.7 The MicroBlaze MCS UART_txd line.

At this point, our MicroBlaze processor has been successfully configured and set within our UART demonstration project. If you want to add another processor, you would just go through the same process, look at IP, and add another MicroBlaze processor. You can add as many processors as your FPGA's resources will allow. As mentioned earlier, each MicroBlaze core consumes resources. In our case, we've only assigned 8KB of memory. If your program becomes larger or performs more complex tasks, it's going to require more memory, and you're going to run out of that. So, you can put as many as your resources can handle. This section explains the MicroBlaze softcore processor within the Xilinx device.

9.3 The Xilinx Software Development Kit

This section is about the Xilinx SDK. In this section, we're going to explain what the Xilinx SDK is and how we'll be using it for a project. The Xilinx SDK, or X-SDK, which stands for Xilinx software development kit, is an integrated design environment that allows you to create embedded applications to run specifically on Xilinx microprocessors. If you're working with the Texas Instruments or any other type of microprocessor, you can't use this tool. It's specifically for Xilinx microprocessors, and in our case, we're using the MicroBlaze softcore processor, so it'll work for that. However, you can't use it for anything else, you can download this as a standalone tool, but if you download Vivado, it is by default included with Vivado webpack, or the free version. So, by default, this is included and it's free. You should be able to use it without any problems. This is what we're going to be using to write the code to run on the MicroBlaze softcore processors. For example, we'll be using it in this UART demonstration project as well as in future projects in this book. This tool is based on the Eclipse development environment. If you're familiar with a lot of code development tools that are based on eclipse, this will look very similar to those, with many of the same features, menus, and everything. It's just Xilinx took the eclipse development environment and modified it to match the tool that they need. So, in our projects, when we create a MicroBlaze softcore processor, we instantiate it in our design and it builds all the logic for it. But if you don't program it, if you don't write any code in the SDK to run on there, you'll put it on your board, and once you start running, you'll realize that nothing's happening. That's because it's just like having a microcontroller or microprocessor on a development board: if you don't program it power it on, it will seem like nothing is working because it's not doing anything, and that's because the effect is the same. You can successfully program your board, but if you don't load any code into the MicroBlaze, it won't do anything either, and it'll just sit there. The SDK has some pretty cool features; it's very powerful and allows you to do a lot of different stuff. It even has the ability to include real-time operating system. In this book, we will not be using that, but just letting you know a little background on it. They've got a free RTOS, which is a free real-time operating system that you can use and run inside the MicroBlaze softcore processors. So, if you're working on a project that requires a real-time operating system, you can include this free RTOS, and it gives you the framework and structure to create a project based on that. You just create your tasks and have them work in a real-time environment. As mentioned earlier, it's a very powerful tool, and we'll be using it a lot. Right now, we're going to go ahead and work with our UART demo project to see what we need to do to get started with our Xilinx SDK. So, if you open up your UART demo project, you'll see that we have the MicroBlaze MCS, which we've already configured. Now, we need to add the IO and create a top-level VHDL design. After that, we'll need to run the synthesis, run implementation, and generate a bitstream so that we can then go ahead and pour everything we need into the Xilinx SDK environment. So, let's create our IO. We'll need to create an IO for our clock and call it the input clock. If we right-click and select Create Port, it brings up a window where you can create inputs and outputs. In this case, we're going to create a clock. So, we're going to call it Input_Clk, select the direction we're going to select, which is an input, and set the type to clock. Then, we'll set the frequency to 100MHz and click OK, as shown in Figure 9.8.

So, we now have an input clock that we can drag and place right in line with the clock (Clk). If you hover over it and your mouse gets close enough, you'll see a pointing finger. Once you get close enough, the cursor will change to a pencil. If you click when you see the pencil, it'll show you what's in place and where you can actually put it too. When we connect it, we can see a little green check next to the clock, that's because we recognize it's a clock signal. Since it's a clock signal, it can be tied together, as depicted in Figure 9.9.

We could tie this Input_Clk to the Reset, but if we select a Reset signal, it'll recognize that this should be connected to a Reset, not to Clk, and vice versa. So, now let's create our reset signal. We'll right-click and select the Create Port from the bottom and give the port a name. We're going to call it Reset_Top; the direction is going to be Input, and for the type, we'll select as Reset. Then, you

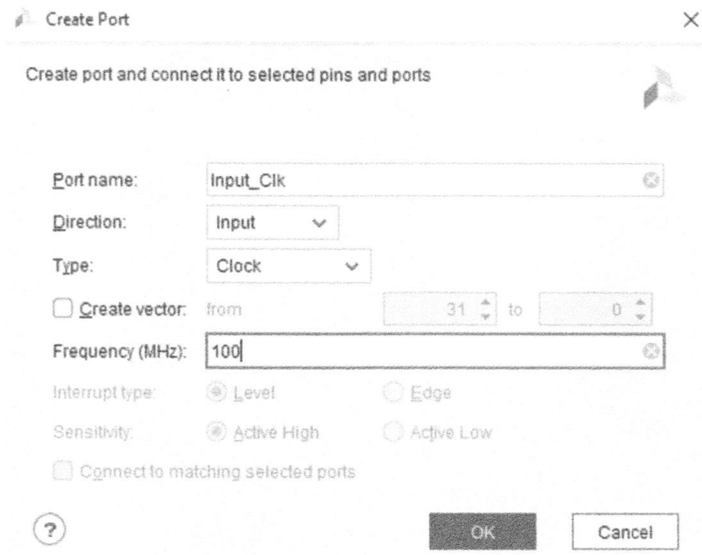

Figure 9.8 The Create Port window for clock.

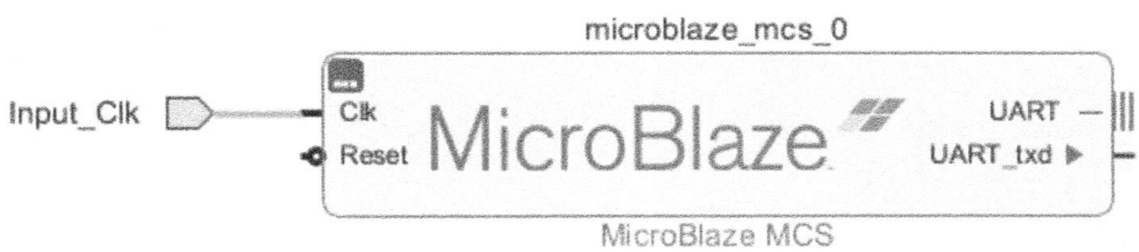

Figure 9.9 Connecting Input_Clk to the Clk of MicroBlaze MCS.

need to select whether it's an active high or an active low. So, in this case, we're going to select the polarity as Active High reset and click OK, as illustrated in Figure 9.10.

Once we see the little finger icon, hand icon, we'll click and drag the line down toward the reset input, and we'll get the pencil icon, and as we start to drag, we can see a green checkmark indicating a valid connection, so we'll drag it there, as shown in Figure 9.11.

Now that we have our Input_Clk tied to clock input and our Reset_Top tied to the reset on the MicroBlaze, there's just one more IO to place. This is called the UART_txd or the transmit. For this UART demo, we're not going to be receiving any data from the computer; we're only transmitting data from the MicroBlaze to the computer. When we open our serial port and see data coming in, we'll know that the MicroBlaze is working correctly. To do this, we right-click and select the Create Port, and we're going to give it a name UART_Tx. For the direction, we are going to make sure it's an Output and the type we will just have it as Other and click OK, as depicted in Figure 9.12.

We'll go ahead and drag this down to the UART_txd, and if we get the pencil icon or drag it to there, we can have a MicroBlaze MCS, where we have the Input_Clk, Reset_Top, and UART_Tx all connected, as illustrated in Figure 9.13.

Now, with our block diagram selected, we want to click Save. At this point, we need to create our top-level VHDL design. To do this, you want to, within the Sources window, have the UART_Demo.bd

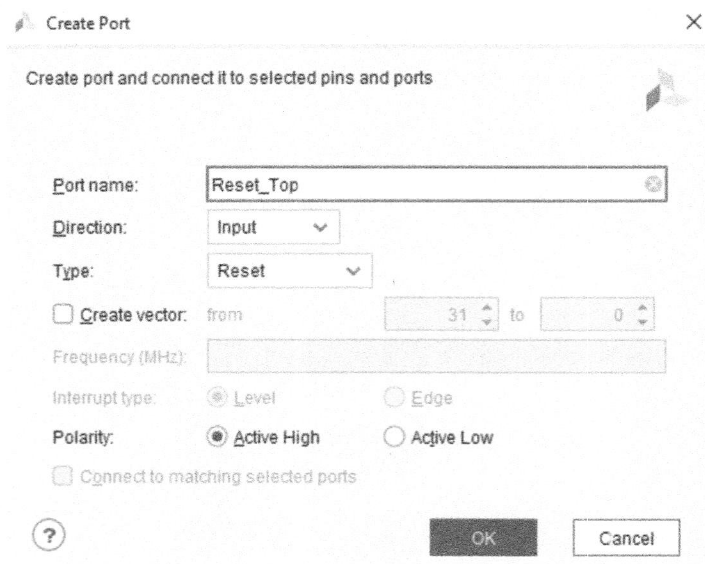

Figure 9.10 The Create Port window for reset.

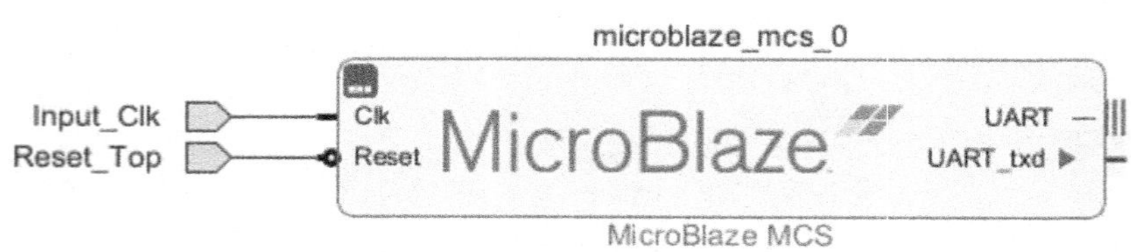

Figure 9.11 Connecting Reset_Top to the Reset of MicroBlaze MCS.

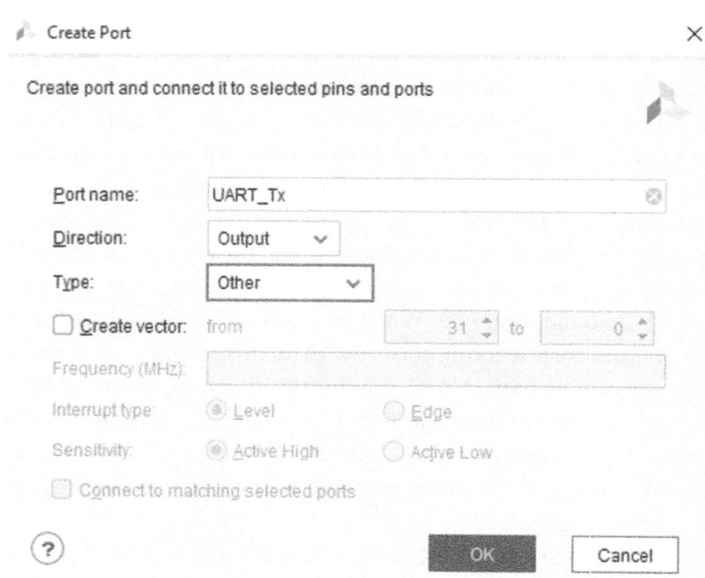

Figure 9.12 The Create Port window for UART_txd.

Figure 9.13 Connecting UART_Tx to the UART_txd of MicroBlaze MCS.

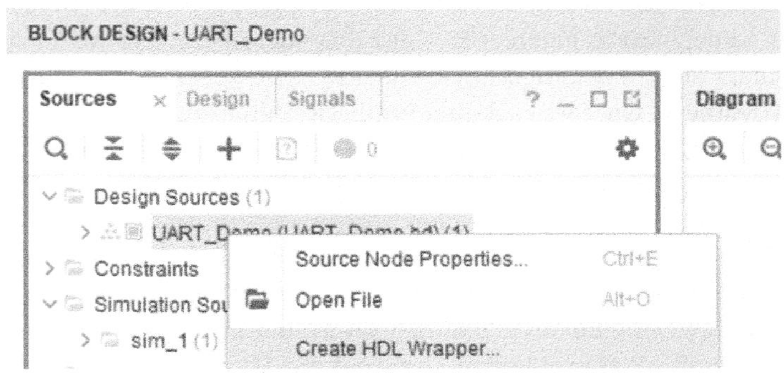

Figure 9.14 Selecting the Create HDL Wrapper.

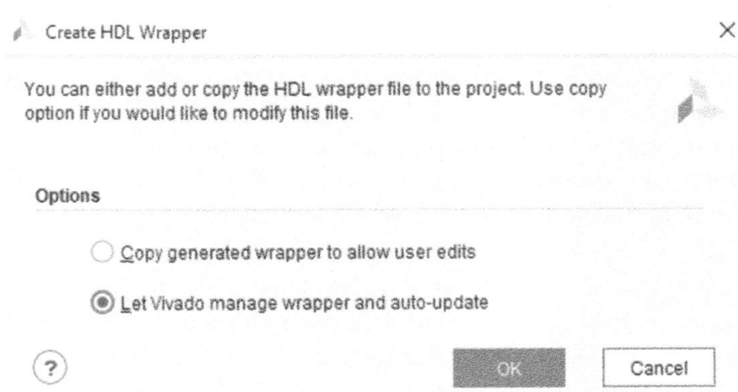

Figure 9.15 Letting the Vivado manage the wrapper and auto-update.

block diagram selected, right-click, and select Create HDL Wrapper that stands for creating a hardware descriptive language wrapper, as shown in Figure 9.14.

So, we're going to click this and let the Vivado manage the wrapper and auto-update, as shown in Figure 9.15.

That's important because if we make any changes to our block diagram, we want Vivado to update the wrapper. We don't want to physically go in and modify the code, so we will go ahead and click OK, and what the Vivado is going to do is take this block diagram and turn it into a VHDL file, because you can't take a block diagram and turn that into a bitstream.

So you have to generate a VHDL file so we can see our top-level design. After doing this, we can view the UART_Demo_Wrapper. If we click down, we can see that it contains UART_Demo.bd block diagram. By double-clicking on that UART_Demo_Wrapper, we can actually open the wrapper that Vivado has generated. If you see, it is very similar to the way in which we would code. So, they have a top-level design where they're giving some sort of comments at the top. They have the library, and the entity is defined, which includes the inputs and outputs we've defined: Input_Clk, Reset_Top, and UART_Tx. To go back to our diagram in Figure 9.13, this is where they're pulling their entity. This is effectively our top-level design. They just generate VHDL to represent that, and also you can see they've got a component. So, they're doing a component instantiation of our block diagram, which we've named UART_Demo. So, they're calling this component UART_Demo and has the port Input_Clk, Reset_Top, and UART_Tx. For their architecture, all they're doing is taking the top-level design of this and mapping it to UART_Demo. So, effectively, this is our design, as depicted in Figure 9.13 – just this MicroBlaze, and we don't have anything else going on, so that's our entire design. So now that we have our UART demo wrapper setup, we can click the Run Synthesis. Once we click Run Synthesis, a little green circle appears on the top to indicate that it's running synthesis. Once we complete our synthesis, we're going to run the implementation and then generate the bitstream. At that point, the reason we want to do this is because we want to make sure that our design is working correctly before we go ahead and write our software for the MicroBlaze. If we have an issue with our design, like the synthesis fails or any type of major failure, we want to fix that before we import it into the SDK development tool. Also, if we go back to a wrapper, we can kind of see that they give you some comments here. So, what's nice about it? If you have multiple MicroBlaze, we only put one MicroBlaze in there. But if we place a second MicroBlaze or a third MicroBlaze, it would instantiate all those components and we don't have to manage it. So, all we have to do is let Vivado update it, and it'll take care of all that for us. So, that's a nice feature to make sure that Vivado takes care of that. We'll go back to our diagram in Figure 9.13, and depending upon how fast your computer is, this running synthesis design can sometimes take a little bit. It's just going through to verify there're no issues with the code that everything is working as expected. So, we wait for that guy to finish up, so right there, our synthesis is complete. So now we want to go ahead and do the run implementation. Now that our synthesis is complete, it's telling us there're no errors in our design that it has synthesized. We're going to run the implementation, which will take our design and actually physically figure out how it's going to implement it on the board itself. Once that's complete, we want to generate the bitstream, and then we're going to have to go to our file and export our hardware. Once we do that, that will take our everything in our hardware and export it so that we can open it in the SDK environment, because then from here, we want to launch our software development kit, or SDK, they call it. So, wait for our implementation to complete, and now that's complete. We'll go ahead and generate a bitstream. When you're running the route design, one of the big things that will come into play is the amount of memory you're using. So, we've only specified eight kilobytes. If you specify more memory, that has to pull it the way our design is set up. We don't have any interface for external memory on the board itself. There's external memory, but we're not using that. We're only using the block RAM or the internal memory. So, it's going to take a little bit to generate all that, so our bitstream generation has failed, as illustrated in Figure 9.16.

So, we say, OK, and that's probably because we haven't defined our IO and that's OK, we'll get to that later, as shown in Figure 9.17. At this stage, we know we passed our synthesis and implementation.

We'll be OK from this point on. We can show you how to take this design and export it into SDK. We will then develop the code, map our IO, and run the thing on the development board. Now, we're going to export our hardware design into the Xilinx SDK, and from there, we are going to take that to write our code, to create a project, and write the code that we're actually going to use on our the MicroBlaze processor. Now that our design has been synthesized and top-level design (UART_Demo_Wrapper) generated, we want to click on the file and export the hardware. This

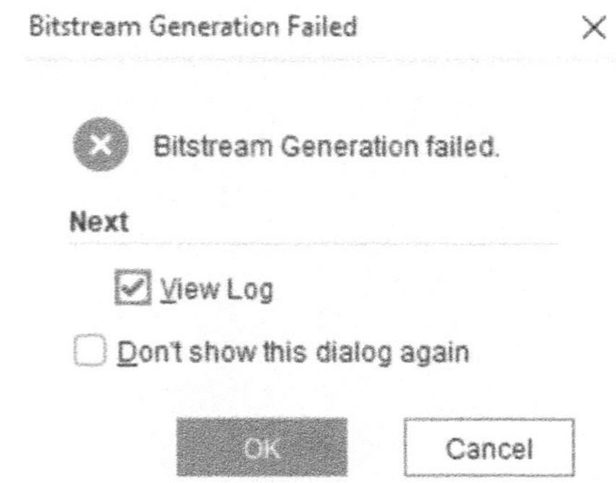

Figure 9.16 The failed bitstream generation window.

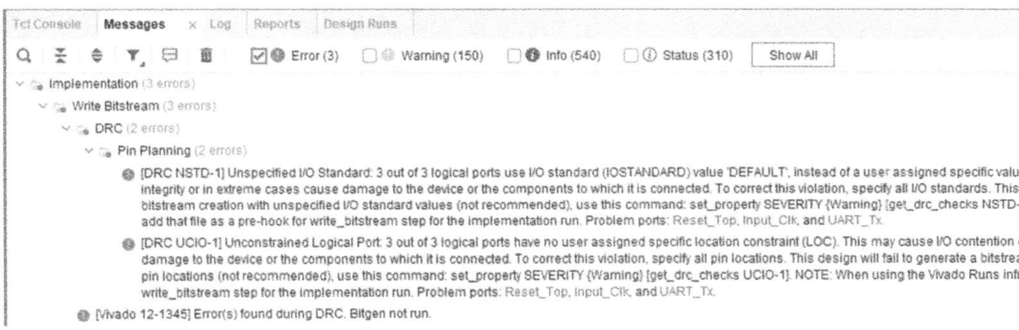

Figure 9.17 The error messages window.

will bring up a little dialogue box that says export the hardware for software development tools and we're going to leave to Include Bitstream unchecked, since we have not generated a bitstream file, so we cannot include it here, which is OK. So, we're just going to export to local project and you could export it to a different location. However, we want to keep as many things on the local project as possible, which makes it much easier when you're working with different tools, especially the software development tools. Therefore, you'll know where to find the different files required to work with the software, as depicted in Figure 9.18.

Next, click OK to export our tools to the hardware. Now, what we want to do is go to the File and Launch SDK. Once this is launched, you'll be prompted to specify the software development tool export location. If you selected a location other than local to project, you'll need to navigate to that location and tell SDK where to find the hardware you've developed. For example, we want to keep all this local to the project, and what this will do, the workspace, is store the project for your SDK: all the settings for that project, the code for that project, where you want that project to be located, and we're going to want it to be located to our local project. So, click OK, as illustrated in Figure 9.19.

This will open up the SDK for Xilinx, and once it opens, you will see a window displaying the UART demo wrapper hardware platform. So, we need to create a project to do that. This gives us all the design information about our MicroBlaze to target FPGA device, as shown in Figure 9.20. It gives us what it's created with, as well as a base address, high address, different memory types of information, and all the IP blocks present in the design. So, within that MicroBlaze block, there're a lot of things in there, like our block memory generator, controller, and IO module. You can look all these and see exactly what they're doing. But this gives you an understanding of what's going on inside the system.hdf. Now, we can close it.

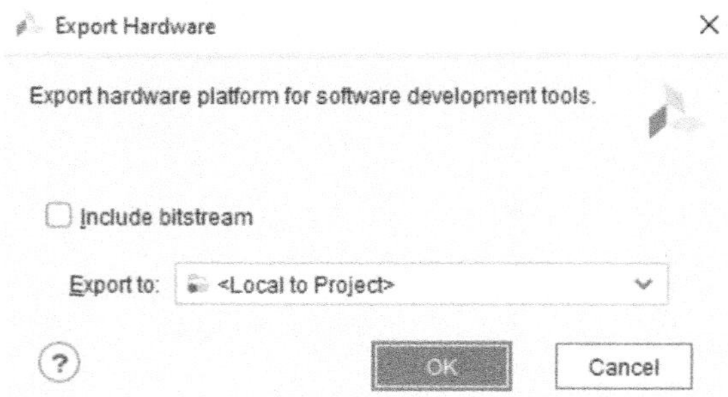

Figure 9.18 Exporting the hardware.

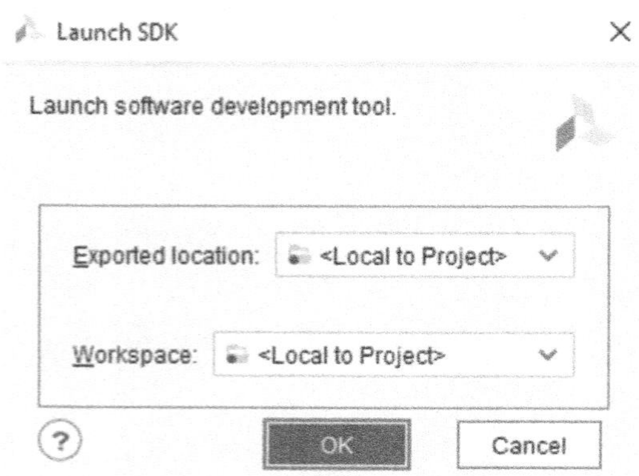

Figure 9.19 The Launch SDK window.

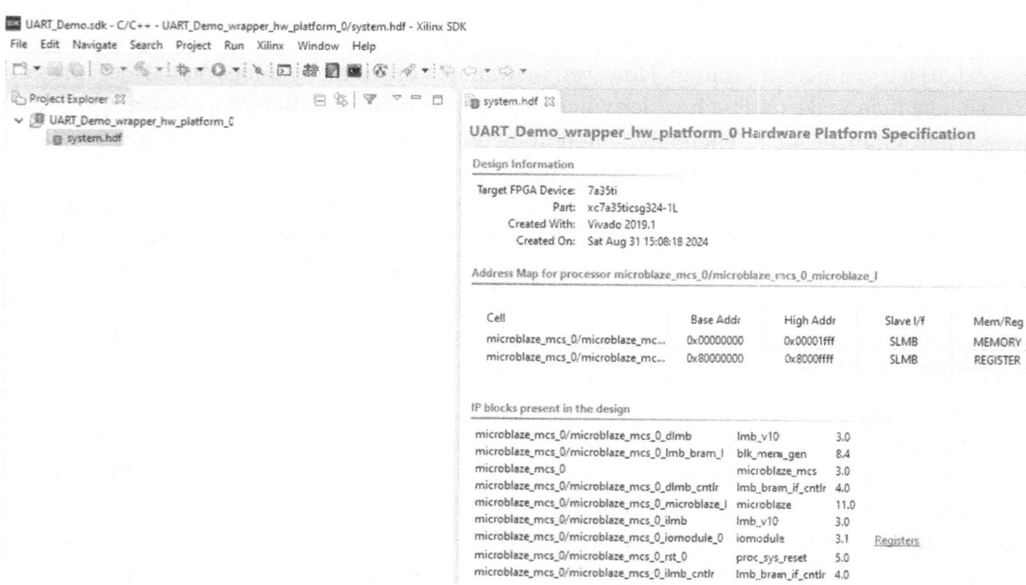

Figure 9.20 The design information about our MicroBlaze.

Within the Xilinx SDK, we want to generate a new application project. To do this, we'll go to File > New > Application Project. From here, we want to give our project a name, Hello_World, and we want to check the use default location. For the OS platform, it gives you different options. We're going to go standalone, but you can do the free real-time operating system, or you can go to Linux or other types of operating systems. But we're going to have a standalone operating system. Our hardware platform is UART_Demo_wrapper_hw_platform_0, which exports our design from the Vivado, and all the required files that the SDK needs to understand what the target processor is. So, it realizes that the target processor, which makes sense, is microblaze_mcs_0_microblaze_1. If we had multiple processors, we created this project, we could do this pull-down menu and select a different processor that we would want to target. But since we only have one processor, we're going to target the only one that's in there. For the target software, you can go with C or C++; in this case, we want to go with C, as it uses less memory. We only have eight kilobytes, so we cannot use a large amount of memory, so C is going to be our best option. The C++, which enables object-oriented programming, can use more memory to do so, and since we're in an embedded type of environment, the memory is very crucial. For a broad support package, we're going to have it create a new one and leave it to the default. You can change the name if you want, but we think it makes sense just to leave it as Hello_World_bsp. So, let's just leave it at that and we'll click the Next, as depicted in Figure 9.21.

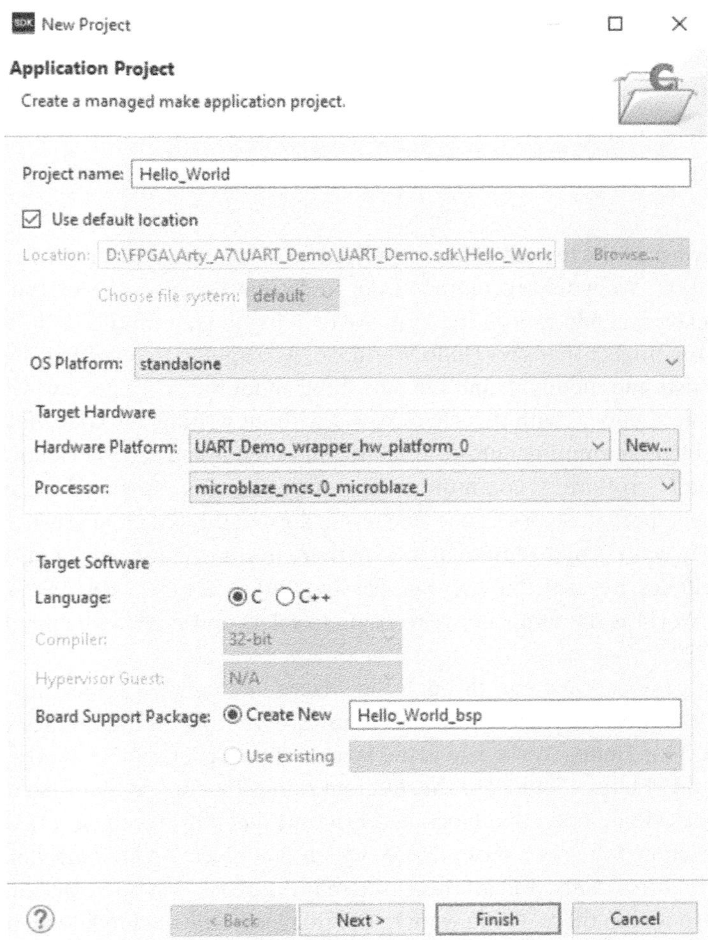

Figure 9.21 The New Project settings.

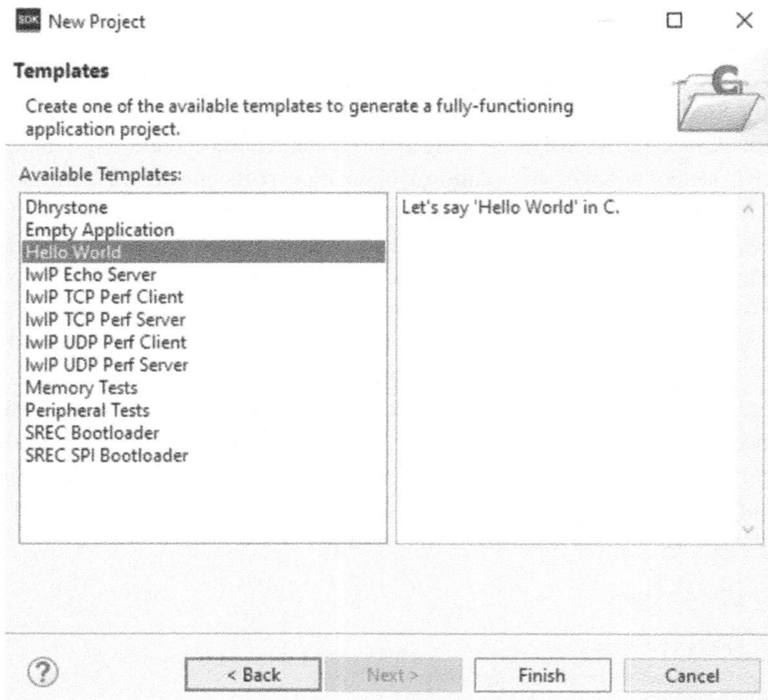

Figure 9.22 Selecting the Hello World template.

In Figure 9.22, we pick the template for our project. So, when you start a project, you can go with an empty application. We would recommend not to do that because it gives you a project, then you have to associate and add everything to it. We're going to go with the Hello World template, so this will give us a project that says Hello World. So, we're going to modify it slightly. We'll take the project that exists and modify it, and the nice thing about this, it's a good way to test, so when you have your project, you go with this Hello World without writing any code, build your project, verify if it works by programming onto your board, and then if you see the Hello World show up, you know that the MicroBlaze is communicating your memories associated correctly and everything is working as expected. So, it's a good first test to do, and this is a great way to test your UART. After that, you may want to add additional tests in there to connect to your IO, such as your LEDs and your slide switches, to verify that they are working. But right now, we're just working with the UART. So, Hello World is the template we're going to select and then we'll click Finish, as illustrated in Figure 9.22.

So, it's going to go ahead and run through the wizard and create a project for us, so we have our UART demo wrapper. This just defines our hardware; the system.hdf defines what's going on inside our design. This Hello_World_bsp is the board support package. So, it gives you the documentation that kind of talks about all of the different things that are going on, so we can open the BSP documentation. If you open that block RAM (bram) file, it brings up an HTML file that tells you about the different things we have: brand, which is a block RAM controller, and they give you some data structures, APIs, and all these different functions that you can call. So, this shows how to use the help within there. If you want to use this iomodule or standalone, they have different HTML documentations. If you want to learn how to use these functions, you'll have to read through this documentation. We're going to show you how to do the basic ones: the UART to write data out to our UART port. But if you want to understand different APIs here, some of them are for how to do so, and that's to show you the board support package, and they have different stuff in here, in this microblaze_mcs_0_microblaze_1 folder, they have some code library and everything inside that. So, the Hello World is our actual project, where we'll be developing our code. So, we have this helloworld.c and we're going to close the system.mss. If we go to helloworld.c and double-click on it, it brings up the helloworld.c, which generates this big header file that explains copyright and all this information that Xilinx provides about the project. What we care about is right here. At this point, we have a main code, an initialization of platform, and we're going to print Hello World, clean up the platform, and return zero, as shown in Figure 9.23.

It is a very basic Hello_World project. But it's very useful and it helps us get a starting point for what we want to do. So, what we want to do is go ahead and add that right now. So, before we do our initializing platform, we are going to create integers i and j. This is because in our project for this UART demonstration, we want to loop through and print Hello World 10 times, and then we'd like to have something that prints to indicate the code is complete. So, we're going to have integers i and j, initialize our platform, and then create a for loop. We'll go from zero to nine, which runs a total of 10 times. Inside that loop, we want to print Hello World, followed by "\n\r" for a newline and carriage return. So, we print those 10 different times, and then we're going to have a second for loop, which is going to act as a delay for us – and that for loop is empty. We're not going to do anything; all we're going to do is just wait. This is a quick and basic way to create a delay. Without it, we'll loop through so fast that we won't be able to see the Hello world come up multiple times. It will all show up at once because we're using a high baud rate of 115200. Once the for loop is complete, we then want to print something that says our code is done. So, we will print a code for complete statement with an exclamation mark, followed by a new line return. This is the code that we'll use to run on our actual FPGA, as depicted in Figure 9.24 and the following codes.

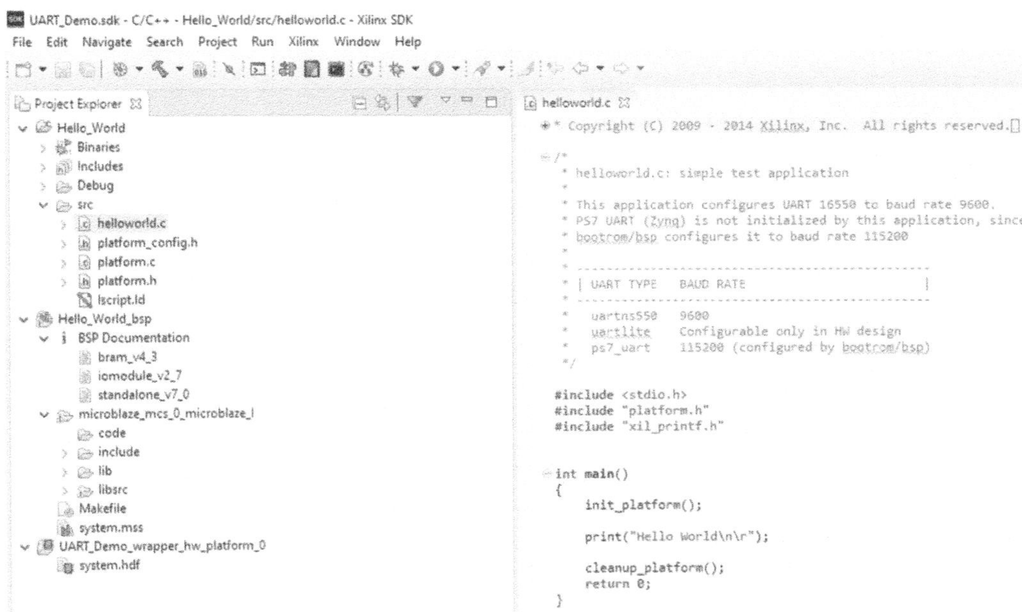

Figure 9.23 The helloworld.c codes.

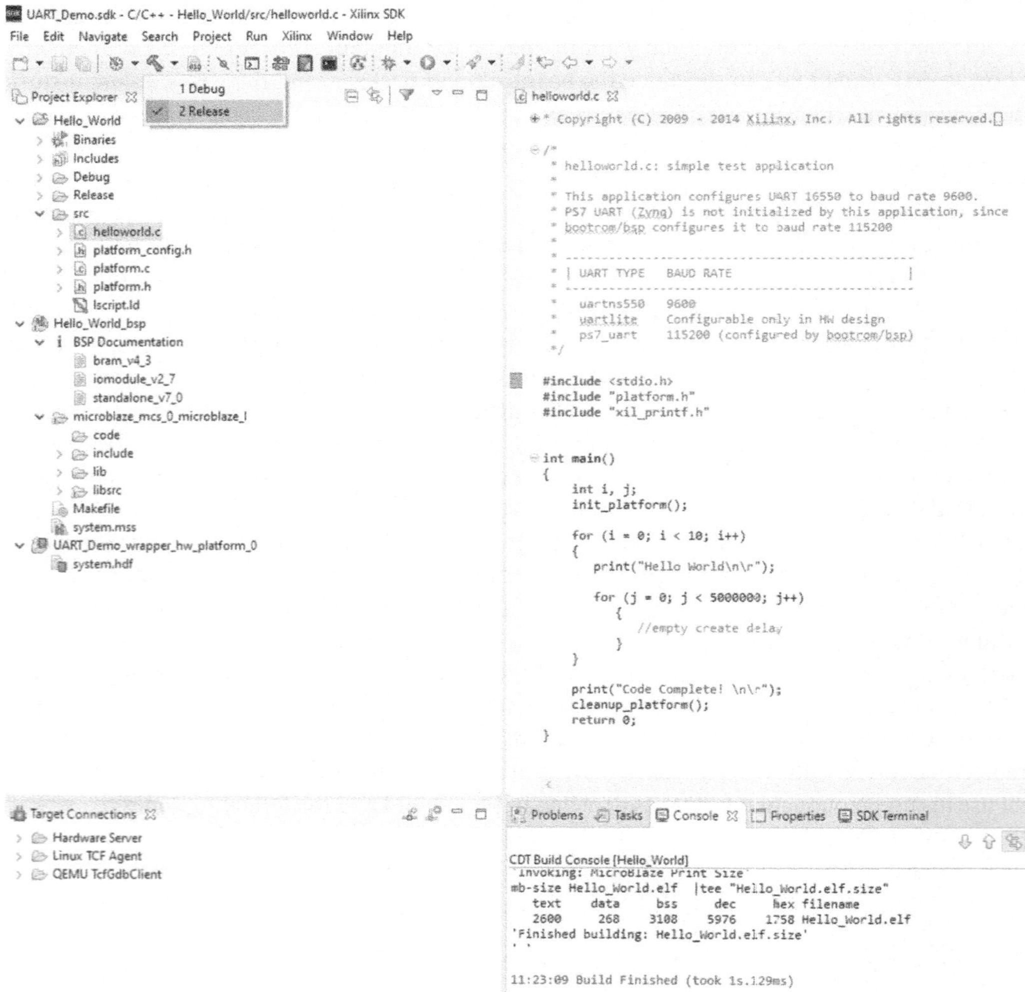

Figure 9.24 The code to run on our actual FPGA.

```
#include <stdio.h>
#include "platform.h"
#include "xil_printf.h"


int main()
{
       int i, j;
    init_platform();

    for (i = 0; i < 10; i++)
    {
      print("Hello World\n\r");

      for (j = 0; j < 5000000; j++)
```

```
            {
             //empty create delay
            }
        }

        print("Code Complete! \n\r");
        cleanup_platform();
        return 0;
}
```

So, from this point, we'll save the project and then build it. We're going to choose build 'debug' for the project Hello_World. Once we start building our project, it'll give us some information, and once completed, it says Build Finished. Once the build is successfully completed, the key thing we want to look at is the .elf file and that's what we're going to use to load when we generate our bit-stream. That is going to contain the code that gets loaded inside the MicroBlaze. To find the .elf file, navigate to the directory where your project is located. In this case, it's located under: Arty_A7 > UART_Demo > UART_Demo.sdk > Hello_World > Debug > Hello_World.elf.

This file is important because that's what we want to use when we go back to Vivado to program our actual board, and we want to use this file to associate with it. At this point, we need to check the Hello_World.elf.size file. You can open it using Notepad++ or any other text editor. This just tells you the size of your file, but more or less the size of the code, as illustrated in Figure 9.25.

So, this decimal value (6192 bytes) is the size of Hello_World.elf. That means we're using 6192 bytes for this code. Now, if you remember, we only allotted eight kilobytes, so we have about two kilobytes left. As you can see, we used quite a bit of memory, and we didn't even do much. That's why eight kilobytes is not a lot of memory. If we had used C++ or written a bunch of code, we would've gotten an error when trying to build it. It would say, no, this won't work because you're using too much memory. In order to associate the .elf file with MicroBlaze, we want to go back to the UART_Demo Hardware project. Then, go to Tools > Associate ELF Files. For our design sources, we have microblaze_mcs, and currently, we have mb_bootloop_le.elf, as shown in Figure 9.26. That is just a basic piece of code – it doesn't really do anything. So, we want to click the arrow, select the file, and add it. To do that, go to Add Files, navigate to the directory where Hello_World.elf file is located, select it, and click OK. Then, click OK again, as depicted in Figures 9.27 and 9.28, respectively.

So now, when we look at our design sources, we have this Hello_World.elf file that is associated to our MicroBlaze MCS processor, as illustrated in Figure 9.29.

So, this section is demonstrating how to develop using the Xilinx SDK to write code that will then run on the MicroBlaze.

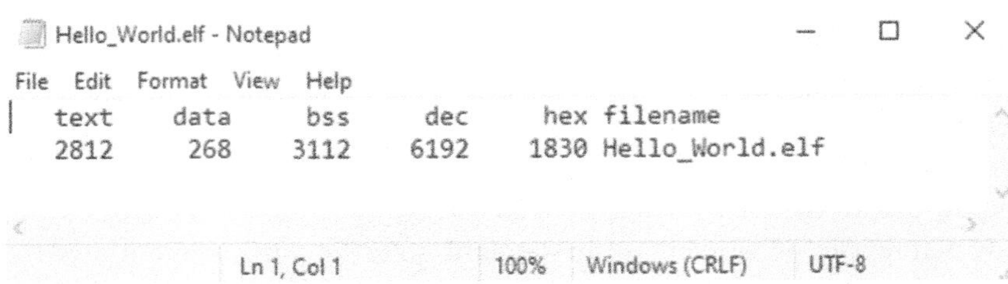

Figure 9.25 The content of Hello_World.elf.size file.

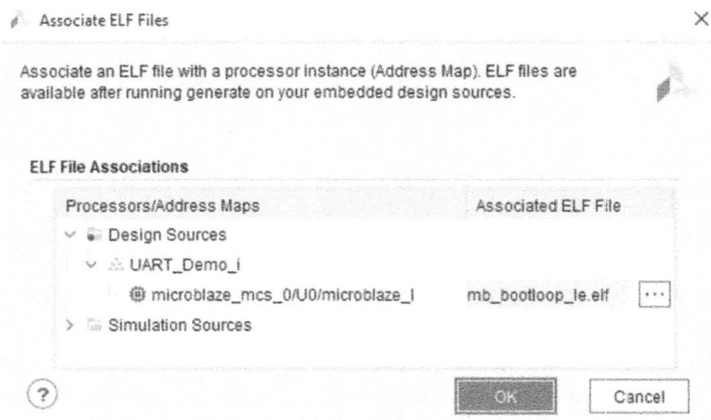

Figure 9.26 The Associate ELF Files window.

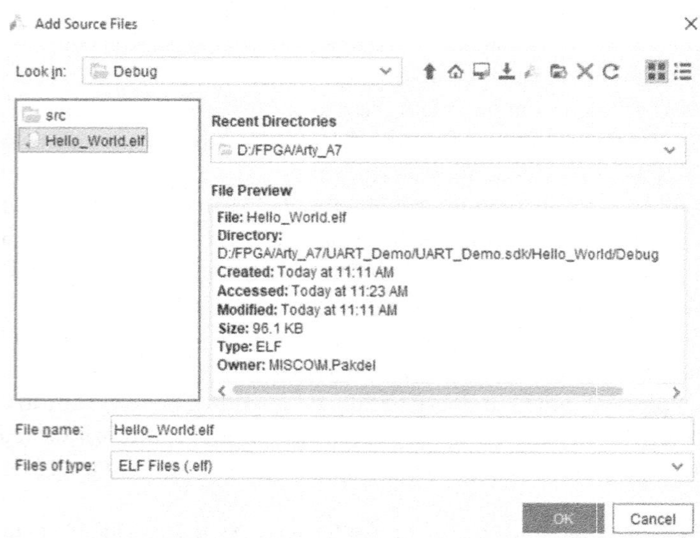

Figure 9.27 The Add Source Files window.

9.4 The UART Demonstration Arty A7 IO Placement

This section focuses on the UART demonstration Arty A7 IO placement. Here, we're going to show you how to place the IO for the Arty A7 on the UART demonstration project. So, if you open the UART_Demo project and the Arty A7 master ".xdc" file, which is the constraints file for this project, we need to map the IO. This is formed on our block diagram in Figure 9.13; we have three different IOs that we need to map: Input_Clk, Reset_Top, and UART_Tx. For the first IO, the Input_Clk, in the Arty A7 master .xdc file, we want to set the clock signal. The package already defines pin E3, as well as the standard, so to get ports where it says 100MHz, we want to replace that with Input_Clk. That gives us the input clock and that maps it to pin E3, which is the pin where the oscillator on the board is located. We'll be sending the clock signal into our design from there. We then have the next line where it says create_clock, and

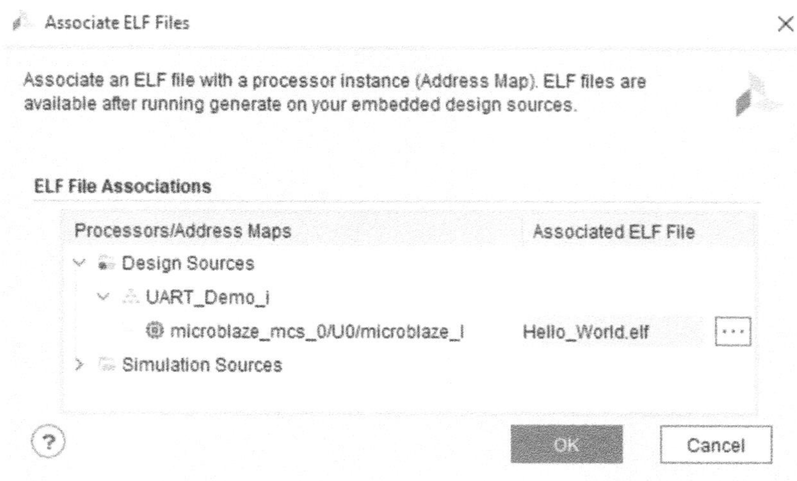

Figure 9.28 The added Hello_World.elf file.

Figure 9.29 The associated Hello_World.elf file.

we want to go to the part where it says get_ports. We want to copy our Input_Clk and paste this there. So, when we run our synthesis and design, it will use these parameters to verify that our timings constraints are being met. So, we save our design and then set the Reset_Top. We're going to replace sw[0] with Reset_Top, so uncomment and save that. Now, we need to place the UART_Tx at the bottom of the .xdc file where we have a USB–UART interface. We want it to be on package pin D10, which is labeled in schematic as uart_rxd_out. So, this is where we have to be careful, as it may confuse you. We'll call this UART_Tx because our Tx is looking at it in relation to the actual computer or USB. So, it's saying this is the Rx, so we want to tie our Tx to that. We want to apply that to pin in D10, as shown in Figure 9.30.

Figure 9.30 The Constrains file for UART_Demo project.

We now save and go back to the diagram. At this point, all our pins are mapped: Input_Clk, Reset_Top, and UART_Tx, on the UART_Demo project. You should now be able to run the synthesis, run the implementation, and generate the bitstream. Once that's complete, you can program it on the actual Arty A7 development board and make sure you have Tera Term or any other terminal emulator program open so that you can view the messages being sent from this MicroBlaze softcore processor to your computer screen.

10

The UART IO and Register Access

10.1 Introduction

This chapter is about the use of MicroBlaze GPIOs. In this chapter, we're going to show you how to use the general-purpose inputs and outputs with the MicroBlaze softcore processor. At this point, we have used the MicroBlaze softcore processor to simply transmit data from our development board to the actual computer and read the message, which was just displaying "Hello World," and the code was complete. Now, we're going to take that a step further, where we must be able to transmit messages from the board to the computer as we did before, as well as be able to receive messages from the computer back to the actual development board and process those messages inside a MicroBlaze processor. For this project, we want to be able to receive commands to turn on the LEDs, read the slide switches, and read the pushbuttons. This will require us to do various things. First, we need to receive the message, which means that messages are going to come in from our UART interface. So, we need to enable our receiver on the UART side so we can actually receive the messages. Then, we want to process them to determine what we want to do next, and in that case, we're going to have outputs to send to the LEDs. So, if we tell it we want to have a value of four, we display that four on the LEDs or light up the specific LEDs corresponding to the command we give it. We also need to be able to read values, which in this case are slide switches or pushbuttons. If they send us a command saying they want to know the status of the pushbuttons, we have to read the values of those pushbuttons and be able to display that information back out on the UART port, which we will then review using Tera Term or any other type of terminal emulator, whichever you decide to use. The big thing for this lab and what stands out here is general-purpose inputs and outputs, which are used in a lot of different ways, and just like the name suggests, they are general-purpose. So, you can use them for various different things. In our specific case, we're using them to put data out on the LEDs and read data in. So, let's go ahead and look inside the Vivado project and examine the MicroBlaze processor. So, we have our MicroBlaze MCS processor, and we want to show you all the properties of that processor that you want to have. So, we're just using MicroBlaze MCS version 3.0, and for the general properties, we want to have the following: for the memory size, we selected 32 KB, and you can go with a larger size if you want. We think it should be more than enough to perform the operation we're working with. But you can go larger, and they recommend you go with sizes that are powers of 2, which are 4, 8, 16, 32, 64, or 128 KB. For the optimization, we're going with area optimization, and that's because we're not doing any sort of intensive processing where we're dealing with an audio signal, image processing, or anything that requires intense CPU usage. However, we're just implementing and reading UART signals and outputting on the UART. So, we're going to go with area optimization so that we don't

Fundamentals of VHDL for FPGA Programming Using Vivado, First Edition. Majid Pakdel.

Companion Website: www.wiley.com/go/pakdelVHDL

use as many resources on our FPGA, and we don't have any IO bus or trace bus, and we're not going to enable either of those. For debugging, we don't have any debugging support, and the software and the program we're going to be running are relatively simple enough that we should be fine without having to have an actual JTAG-supported debug interface, as shown in Figure 10.1.

For the UART, we want to enable the receiver and transmitter. The transmitter will allow us to send messages to the computer so that we can read them, and the receiver will allow us to receive messages typed in the terminal emulator so that we can read and process these data. For the baud rate, we're going to set it to 115200, and we'll leave it as non-programmable. We could make it programmable, but since we're not going to change the baud rate, we don't want to make it more complex than it needs to be. For the number of data bits, we'll set that to eight, and we're not going to check for parity or use parity. Then, two stop bits are included by default, so we don't need to have a parameter for that, as depicted in Figure 10.2.

For the fixed interval timers, they have FIT 1, FIT 2, FIT 3, and FIT 4. We're not using any of those, so we don't have any sort of fixed interval time or so; we have all of those unchecked, and by default, these are all unchecked. So, you can just leave them as they are. For our programmable interval timer (PIT), we're not going to use any of these either, and by default, these are all

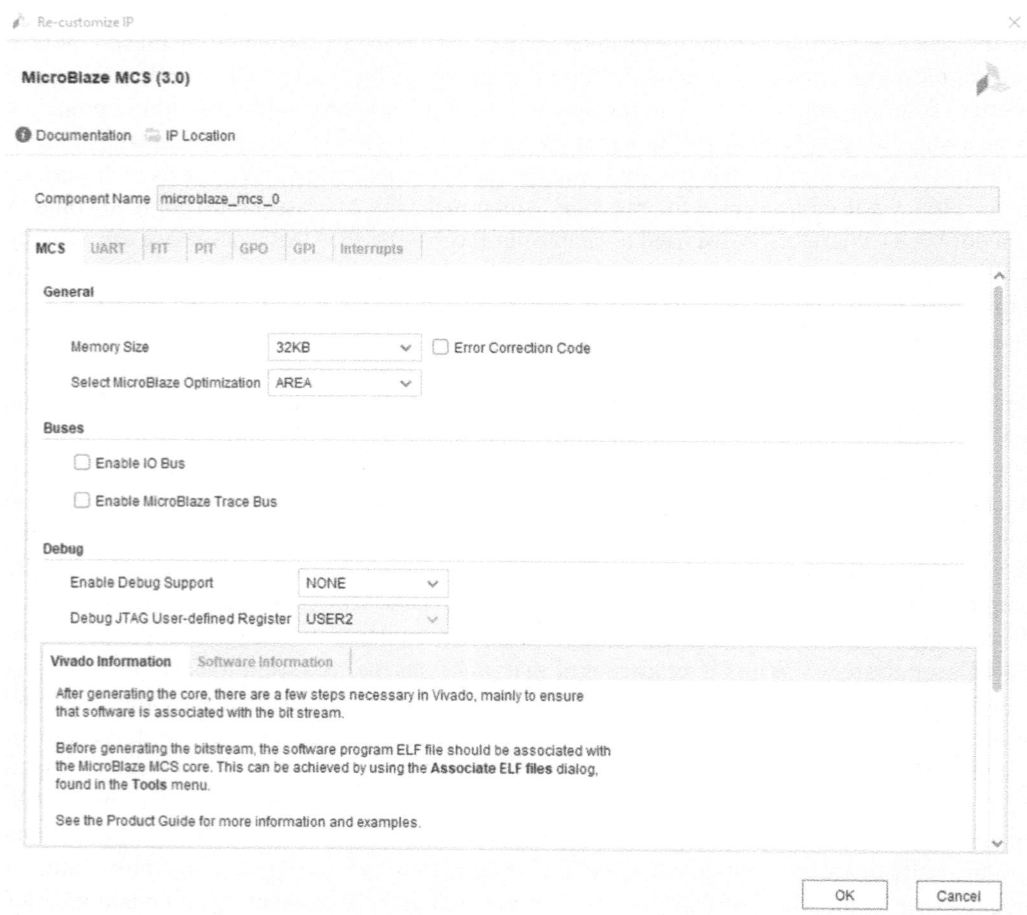

Figure 10.1 The MCS tab settings of MicroBlaze MCS.

Figure 10.2 The UART tab settings of MicroBlaze MCS.

unchecked. So, for the PIT 1, PIT 2, PIT 3, and PIT 4, we will leave all of them unchecked. For our GPO or general-purpose output, we have GPO 1. We're going to select that, and we want to set the number of bits to four because this will be connected to the four actual LEDs on our output. We have four RGB LEDs and four regular LEDs, and we're going to tie this general-purpose output to our four standard LEDs, so we'll set it to four bits wide, as demonstrated in Figure 10.3.

For our general-purpose inputs, we're going to have two. We're going to have our pushbuttons as well as our slide switches. For GPI 1 or general-purpose input one, it will be four bits wide, and it'll be tied to our four pushbuttons. For GPI 2 our general-purpose input two, we're going to set the number of bits to three because one bit is tied to the reset, which is directly connected to MicroBlaze. Thus, we only have three bits, which we'll use for the switches, as illustrated in Figure 10.4.

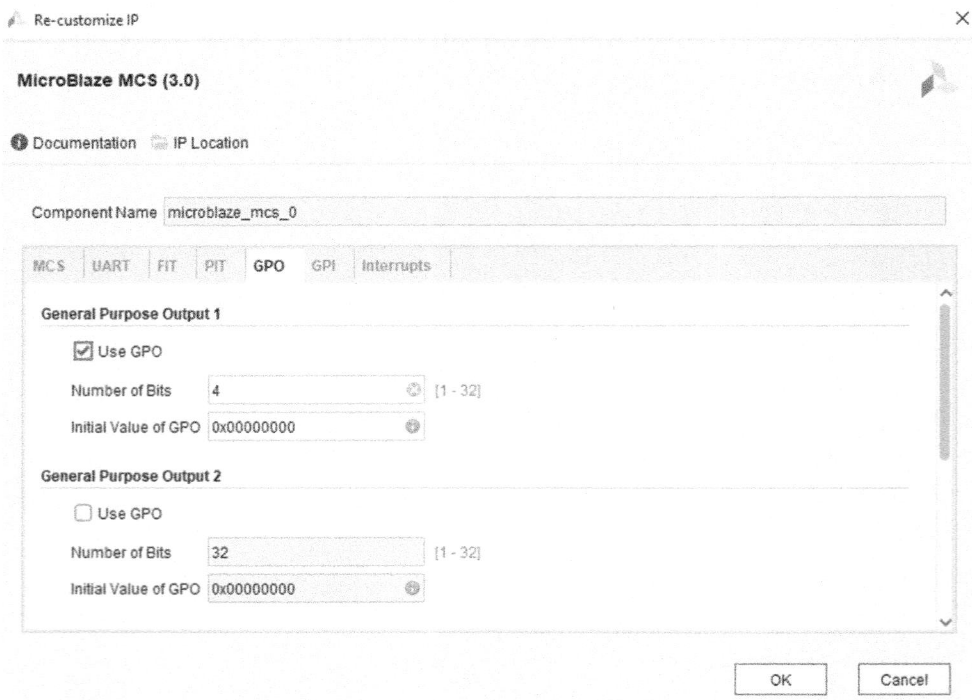

Figure 10.3 The GPO tab settings of MicroBlaze MCS.

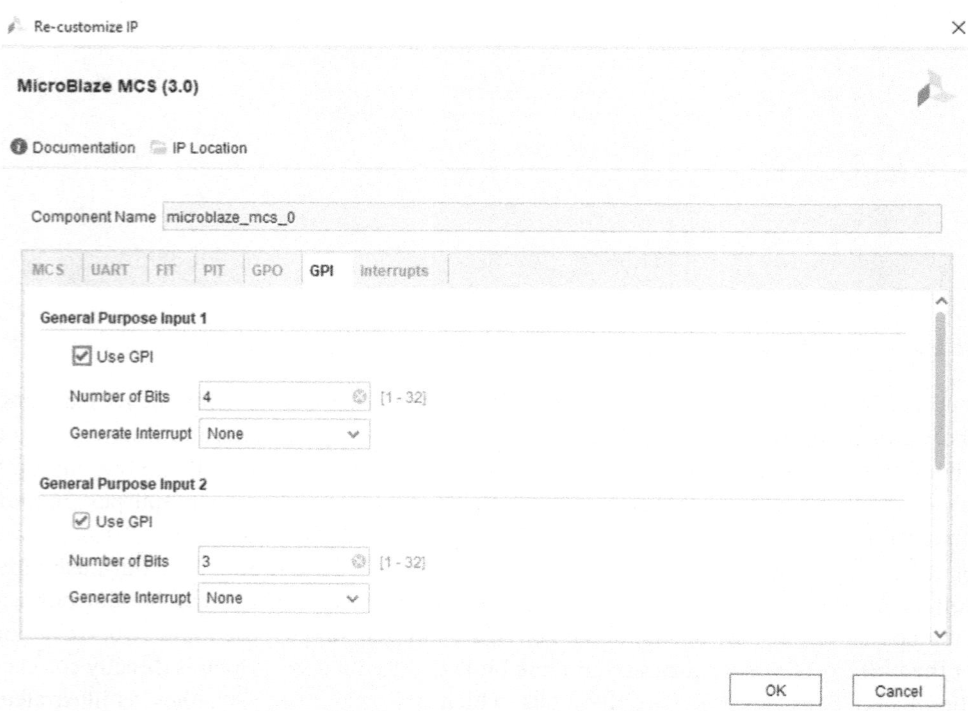

Figure 10.4 The GPI tab settings of MicroBlaze MCS.

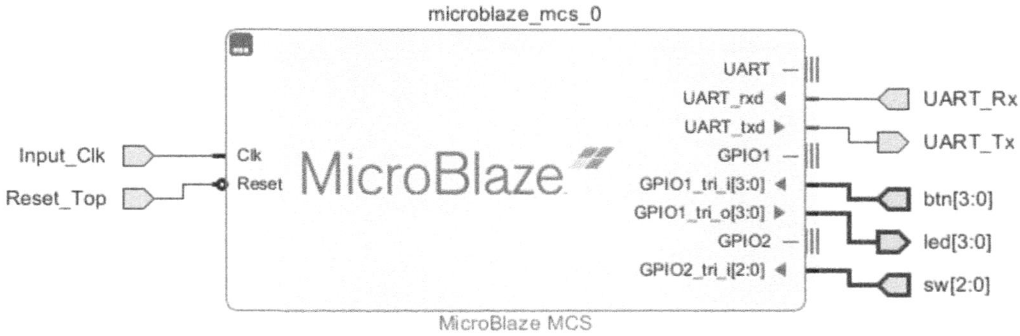

Figure 10.5 Creating ports for the MicroBlaze MCS.

For interrupts, we are not going to use any of them. We don't have any sort of interruption, and we'll just keep looping and waiting for a response from UART. Once we get that response, we'll go ahead and process it so we don't have any need for an interrupt. We're not going to use any sort of external interrupts, and at this point, you'll click OK, which will generate a MicroBlaze MCS. Next, we create a port, where we have Input_Clk, Reset_Top signal, and for UART, we have UART_Tx (UART transmitter) and UART_Rx (UART receiver). For general-purpose inputs and outputs, we will have an input that's four bits wide for different buttons and an output that's four bits wide, which will be connected to our LEDs. From GPO 2, we have an input that is three bits wide, which is tied to the slide switches, as shown in Figure 10.5.

So, that is how to configure the MicroBlaze MCS for the general-purpose input and output usage in the UART_IO project.

10.2 The UART IO Complete Design

This section is about the UART IO complete design walk-through. In this section, we're going to walk you through the completed design used for UART IO project to run on the Arty A7 development board. So, we're going to walk through the design using the Vivado tool and open it up. We have a block diagram, and this block diagram contains the MicroBlaze MCS soft-core processor, along with multiple IOs, as depicted in Figure 10.5. We've got our Input_Clk, Reset_Top, and UART communication – our receive and transmit signals – which allow us to communicate with the computer in order to transmit and receive commands. We have three different general-purpose input/output (GPIO) interfaces. GPIO_0 is an input from pins 3 down to 0 and corresponds to the buttons on the Arty A7 development board. GPIO_1 is an output from pins 3 down to 0, connected to four LEDs controlled by the MicroBlaze processor. GPIO_2 is also an input, connected to switches 2 down to 0 – giving us three switches used to read switch status. Although the board has four switches, one is reserved as the system reset (Reset_Top), and the remaining three are used for input. So, if we look at the block diagram, it contains our MicroBlaze with several inputs and outputs. What we do is take UART_IO.bd, generate a wrapper, and then let Vivado manage the wrapper and auto-update it, as illustrated in Figure 10.6.

If we navigate to our source file, we can see UART_IO_wrapper, which is simply the name of our block diagram (UART_IO) with an underscore wrapper to it. We can open UART_IO_wrapper, which is a VHDL file automatically generated by Xilinx based on the block diagram we created. So, if we look at the diagram in Figure 10.5, we can see that the entity of this UART_IO_wrapper contains all

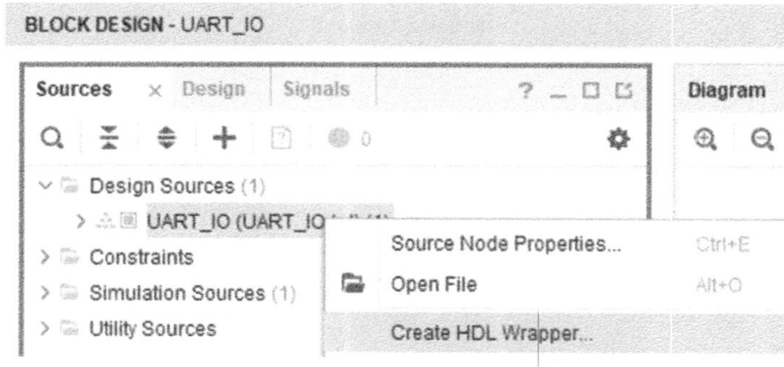

Figure 10.6 Creating the HDL wrapper.

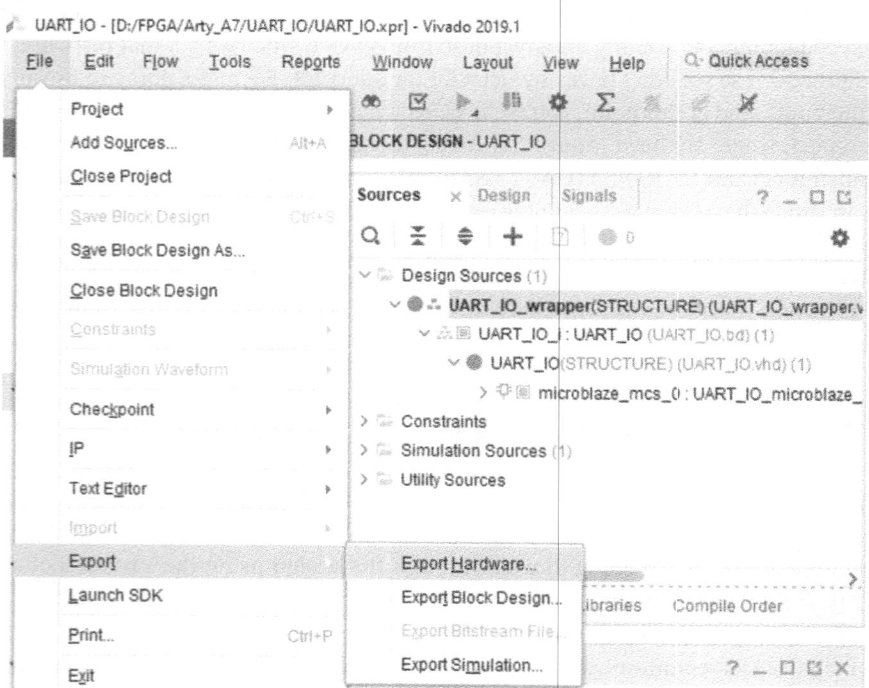

Figure 10.7 Exporting the hardware.

the inputs and outputs we've defined in our diagram. Then, we run synthesis, implementation, and bitstream generation, respectively. If any errors occur during bitstream generation, we ignore them and proceed by clicking the file. The next support is to export the hardware, as shown in Figure 10.7.

This will bring up a dialog box that says "Export hardware platform for software development tools." We're going to leave the "Include bitstream" option unchecked since we have not yet generated a bitstream file, meaning we cannot include it at this stage, and that's OK, as shown in Figure 9.18. Next, we open the Xilinx SDK and walk through the code used to read and send data. So, we create a new application project, as depicted in Figure 10.8.

We apply the following settings for the new project window, as illustrated in Figure 10.9.

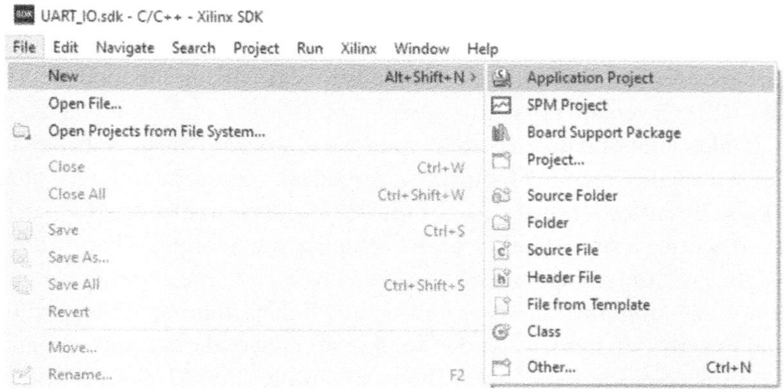

Figure 10.8 Creating a new application project.

Figure 10.9 Settings for the new project window.

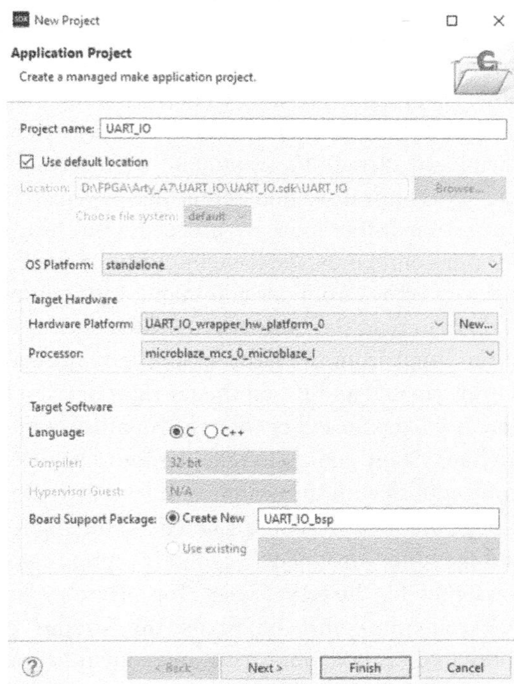

Then, we rename the file helloworld.c to testperiph.c. Inside the main function, we initialize the variables we will use. Above this, we have the libraries and the header files: stdio.h, string.h, xparameters.h, xil_cache.h, and xiomodule.h. Then, we define some integers, characters, and unsigned data types, and then we just have data in this XIO module, and there we have our Xilinx cache enable and Xilinx D cache enable, and then we do our module initialization. So, with this initialize module, we set and tell it to this IO module, which we have as an XIO module. It just gives us a handle that we can then refer to when we want to write values. So, if we look at this X parameter device (XPAR_IOMODULE_0_DEVICE_ID), it tells us we have our IO module.

For example, in this case, if you have multiple IO modules, you'll have to initialize each one of them with a different ID. However, in this case, we only use one of them, so we can initialize it here, and this goes for just those three commands for initializing the module. The xil_printf function is just a Xilinx's version of printf. So when you're using a printf, if you use the standard C/C++ printf, it takes a lot of data to do that. Since we're working on an embedded application, the ability to have a smaller memory footprint is very ideal. So, for the initialize module or after we do the module initialization, we can say that CFInitialize returned (0 = success), which means success. If we see a one, then we didn't initialize successfully. Then, we go ahead and print a message to say, "Enter a command," followed by a new line and return character, After that, we enter a while loop until the user enters the "finish" command. So, while our "done" variable is equal to zero – up here, we know we have an integer done and it's equal to zero, and the only time it ever equals a one is when the command is "finish." Inside this loop, we have other operations such as execute command, memory set, and byte count, which we set to zero. Here's where we build up the message from our UART terminal. While our execute command is equal to zero, we are going to loop through. If our received data from the XIO module is equal to zero, our buffer is going to be equal to our temporary buffer. If our buffer byte count is equal to "\n" or a new line, then we are going to say execute command is equal to one. We keep looping through here until we recognize this "\n" or newline character, and what that does is when we send a command, we want to make sure that we send the command followed by "\n" so that we can give it the command "\r\n". This way when we see that "\n" or the new line, we know to go ahead and execute the command we received as needed. So, then we go ahead and increment our byte count if we don't enter that execute command (if received buffer is not equal to "\n"). Once we exit that loop, we proceed with these codes, where we build up our command arrays. For our commands, we have in our buffer zero, which we copy, and we perform three different memory copies from our command, LED command, and LED numbers. Now, what we're going to do is to read the button and switch status. The button data will be a value that corresponds to the input from the button. So, if all buttons are zero, the value will be zero. If we have one button pressed and it's on the far right or what could be the least significant bit, then we'll get a one. If it's the far left or the most significant bit, then we would have a four or the highest value we have. So, it's going to take the four buttons we have and treat them as four binary bits. The same applies with the switch data – it's going to take that data and treat it as binary values. Even though we have it as an unsigned 32, the highest number it can represent is a total of four bits, and the other unsigned 32 can only represent a total of three bits because we use only one bit of the switch for the reset (Reset_Top). Here's where we actually execute the command we receive. To execute the command, we use the "strcmp" command, which is a string compare, so what it's going to do is take one string and copy it to another, and if they match, the value will be equal to zero. Otherwise, it'll give you a value of one or negative one, indicating that the strings do not match. The first thing we check is whether our command is equal to "led"; if it is equal to "led", then what we'll get is zero, and our led_date is going to be equal to this led_num[1] minus 48, and this is because the numbers we send them are in ASCII. So, if you subtract 48 from that number, you'll get the actual corresponding number. For example, the ASCII representation of the number zero is 48. So, when you subtract 48 from 48, you'll get your led_data to be equal to zero. Next, we perform a discrete write operation using the XIO module. We send the led_data to the XIO module's discrete write output. This logic is implemented within a series of if-else if statements. First, we check whether the command is "led." If it is, we take the led_num specified in the command and write it to the output port. Otherwise, if the command is "button," we print the button status by outputting the value of button_data, which was previously read using the XIO module's discrete read command. This read operation captures the button and switch data. Every time we read these data, depending upon the command, the users ask to print either the button data or the switch data. When we do a Xilinx printf command, this is actually taking the value and printing it out onto the UART terminal. If we're using Tera Term, which is what we used in this project (or any other terminal emulator), it'll report the button or switch data. Also, if you type in the command "finish", then all we do is set "done" equal to one. So, when we go down and re-enter our while loop, done will be equal to one. Since the condition while done is equal to zero (while(done == 0)) is no longer true, we exit that loop. Once outside of that loop, we will print a message saying "Exiting main", then disable Xilinx D cache and Xilinx I cache, and finally return zero, signaling the processor to stop running. We then return and exit our main loop, and that is the softcore MicroBlaze MCS processor. The complete code for this project is shown as follows.

```
#include <stdio.h>
#include <string.h>
#include "xparameters.h"
#include "platform.h"
#include "xil_printf.h"
#include "xil_cache.h"
#include "xiomodule.h"

int main()
{
      int done = 0;
      int byte_count = 0;
      int execute_cmd;
      char command[6];
      char led_cmd[3];
      char led_num[2];
      u8 tmp_rx_buf;
      u8 rx_buf[40];
      u32 button_data = 0;
      u32 switch_data = 0;
      u32 led_data = 0;
      u32 data;
      XIOModule iomodule;
      Xil_ICacheEnable();
      Xil_DCacheEnable();

      //Initialize module
    init_platform();
    data = XIOModule_Initialize(&iomodule, XPAR_IOMODULE_0_DEVICE_ID);
    data = XIOModule_Start(&iomodule);
    data = XIOModule_CfgInitialize(&iomodule, NULL, 1);
    xil_printf("CFInitialize returned (0 = success) %d\n\r", data);
    print("---Enter a command---\n\r");

    // While loop until user enters "finish" command
    while(done == 0)
```

```
    {
            execute_cmd = 0;
            memset(rx_buf, 0, sizeof(rx_buf));
            byte_count = 0;

            //Build up message from UART Terminal
            while(execute_cmd == 0)
            {
                    //Read UART data
                    while((data = XIOModule_Recv(&iomodule, &tmp_rx_buf, 1)) == 0);
                    rx_buf[byte_count] = tmp_rx_buf;
                    if(rx_buf[byte_count] == '\n')
                         execute_cmd = 1;
                    byte_count++;
            }

         //Build the command arrays
            memcpy(command, &rx_buf[0], 6);
            memcpy(led_cmd, &rx_buf[0], 3);
            memcpy(led_num, &rx_buf[4], 2);

            //Read the button & switch status
            button_data = XIOModule_DiscreteRead(&iomodule, 1);
            switch_data = XIOModule_DiscreteRead(&iomodule, 2);

            //Execute the received command
            if(strcmp(led_cmd, "led") == 0)
            {
                      led_data = led_num[1] - 48;
                      if((led_num[0] - 48) == 1)
                              led_data += 10;
                         XIOModule_DiscreteWrite(&iomodule, 1, led_data);
            }
            else if(strcmp(command, "button") == 0)
                       xil_printf("Button Status: %d\n\r", button_data);
            else if(strcmp(command, "switch") == 0)
                 xil_printf("Switch Status: %d\n\r", switch_data);
            else if(strcmp(command, "finish") == 0)
                       done = 1;
            else
                       print("Command is invalid, re-enter below:\n\r");
    }

    print("---Exiting main---\n\r");
    Xil_DCacheDisable();
    Xil_ICacheDisable();
    cleanup_platform();
    return 0;
}
```

So, we go ahead and build this, which generates the UART_IO.elf file we were talking about inside our Vivado design, as shown in Figures 10.10, 10.11, and 10.12.

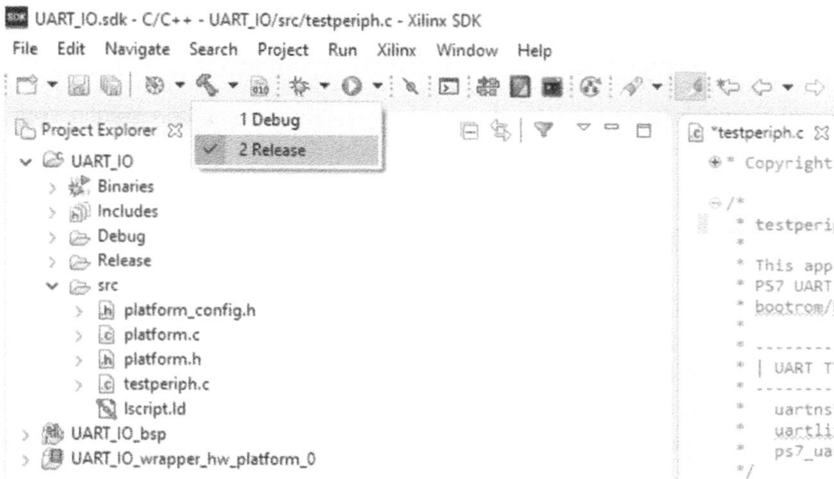

Figure 10.10 Building the UART_IO.elf file.

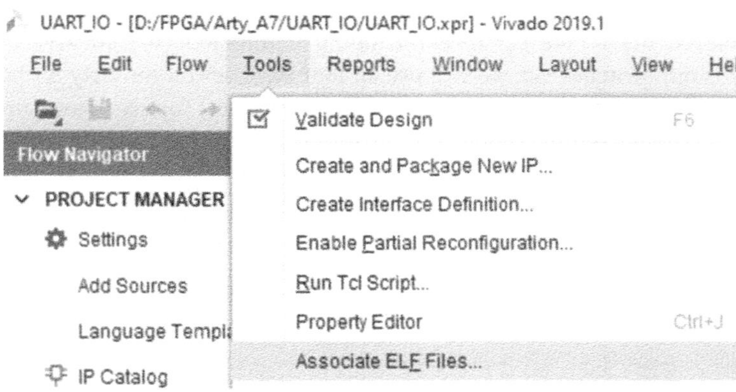

Figure 10.11 Selecting associate ELF files.

Figure 10.12 The UART_IO.elf file in design sources of UART_IO project.

That is effectively the entire design of the UART_IO for the Xilinx development board. So, there's a complete design walkthrough from start to finish to give you an overview – here is everything, and here's how it all works together. There're all ports and bits, and at this point, all you'd have to do is, once you assign the IO, go ahead and generate your bitstream and program it onto the board to make sure it's actually doing what we intended it to do.

10.3 The UART IO Arty A7 IO Placement

This section is about the UART IO Arty A7 IO placement. In this section, we're going to show you where to place all of the IOs for the UART IO Arty A7 project. On the Arty A7 board, we have a master .xdc file. Within this file, we want to go ahead and place all of the IOs, and in this case, we have Input_Clk, Reset_Top, UART_Rx, UART_Tx, our btn from three to zero, led from three to zero, and our sw from two to zero. If we go to our Arty A7 master .xdc file, the first thing we do is set property CFGBVS VCCO in current_design and set property CONFIG_VOLTAGE to 3.3 in current_design. Now, for our clock signal, we have this package pin set to E3, and we set that to our Input_Clk. Then, we use this create clock add, and we name it Input_Clk with a period of 10 and a waveform from 0 to 5 and get_ports Input_Clk. That's just going to help us when it goes through and does our synthesis, as well as the implementation, and it will generate and ensure that our Input_Clk, which is defined there, can meet the timing requirements of a 100MHz clock. The next line we will use is our switches. So, we will use all four switches on the Arty A7 board. The first switch, which, if you look at it, is switch zero, is our reset (Reset_Top). So, based on whether this is high or low, it is going to either allow our MicroBlaze softcore processor to run or not. It'll put it in a reset state or run state. For the next pins, we have what would be our switches 1, 2, and 3,

```
## Voltage Constraints
set_property CFGBVS VCCO [current_design];
set_property CONFIG_VOLTAGE 3.3 [current_design];

## Clock signal

set_property -dict { PACKAGE_PIN E3    IOSTANDARD LVCMOS33 } [get_ports { Input_Clk }]; #IO_L12P_T1_MRCC_35 Sch=gclk[100]
create_clock -add -name Input_Clk -period 10.00 -waveform {0 5} [get_ports { Input_Clk }];

##Switches

set_property -dict { PACKAGE_PIN A8    IOSTANDARD LVCMOS33 } [get_ports { Reset_Top }]; #IO_L12N_T1_MRCC_16 Sch=sw[0]
set_property -dict { PACKAGE_PIN C11   IOSTANDARD LVCMOS33 } [get_ports { sw[0] }]; #IO_L13P_T2_MRCC_16 Sch=sw[1]
set_property -dict { PACKAGE_PIN C10   IOSTANDARD LVCMOS33 } [get_ports { sw[1] }]; #IO_L13N_T2_MRCC_16 Sch=sw[2]
set_property -dict { PACKAGE_PIN A10   IOSTANDARD LVCMOS33 } [get_ports { sw[2] }]; #IO_L14P_T2_SRCC_16 Sch=sw[3]

##LEDs

set_property -dict { PACKAGE_PIN H5    IOSTANDARD LVCMOS33 } [get_ports { led[0] }]; #IO_L24N_T3_35 Sch=led[4]
set_property -dict { PACKAGE_PIN J5    IOSTANDARD LVCMOS33 } [get_ports { led[1] }]; #IO_25_35 Sch=led[5]
set_property -dict { PACKAGE_PIN T9    IOSTANDARD LVCMOS33 } [get_ports { led[2] }]; #IO_L24P_T3_A01_D17_14 Sch=led[6]
set_property -dict { PACKAGE_PIN T10   IOSTANDARD LVCMOS33 } [get_ports { led[3] }]; #IO_L24N_T3_A00_D16_14 Sch=led[7]

##Buttons

set_property -dict { PACKAGE_PIN D9    IOSTANDARD LVCMOS33 } [get_ports { btn[0] }]; #IO_L6N_T0_VREF_16 Sch=btn[0]
set_property -dict { PACKAGE_PIN C9    IOSTANDARD LVCMOS33 } [get_ports { btn[1] }]; #IO_L11P_T1_SRCC_16 Sch=btn[1]
set_property -dict { PACKAGE_PIN B9    IOSTANDARD LVCMOS33 } [get_ports { btn[2] }]; #IO_L11N_T1_SRCC_16 Sch=btn[2]
set_property -dict { PACKAGE_PIN B8    IOSTANDARD LVCMOS33 } [get_ports { btn[3] }]; #IO_L12P_T1_MRCC_16 Sch=btn[3]

##USB-UART Interface

set_property -dict { PACKAGE_PIN D10   IOSTANDARD LVCMOS33 } [get_ports { UART_Tx }]; #IO_L19N_T3_VREF_16 Sch=uart_rxd_out
set_property -dict { PACKAGE_PIN A9    IOSTANDARD LVCMOS33 } [get_ports { UART_Rx }]; #IO_L14N_T2_SRCC_16 Sch=uart_txd_in
```

Figure 10.13 The constraints file for Arty A7 UART_IO project.

which become sw[0], sw[1], and sw[2]. So, we're only reading the three switches from zero to two, and we have these as our PIN C11, C10, and A10. And just like that comes with the Arty A7 master .xdc file, these are all 3.3V standard. So, just leave those as such. We will then uncomment and just change the name of the switches to say "sw" and order them from 0 to 2. If we return to our diagram in Figure 10.5, it covers our Input_Clk, our Reset_Top, and our switches (sw) from 2 down to 0. If we return to our Arty A7 master .xdc file, then for LEDs, we have led[0] through led[3], and if we look at the diagram, that's from three to zero, so we got those covered. Those are just the four standard LEDs that are on the board. There's a total of eight LEDs. There are also LEDs from zero to three, which are the RGB LEDs on the Arty A7 development board, but we're only worried about the standard LEDs. So, they call those four, five, six, and seven. So, we just rename them to zero, one, two, and three so that we can match up with what we have on our diagram for our output there. If we go back to our .xdc file, there are four buttons, and we're using all four of them in schematic (buttons zero through button three or btn[0] to btn[3]). Therefore, wewill uncomment those and use button zero through button three, and that's because our naming convention is the same btn[3] to btn[0]. Now, finally, we have our UART_Tx and UART_Rx. So, if we go down further to where we have this USB-UART interface, we have the two signals that are UART_Tx and UART_Rx, and that is where we want to set our UART_Tx and UART_Rx. Now, note that this is where it might be slightly confusing, but we have our UART_Tx set to where the schematic shows uart_rxd_out. If we look at our diagram, our UART_Tx is from our MicroBlaze or UART, which transmits that data out. If we look here in .xdc file where it says we're transmitting it out, this is where it's received – uart_rxd_out. Then, UART_Tx is pin D10, or UART_Rx is pin A9 at that point. That is all of the IOs we're using for this UART_IO project, and that is the placement location for each IO, as shown in Figure 10.13.

11

The ADC Processing Project

11.1 Introduction

This chapter focuses on temperature sensor ADC processing. We'll explain how various temperature sensors work, with a specific focus on the temperature sensor used in our ADC processing project. In this project, we aim to process an analog signal and convert it into a temperature value that can be displayed on a serial port and viewed on a computer, allowing us to monitor the actual temperature. There are various types of temperature sensors, which typically fall into one of two categories: digital temperature sensor and analog temperature sensor. Both types perform the same function – measuring and sensing temperature. However, a digital temperature sensor has a digital interface, where you simply read the bits and they tell you what the temperature is. This is usually done with an SPI or an I^2C type of interface. On the other hand, an analog temperature sensor outputs an analog voltage that correlates to a specific temperature, at which point you need to take the analog voltage, perform an analog to digital conversion, process it, and convert it to the temperature. In this chapter, we're using an analog temperature sensor, specifically the TMP36 from analog devices. The TMP36 has three pins: a voltage source pin, an output pin, and ground. We'll apply 3.3 V to the voltage source, connect ground appropriately, and the output voltage is where we'll perform our analog to digital conversion. That output voltage signal will be used to determine the temperature. So, the way the sensor works is straightforward: as the temperature increases or decreases, the output voltage of the sensor will increase or decrease in correlation with the temperature. From there, we just have to perform the manipulation, where we're going to use the MicroBlaze or the softcore processor to go ahead and perform the processing of that information. Also, with the analog to digital conversion, the way it works is: you take an analog voltage, and you output the digital representation of that voltage because we can't actually process an analog value inside the MicroBlaze. We're working in a digital world, so we need actual bits or values to process and work with. So, all that analog to digital conversion is doing is taking the analog value and converting it into a digital value, specified by the number of bits you have. With the Arty A7 development board, using the Artix FPGA, we have what's called an XADC core built into the actual FPGA. What that means is that certain pins can allow for analog signals to come directly in, and the FPGA already has the analog to digital conversion built-in. So, within Vivado, all we have to do is drop in the XADC IP core into a block design. Then, we can figure that XADC core to operate on whatever channel we're bringing our temperature analog signal into. We just have to tell it which channel we're using and set the parameters accordingly so that we can then process that data with the MicroBlaze, and so it's really nice. In fact, we're going to create one design where we have a MicroBlaze softcore processor and our XADC core. They're going to interface with each other, and all they're going to do is: the MicroBlaze will continually read the analog to digital conversion

Fundamentals of VHDL for FPGA Programming Using Vivado, First Edition. Majid Pakdel.

Companion Website: www.wiley.com/go/pakdelVHDL

value from that XADC core, and that XADC core will just take that analog value coming from the temperature sensor, convert it to a digital value, and continually do that with a MicroBlaze processor, which will determine the temperature and send it out to the serial port. So, the next thing is to take a look at the actual datasheet of the analog to digital devices, analog devices ADC converter CHIP, the TMP36. We'll take a look at the datasheet and kind of run through it to get an understanding of what this device is and how it actually operates. So, the datasheet shows us it is called a low-voltage temperature sensor. They refer to it this way because it can operate anywhere from 2.7 to 5.5 V. On most circuit boards, the voltages range from 1.8 to 5 V. So, 2.7–5.5 V covers a pretty wide range of voltages you'll typically see on these development boards. Since we're working with 3.3 V, we are right in the middle of this operating range. This means the temperature sensor should work just fine with our FPGA development board. We're using the TO-92 package, which has three pins. Pin 1 is the positive voltage input, so we'll connect that to 3.3 V. Pin 2 is V_{OUT}, which is the actual analog signal we're going to process and interpret to determine the temperature. Pin 3 is the ground. So, an important thing to note here is pin 2 is V_{OUT}. When you're connecting this sensor to your board, you'll have to make sure you look at this package, understand which pin is which, and make sure you connect it correctly, as shown in Figure 11.1.

The next thing we'll take a look at – with importance – is a chart included in the datasheet that is very important. It shows the temperature versus output voltage for three sensors: TMP35, TMP36, and TMP37, as depicted in Figure 11.2.

We're using the TMP36, so we want to focus on line b. If we look closely, we'll see that this is a linear line – actually, all these are linear lines – and they provide a clear relationship between temperature and output voltage. We're going to read the output voltage, and based on this voltage, we can determine the temperature using the equation for line b, which we'll show you how to solve. This will give us the temperature in degrees Celsius. From that point, we can perform a simple Celsius to Fahrenheit conversion in order to determine the temperature in Celsius or Fahrenheit. An important point to note when figuring out the actual equation for this linear line is found in the functional description section of the datasheet. They have a table showing the output characteristics, as illustrated in Table 11.1. The key values we want to look at are the offset voltage, which is half a volt (0.5 V), and output voltage scaling, which is 10 mV/°C. This means that at 25 °C or your room temperature, the output voltage for the TMP36 should be about 750 mV. So now, let's take some notes and understand what exact equation we need to operate in order to perform this calculation.

So, if we get a voltage output in millivolts (mV), how can we actually convert that to a temperature? Let's go ahead and take some notes that we'll be using. Since we know this is a linear line, we can define it using the basic equation as follows.

$$y = mx + b \tag{11.1}$$

That b in Equation (11.1) is obtained as follows: we looked at Table 11.1, which told us that our offset voltage is 0.5 V or 500 mV, so that would be our b term. So, we know our b is equal to 500 mV, and that's our offset. Also, it gave us the scaling factor, and it was 10 mV/°C, and that was equivalent to our m value, that's like our slope, so our m is equal to the 10 mV/°C, and that would be our slope.

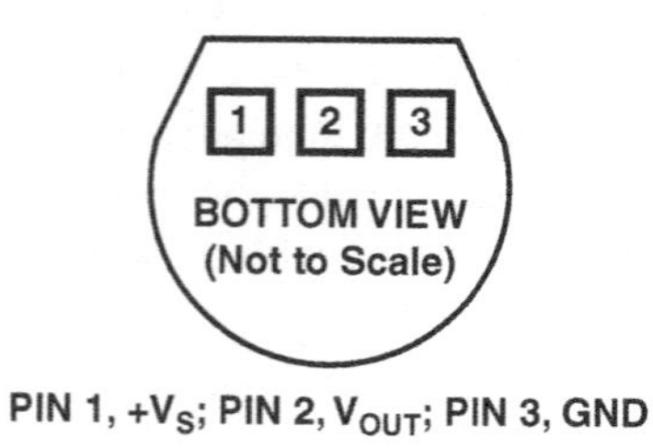

Figure 11.1 The TO-92 package of TMP36.

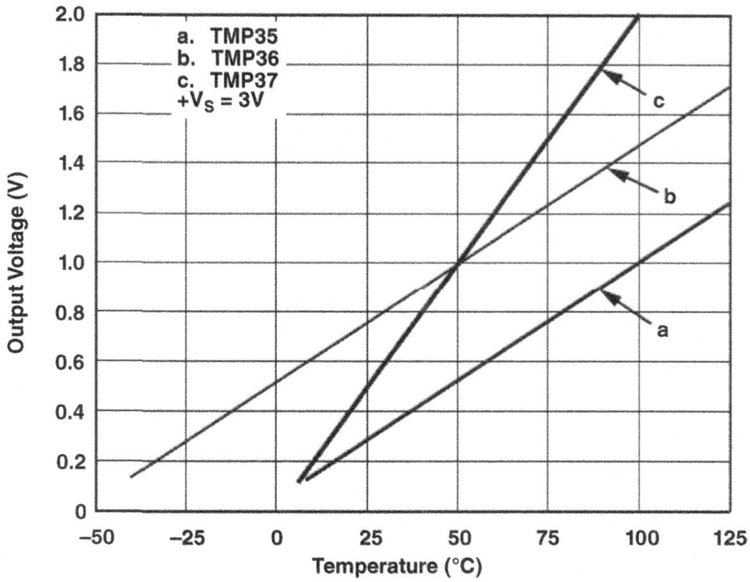

Figure 11.2 The output voltage versus temperature.

Table 11.1 The TMP35/TMP36/TMP37 output characteristics.

Sensor	Offset voltage (V)	Output voltage scaling (mV/°C)	Output voltage at 25 °C (mV)
TMP35	0	10	250
TMP36	0.5	10	750
TMP37	0	20	500

Then we also have our *x* value, which is going to correlate to a temperature in °C, and the reason for this is because we had Table 11.1 in there, we looked at our output voltage versus our temperature in °C. Temperature was displayed on the X-axis and output voltage on the Y-axis, as shown in Figure 11.2. So, we have a linear line, and we are going to know our output voltage. So, we're trying to find our temperature. So, our *x* is what we're trying to actually find, since *x* is equal to our temperature – call it temp_C – that is our *x*, and *y* is equal to our output voltage and we'll put it in mV because we have b in mV, so y is equal to our voltage_mV, and this is the actual reading we're going to take. We're going to take a reading of this voltage, which is going to be continually changing, and we need to solve for *x*. So, in order to solve for *x*, if we just solve that linear Equation (11.1), we know that we will have the following equation.

$$x = \frac{y - b}{m} \tag{11.2}$$

So, to write that out, our temp_C is going to be equal to the following equation.

$$\text{temp\_C} = (\text{voltage}_{\text{mV}} - 500\ \text{mV})/10\ \text{mV} \tag{11.3}$$

That's going to give us our temperature in °C, and that's how we're going to calculate our Celsius temperature. Then, when we want to go to Fahrenheit, we just perform the Celsius to Fahrenheit conversion equation to get our temperature in Fahrenheit. Now, another important thing to note when you're working with this or any other type of analog to digital conversion is resolution, which is important because this will tell you how accurate you can actually be. This is going to be based on the number of bits you have, as well as the voltage range you're trying to sense. So, the way you calculate your resolution is by taking the voltage range, which in this case is 3.3 V, and then you divide it by the number of possible positions you have. So, in this case, we have 12 bits. So, if we have 12 bits, we take 2 to the power of 12, and that gives us 4096, and that's 4096 possible combinations. So, the following equation would give us our resolution:

$$\text{Resolution} = \frac{3.3\ V}{2^{12}} = \frac{3.3\ V}{4096} \approx 0.81\ \text{mV} \tag{11.4}$$

What that means is that our temperature we're reading, or our voltage reading, is approximately accurate to 0.81 mV, and if we look at our m or 10 mV/°C, then we are going to be more accurate than a degree Celsius. So, we're within a very high level of accuracy because 10 mV is equal to a degree Celsius, and we're even less than 1 mV, which is less than one-tenth of a degree. So, we have a high resolution of accuracy. Now, taking this concept, if we have the same number of bits and a higher voltage range – so, for example, let's say we have another common case, which would be 5 V, and we're still using our 12-bit analog to digital conversion, and we still have 4096 possibilities – then the resolution would be as in the following equation:

$$\text{Resolution} = \frac{5\ V}{2^{12}} = \frac{5\ V}{4096} \approx 1.2\ \text{mV} \tag{11.5}$$

The resolution is still really pretty good, and that's still really quite accurate. We're still accurate to within a degree – maybe not greater than a degree. But if you look at using the same analog to digital conversion, whether you're using a 3.3 V signal on your input or a 5 V signal, you can see that the 3.3 V gives you 0.81 mV resolution versus 5 V, which gives you 1.2 mV resolution. So, we get a decent amount better with the 3.3 V. Obviously, if you take this even higher to 10 V or any higher voltage, you can see that as you have a higher range, your resolution is going to decrease, which makes sense because the bits are basically telling you how many different possibilities you have in between each voltage range. So, that's another thing to note with resolution, and that comes into play when you're working with accuracy. If you have a circuit or something you're working with where you're trying to measure the temperature to a certain degree of accuracy, you'll need to use this calculation to determine how accurate you can actually be. So, that is just an excellent explanation of the temperature sensor circuit and how we're going to use it in our project.

11.2 Using the Xilinx XADC and MicroBlaze MCS

This section is about using the Xilinx XADC and MicroBlaze MCS. Here, we're going to walk through how to set up the XADC core, which is a core provided by Xilinx to use the internal analog to digital converter on certain Xilinx chips. The chip we're using on the Arty A7 board has one of these cores. So, within Vivado, we will drop in an XADC core and configure it so that we can read this analog to digital conversion values. We want to have it set up in such a way that we can interface with it using a MicroBlaze MCS. So, in this section, we're going to show you how to set up the XADC core, and in the next section, we'll show you how to set up the MicroBlaze so we can get them to interface together. Because on the chip, we have the capability of doing both processing and analog to digital conversion, and we're going to show you how to get all these working together on the same chip. So, if you open up your project – the temperature sensor project – we're going to place the XADC Wizard core. Go ahead and open it up, and we're going to show you all the properties you'll want to set on this core. The first thing is we're going to go to the Basic tab, where we have different interface options: AXI4Lite, DRP, or None. We want to select DRP, which stands for dynamic reconfigurable port, which allows us to interface externally. If we click on the plus buttons (s_drp and Vp_Vn), with "Show Disabled Ports" unchecked, then as we make changes through our Basic ADC Setup, Alarms, Single Channel, and Summary tabs, we make these changes, and we'll see the different ports that appear and disappear. So, since we're using this dynamic reconfigurable port, this is going to give us daddr_in[6:0], or the address to select which analog input we want to read, and whichever one we select is going to come out on this do_out[15:0], or our digital output, which is a 16-bit value. Then, for a startup channel selection, we just want to select a single channel, and for the AXI4STREAM options, we're going to leave that unchecked. We're going to have our reset_in in control/status ports, and this just allows us to have this reset configurability, which we'll tie to our MicroBlaze, and we'll have that tied off to a slide switch on our development board. For the timing mode, we'll just have it set to continuous mode, and for the DRP timing options, our DCLK frequency is 100 MHz, and that's because we're going to just bring in the clock, our system clock, which comes in externally from the board, and that's 100 MHz. So, we're just going to leave that as 100 MHz. For the ADC conversion rate in kilo samples per second, we'll set it to 1000. We're going to set it to a really high sample rate, and you could reduce this if you want. It's not required to go this high where we're sampling something like temperature, which shouldn't be changing super-fast. So, you could have this at a lower sampling rate if you want, or you can just leave it at 1000, which is what we have. For the acquisition time, we're going to say that's 4. So, based on those two parameters (ADC conversion rate and acquisition time), it gives you how many samples you'll actually be getting. So, we're trying to get 1000 kilo samples per second (KSPS), but it's going to actually give us 961.54 KSPS, which is still more than we need. So, this is fine, and it meets the criteria for sampling a temperature signal, as illustrated in Figure 11.3, and we'll go ahead to the ADC

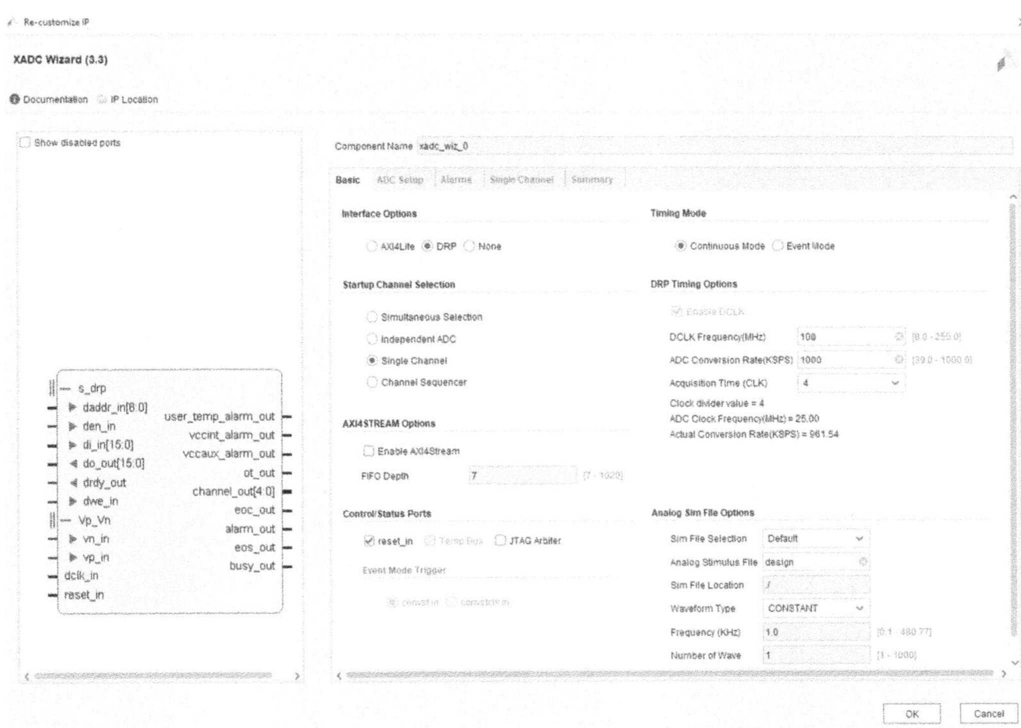

Figure 11.3 The Basic tab of XADC Wizard.

Setup. For the ADC Setup, we're going to leave off the ADC offset, the ADC offset and gain, sensor offset, and sensor offset and gain calibrations.

So, you can have these where, if you're wanting a really precise, high-quality signal that you're trying to make super accurate, you can use these calibrations, which gives you the offsets and gains. But we're not going to worry about that for this section. Since we're doing just a basic temperature sensor, we don't need to worry about all these types of precision. For external multiplex setup, we're going to check the external multiplexer, and we want to use VAUXP4 VAUXN4. We're going to select channel 4 for our auxiliary analog input, and we'll show you, once we go through here, why we're choosing 4. The reason we're choosing 4 is because it ties to analog zero, or the A0 pin. If you look on the actual development board where they have the ChipKit or the Arduino headers, there's an A0 pin, which is tied to VAUXP4. We'll show you that as we go through, as shown in Figure 11.4.

For the alarms, we're going to uncheck all of them, and we're not going to have it give us any type of alarms. You can set these so that if your board reaches a certain temperature – because inside the actual chip, there are little temperature diodes or transistor demonstrators in there – they will give an indication. When it reaches a certain voltage, it indicates by a certain temperature, so you can have your board shut down. If it's in a situation where it's overheating, it can trigger an alarm to shut the system down or turn a fan on or take whatever action is needed to cool the board down, as depicted in Figure 11.5.

Then, we have our Single Channel tab. We're selecting VAUXP4 VAUXN4, which enables the specific channel we want. Options Bipolar and Acquisition Time are available, but we're going to leave those unchecked, as illustrated in Figure 11.6.

Then, we go to the Summary tab, which shows us a quick summary of everything. In summary, we have the DRP interface selected, the XADC operating mode set to single_channel, and the AXI4Stream Interface set to false since we're not using it. Our Timing Mode is Continuous, the

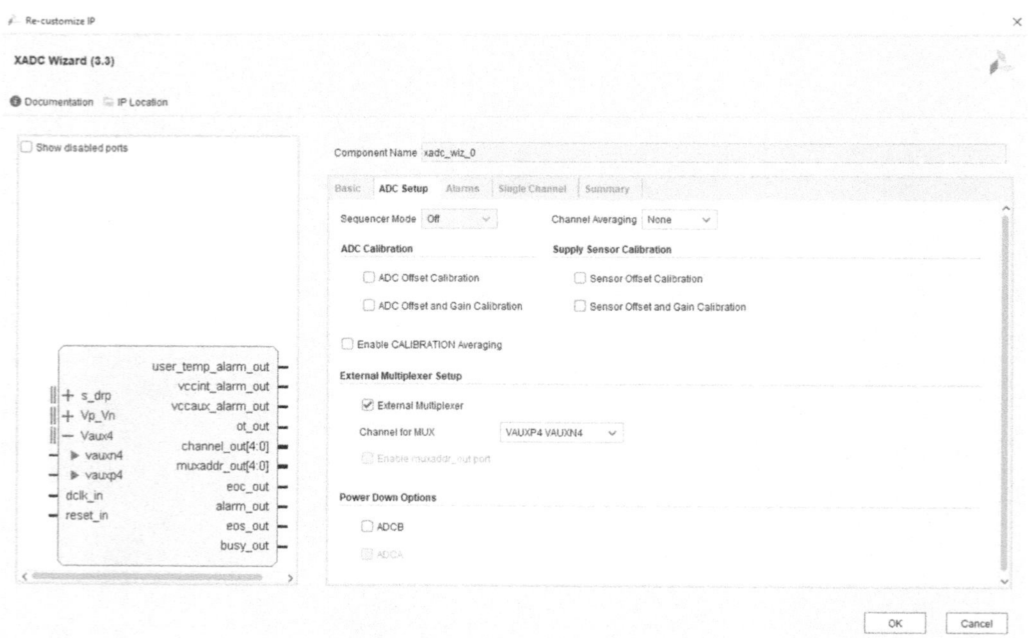

Figure 11.4 The ADC Setup tab of XADC Wizard.

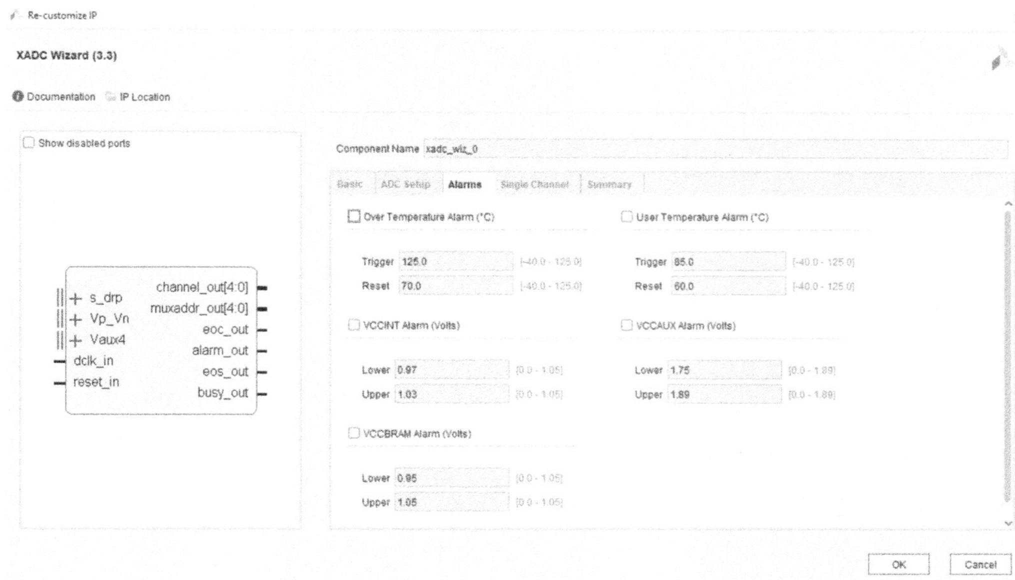

Figure 11.5 The Alarms tab of XADC Wizard.

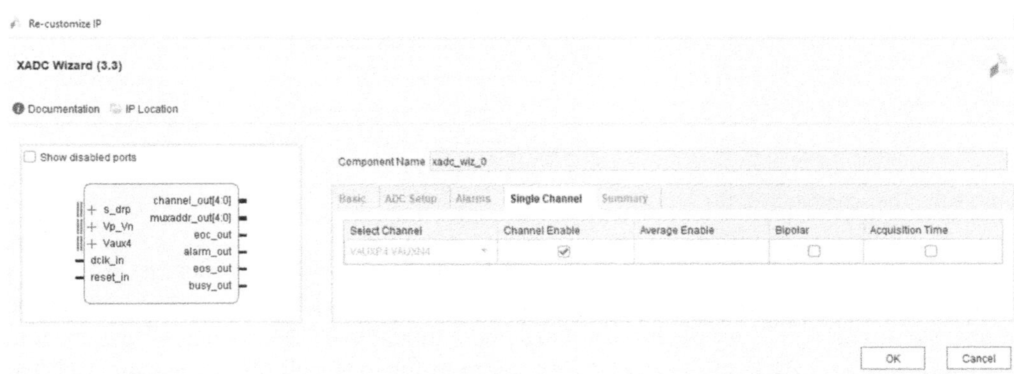

Figure 11.6 The Single Channel tab of XADC Wizard.

clock frequency is 100 MHz, the Sequencer Mode is off, and Channel Averaging is set to None. We also have Enable External Mux set to true, as shown in Figure 11.7.

So, then you go ahead and click OK, and once you do, you should see that even though we're only using several of these outputs from the XADC, such as daddr_in, den_in, and do_out, they're sufficient for our configuration. For the ADC address, we're having this set to a constant. So, if we double-click and pull it up, we're just placing a constant, so if you search for one of those in the IP core, you can just drop that in. We want to set the width to 7 and the constant value to 20. The reason we want to set that to 20 is because if we look on the schematic for the Arty A7, we can see that CK_AN0_P and CK_AN0_N are tied to the single-ended analog input signal A0. Looking further into the IO name on BANK 35, you'll find IO AD4P and AD4N. Here, P and N stand for positive and negative, and AD refers to analog to digital value. This indicates that the signal can be used for analog input and it's tied to the XADC core, as depicted in Figure 11.8. So, now that we know we want channel 4 (AD4, which is tied to AN0), how do we know what that actual address is?

Figure 11.7 The Summary tab of XADC Wizard.

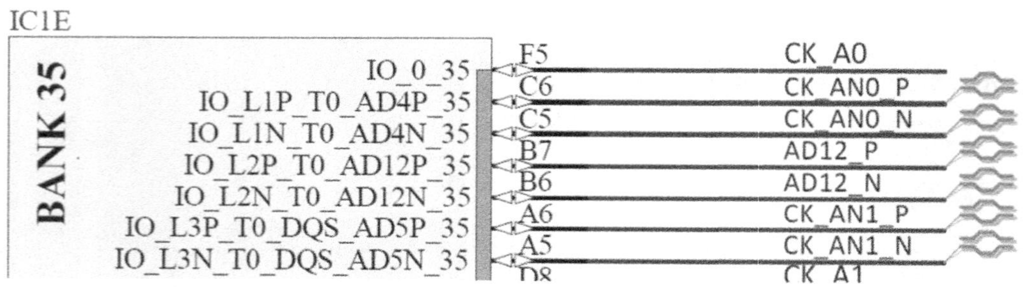

Figure 11.8 The schematic for the Arty A7.

If we look at the user guide for XADC (UG480), it tells us the actual address locations for the column status registers. But for this DRP interface, it tells us that DADDR[6:0] from 00h to 3Fh contains the read-only status registers. If we go through these read-only status registers, when we get down to the VAUXP[15:0]/VAUXN[15:0] addresses, they range from 10h to 1Fh. So, the address for VAUXP4/VAUXN4 will be 14h, which, when we convert that to decimal, gives us the constant value of 20. Therefore, we're going to set it to 20, and that's how we're selecting the actual signal that will appear on our output, do_out[15:0]. This output is a digital signal that will then be fed into our MicroBlaze MCS processor. We have den_in, which enables our dynamic reconfigurable port (DRP). We'll tie that to a signal, which then gets tied to a slide switch to enable or disable our XADC. The vauxn4 and vauxp4 are our differential analog signals, which we will tie to specific IO, and these correlate to the actual single-ended analog input we have on the A0 pin for the Arduino header. Then, we have our clock and reset, and those are all the different IO we're using for this XADC Wizard. There are several ports that are not being used, which is fine – we don't have to use all of them. Even though we're not using everything on this XADC Wizard, when you go ahead and synthesize, it might give you some warnings. However, you can disregard them as it should optimize some of them away, but a lot of them will still be present, but that's fine. Even though we don't have them connected, we can leave them disconnected, and our project will still work and be able to configure just fine, as illustrated in Figure 11.9. So, that's essentially how you use the Xilinx XADC Wizard within Vivado to start reading the analog signal and its output from the temperature sensor.

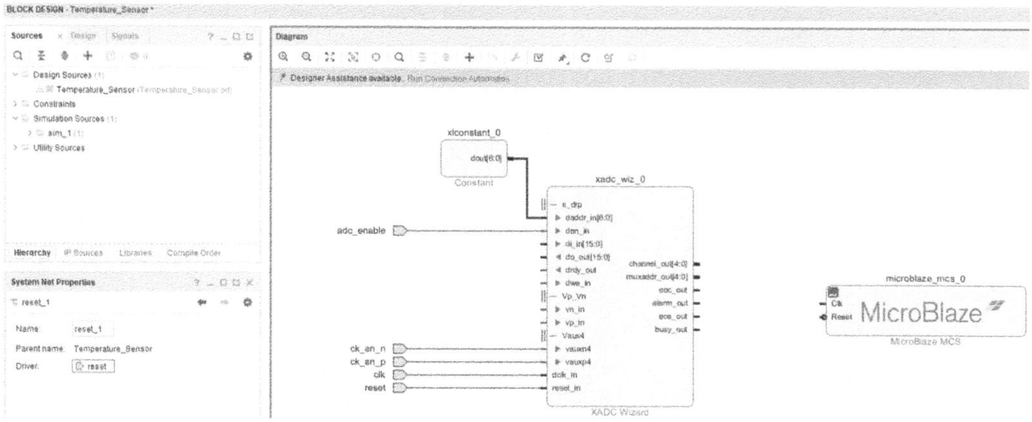

Figure 11.9 The XADC Wizard for Temperature_Sensor block diagram.

11.3 The MicroBlaze MCS ADC Configuration

This section is about the MicroBlaze MCS ADC configuration. In this section, we're going to show you how to take the MicroBlaze MCS core and configure it so that we can do analog to digital conversion readings. Now, the MicroBlaze MCS core from Xilinx doesn't have an analog to digital conversion built in. However, it's on the same chip as the XADC. First, we're going to see how to set up the XADC, so the XADC has been placed and configured. Now, we want to read that data from a parallel port. We've got these general-purpose inputs and outputs (GPIOs). We'll take a 16-bit general-purpose input and use it to read the data out coming from the actual specific analog channel. So, we'll go ahead and place the MicroBlaze MCS on the block diagram, and then configure it as follows. For memory size, we're using 64 kB – you can increase this if needed. If you go beyond 64 kB, we suggest you go to 128 kB; however, that depends on how large you want to go. For MicroBlaze optimization, we're going to optimize it for area since we're only working with temperatures, which don't change super often, so we don't need any sort of super high performance. We're just realistic, and all we want to do is to update our temperature once every roughly half a second or every second, every so many seconds. Because the temperature is not going to change that fast anyway, we can optimize it for area just to save on resources. We'll leave both of the buses unchecked, and for debug, we'll select NONE, as shown in Figure 11.10.

For UART, we want to enable both the receiver and the transmitter. We don't need the receiver right now, but we're going to enable it anyway – just in case you want to add something later to the project that involves sending data. But right now, we're just going to enable the transmitter, and we're going to pull both of them, since we're going to be transmitting the temperature data that we read from our board to the Tera Term terminal emulator, as depicted in Figure 11.11.

For our FIT or fixed interval timer, we're not going to be using any of those, so FIT 1 through FIT 4 can all be left unchecked. The same goes for programmable interval timers, PIT 1, PIT 2, PIT 3, and PIT 4; we're not going to be using those either, so leave them unchecked. For the GPOs, we're not going to be having any general-purpose outputs since we're only reading data. The only output we're using is the UART output, so we'll leave all the GPOs unchecked. For general purpose inputs (GPIs), we want GPI 1 to be our analog to digital conversion value. We're going to take that ADC output value from the XADC and read it into GPI 1. For GPI 2, we'll leave

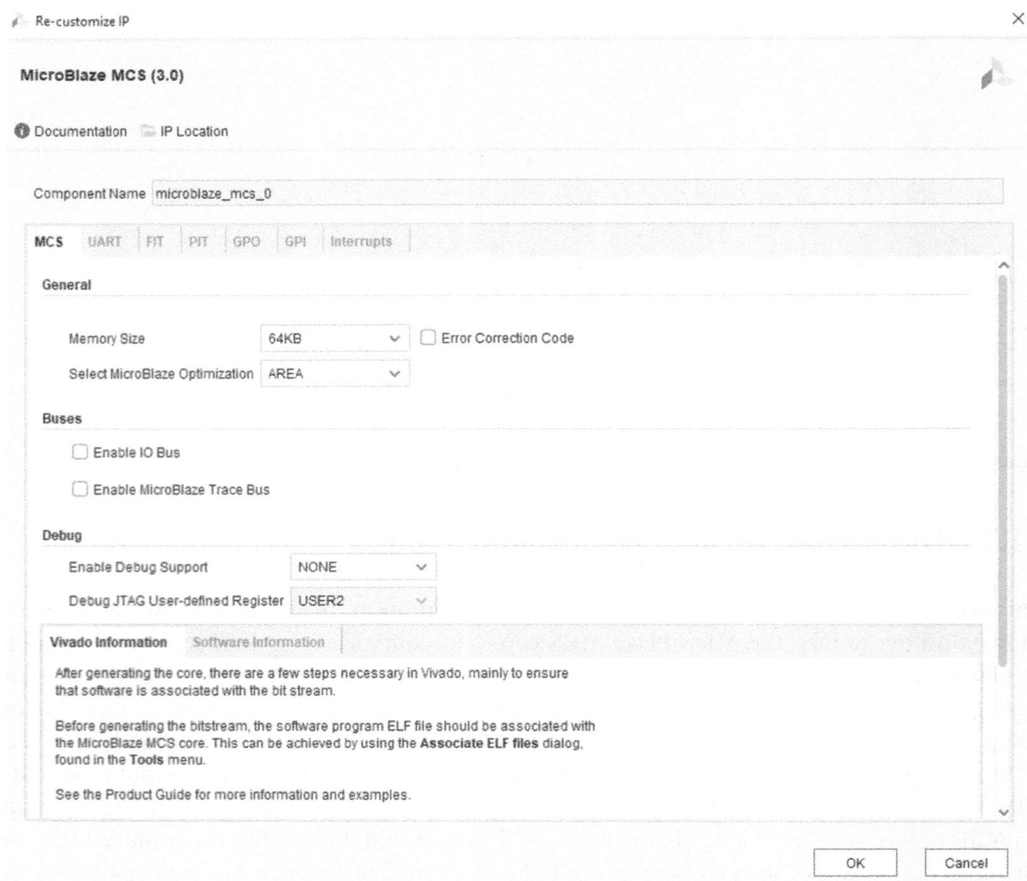

Figure 11.10 The MCS tab of MicroBlaze MCS.

that as a one bit, which will be tied to a slide switch that determines whether we're interpreting the temperature in Fahrenheit or Celsius (1 for Fahrenheit, and 0 for Celsius). We're going to calculate both, but we'll see that one amount, as illustrated in Figure 11.12.

Then, for our interrupts, we're going to leave that blank, and we're not going to be using any of the interrupts, so we'll leave that unchecked at that point. You can go ahead and click OK to configure your MicroBlaze MCS so that it can interface with the XADC core, as shown in Figure 11.13.

11.4 Interfacing with the XADC Core

This section is about interfacing with the XADC core. Here, we're going to show you how to write the software to interface with the XADC IP core. What we do is take the Temperature_Sensor.bd, then generate a wrapper, and let Vivado manage the wrapper and auto-update it, as illustrated in Figure 11.14.

Then, we run synthesis, run implementation, and generate bitstream, respectively. If we encounter errors during bitstream generation, we ignore them and click the file. Next, we want to export the hardware as shown in Figure 11.15.

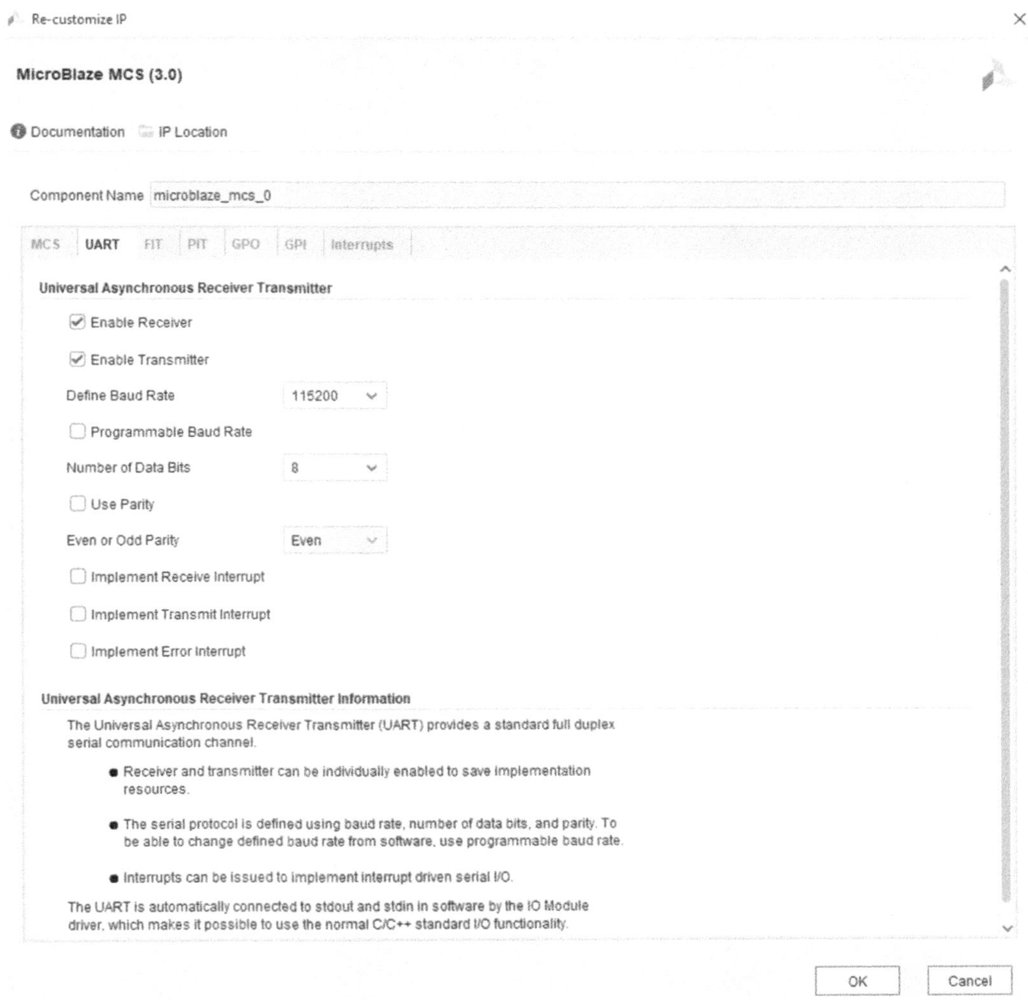

Figure 11.11 The UART tab of MicroBlaze MCS.

This will bring up a little dialogue box that says "Export the hardware for software development tools." We're going to leave "Include bitstream" checked off, since we have not generated a bitstream file, so we cannot include it here, and that's OK. Then, we want to open the Xilinx SDK by going to File and selecting Launch SDK, as depicted in Figure 11.15. Once you open up the Xilinx SDK, go to File > New > Application Project. From here, we'll give our product a name – we'll call it Temp_Sensor for temperature sensor – and keep the "Use default location" option checked. For the OS platform, we'll select standalone. Our target hardware is defined by the Temperature_Sensor_wrapper_hw_platform_0, and the processor is the MicroBlaze MCS zero, which is the MicroBlaze MCS processor we placed in our Vivado project. For Language, we'll use C, and for the Board Support Package, we will use Create New with Temp_Sensor_bsp, as illustrated in Figure 11.16.

Then, we'll click Next, and for our application template, we are going to click on the Peripheral Tests, and then click Finish, as shown in Figure 11.17.

This is going to generate a project for us. Now, if we expand our Temp_Sensor and navigate to our source (src), we'll find testperiph.c. If we double-click it, it'll open up the testperiph.c,

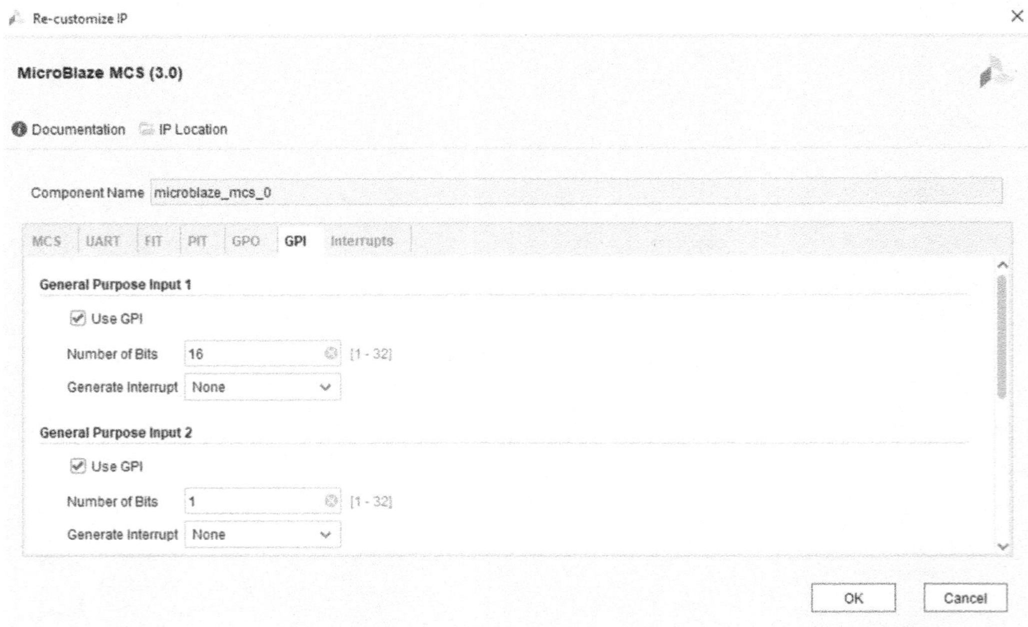

Figure 11.12 The GPI tab of MicroBlaze MCS.

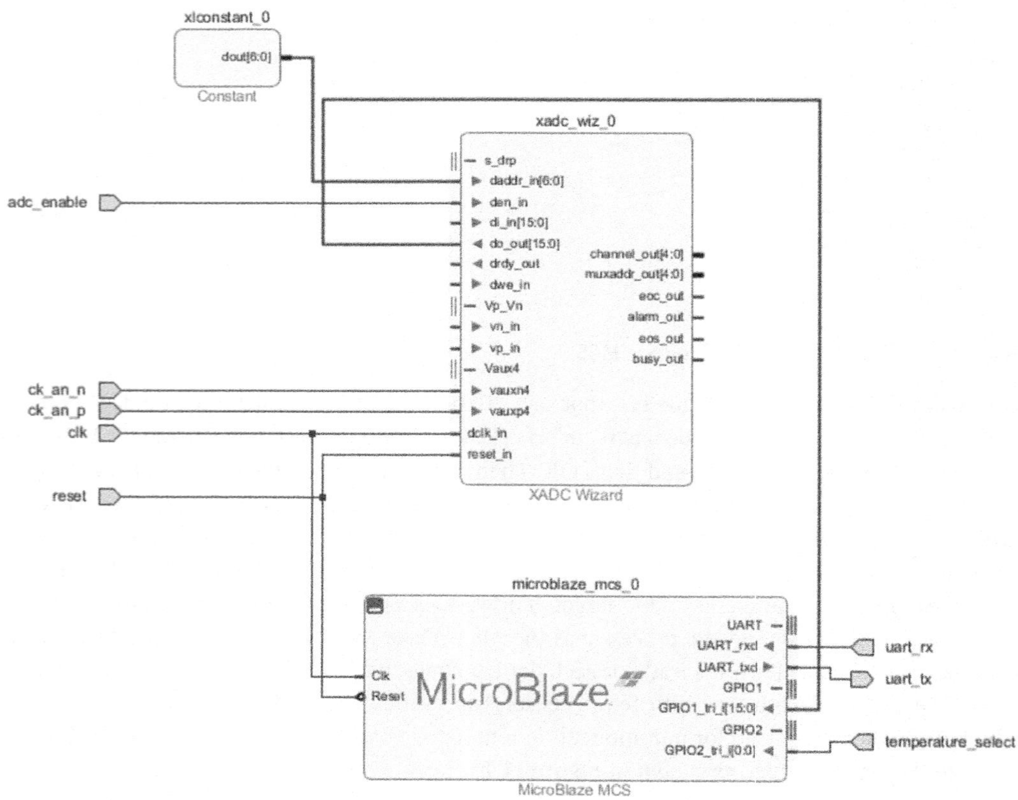

Figure 11.13 MicroBlaze MCS interface with the XADC core.

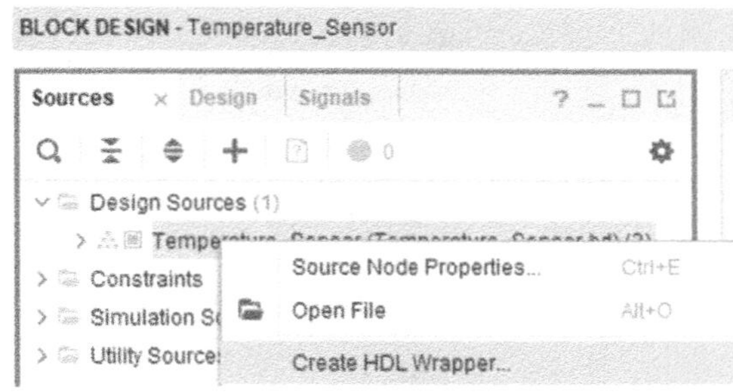

Figure 11.14 Creating the HDL wrapper.

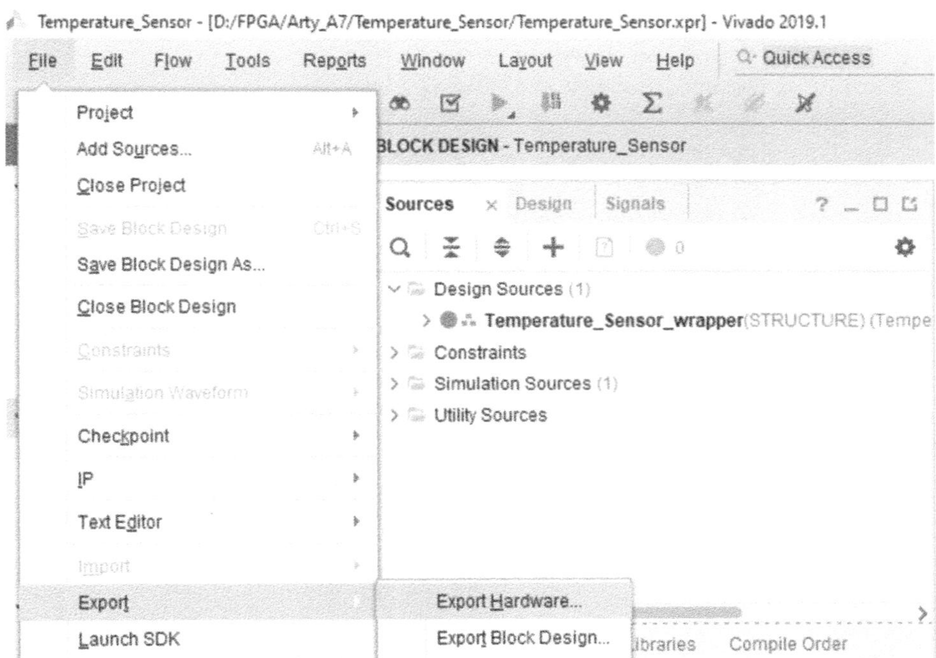

Figure 11.15 Exporting the hardware.

which contains a basic little program that's going to test the print and is testing UART. From this point, we now want to add to this so we can read our temperature, as well as process that information and perform the calculation that we had determined to figure out what the temperature is. So, from within here, we are going to have to add several more header files. We have our standard IO (stdio.h), X parameters (xparameters.h), Xilinx cache (xil_cache.h), XIO module (xiomodule.h), and IO module header (iomodule_header.h). Now we can go inside our main program. So inside our main program, we are going to define our parameters. These are just different values we will be using throughout the project parameters. To begin, we're going to have an unsigned 32-bit data type, and we'll call it data. We'll have an ADC_reading, and we will have Selected_Degree. The Selected_Degree is what we're going to use to determine whether we display our value in Fahrenheit or Celsius. The ADC_reading is the initial analog to digital conversion, which we then need to convert to a voltage in millivolts, and the data is what we're going to use when we initialize and instantiate our general-purpose inputs and outputs so that we can read from and write to them. Next, we want some floating-point or 32-bit floating

Figure 11.16 The Application Project settings.

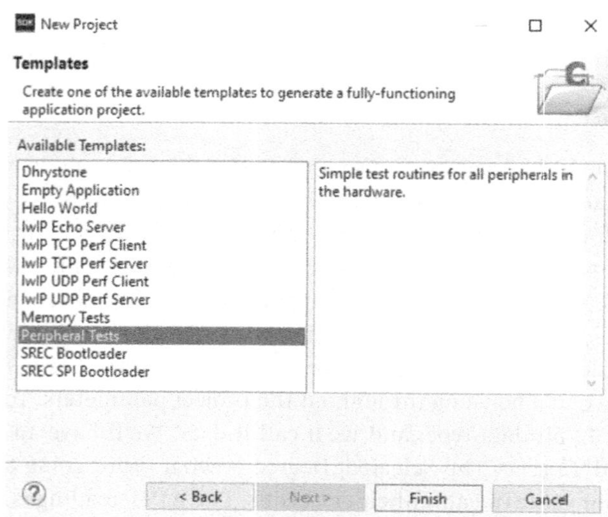

Figure 11.17 Selecting the Peripheral Tests for our application template.

point values. This will be our Voltage_mV. We want that to be a floating point. We'll also have our temperature in degrees Celsius (Temperature_C) and degrees Fahrenheit (Temperature_F). We also want to have another floating-point value, which will be Temperature_out, and this is going to be the value that we display on our output. Then finally, we want to have some integers, so we're going to have a delay_count. That's because we're going to have a delay from when we process. We don't want to keep printing this temperature value so fast on our terminal that we don't have any time to read it. We want to have it slow enough that we can see it as it is printed out, as well as Temperature_whole, which is going to be the whole digit number. So, we're going to be using the Xilinx printf command, which will allow us to print decimals, but we cannot print a floating-point number. So, we're going to have to take the floating-point value of our Temperature_out and convert it to a whole as well as a decimal portion. We'll call it Temperature_hundreds, and that's just because we're going to show two decimal places, temperature hundreds. Now, we want to go ahead and have our XIO module, so we'll call it XIO module, and this is just like we did in the UART_IO project. So, we're following the same scheme, except we only have a general-purpose input, and we don't have any general-purpose outputs. Now, we're going to initialize our GPIOs, and we'll do that by using our data, which is equal to XIOModule_initialize(&gpi, XPAR_IOMODULE_0_DEVICE_ID). So here we're initializing our general-purpose input. Now, we want to go ahead and start that, so we will do that with the XIOModule_Start(&gpi) command put right underneath there. So that's going to initialize all of our general-purpose inputs – in this case, just our inputs. We are going to comment out all of the remaining codes. So, we'll just comment all of that out – if we want to use it later, we can. But right now, we're not worried about it, and we don't need to return anything because our call to main doesn't have anything, and we're never going to get there. So to have a return doesn't really matter. We're going to get into a while loop, where we never have to return. Now, we want to enter our infinite while loop. In this while loop, we are going to be doing various things: perform the ADC reading, calculate the temperature values, and determine whether to display it in Fahrenheit or Celsius. The first thing is to read the ADC value by taking the XIO discrete read and passing it to the handle of our general-purpose input, which we've already defined in the following code.

```
ADC_reading = XIOModule_DiscreteRead(&gpi, 1);
```

So that's the actual raw analog to digital conversion reading, and it's going to be the 16-bit value that we're reading from our XADC module. Now we want to shift out the four least significant bits (LSBs), and that's because our ADC rating is 12-bit wide. However, they give you 16 bits, and they tell you inside the XADC user guide to go ahead and shift out the four LSBs. So, we're going to shift those out, and let's make a comment so we know what we're doing. The next thing we want to do at this point is have our actual ADC reading at this stage right here, and that is the actual raw ADC reading that we have and will work with. So, the next thing we want to do is use our slide switch to determine if we're displaying our results in Fahrenheit or Celsius. That takes place using this variable to hold the selected degree, and we're making that equal to our XIO module discrete read on channel 2, because that's the slide switch that our general-purpose input is connected to, as shown in the following code.

```
Selected_Degree = XIOModule_DiscreteRead(&gpi, 2);
```

Then, right below that, we want to calculate our voltage in millivolts, and this is a floating-point value. So, in order to calculate this, we use the following code.

```
Voltage_mV = (float)(3.3/4095) * (float)ADC_reading * 1000;
```

We're going to multiply that value by a thousand, and that is so we put it in mV. If we don't multiply it by a thousand, the value would remain in volts. But since we want the millivolt reading, we're going to multiply by a thousand. So, this whole equation is used to determine the voltage in millivolts. Now, we want to determine the temperature in degrees Celsius. We have to calculate this every time, even if the user has selected to display the temperature in Fahrenheit, because our formula to determine temperature is based on Celsius. Once we've calculated the temperature in Celsius, if the user has selected Fahrenheit, we'll go ahead and perform the conversion. If not, we'll just use the Celsius temperature directly. This logic is written as the following code.

```
Temperature_C = (Voltage_mV - 500) / 10;
```

So now, at this point, we're going to set Temperature_out because Temperature_out is the value that we're going to display via the UART terminal. So, Temperature_out is initially equal to Temperature_C. Now, what we need to do is check and determine what our Selected_Degree is. Selected_Degree, which we've already read in, tells us whether we're reading in Fahrenheit or Celsius, so we need to use an if statement as shown in the following codes.

```
if(Selected_Degree == 1)
{
    Temperature_F = (Temperature_C * 1.8) + 32;
    Temperature_out = Temperature_F;
}
```

So, if the selected degree is 1 – meaning your slide switch is set to high, which means we want to display the output temperature in Fahrenheit – then our Temperature_out is now equal to the temperature in Fahrenheit. If not, the Temperature_out stays at the value in degrees Celsius. So, at this point, we've calculated the temperature in both degrees Celsius and degrees Fahrenheit, based on whether or not the user selects to display one or the other. Now, the final thing we need to do is write the output temperature so that the user can actually see what the temperature value is. If we don't display it anywhere, we won't know the temperature is changing, and we can't verify that our design is working correctly. In order to display it, we can only print out whole decimal values. But we have a floating-point value, and we want to show the accuracy – specifically the changes in the hundredths position. So, to do that, we'll say Temperature_whole is equal to Temperature_out, and Temperature_out is a floating-point value, and Temperature_whole is an integer value. This will take everything to the left of the decimal point and will be represented in Temperature_whole. So, let's say our temperature is 46.87 degrees. At this point, temperature is going to be equal to 46. Now, what we need to do is get that 87 to display. To do that, we want it to be stored in Temperature_hundreds, and that Temperature_hundreds is equal to Temperature_out, which is 46.87, and we're going to subtract Temperature_whole, which is 46. That gives us 0.87. We then multiply that by 100 to get 87. So now, Temperature_whole is equal to 46 and Temperature_hundreds is equal to 87. When we display this, we need to make sure that we put a decimal point between them to display 46.87. Finally, we need to determine whether we're in Fahrenheit or Celsius because based on the selection, we want to print either a C or an F to let the user know that it's degrees Fahrenheit or degrees Celsius. So, we'll just use an if-else statement in the following codes.

```
if(Selected_Degree == 0)
   xil_printf("Temperature: %d.%d C\n\r", Temperature_whole, Temperature_
hundreds);
```

```
else
   xil_printf("Temperature: %d.%d F\n\r", Temperature_whole, Temperature_
hundreds);
```

After this, we want to create a delay, and the reason we want to create a delay is that this is going to happen so fast that if we don't have a delay, your screen will fill up so quickly that you won't even be able to read or see what the temperature is. So, as the temperature changes, you won't be able to see the rising or the lowering of the temperature. So, to create a delay, we're just going to use a simple count. It's a really simple and basic way to create a delay in an embedded type of system.

```
delay_count = 0;
while(delay_count < 50000000)
     delay_count++;
```

So, we're going to keep counting, so something is going to count to a high number. In a way, you can change that number – the delay will change accordingly. If you make it a higher number, you'll get a longer delay. If you make it a lower number, you'll have a shorter delay. While it's less than that number, we are going to increment our delay count, and then we're going to continue in the outer while loop. So, our while loop will continue on, and we have this while(1) loop. So, while it is running, we'll never actually get to our main, and actually, in our main, we are going to have our return 0, and we'll uncomment it. We'll go ahead and save the code, and the complete code for this project is shown as follows.

```
#include <stdio.h>
#include "xparameters.h"
#include "xil_cache.h"
#include "xiomodule.h"
#include "iomodule_header.h"
int main ()
{
   //Parameters
   u32 data, ADC_reading, Selected_Degree;
   float Voltage_mV, Temperature_C, Temperature_F;
   float Temperature_out;
   int delay_count, Temperature_whole, Temperature_hundreds;
   XIOModule gpi;
   //Initialize our GPIOs
   data = XIOModule_Initialize(&gpi, XPAR_IOMODULE_0_DEVICE_ID);
   data = XIOModule_Start(&gpi);
   while (1)
   {
          ADC_reading = XIOModule_DiscreteRead(&gpi, 1); //Perform ADC Reading
          ADC_reading =  ADC_reading >> 4;  //Shift right 4 LSBs
          Selected_Degree = XIOModule_DiscreteRead(&gpi, 2); //Read the slide
switch to determine C or F
          Voltage_mV = (float)(3.3/4095) * (float)ADC_reading * 1000; //Voltage
reading in mV
          Temperature_C = (Voltage_mV - 500) / 10; //Calculate temperature in C
          Temperature_out = Temperature_C;
          if(Selected_Degree == 1)
          {
                 Temperature_F = (Temperature_C * 1.8) + 32; //Calculate
temperature in F
```

```
                    Temperature_out = Temperature_F;
            }
            //Write the output temperature
            Temperature_whole = Temperature_out;
            Temperature_hundreds = (Temperature_out - Temperature_whole) * 100;
            if(Selected_Degree == 0)
                  xil_printf("Temperature: %d.%d C\n\r", Temperature_whole,
Temperature_hundreds);
            else
                  xil_printf("Temperature: %d.%d F\n\r", Temperature_whole,
Temperature_hundreds);
         //delay
            delay_count = 0;
            while(delay_count < 50000000)
                 delay_count++;
    }

         /*
    Xil_ICacheEnable();
    Xil_DCacheEnable();
    print("---Entering main---\n\r");
    {
       XStatus status;
       print("\r\nRunning IOModuleSelfTestExample() for microblaze_mcs_0_
iomodule_0...\r\n");

       status = IOModuleSelfTestExample(XPAR_MICROBLAZE_MCS_0_IOMODULE_0_
DEVICE_ID);

       if (status == 0) {
          print("IOModuleSelfTestExample PASSED\r\n");
       }
       else {
          print("IOModuleSelfTestExample FAILED\r\n");
       }
    }
    print("---Exiting main---\n\r");
    Xil_DCacheDisable();
    Xil_ICacheDisable();
    */
    return 0;
}
```

Then, we go ahead and build, and it gives us this information right here on our output terminal. We have that Temp_Sensor.elf file, which we'll then want to associate in Vivado, as depicted in Figure 11.18.

We'll want to take and associate the .elf file with our actual MicroBlaze MCS softcore processor, so that we can reprogram the device with the software already embedded in the MicroBlaze MCS. This way, we don't have to program the board twice, but it gives you the size of that .elf file in decimal, as illustrated in Figure 11.19.

You can see we're using approximately 26 kB, and we specified about 64 kB, so we're using less than half of what we specified. That means we have more than enough memory to work with. This wraps up everything needed to interface with the XADC core in Vivado from the software side.

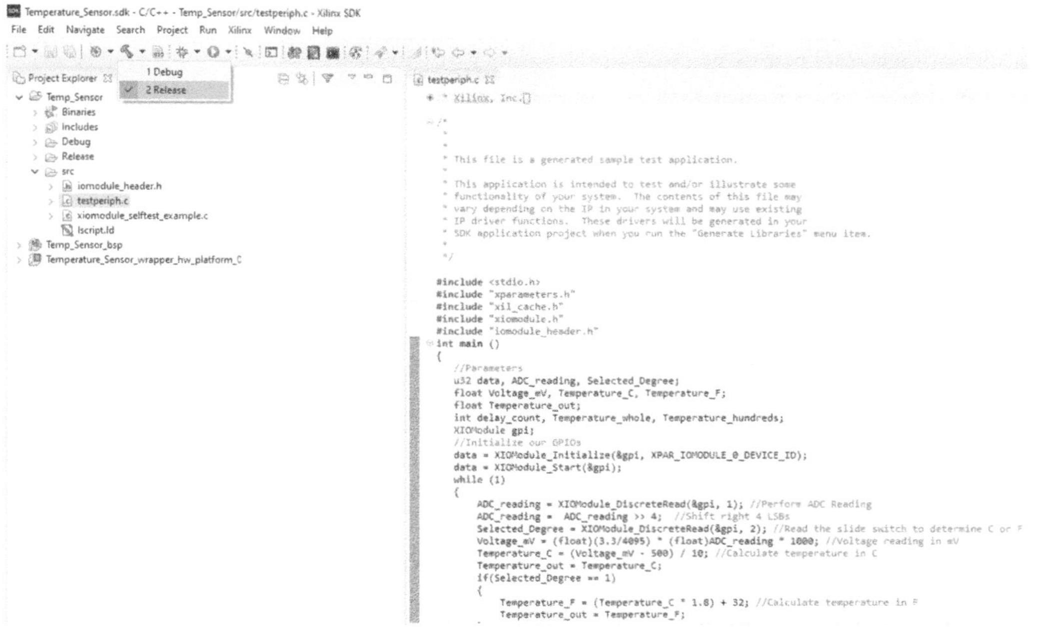

Figure 11.18 Building the Temp_Sensor project.

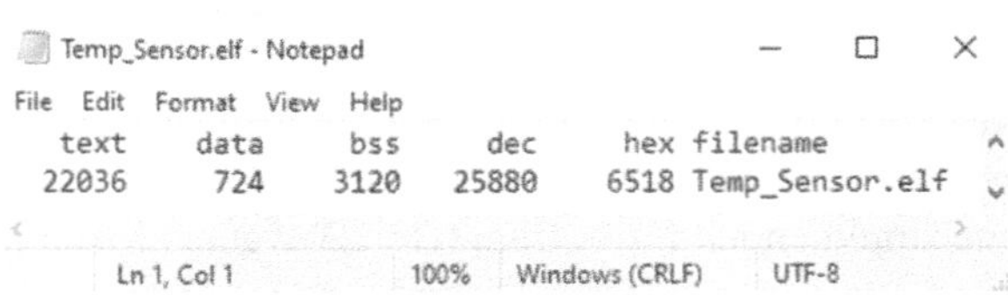

Figure 11.19 The content of Temp_Sensor.elf.size file.

11.5 The ADC Processing Arty A7 IO Placement

This section is about ADC processing Arty A7 IO placement. Here, we'll walk through all the different IO placements for the temperature sensor project. For this project, we have various inputs and outputs. Our inputs include adc_enable, reset, and temperature_select, which are tied to slide switches. If we go to our Arty_A7_Master.xdc file and look at the top, we can see the slide switches listed first. So, our switch zero, which we'll tie to adc_enable, is going to enable or disable the analog to digital conversion. Then, we have reset, which is tied to switch two, and that's going to determine whether our MicroBlaze and analog to digital conversion are held in a reset state. Our temperature_select is tied to switch three, and this will determine whether we're going to output the temperature in Celsius or Fahrenheit. For our clock (clk), which goes into our XADC and our MicroBlaze MCS, we're going to tie that to package pin E3, which is connected to an external oscillator on the actual development board. That's just a 100 MHz clock that we're bringing into the FPGA. We use the next line to create a clock and we're just telling it the period and the waveform so that when it does an analysis,

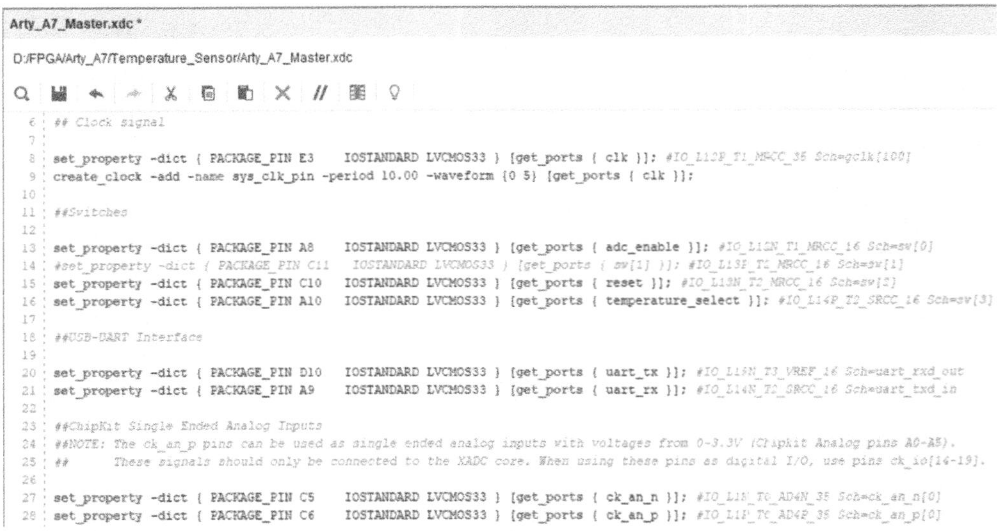

Arty_A7_Master.xdc *

D:/FPGA/Arty_A7/Temperature_Sensor/Arty_A7_Master.xdc

```
## Clock signal

set_property -dict { PACKAGE_PIN E3    IOSTANDARD LVCMOS33 } [get_ports { clk }]; #IO_L12P_T1_MRCC_35 Sch=gclk[100]
create_clock -add -name sys_clk_pin -period 10.00 -waveform {0 5} [get_ports { clk }];

##Switches

set_property -dict { PACKAGE_PIN A8    IOSTANDARD LVCMOS33 } [get_ports { adc_enable }]; #IO_L12N_T1_MRCC_16 Sch=sw[0]
#set_property -dict { PACKAGE_PIN C11   IOSTANDARD LVCMOS33 } [get_ports { sw[1] }]; #IO_L13P_T2_MRCC_16 Sch=sw[1]
set_property -dict { PACKAGE_PIN C10   IOSTANDARD LVCMOS33 } [get_ports { reset }]; #IO_L13N_T2_MRCC_16 Sch=sw[2]
set_property -dict { PACKAGE_PIN A10   IOSTANDARD LVCMOS33 } [get_ports { temperature_select }]; #IO_L14P_T2_SRCC_16 Sch=sw[3]

##USB-UART Interface

set_property -dict { PACKAGE_PIN D10   IOSTANDARD LVCMOS33 } [get_ports { uart_tx }]; #IO_L19N_T3_VREF_16 Sch=uart_rxd_out
set_property -dict { PACKAGE_PIN A9    IOSTANDARD LVCMOS33 } [get_ports { uart_rx }]; #IO_L14N_T2_SRCC_16 Sch=uart_txd_in

##ChipKit Single Ended Analog Inputs
##NOTE: The ck_an_p pins can be used as single ended analog inputs with voltages from 0-3.3V (Chipkit Analog pins A0-A5).
##      These signals should only be connected to the XADC core. When using these pins as digital I/O, use pins ck_io[14-19].

set_property -dict { PACKAGE_PIN C5    IOSTANDARD LVCMOS33 } [get_ports { ck_an_n }]; #IO_L1N_T0_AD4N_35 Sch=ck_an_n[0]
set_property -dict { PACKAGE_PIN C6    IOSTANDARD LVCMOS33 } [get_ports { ck_an_p }]; #IO_L1P_T0_AD4P_35 Sch=ck_an_p[0]
```

Figure 11.20 The constraints file for Arty A7 Temp_Sensor project.

it can verify that everything connected to this clock can run at 100 MHz. We also have ck_an_n and ck_an_p, and this is the differential analog signal that gets fed into our XADC core to perform the analog to digital conversion. If we scroll down to the bottom of the Arty_A7_Master.xdc file, we'll find a section titled ChipKit Single Ended Analog Inputs, and then it goes on. The ck_an_p pins can be used as single-ended analog inputs with voltages from 0 to 3.3 V (Chipkit Analog pins A0–A5). These are six different analog channels, and we're using A0. So, if you're working with the ChipKit or Arduino, this would be your analog A0 input is, which is what this would be tied to. We can bring a signal from 0 to 3.3 V in. It takes that signal and turns it into a differential signal. So, one single-ended input from 0 to 3.3 V gets broken into two actual pins on the FPGA cell. By using analog input A0, we're going to require two inputs: ck_an_n and ck_an_p. If we look, those are tied directly to vauxn4 and vauxp4. On XADC, that's just one of the channels for analog to digital conversion. So, we'll go ahead and tie those there. Next, we have our UART–USB interface, which we'll use to transmit the converted temperature values to our computer screen. That's just a standard UART interface – uart_tx and uart_rx – and we'll place those in the UART–USB interface on pins D10 and A9. Once we do that, that is all we have for the temperature sensor project. So, we're connecting ck_an_n and ck_an_p, which go out to A0 on the ChipKit header. This will connect and bring in the output signal from our temperature sensor, which is fed to the XADC. Then, it performs the conversion. The MicroBlaze reads that conversion, determines the voltage value, converts it to a temperature, and then reports that temperature on the serial port terminal or UART terminal. That is the ADC processing Arty A7 IO placement for Temp_Sensor project, as shown in Figure 11.20.

12

The SPI Interface Project

12.1 Introduction

This chapter is about using the SPI interface with the external flash memory. First of all, we're going to explain and briefly talk about the external flash memory chip that is used on the Arty A7 development board. For the Arty A7 development board, there is an external QSPI flash chip located on the board, and typically what this is used for is to store the configuration of the FPGA. When you power up the board, if you think of an FPGA, a lot of times in our labs, we have to program it using the JTAG cable. Every time we power it off and then power it back on, we have to reprogram it. So, what these QSPI external flash memory chips are commonly used for is to hold that configuration. That way, as soon as the board powers up, it reads the configuration from that external flash memory and configures the FPGA automatically. Out in the field, you wouldn't want to reprogram the board every time it is powered off – that would not be feasible. So, that's typically what these QSPI chips are used for. The chip used on the Arty A7 board is a 16 MB flash chip, and the Arty A7 configuration file takes approximately 2 MB. That leaves about 14 MB free to work with. Now, we are only going to be using a few bytes since we're just going to store a name and some additional information. But you could store up to 14 MB of data, so there's a lot of information you could store on there. The whole idea of this lab in this project is to help you understand how to use the SPI interface to communicate with a chip external to the FPGA. In this case, it's a memory chip manufactured by Micron, which is an IC manufacturer that commonly makes memory devices. This is a nonvolatile external flash memory chip. They also manufacture RAM, SD-RAM, and various other types of integrated circuits you can work with. Don't worry about overwriting anything in the QSPI flash during this project, because when we are working with the boards and all the projects we're doing, we're just loading them directly into the FPGA from the actual JTAG interface. So even if we write over something in the QSPI flash, we can always rewrite it later. If you visit the digital website, they provide you with the initial file that was preprogrammed on the Arty A7, so we can always reprogram or restore the original contents. Whatever you write to the QSPI flash, don't worry – we can always erase or replace it. Now, we're going to take a look at the Micron datasheet for the actual flash part that we'll be interfacing with. On the first page, it gives us a quick summary of the device. It's a Micron serial NOR flash memory – NOR refers to the type of architecture it uses. They can have NOR or NAND flash, and that's just based on the gates or the architecture of how they construct the actual memory. It's a 128 Mb device, which equals 16 MB (128 Mb/8 = 16 MB). It supports 4 kB sector erase, which means you can erase a whole 4 kB block at once, instead of erasing byte by byte, which would be much slower.

Fundamentals of VHDL for FPGA Programming Using Vivado, First Edition. Majid Pakdel.

Companion Website: www.wiley.com/go/pakdelVHDL

So, it tells us that's an SPI-compatible serial bus interface, which is the interface we're going to use to communicate with this flash memory chip.

Also, it gives us a max clock frequency of 108 MHz, so this thing can run very fast. If you think about it, the Arty A7 development board has a 100 MHz clock coming in. So, no matter how fast we want to run this SPI device, we can run it super fast – it can go up to 108 MHz, and we're only coming in at 100 MHz. We're not going to use a Phase-Locked Loop (PLL) to increase the frequency on the actual Arty A7 device. So, any speed we want to run at will be well within specifications. For example, if we run it at 50 MHz, we'll be well within the specifications of 100 MHz it can handle. Also, it gives you additional details, such as the voltage range (2.7–3.6 V), which is important when you're working with different flash memories. The main thing we need to worry about is making sure we understand the actual SPI characteristics in terms of voltages and currents. The team that designed and built the board already went ahead and verified all of that to make sure that everything works just fine. We just want to focus on understanding how to communicate with the chip – what we send down and what we can expect back. So, let's go ahead and take a look at the description of the memory organization and the block diagram, as shown in Figure 12.1. The block diagram shows us the different signals: hold, write/Vpp, S, C, and DQ0, DQ1 to DQ2 and DQ3. These are used and fed into the I/O shift register. When you think about SPI, which we'll discuss later, it's actually a big shift register. You're sending data and controlling the clock that shifts data in and out. So, we're going to shift data into this I/O shift register, and based on the information we're shifting in, it's going to set the address register. There's also a 256-byte data buffer and a status register. We're going to tell it where we want to write data or read data from, and it's going to process that and send data back to us. This memory allows us to access and write single bytes at a time, or we can write in chunks. The next thing we will look at is the protocol, but then we'll see the commands, which shows us what we can actually write down and the data that we can read back. If we look at the supported protocols, they list extended, dual, and quad SPI protocols. This extended protocol is the one we're going to use. In this mode, we just have one data bit on every clock cycle. In dual, you use DQ1 and DQ0, and in quad, you send four data bits on every clock cycle using the DQ3 down to DQ0. For our purposes, we're just using DQ0 and they give you a quick description. So, this is the device default protocol from the factory and additional commands extend the SPI protocol. If we want to use these other protocols, we have to send down specific commands in this extended format so that we can use the dual or quad interface. But we are sticking with the extended protocol. So, when you look through the datasheet, if you want to understand how it works, just look at the extended protocol. Next, let's take a look at some of those supported flash commands. They list different operations that we can do. If you notice, in the left-hand column, they provide the command, the code, and indicate whether it is extended, dual I/O, or quad I/O, along with the data bytes. So, we are using the extended mode. If we look at the table, it shows whether it's supported or not. Some of these operations are only supported on certain interfaces. For example, this multiple I/O read ID is not supported in extended mode, but it's supported in dual or quad modes. For our purposes, we only need to perform read and write operations. The read operation is supported in the extended mode, so we can proceed with that. Similarly, write operations are also supported. For example, the write enable command is yes, and it supports that. So, everything that we need to do is supported in the extended mode. So, that code column shows the actual data that we're going to send. So, if we want to enable the reset, we need to send down the 66 h or this 8-bit value, so it's in a hexadecimal format. We're not going to go into detail on all these commands because there are quite a few of them here and their descriptions can be complex. But once we get into writing the actual software and designing our SPI core, we'll take a look at all these and make sure that we can support and send all these down correctly. One helpful

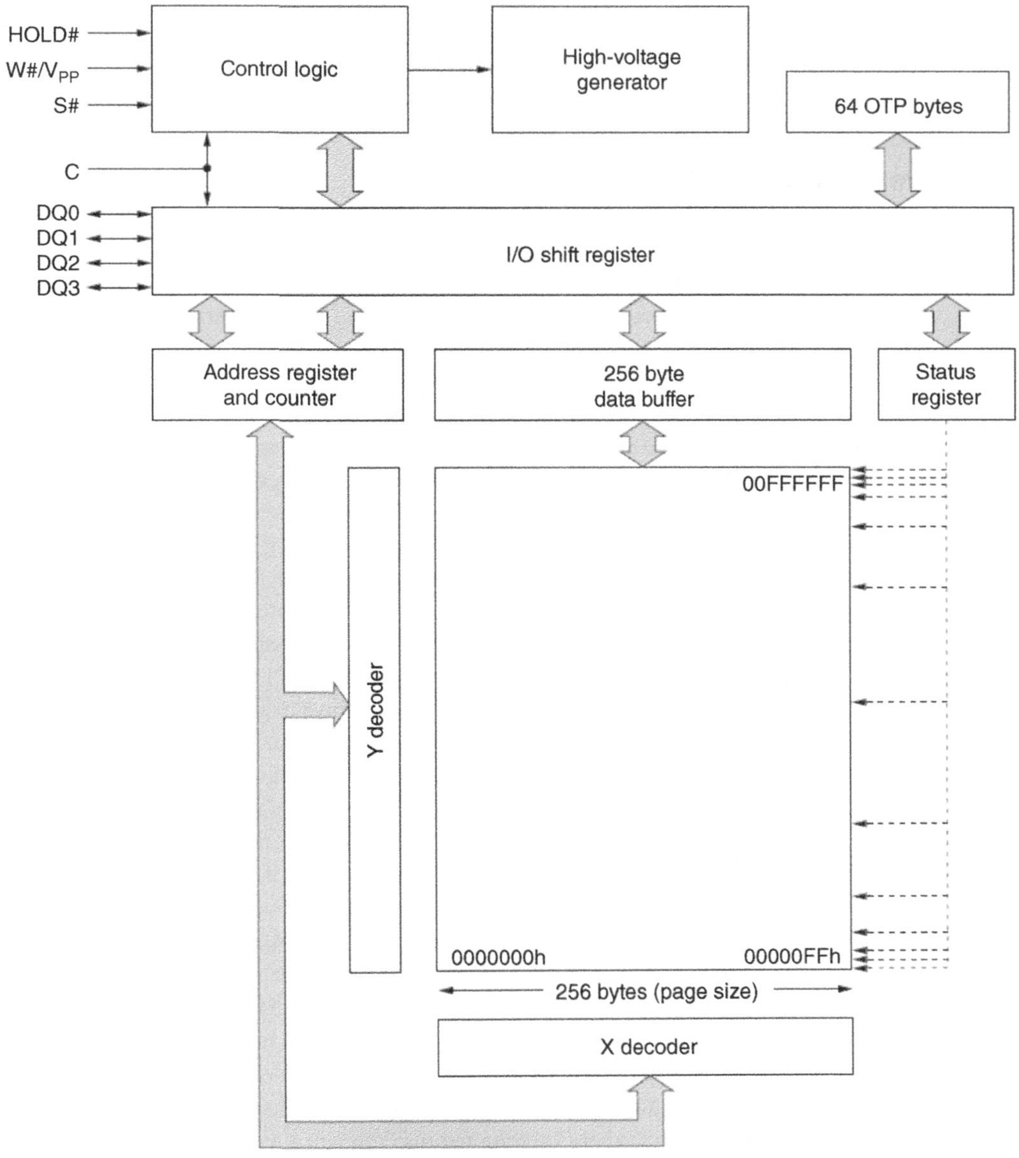

Figure 12.1 The block diagram of Micron serial NOR flash 16 MB memory.

aspect of the datasheet is that it provides detailed descriptions of each command. If you scroll further down, it explains the data format for transmissions and gives in-depth information about each operation. For example, the "write enable" command is thoroughly described, including what it does and how it works. So, the datasheets are very useful, and when you're working with these types of devices and you want to interface with them using specific protocols such as SPI, it's essential to read, understand the commands, know what data to send, and know what responses to expect. It's very important to look through this here and understand that all of these commands make sense – both in terms of how they are sent and how responses are received. As we move forward in designing the software, we'll explain what we're sending and what we're reading back.

We won't need every command listed, but we'll use many of them – especially the read and write commands, which will be our main focus. So this is just an overview of the external flash memory we'll be using in this project, specifically the one on the Arty A7 development board.

12.2 Overview of the SPI Interface

This section is about the SPI interface explanation. Here, we're going to explain the SPI interface, specifically how we'll be using it in a project to communicate with an external memory chip. First off, what is SPI? It stands for serial peripheral interface, and SPI is a serial communication protocol used by different integrated circuits (ICs) on a circuit board to communicate with each other. When you think of UART, it's another protocol ICs can use to communicate. However, SPI is another communication protocol. For example, if you have two microcontrollers on a circuit board, they can use an SPI interface to communicate with each other. Or in our case with an FPGA, you can have two different FPGAs on the same board communicating over SPI. Typically, SPI is not used for long-distance communication. Most of the time, SPI is used when two devices are located on the same physical PCB board. If you need to communicate over long distances, you'd have to put it in a different format, such as RS-45, or an ethernet-style format. So, the way we're going to be using it is to have our device or our FPGA and processor to communicate with an actual memory chip. One advantage of SPI is that it allows for very fast data transmission. The disadvantage is that it uses more pins, but you can transmit data very quickly with an SPI interface. There are four common signals used in SPI communication: MOSI (master out, slave in signal), MISO (master in, slave out signal), SCK (serial synchronous clock signal), and the CS (control select). The CS signal is used when you have multiple devices on the SPI bus. So, when you have an SPI bus, you might have one device – your microcontroller, or whatever you want to call the master – and it will have several other slave devices attached to that same bus. They'll have a separate CS line, but share the MOSI, the MISO, and the SCK lines. So, the master picks the device it wants to talk to and sets that CS line low. All the others CS lines will be high. That device knows, once the CS goes low, that the master is talking to him, so it starts listening on the bus. The MOSI is the signal that the master sends data out on, and it's synchronous with the clock. Once the data is fully received and the command is sent to the slave, that slave returns the data, or responds based on whatever that message was. So, for example, in the case of the memory, if we want to read at a specific address, we'll send that data down on the MOSI line. Then the actual memory device will just keep sending our synchronous clock and the memory line on the MISO line. It'll send the data back to us, and we'll be able to read and interpret what that data means. You can run SPI very fast depending on the device. As long as both the slave and the master can operate at a given frequency, you can run your SPI at that speed. With SPI, there are several different versions: standard SPI, which is just a single bit of data transferred per clock cycle, and that's where you send a synchronous clock, where you send one bit of data down to that same device. They have a dual SPI, where in every clock cycle, you send two bits of data to the same device, and when it responds, it'll send two bits of data back. Then, they have the quad SPI (QSPI), where you send four bits of data every clock cycle. So, the quad SPI is obviously the fastest because you're sending more data for less clock cycles. If you think about it, you have eight bits of data or one byte to transfer to the slave device. If you're using a standard SPI, it's going to take you 8 clock cycles to get all 8 bits of data down to the slave device. If you're using dual SPI, it'll only take you four clock cycles to get those 8 bits to the same device, whereas if you're using a quad SPI, you're only going to use two clock cycles to get that full byte of data down to the same device, So, it's much faster, and it just takes up more IO lines. Typically, when you're working with FPGAs, you'll have devices where the QSPI is very common because you have a big configuration bit file, so that when you power on, the FPGA is going to look into that key SPI device and pull a configuration file in, and you want that to happen pretty fast. You don't want to turn your device on and have to wait for a couple of seconds for the JTAG type of programming to take place. If you use the QSPI, it'll happen very quickly in a matter of milliseconds or microseconds in any type of situation. When you're working with SPI bus, you can connect multiple devices to an SPI bus. So, you'll have a master, and you can have multiple slaves. Each slave device you add requires a new CS pin in order to select that individual device. If you look at it from a high level, in the SPI interface, we have a master that selects a slave device it wants to talk to, and then the master controls the serial clock (SCK) and the MOSI line and transmits data down to the selected slave device, which it has selected from the control select line, and then slave responds by transmitting data back on that MISO line, so the master sends and it initiates the conversation. The slave only responds when the master sends information to them and requests something back. So, even if you have everything, depending on a protocol that's been established or the actual device, you have to look at the datasheet; for example, we're going to refer back to the memory chip again. If we want to read a memory location back, we're going to send down the command, and we'll send down the address, and then the chip is going to respond back with the actual contents in that address. If we go to a write command, we're going to tell the chip that, OK, we want to do a write. We specify the address and the data we're writing, and a lot of times, depending on the chip, you have to look at the datasheet. But they'll tell you that once you send the write command down, it will either do nothing, or it'll respond with an acknowledgment or give you some sort of response to say that write command has finished or has been successfully interpreted as a write command. In this SPI interface project, the FPGA is going to be the master, and the external memory chip is going to be the slave. So, we're going to initiate communication with the external memory chip, where we're going to ask the chip to read us back data at a specific address location, and we will write data down to the actual chip to where we want to write any actual data we want to write on that chip. So, let's go ahead and see how the SPI interface actually works, and first off, we will talk about what the actual physical lines look like. So, if we have our master device, we'll call it the master, and this could be anything. This could be a microcontroller, an FPGA, or any sort of chip. So, we'll have a CS line. Also, we'll have a SCK line, and we have our MOSI line, or master out, slave in data line, and our MISO line. We're just assuming a standard SPI interface where we're sending down one bit of data for each clock cycle. So, if we have a slave device, we have a CS as well. Also, we have a serial clock (SCK) line, a MOSI line, and a MISO line. So, the CS is driven by the master – that's an output of the master and an input to a slave. Our serial clock (SCK) is the same – it's an output of our master and an input to the slave, and that's driven by the master as well. Our MOSI lines are master output, slave input, so that's the data the master device transmits to the slave device, which is driven by the master as well. Then, the MISO lines come from the slaves, so it's the master input, slave output that goes back to the master. So, when the slave responds, it sends the information back on the MISO line. So, that's the connection that you have between your master and your slave devices, as depicted in Figure 12.2.

Next, we're going to talk about what the actual waveforms would look like. So, the master – the way it's going to work – has that control select (CS) line, and when it's idle, it'll be high. Then it's going to drop low. When it drops low is when the slave starts looking at the serial clock line and the MOSI line to determine the data that's coming in. So, we'll then have our SCK clock, or synchronous clock, that's high. Then, all of a sudden, when our control select line goes low, we'll see when it occurs at that point. Our SCK clock will then go low, and it's just going to toggle at

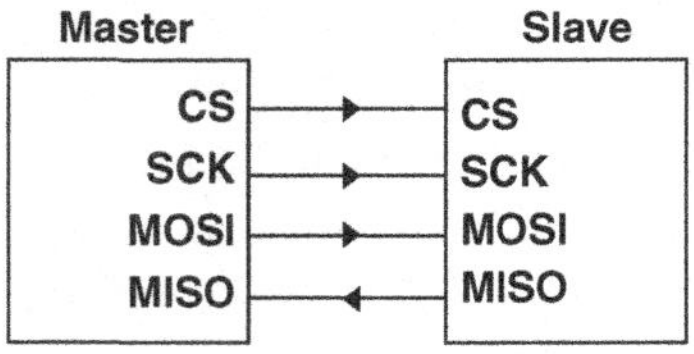

Figure 12.2 The SPI interface.

whatever clock frequency you're sending. This determines the data rate at which data is sent. In this case, we're going to send out one bit of data every clock cycle. Depending on the devices, and typically on the rising edge of the clock, though you'd have to look at the datasheet of this specific device, you'll know when the data is being read. Once the data transmission stops, the serial clock (SCK) line will run high. Now we have our MOSI line, which determines the data being sent from the master to the slave device. This could be any type of command depending on what the slave device is doing. By checking the datasheet, it tells you the type of commands to send. So, it'll be high as it's idle, and then it'll drop low or high based on whether it's a one or a zero. Since we're low in this situation right here, the first rising edge is when we're going to look at our data. Within one cycle, and between the rising edges of SCK, nothing's happening. But at the rising edge of our SCK clock, the slave device sees the status of this MOSI line and determines whether it's a one or a zero. At the rising edge of this SCK clock, our MOSI line is low. That tells us we have a zero. So right here is a zero. At the next rising edge, it checks the MOSI line again – it's still low, so we have another zero. Then the MOSI line goes high for the next two clock cycles. That gives us two ones. Finally, we're going to stay high, which makes that a one as well. We can keep this out, but we're just showing that, in this case, you would have 00111 transferred on the MOSI line. That's because on every rising edge, or in the falling edge, it's going to read the data based on the data associated with the slave. In this case, we're assuming the rising edge, so on every rising edge of this SCK line, we are going to read this MOSI line. If it's low counted as a zero, and if it's high counted as a one. So, then this MOSI line will just stay high or idle at this final point once the full data transmitted, and that's going to be based on once again it really comes down to what the data shit in that slave tells it, you may send 8 bits, 16 bits, 4 bits or 32 bits just depending upon which protocol you have implemented, and then we have our MISO line, and this is the line that the slave responds back to the master on. So, the whole time this is going on, and the MOSI line is transmitting, our MISO line is going to just be idle and stay high and it's going to stay high until our SCK clock starts clocking again. So, after we transmit everything on our MOSI line, we stop our clock and then we're going to go ahead and start our serial clock up again, which is going to then data is still going to be transmitted on the rising edge of the clock, except instead of on the MOSI line, it's on the MISO line this time, and in the whole time while this is going on, our MOSI line is going to just stay idle or stay high because we're not going to be transmitting at this point, and we're waiting for the response. This will just stay high throughout this entire time period and then our MISO line will come here and for our first rising edge, we're high, so we know that is a one and then we'll drop low, and go low for to still be a zero, and we will have two zeros, then we'll say we go high again and then it's as high as it's idle and that becomes a one. So, in this case, our MISO would respond with 1001, and this is really going to all come down to just depending upon what the datasheet tells you. But what the way that slave device is going to respond, our master is going to on the rising edge of the clock, it's going to look at this MISO line and determine if the MISO line is high, that's a one, and the MISO line is low, that's a zero, and so that our control select is going to stay low until around there once the master receives a response back from a slave and no longer wants to talk to it or talk to a different slave, it's then that control select line will be high. So, at this point, if the master sends

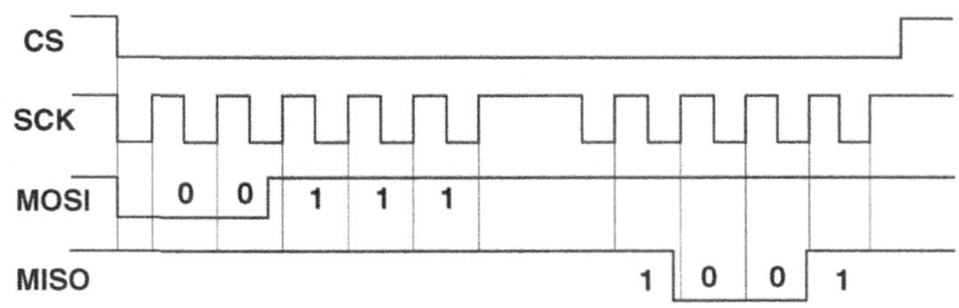

Figure 12.3 The SPI interface signals waveform.

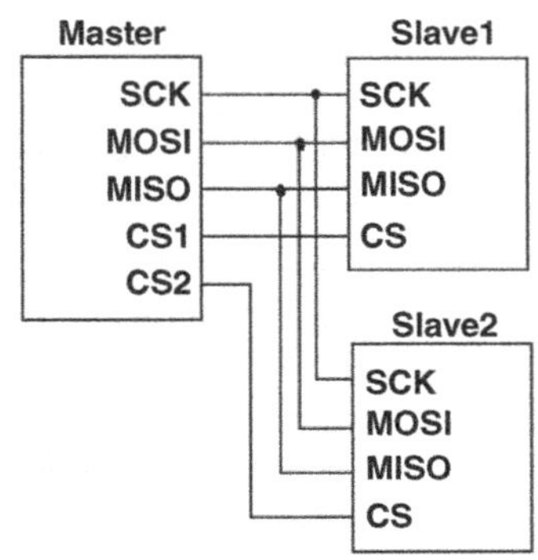

Figure 12.4 The SPI interface between a master and two slaves.

another message out in this MOSI line and the control select is high, the slave is going to ignore that entire command as illustrated in Figure 12.3.

So, the next thing is, if we want to have a master and two slaves, what would that look like? If we have a single master, we're going to use a synchronous clock (SCK), a MOSI line, and a MISO line. Then, depending on the number of slave devices we want to connect to the bus, that will determine the number of CS signals we need. Since we have two slaves, we'll have CS1 and CS2. For Slave1, we'll represent that as a device, and it will be connected to SCK, MOSI, MISO, and that is just going to have a CS.

Then we have our second slave, Slave2, and these could be various types of devices, so you could have a memory chip, an analog to digital conversion chip, or any type of integrated circuit on your PCB board. We'll have the SCK, MOSI, and MISO lines, and each slave will also have a CS. The way this will be connected is: CS2 from the master goes to Slave2, and CS1 from the master goes to the CS of Slave1, as shown in Figure 12.4. The SCK clock, MOSI, and MISO lines are shared between the master, Slave1, and Slave2 – Slave 1, Slave 2, and any additional slaves. The CS lines determine which slave is currently active. If we have a third slave, we'd introduce a third CS line. That's how you connect multiple devices together using the SPI interface.

12.3 The SPI Interface Design Strategy

This section is about the SPI interface design strategy. In this section, we're going to discuss the strategy that we're going to use to implement our SPI interface project. The SPI interface design will be used to interface with external memory. We want to use the Micro Blaze MCS processor. However, it doesn't contains a dedicated SPI port. Typically, if you'reworking with certain micro-controllers or microprocessors, a lot of times they'll have dedicated hardware to implement the SPI interface. But in our case, we are going to implement SPI interface using VHDL. So, for our Micro Blaze, we want to have a general-purpose input as well as an output to interface with our dedicated SPI controller, which we're going to implement in the actual FPGA logic. We'll have an SPI interface in VHDL, as well as our Micro Blaze, and we're going to interface the two together so they can communicate with the actual memory chip located on the board itself. Then, we're going to use the UART interface to display the data we read from the memory chip and display it on the computer. Additionally, we'll send transmissions from the computer to the MicroBlaze MCS to determine what to write to the actual memory chip. It's always good to have a strategy up front before diving into the design – to know, at a high level, what you're wanting to do. At a high level, we are going to have our MicroBlaze MCS processor, which we've used for various different projects in this book. That's our softcore processor, and this will be implemented into the block design. We're also going to create our VHDL module – we'll call it our VHDL SPI design, and we're going to have to create this as an IP block that we then drop into our block design. From there, we want to have an interface from the MicroBlaze to our VHDL SPI design, and we're going to do this using two different GPIOs. So, we'll go ahead and create two of them here that will allow us to communicate back and forth. These will be implemented just using general-purpose inputs and outputs, because once we developed this VHDL SPI design, we'll know specifically how many inputs and outputs we need, and what bit widths are required for the MicroBlaze MCS to communicate with the SPI design. As for our SPI design, we are going to implement it later. The most important is implementing this VHDL SPI design interface, so we can talk to the memory. We'll just make a block here and call it memory. This is external to the chip on the board. This SPI interface will either write data to the memory or read information from the memory and feed it back into the MicroBlaze MCS softcore processor. We are also going to interface to a computer, so we'll have our computer using a COM port where we'll be using Tera Term or any other type of terminal emulator. That's just a UART interface between the computer and the MicroBlaze MCS. We know the MicroBlaze MCS supports UART. So, we'll have to transmit data to our computer and receive. Those commands will be sent down to the MicroBlaze processor to determine what values to write to memory. We'll then use our general-purpose inputs and outputs to interface with the VHDL SPI design, and that, in turn, will communicate with the memory. So, it's kind of a whole block chain, as depicted in Figure 12.5.

We've got our computer, which we're going to use to send information to our processor. The processor will then send that information to our SPI design, which will forward it to the memory, and then obviously, we're going to have other generics, such as a reset and a clock that'll come in there, and these are just kind of a global reset that we'll probably just tie to a slide switch. We'll have our clock as well, and our SPI design here will have a clock, and we'll try to pull that same clock from there, which will then come into our SPI design. Next is UART, and we might also include another slide switch here that uses a GPIO to indicate whether we want to read from or write to the memory. So, we can do this in different ways, but what we will have here is a general-purpose

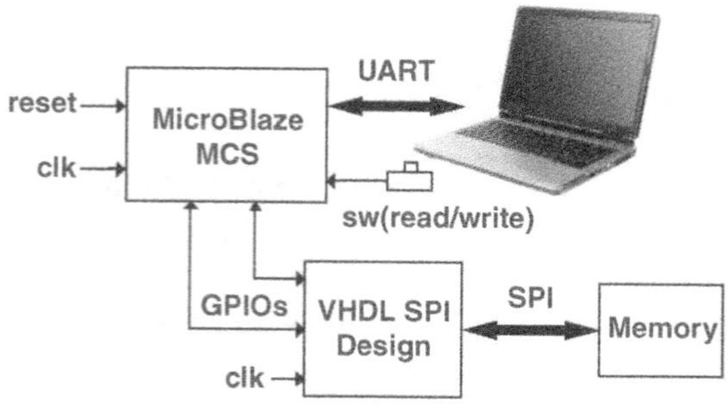

Figure 12.5 The whole block diagram.
Source: Fatman73/Adobe Stock.

input that will determine, by reading the slide switch, whether we are going to read data from the memory or write data to the memory. If we're going to read or write data to the memory, we then need to ask the user via the UART port if they want to read or write. So, that is the overall design strategy for this SPI project. We'll basically use the VHDL to design this SPI and the MicroBlaze; we already know we can drop in the IP core and we'll have to write software to interface with the GPIO as well as the UART. So, that is the SPI design strategy from a high level.

12.4 External SPI Flash Chip Memory Organization

This section is about the external SPI flash chip memory organization. In this section, we're going to explain and help you understand how the Micron SPI chip that's located on the Arty A7 development board is structured. So, we have the Micron datasheet for the flash chip on the Arty A7 development board, and they have a block diagram that's located in their memory organization structure, as shown in Figure 12.1. So there, you've got a block that they call control logic, and this is where you have the external interface, which is shown with this hold#, write#, S#, C, and your DQ0 through DQ3. This is the SPI interface. So, we know the memory chip supports the extended, the dual, and the quad SPI, and we were working with the extended SPI, which just uses your clock, your CS, your MOSI line, and your MISO line. These are used to interface with your control logic and specifically interface with the I/O shift register. The I/O shift register is what we use to come into here, where we have the actual block, this main block here, where they have 256 bytes (page size) from the 0000000h all the way up to your 00FFFFFF. That is the actual memory itself, and that's what we're trying to access.

So, we're just shifting in command and information into this I/O shift register that we send down to this address register and counter, as well as a 256-byte data buffer, and we send down the data in multiple shots. So, this is a good thing – when you look at the Micron datasheet, this is something to review to get an understanding of what you need to do when working with this SPI flash chip, both in terms of how it works at a high level and the different commands you need to send down if you want to interface with it and get this type of information. So, what we're going to do is show what this means for us. We'll just call this the external flash memory, and we are going to give a high-level explanation of what we need to know or what's important for us. If we have our external flash memory chip, let's just take a look at it from a block diagram view, which is helpful when you're working with memory or similar components. It's good to get an understanding by looking at it from a block memory view. So, we will say that from up here we have our address starting at 0x000000h (0 MB), and this is our address zero, all the way down to 0xFFFFFFh, which is like our 16 MB. Now, what's important is if you look up, there's an app note (XAPP951 v1.3) that specifies when you load a configuration file – or in this case, take a project we created in Vivado and load it into this external flash memory – it's going to start from address zero to wherever it sees a break. So, what that means is, if we want to store additional information or use this external flash or anything else, we need to set aside the space from zero to however large our .mcs file, or our configuration file, is. For the Arty A7, they say it's approximately 2 MB. Now, that's kind of an approximation based on your design and how much memory it consumes. In this case, we would just go ahead and put a line there at 0x1FFFFF ($\approx$ 2 MB), and that marks approximately the 2 MB line. This is where we store the .mcs file. We get this file from the .bit files we've generated. So, using a tool (Vivado), we convert that .bit file into a .mcs file, which we then load onto our external flash – this right here is our external flash. If we're going to write to this flash or in the future store any additional memory or information onto this flash chip, we want to stay out of the address regions from zero all the way up to the 0x1FFFFFh address (approximate 2 MB). You want to leave that alone for the configuration because the app note tells us that when the Arty A7 or the Arix FPGA boots up and is configured to read from the QSPI flash, it starts at address zero and reads through the flash until it sees the end of file indicator, signifying the .mcs file is complete. If we were to load this configuration file down here at the bottom, then it won't load into the FPGA. Or if we overwrite any of this information – if our SPI interface stores information on this flash chip and overwrites some of the configuration, then erase and overwrite it – there will be an issue where it tries to reassemble the corrupted data, and your design won't load because the checks don't match, and there'll be a lot of issues. So, the important thing here is you need to know where the configuration file is going to be stored. We approximate this to be 2 MB here. So, we'd still have around 14 MB to store what we want. So, we would probably make a line somewhere at 0x7FFFFFF, which is approximately at the 8 MB mark, and say, "OK, this right here is safe for user flash use," and that's where you could store additional information. This gives you the buffer for safety. You could probably move it up a little closer since you're not likely to use all of that. But just to be safe, give yourself 8 MB and use this space for storing in flash if needed. It's really important to note that the configuration file should start at address zero and just keep writing until it's full. So, you'll want your configuration file to begin at address zero. This is important when loading it onto the flash, as it ensures the FPGA will automatically load the configuration or program we've created – without needing to be connected to a computer each time the board powers on. This becomes even more important if you're working with a board that contains an FPGA along with a microprocessor, microcontroller, or other control logic. In that case, the microprocessor, microcontroller, or even the FPGA itself could take in a new configuration file, a new .mcs file, and write it to this location. When it reboots, it'll take on that new file. So, you could even have something like a multiple configuration file system, where your primary configuration file starts at address location zero. If you have a second configuration file, you might store that one starting at the 8 MB or 0x7FFFFFh address. In this setup, you'd store the old one at 0x7FFFFFh and the new one at 0x000000h. When you boot up, the configuration file that we take gets loaded, as illustrated in Figure 12.6.

It is very important to kind of understand the whole memory organization of the flash chip, because when you're working with it, you want to be careful that you're not erasing specific stuff in here (configuration file (.mcs) area starting at 0x000000h address). This is very safety-critical, and you don't

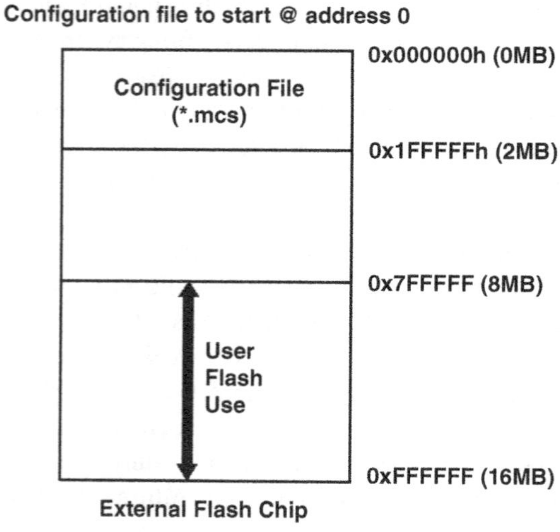

Figure 12.6 The external flash memory organization.

want to erase anything there or cause any type of issues, or all of a sudden, you'll get these problems where your board won't boot up. Now, when you're working with all the boards, the beauty of it is you have it right there on your desk. So, if there is an issue, you're able to just reflash it. But if you're designing a board that you want to go out in the field, or somewhere where you're not going to be, or where somebody is not going to be there if something goes wrong, you want to make sure that you have your protections and safety checks in place, so that this configuration file right here doesn't get overwritten or corrupted in any way. So that's just kind of a quick overview of the external SPI flash chip memory organization and how you can interface with that structure.

12.5 The SPI Configuration and File Generation

This section is about the SPI configuration and file generation. Here, we're going to show you how to go through and take a project to generate the file that we can then load onto the SPI flash memory located on the Arty A7 development board. We have previously completed the UART IO project, and we are going to go ahead and create a .mcs file that we can load onto the external flash. So, in order to do that, we want to go through all the necessary steps: run synthesis, run implementation, and generate the bitstream. We go ahead and rerun Generate Bitstream. We're going to regenerate the bitstream because, once it completes, a pop-up window will appear asking if you want to generate a memory configuration file or a .mcs file. From there, we'll click Yes, because we're going to take that file and load it onto our QSPI flash, as shown in Figure 12.7.

So right now, we have our Tera Term window, which is connected to the COM port associated with the Arty A7 development board. If we go ahead and click the reset button, we'll see that the

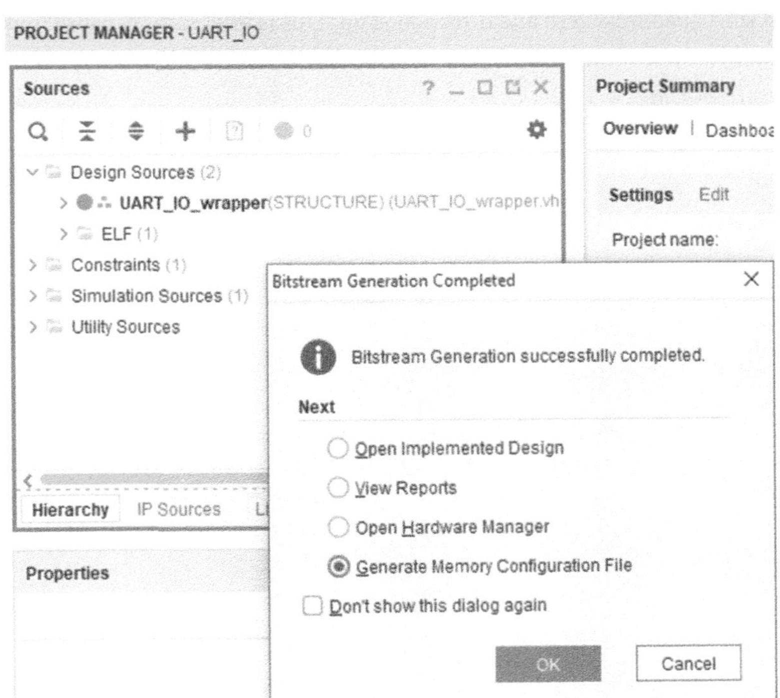

Figure 12.7 Selecting Generate Memory Configuration File.

green LED goes off and the FPGA resets. So, we'll go ahead and try to reboot. But two things are preventing it from working: first, the jumper JP1 is currently set to JTAG, and we need to change that to QSPI. Second, we don't have the file loaded into that SPI device, so there's nothing to boot from. Our bitstream generation has been completed, and we now want to open the Hardware Manager, but we want to select the last option: Generate Memory Configuration File. So, we select that and click OK, which will bring up a window called Write Memory Configuration File. From here, we choose the file format – we can go with a bin or a hex, but we're going to select the .mcs file format. You can select a specific memory part or go with a custom memory size. Since we know the memory part used on the board, we're going to select a memory part. To do this, we click on the Part Chooser, and then we are going to select All, and then Reset All Filters. After resetting the filters, we set the manufacture to Micron, density to 128 Mb, and the type to SPI. This narrows it down to two options. We then select the 3.s V part, and click OK, as depicted in Figure 12.8.

So, it recognizes the memory part, and we're going to select that because that's the part we want to use. Then, we need to specify the filename, which is where we want to store the actual .mcs file. So, we are going to save it in the project directory where the UART IO project is located. We have runs and implementation folders, and this is the same location where the bit file and all other configuration files are stored. We want them to be stored in the same locations. So, we're going to save it in that same project directory and name it UART_IO.mcs, as illustrated in Figure 12.9.

So, it's going to create the file at the specified location, and for our interface, we want to select SPIx1. Currently, we have the device set to SPI with a single interface, meaning we're just using the MOSI and MISO lines – two data lines. So, we want to select SPIx1. Now, we need to load the bitstream file. From here, we go ahead and click the three-dot button and navigate within the UART_IO project to runs > implementation, and then select UART_IO_wrapper.bit and we'll click OK. This is going to take that bit file and use it to generate .mcs file. For the start address, we just want to have it start at address zero and direction is up. This is important because if we look at the

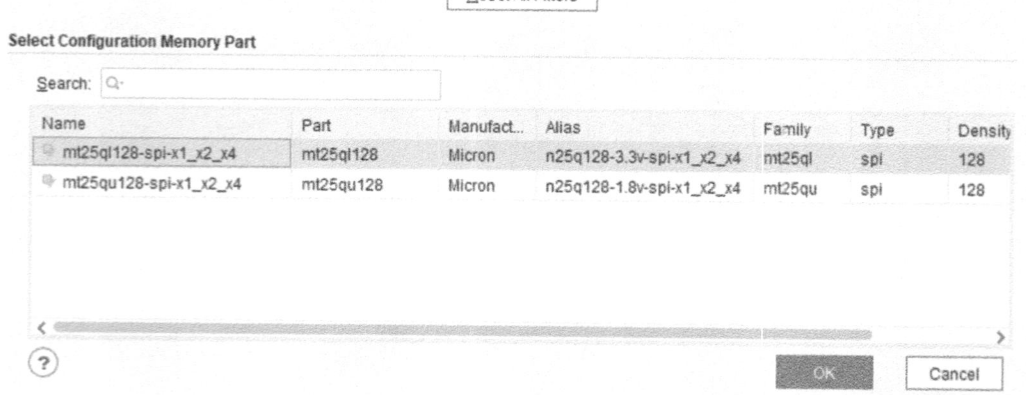

Figure 12.8 Selecting the configuration memory part.

Figure 12.9 Specifying the configuration filename.

user guide from Xilinx, the memory configuration must start at address 0 and count upward. As for the remaining options, we will leave them unchecked: Write checksum, Disable bit swapping, and Overwrite. So, we're going to go ahead and click OK, as shown in Figure 12.10.

So right there, it gives you a little information indicating that the .mcs file is generated, and that gives you the starting address, ending address, and other indicators like Addr1 and Addr2. These fields help you understand the size of the memory block being used and how much of the external QSPI flash memory will be occupied by the configuration file, as depicted in Figure 12.11.

So, from there, we can now go ahead and load that onto our QSPI flash part. We'll select our external memory part, right-click, and choose Program Configuration Memory Device. That is our memory device. For the configuration file, we'll go to our project directory, and we'll see the .mcs file there. We'll click OK. For the address range, we're going to select Configuration File Only. We don't need a PRM file, so we'll leave that set to None, we will also set Pull to None. We want to have address range configuration file only, and check the Erase, Program, and Verify options. Then, we'll click OK. Now we get a little indicator showing us what's happening. The way these flash devices work – and this is very important to understand when working with them or interfacing with them – is that any time you want to write, step one is to erase. Then you can write. It's the erase operation that usually takes a bit more time. So obviously, when we load a design directly onto the FPGA, it's much faster than this operation because the FPGA uses RAM-based memory, which works a lot faster than flash memory. The flash memory is decently slower, but the advantage is that it's nonvolatile. So it stores it in there, and we can power off our board, and when we power it back, it'll take effect immediately. So it's very powerful to use this, and as you can see here, we're almost completely done. So, once we get this programming done here, the only thing we have to do then is set our jumper on our development board and we'll hit that program button to reset it at that point. Then it will load the MCS file and transfer it from the QSPI flash to the FPGA, allowing you to run the UART_IO project – or any other project you're working on. This is part of the typical development flow: you first load your design onto the device using the JTAG cable, as we've done with previous projects. This approach is ideal for debugging, testing, and verifying that everything functions correctly. Once verified, you can then program the design into the QSPI flash for standalone operation. Finally, click OK to complete the process successfully. So, if

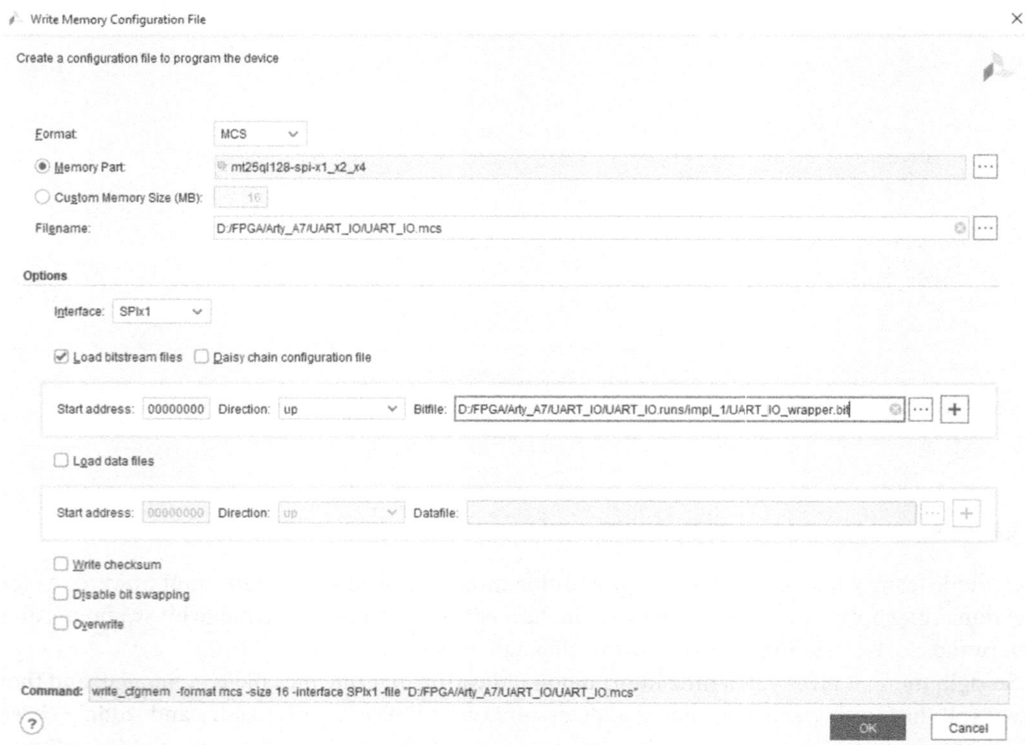

Figure 12.10 The Write Memory Configuration File window.

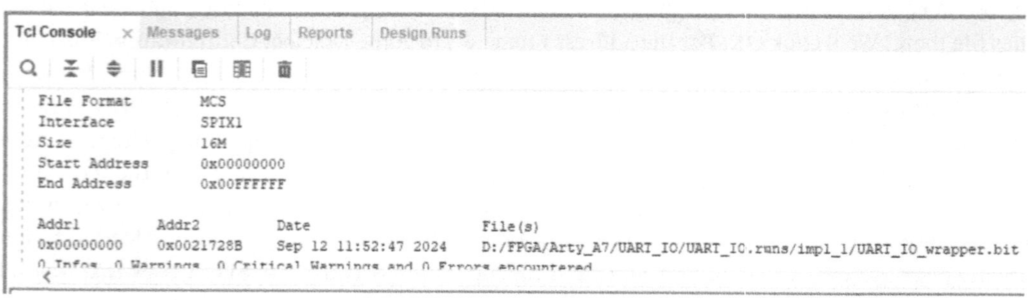

Figure 12.11 A little information about the QSPI flash memory.

we go, we can then go ahead, and we are going to take a jumper and move it over. So now we're going to boot from our QSPI device, and if we hit this Program button right there to reset to the QSPI device, we'll reboot, and we have our terminal window right there that has our information. So we can see all signs that it booted up – we now have this QSPI, this UART_IO project is now running on here. So, you know, if we just type in led 15, we can see the four LEDs light up. So that's how you work with and generate a configuration file and load it onto the SPI device of your Arty A7 development board.

13

Some Miscellaneous Projects with Vivado

13.1 Introduction

In the rapidly evolving landscape of digital design, the Xilinx Vivado Design Suite stands out as an indispensable tool for engineers and enthusiasts. Its powerful features cater to a wide array of applications, from simple field-programmable gate array (FPGA) projects to complex system-on-chip (SoC) designs. This chapter delves into a collection of miscellaneous projects that exemplify the versatility of Vivado, bridging theoretical knowledge with practical implementation. As we embark on this journey through various projects, we'll explore different aspects of hardware design, ranging from basic circuit implementations to more advanced concepts such as signal processing and system integration. Each project showcases the unique capabilities of Vivado, highlighting its user-friendly interface, comprehensive design validation tools, and seamless integration with simulation and synthesis workflows.

13.2 The I2C Interface Project

This block diagram of the project is shown in Figure 13.1.

By right-clicking and selecting "Run Connection Automation" and also "Regenerate Layout," we can easily connect and use the required IP cores. Then, we use the following Address Editor window to configure the data and instruction memory of MicroBlaze, as depicted in Figure 13.2. By right-clicking on microblaze_0, we can also select "Auto Assign Address."

Then, we right-click on the block diagram file name and select Create HDL wrapper..., and then run synthesis, run implementation, and generate bitstream, respectively. In the next step, we go to the File menu, and select Export, and then Export Hardware... options. Finally, we choose Launch SDK from File menu to open the Xilinx SDK software. We create an application project by going to File> New > Application Project. In the New Project window, we enter a project name, select C as the language, leave others as default, and click the Next button. Then, we choose Peripheral Tests as our template and click the Finish button. Then, we copy and paste the following codes into the testperiph.c file.

```
#include <stdio.h>
#include "xil_printf.h"
#include "xparameters.h"
#include "XIic.h"
```

Fundamentals of VHDL for FPGA Programming Using Vivado, First Edition. Majid Pakdel.

Companion Website: www.wiley.com/go/pakdelVHDL

```
u8 readReg(u16 reg_addr)
{
        u8 write_buf[3];
        write_buf[0] = (reg_addr >> 8) & 0xff;
        write_buf[1] = reg_addr & 0xff;
        u8 read_buf;
        XIic_Send(0x40800000, 0x36, write_buf, 2, 0);
        XIic_Recv(0x40800000, 0x36, &read_buf, 1, 0);
        return read_buf;
}


int main()
{

    u8 data;
    data = readReg(0x300b);
    xil_printf("%x\r\n", data);

    return 0;
}
```

The code provided is a C program that utilizes a Simple I2C interface to read a value from a specific register of a device connected to an I2C bus. Let's break it down step-by-step:

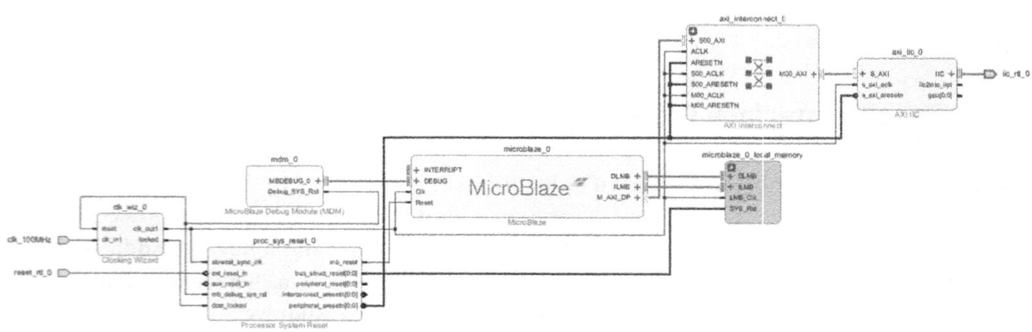

Figure 13.1 The block diagram of I2C interface project.

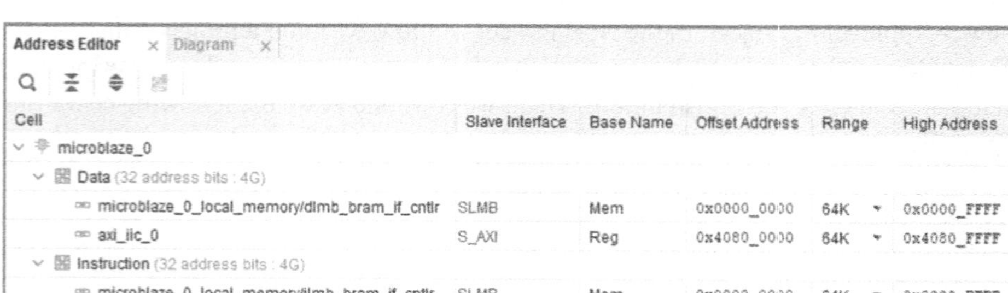

Cell	Slave Interface	Base Name	Offset Address	Range	High Address
microblaze_0					
Data (32 address bits : 4G)					
microblaze_0_local_memory/dlmb_bram_if_cntlr	SLMB	Mem	0x0000_0000	64K	0x0000_FFFF
axi_iic_0	S_AXI	Reg	0x4080_0000	64K	0x4080_FFFF
Instruction (32 address bits : 4G)					
microblaze_0_local_memory/ilmb_bram_if_cntlr	SLMB	Mem	0x0000_0000	64K	0x0000_FFFF

Figure 13.2 The Address Editor window.

1. Header Files:
   - #include <stdio.h>: Includes the standard input–output library for functions like printf.
   - #include "xil_printf.h": A custom library provided by Xilinx for output on specific hardware, allowing formatted output similar to standard printf.
   - #include "xparameters.h": A header file usually generated by Xilinx tools, containing parameters for the hardware system, such as device addresses.
   - #include "XIic.h": Supports the use of I2C interface in Xilinx systems.
2. Function Definition readReg(u16 reg_addr):
   - This function reads a byte from a specific register address (a 16-bit address) via the I2C interface.
   - The variable write_buf is defined with a size of three bytes, which is used to format the register address into two bytes: a high byte (write_buf[0]) and a low byte (write_buf[1]).
   - The address is divided into two bytes using bitwise operations. The high byte is obtained by shifting the address right by 8 bits, and the low byte is obtained by masking with 0xff.
   - read_buf is defined to hold the single byte that will be read from the device.
   - The function then utilizes the XIic_Send function to send the constructed address to the device at I2C address 0x36.
   - After sending, it calls XIic_Recv to receive a single byte from the same device address, which contains the data from the specified register.
   - Finally, the function returns the received byte.
3. main Function:
   - A variable data of type u8 (unsigned 8-bit integer) is declared.
   - The program calls readReg(0x300b), which reads the data from register address 0x300b and stores it in data.
   - It then uses xil_printf to print the hexadecimal representation of data followed by a carriage return and a newline.

In summary, the code is designed to communicate with an I2C device, read a single byte from register 0x300b, and print that byte in hexadecimal format. This code is suitable for embedded systems using Xilinx hardware and the I2C communication protocol.

Finally, we can build and generate the ELF file and then associate it with the Vivado project from Tools menu and Associate ELF Files... option.

13.3 The VHDL CORDIC Sine-Cosine Generator

CORDIC (Coordinate Rotation Digital Computer) is an algorithm used to perform a wide range of mathematical functions, particularly trigonometric, hyperbolic, exponential, and logarithmic calculations. It is efficient for hardware implementation because it requires only shift and add operations, making it suitable for FPGA architectures. In the context of VHDL (VHSIC Hardware Description Language) and Vivado, which is a design suite provided by Xilinx for FPGA and SoC development, CORDIC can be implemented as a VHDL design. This involves creating entities that define the CORDIC algorithm to compute various functions, such as sine, cosine, and tangent, with hardware efficiency. The key features of VHDL CORDIC in Vivado are listed as follows:

1. Algorithm Implementation:
   - The CORDIC algorithm can be implemented in VHDL to achieve various mathematical computations with a minimal number of hardware resources.

   - The VHDL code typically includes iterative structures to handle the rotation operations and convergence of angles.
2. Static versus Dynamic Modes:
   - CORDIC can operate in various modes, including vectoring mode and rotation mode, which can be selected based on the application requirements.
3. Fixed-Point Representation:
   - CORDIC typically uses fixed-point arithmetic, which is well-suited for FPGA implementations to balance precision and resource utilization.
4. Parameterization:
   - CORDIC implementations can be parameterized for different precision requirements (number of iterations, bit-width, etc.), allowing designers to optimize for speed or resource usage.
5. Integration with Other Designs:
   - CORDIC modules can be integrated with other digital signal processing (DSP) components in an FPGA design, such as filters, to enhance functionality.

Implementing CORDIC in VHDL within the Vivado environment provides an efficient way to perform complex mathematical computations with an FPGA. By taking advantage of the CORDIC algorithm's properties, designers can achieve high performance while minimizing resource consumption, thereby making it a popular choice for applications in signal processing, communications, and control systems. In this section, we are going to generate sine and cosine waveforms using the VHDL CORDIC using Vivado. The CORDIC IP core settings window is illustrated in Figure 13.3.

After locating the CORDIC IP core in the IP catalog and configuring the settings as demonstrated in Figure 13.3, click the OK button. This action opens a new window, where we should click the Generate button, as shown in Figure 13.4.

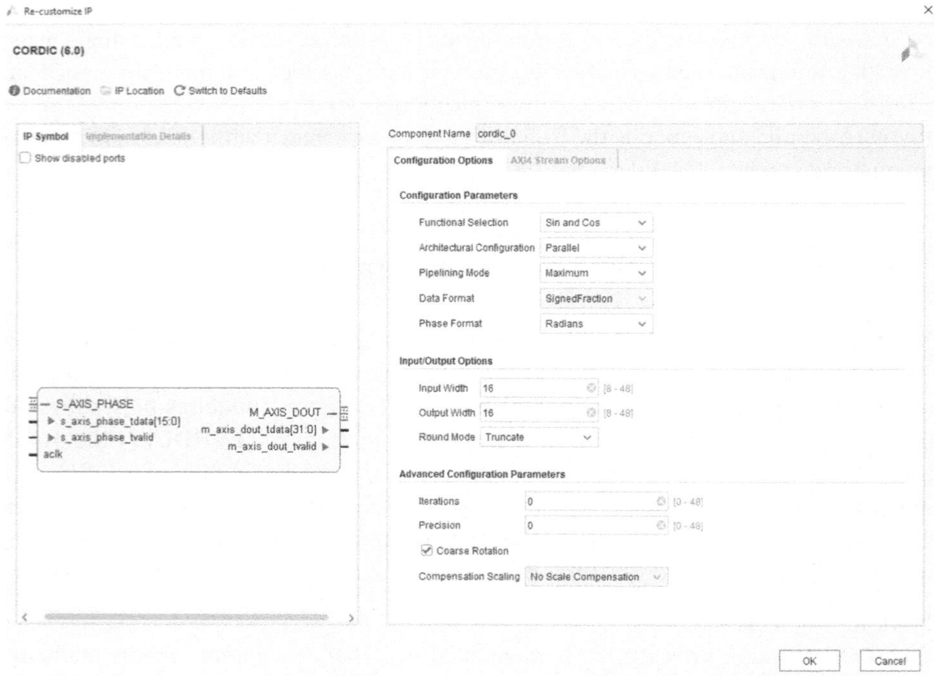

Figure 13.3 The CORDIC IP core settings window.

Then, we create a new text editor file as shown in Figure 13.5.
We give it a name (sincos.vhd) and save it in a folder as depicted in Figure 13.6.
Next, we add the written VHDL code to the Design Sources folder as illustrated in Figure 13.7.

Figure 13.4 Clicking on the Generate button.

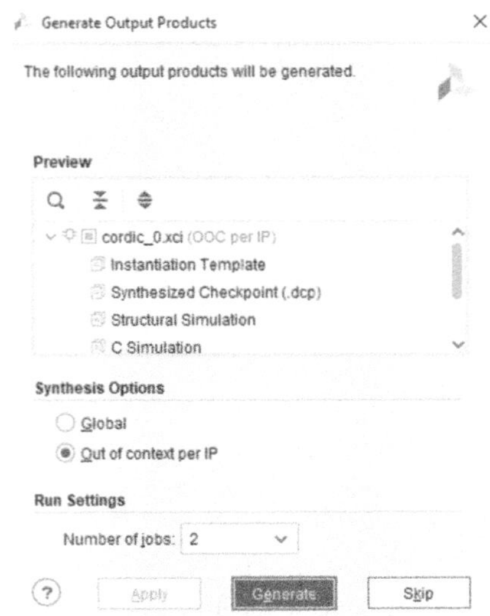

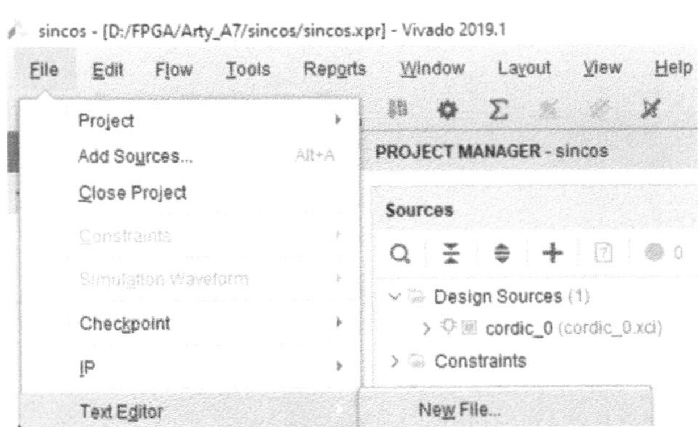

Figure 13.5 Creating a new text editor file.

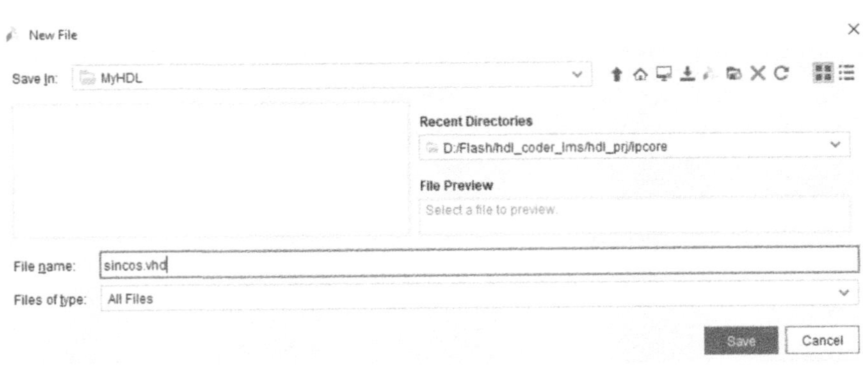

Figure 13.6 Saving the file sincos.vhd in a folder.

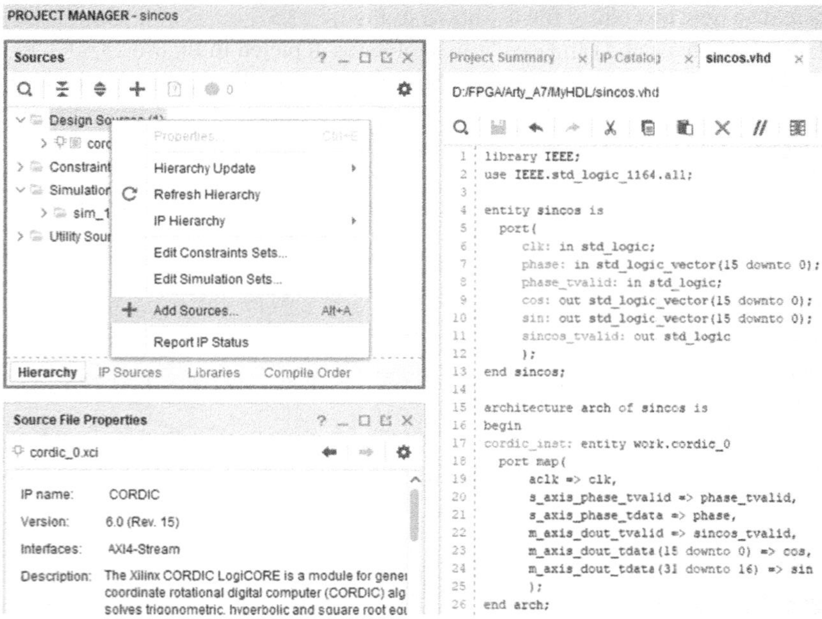

Figure 13.7 Adding the VHDL code to the Design Sources folder.

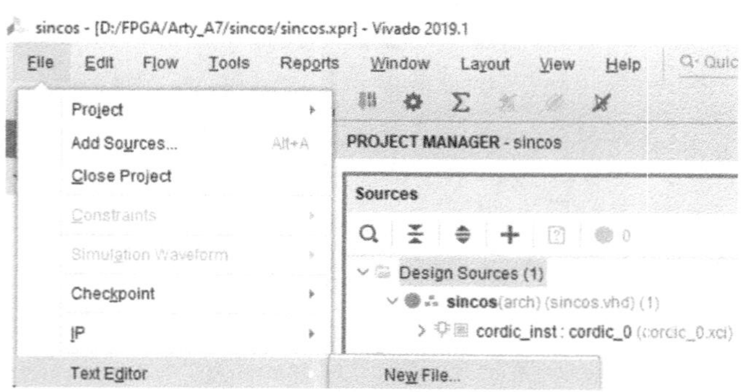

Figure 13.8 Creating another text editor file for our test bench.

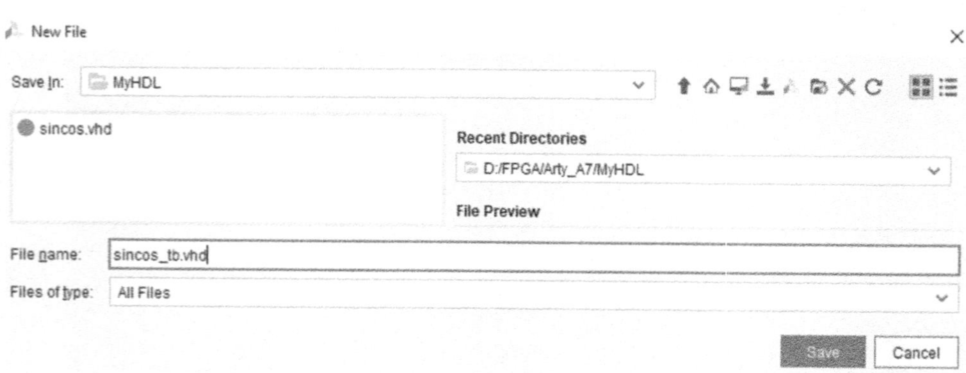

Figure 13.9 Saving the file sincos_tb.vhd in the same folder.

The file sincos.vhd is added to the Design Sources folder, and then we create another text editor file for our test bench as shown in Figure 13.8.

We give it a name (sincos_tb.vhd) and save it in the same folder where the file sincos.vhd was already there, as depicted in Figure 13.9.

We write the following code into the file sincos_tb.vhd.

```
library IEEE;
use IEEE.std_logic_1164.all;
use IEEE.numeric_std.all;
entity sincos_tb is
end sincos_tb;
architecture arch of sincos_tb is
constant CLOCK_PERIOD: time := 10ns;
constant PI_POS: signed(15 downto 0) := "0110010010001000";
constant PI_NEG: signed(15 downto 0) := "1001101101111000";
constant PHASE_INC: integer := 256;
signal clk: std_logic := '0';
signal rst: std_logic := '1';
signal phase: signed(15 downto 0);
signal phase_tvalid: std_logic;
signal cos, sin: std_logic_vector(15 downto 0);
signal sincos_tvalid: std_logic;
begin
sincos_inst: entity work.sincos
   port map(
     clk => clk,
     phase => std_logic_vector(phase),
     phase_tvalid => phase_tvalid,
     cos => cos,
     sin => sin,
     sincos_tvalid => sincos_tvalid
     );
process
begin
    clk <= '0';
    wait for CLOCK_PERIOD/2;
    clk <= '1';
    wait for CLOCK_PERIOD/2;
end process;
process
begin
    rst <= '1';
    wait for CLOCK_PERIOD*10;
    rst <= '0';
    wait;
end process;

process(clk)
begin
   if rst = '1' then
      phase <= (others => '0');
      phase_tvalid <= '0';
   else
```

```
            phase_tvalid <= '1';
            if(phase+PHASE_INC < PI_POS) then
               phase <= phase+PHASE_INC;
            else
               phase <= PI_NEG;
            end if;
         end if;
      end process;
      end arch;
```

Then we add the file sincos_tb.vhd to the Simulation Sources folder, as illustrated in Figure 13.10.

Then, we can run the simulation and view the waveforms window. We select the phase signal, right-click, and choose Radix followed by Real settings, as shown in Figure 13.11.

In the Real Settings window, we select Signed under Fixed Point with 13 binary points, as depicted in Figure 13.12.

We right-click on the phase signal again, then select the Waveform Style, and Analog Settings..., as illustrated in Figure 13.13.

We enter 150 pixels a row height in the Analog Settings window and leave the rest unchanged, as shown in Figure 13.14.

The waveforms of phase, cos, and sin in the Waveforms Window are depicted in Figure 13.15.

13.4 The Trailing Edge PWM in VHDL

In this section, we are going to implement a trailing edge PWM using VHDL in Vivado. So, we create a new VHDL project in Vivado, and add the file PWM_carrier.vhd to the Design Sources folder, as illustrated in Figure 13.16.

Then, we click on the Finish button, as shown in Figure 13.17.

In the Define Module window, we add input and output ports and click on the OK button, as demonstrated in Figure 13.18.

Then, we create an incremental carrier signal inside a process, and from Flow menu, we select Run Behavioral Simulation, as shown in Figure 13.19.

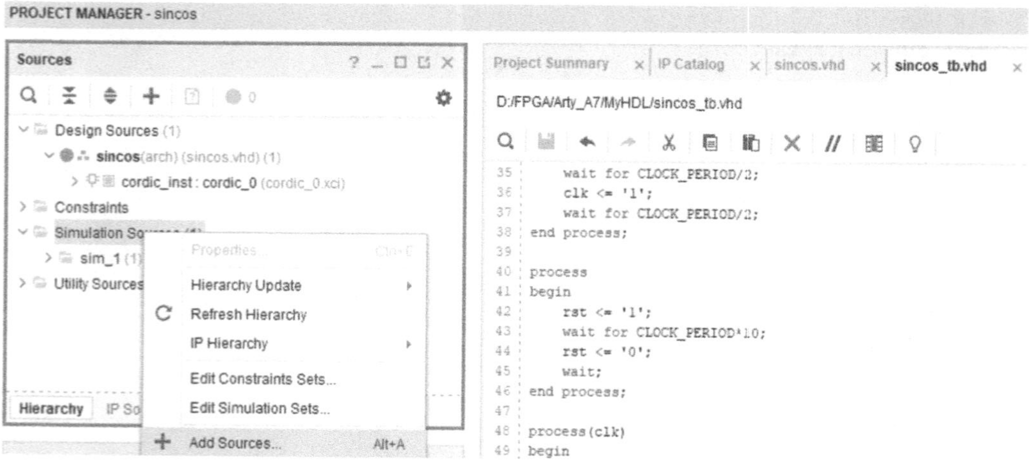

Figure 13.10 Adding the file sincos_tb.vhd to the Simulation Sources folder.

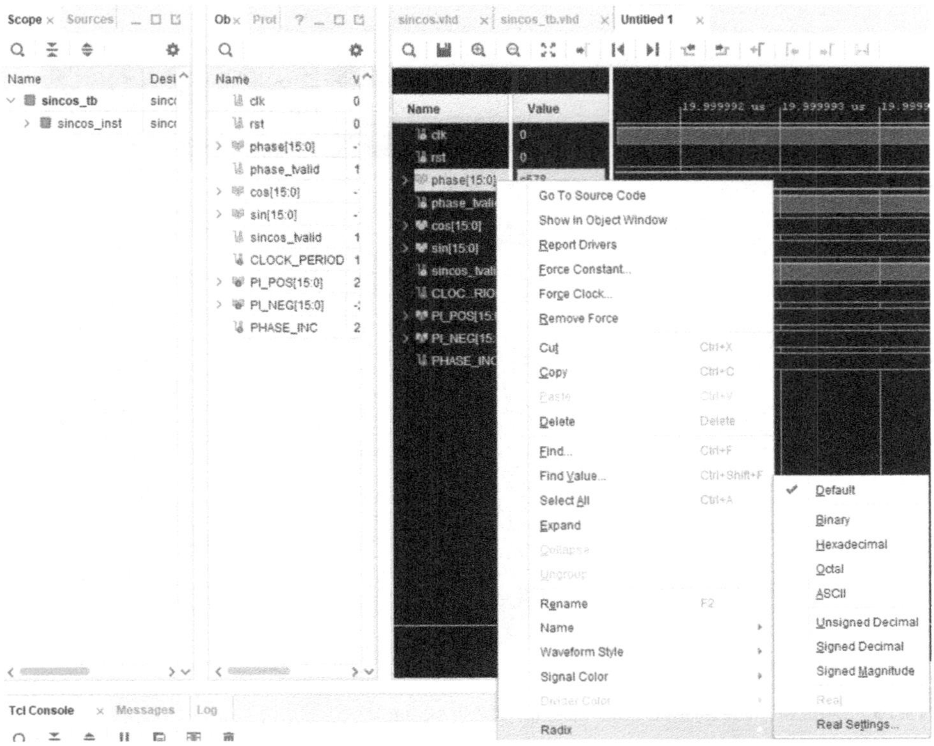

Figure 13.11 Selecting Real Settings for phase signal.

Figure 13.12 Selecting the signed fixed point with 13 binary points.

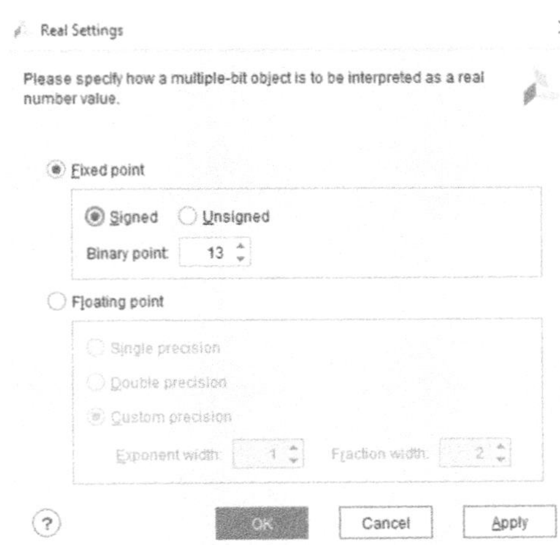

We enter a simulation time step of 2 ms in the boxes and click on the Restart button, then the Time Step Simulation button, and also click on the Zoom Fit button, as depicted in Figures 13.20 and 13.21, respectively.

We right-click on the signal "clk" and select the Force Clock... option, as illustrated in Figure 13.22.

We configure the settings of Force Clock window, as shown in Figure 13.23.

Then we click on the Restart button, the Time Step Simulation button, and then the Zoom Fit button. The waveforms window is depicted in Figure 13.24.

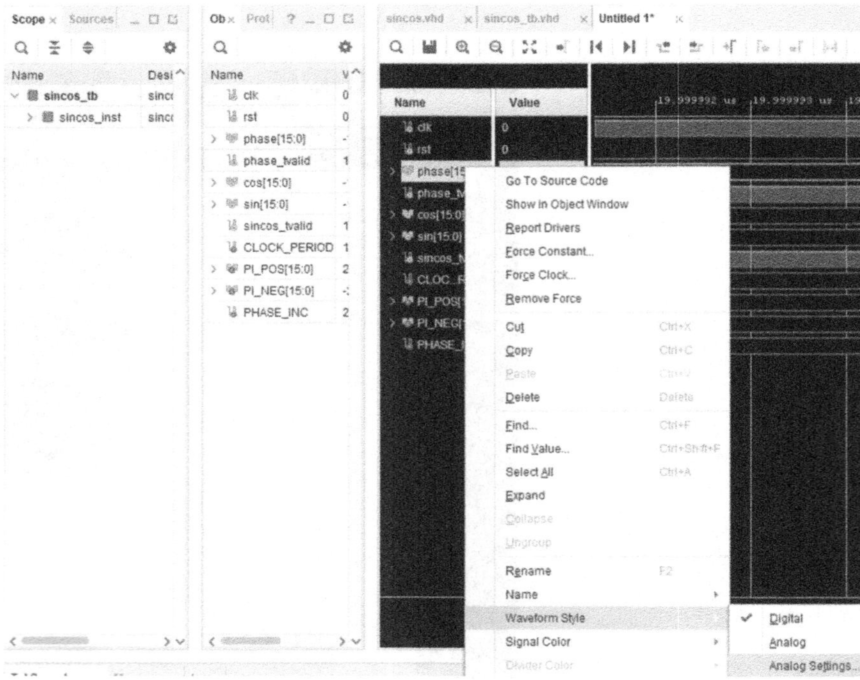

Figure 13.13 Selecting the Analog Settings... for phase signal.

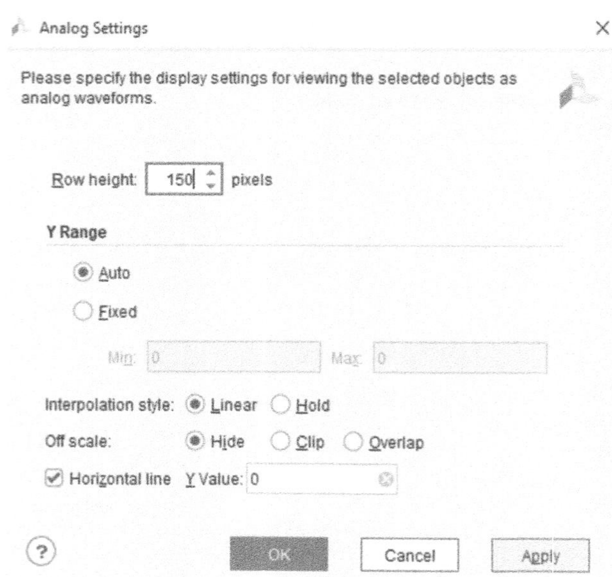

Figure 13.14 The Analog Settings window.

We right-click on the signal "car_sig", and select Waveform Style as Analog, as illustrated in Figure 13.25.

Also, we right-click on the signal "car_sig", and select Radix, and Unsigned Decimal, as demonstrated in Figure 13.26.

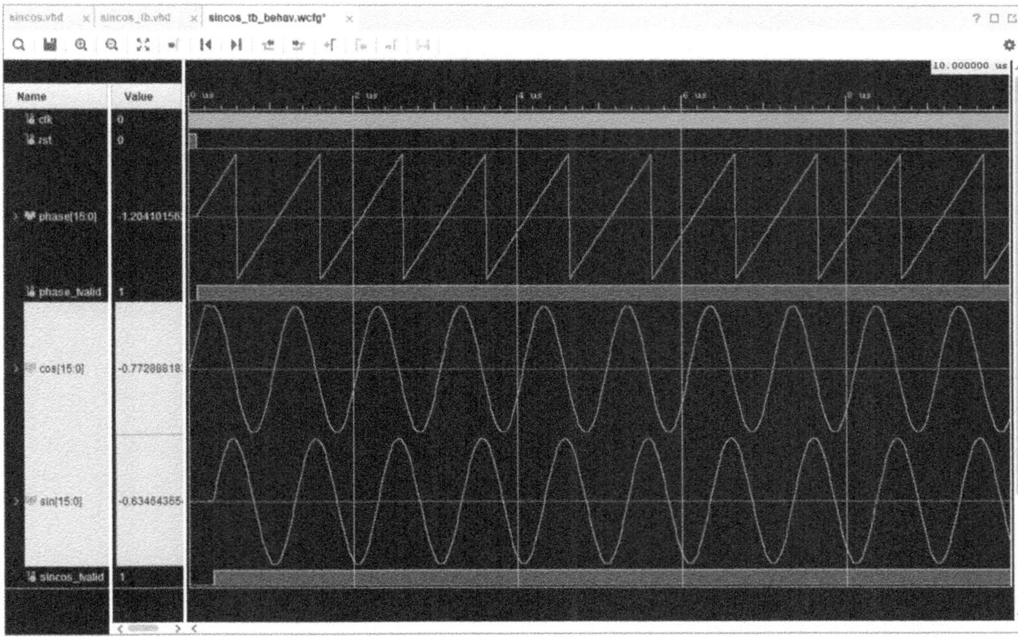

Figure 13.15 The waveforms of phase, cos, and sin.

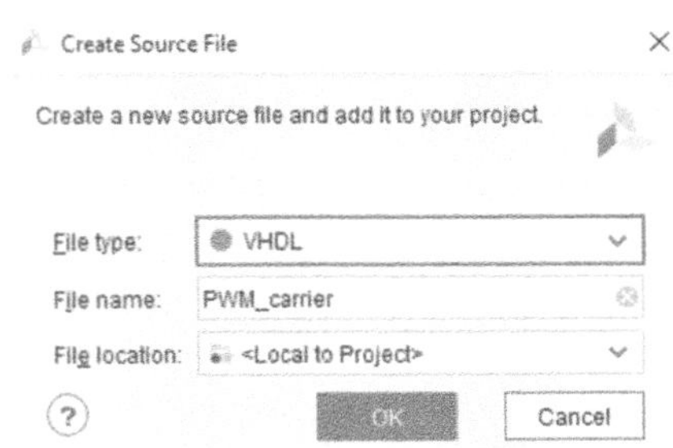

Figure 13.16 Creating the source file PWM_carrier.vhd.

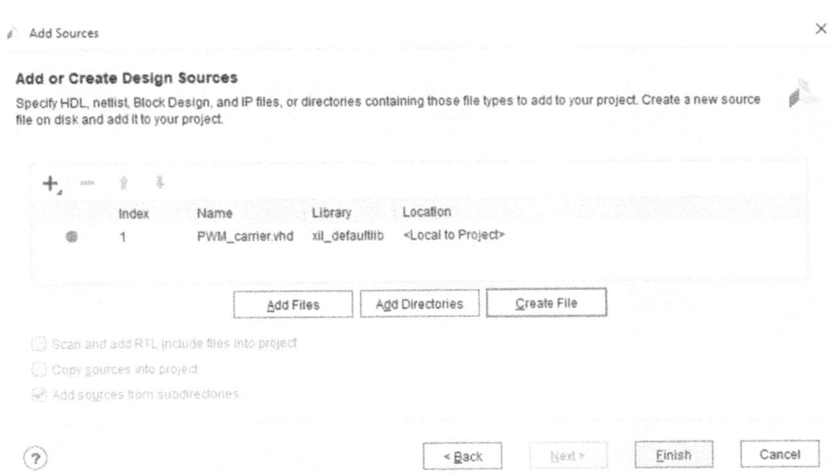

Figure 13.17 Clicking on the Finish button.

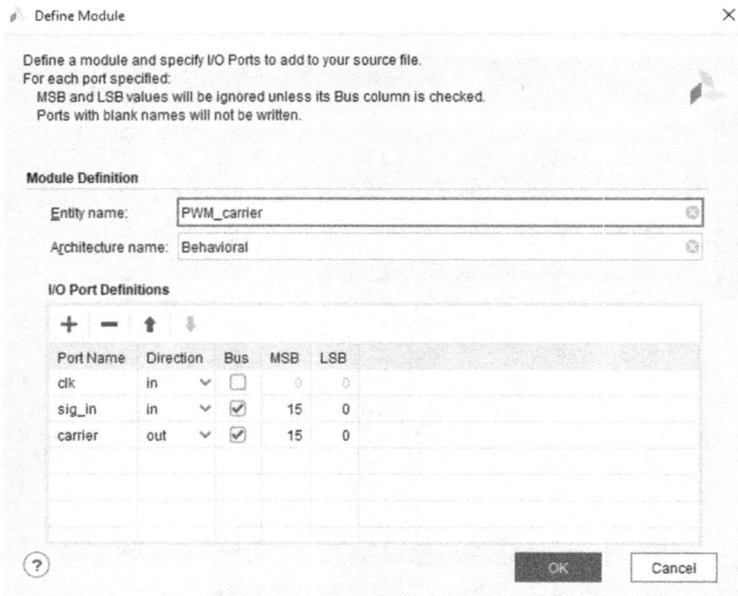

Figure 13.18 Adding input and output ports.

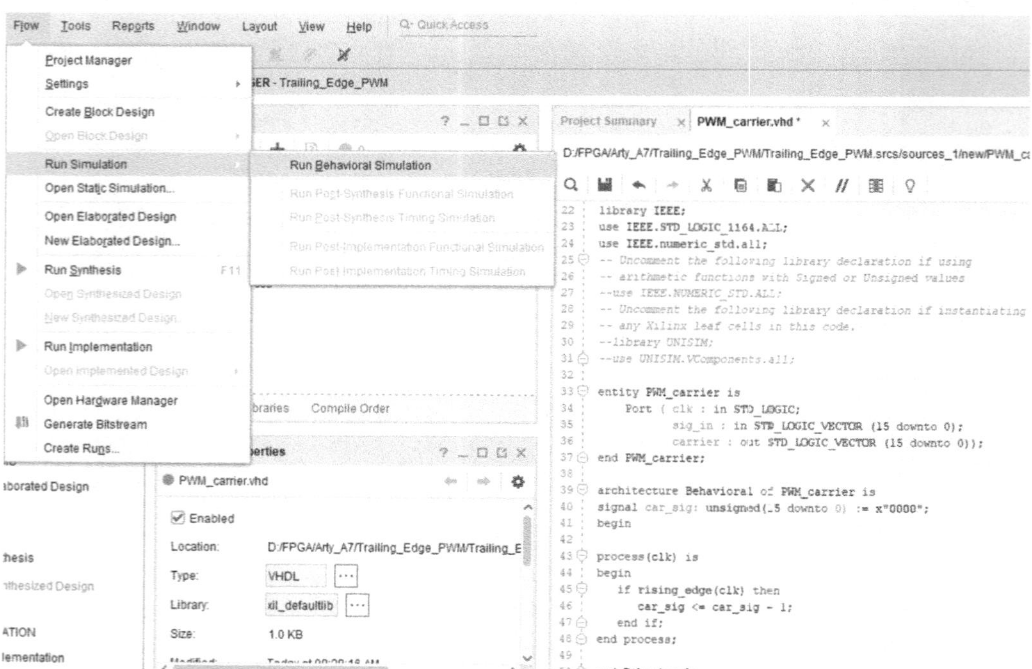

Figure 13.19 The carrier signal simulation.

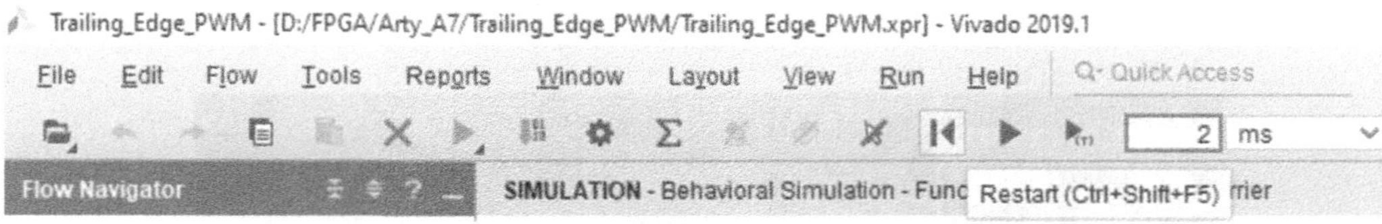

Figure 13.20 Setting simulation time step and clicking the Restart button.

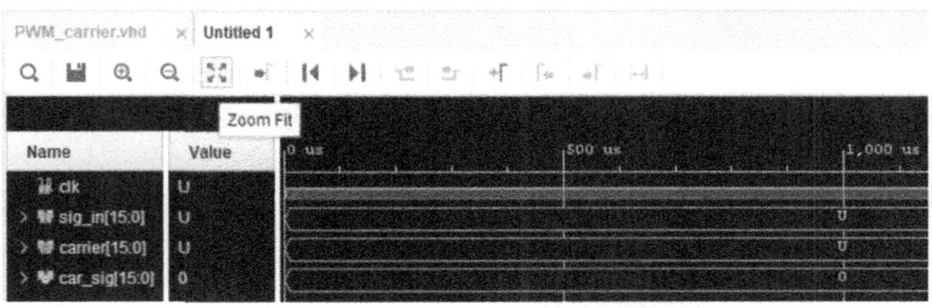

Figure 13.21 Clicking on the Zoom Fit button.

Figure 13.22 Selecting the Force Clock... option.

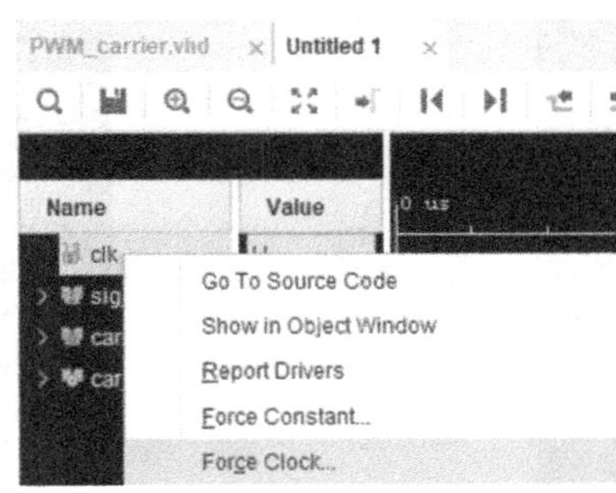

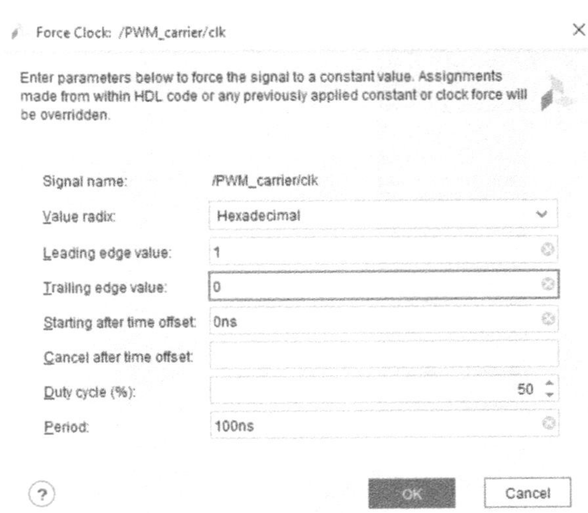

Figure 13.23 The settings of Force Clock window.

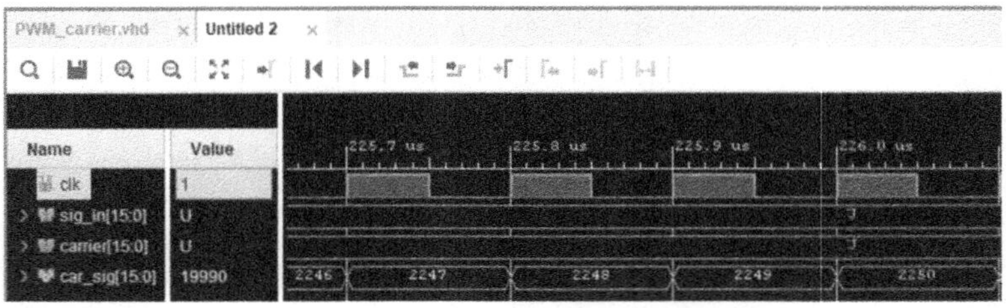

Figure 13.24 The waveforms window.

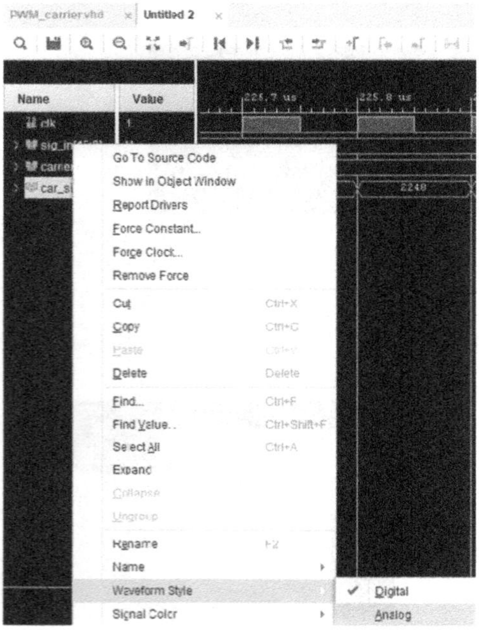

Figure 13.25 Selecting Waveform Style as Analog for the signal "car_sig".

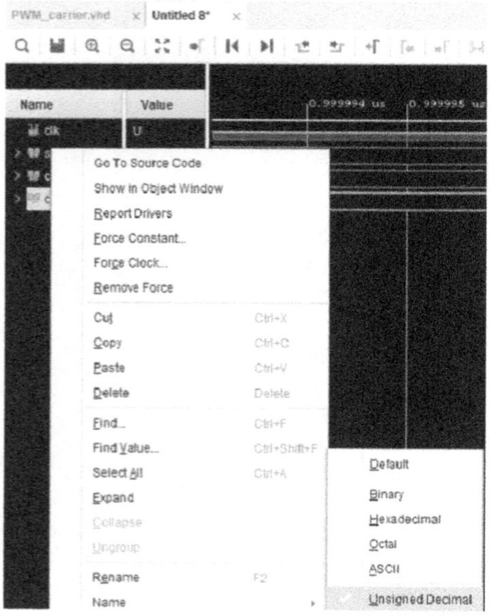

Figure 13.26 Selecting the Radix, and Unsigned Decimal.

We click on the Restart button, the Time Step Simulation button, and also the Zoom Fit button. The analog signal "car_sig" waveform is shown in Figure 13.27.

We want to limit the signal "car_sig" to count up the maximum value of 29999. To do this, we modify the codes and run the behavioral simulation, as depicted in Figure 13.28.

We repeat the same procedures to simulate in the waveforms window. We can see that the frequency of the carrier signal increases as illustrated in Figure 13.29.

Next, we create a PWM signal by comparing a constant signal to the carrier signal and run the behavioral simulation, as shown in Figure 13.30.

We right-click on the signal "sig_in" and select the Force Constant... option, as depicted in Figure 13.31.

We choose 1000 as the unsigned decimal value, as illustrated in Figure 13.32.

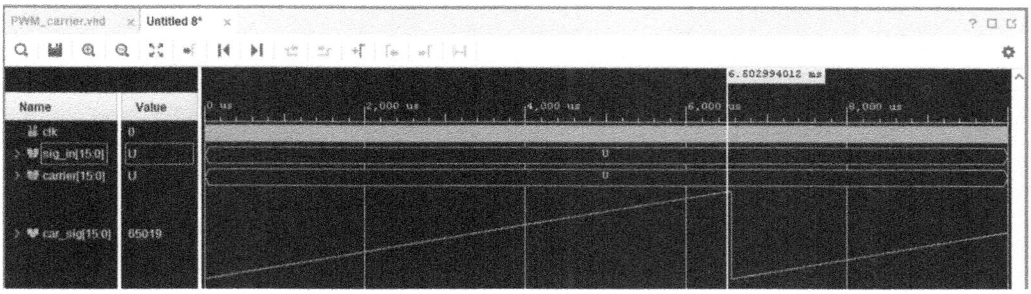

Figure 13.27 The analog signal "car_sig" waveform.

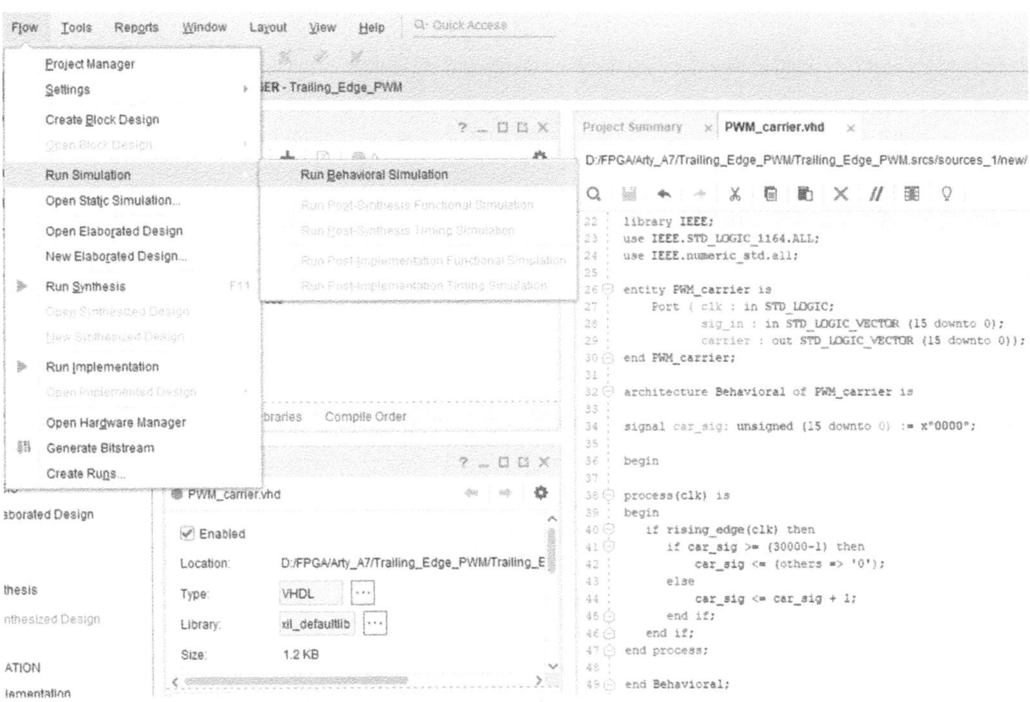

Figure 13.28 Modifying the codes and running the behavioral simulation.

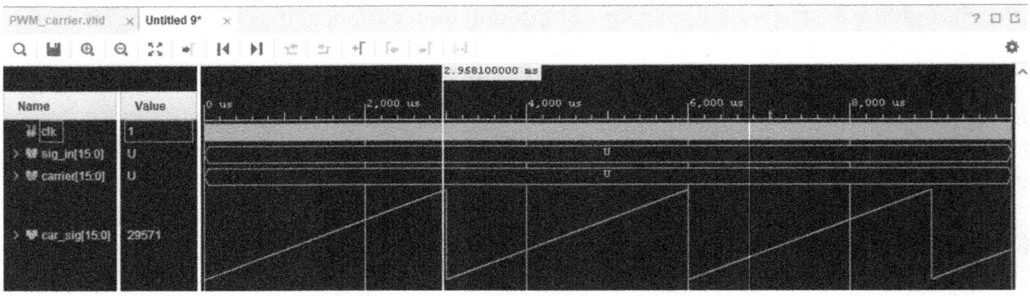

Figure 13.29 The increased frequency of carrier signal.

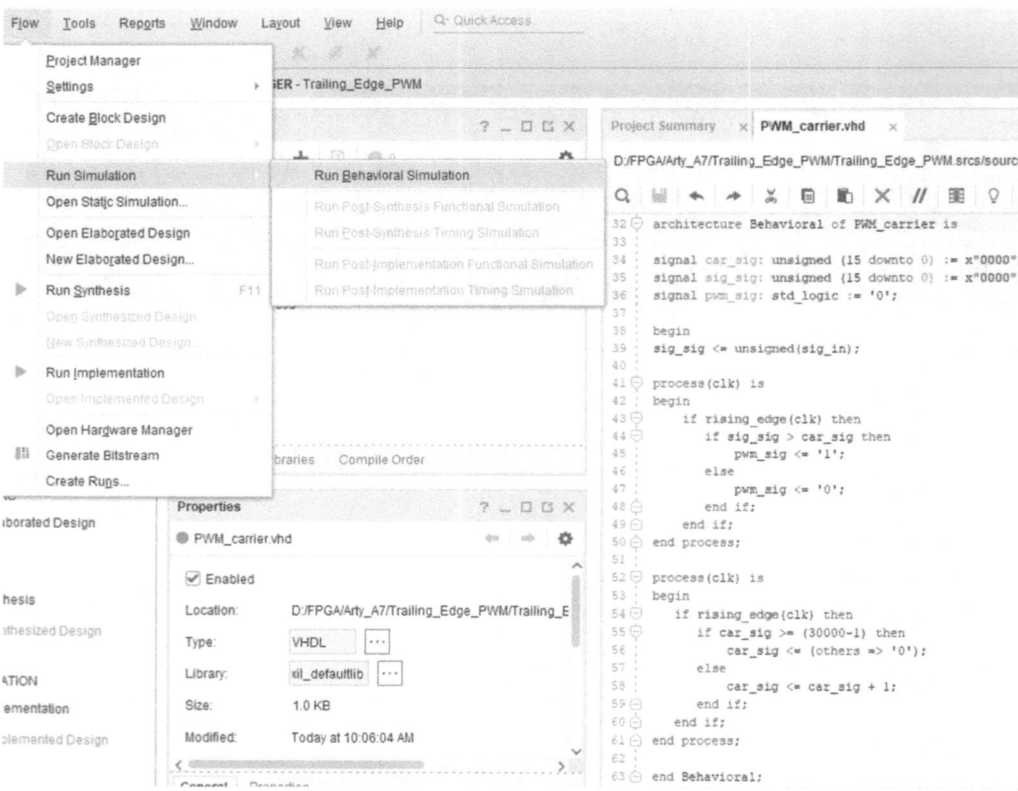

Figure 13.30 Running the behavioral simulation.

The signals on the waveforms window are shown in Figure 13.33.

We set the force value to 5000, and click on the Step Time Simulation button to see signals on the waveforms window as depicted in Figures 13.34 and 13.35, respectively.

We change the force constant value to 15000 and click on the Step Time Simulation button to view the change in PWM signal as illustrated in Figures 13.36 and 13.37.

We also change the force constant value to 25000 and click on the Step Time Simulation button to again view the PWM signal as illustrated in Figures 13.38 and 13.39, respectively.

The signals for total time simulation on the waveforms window are shown in Figure 13.40.

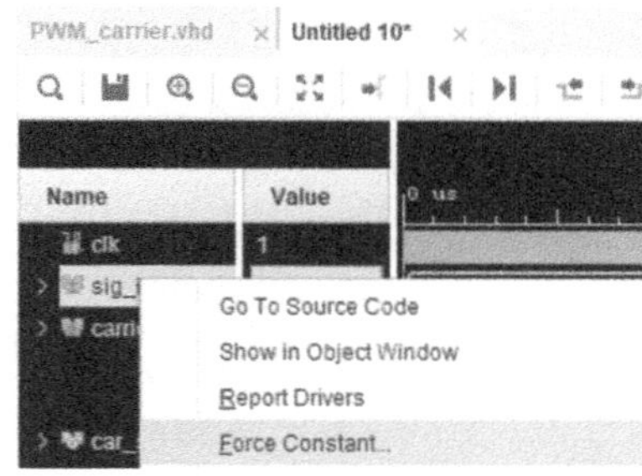

Figure 13.31 Selecting the Force Constant... option.

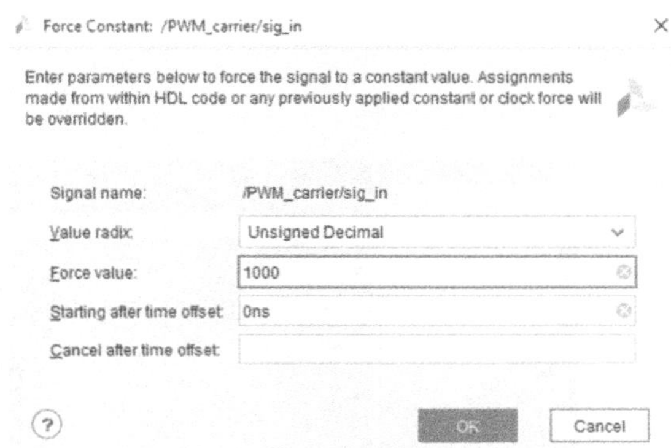

Figure 13.32 Settings of the Force Constant window.

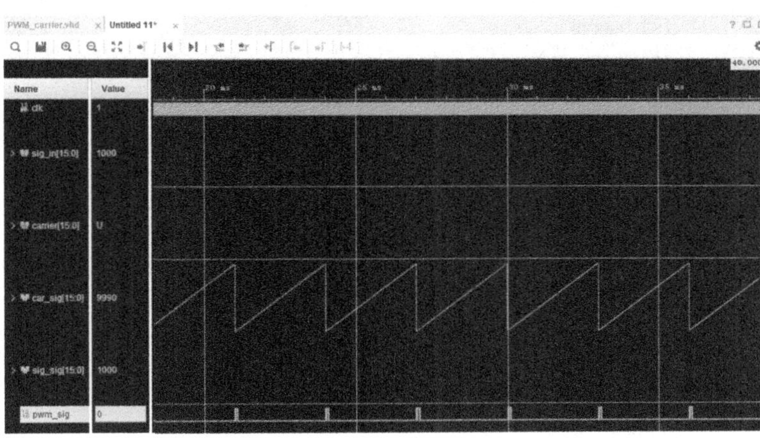

Figure 13.33 The signals on the waveforms window.

We add the "pwm_out" as an output port and assign it to the signal pwm_sig. Additionally, assign the signal "car_sig" to the output port "carrier" and run behavioral simulation as depicted in Figure 13.41.

We change the force constant value to 10000, and click on the Step Time Simulation button to view the total signals on the waveforms window as illustrated in Figure 13.42.

Now, we can create an IP core for this PWM carrier. To do that, from Tools menu, we select Create and Package New IP... as shown in Figure 13.43.

In the opened window, we click on the Next button as depicted in Figure 13.44.

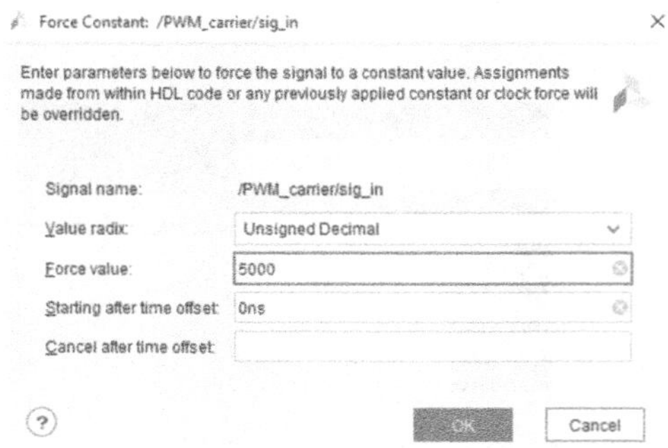

Figure 13.34 Forcing the constant value to 5000.

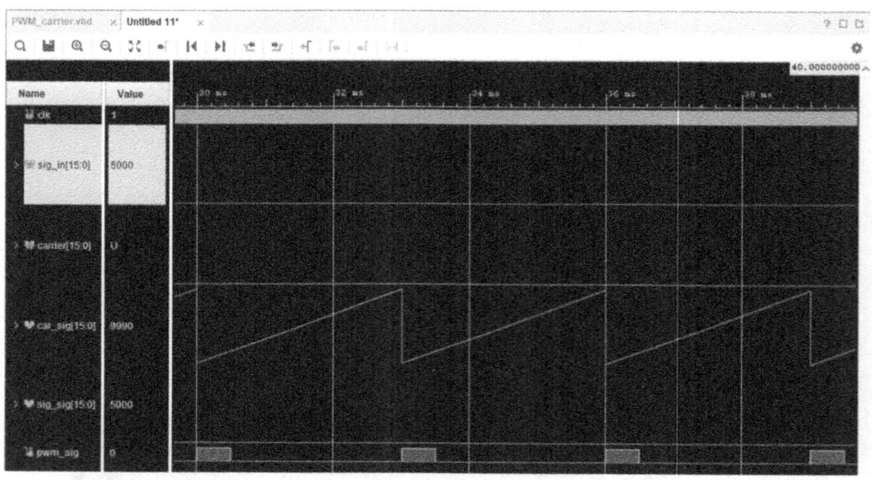

Figure 13.35 The signals on the waveforms window.

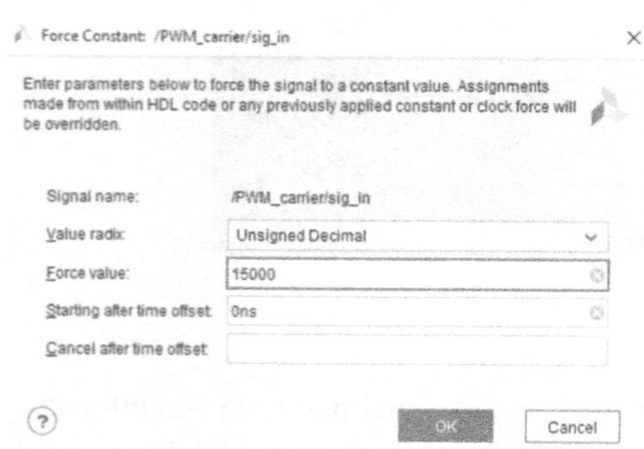

Figure 13.36 Changing the Force value to 15000.

Then, we select Package your current project, and click on the Next button as illustrated in Figure 13.45.

We leave the IP location unchanged and click on the Next button as shown in Figure 13.46.

In the next window, we click on the Finish button as depicted in Figure 13.47.

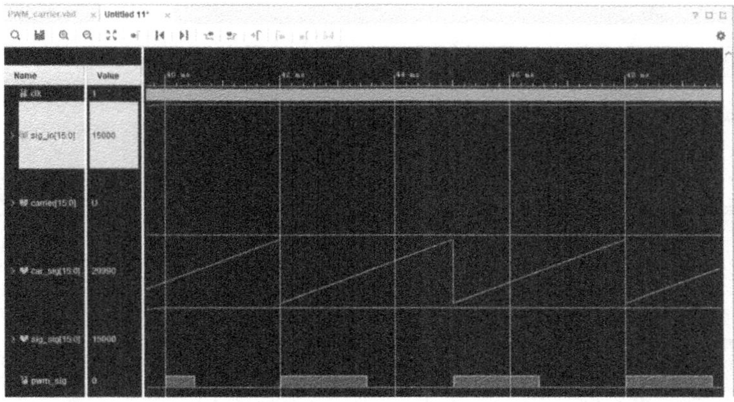

Figure 13.37 The signals on the waveforms window.

Figure 13.38 Changing the Force value to 25000.

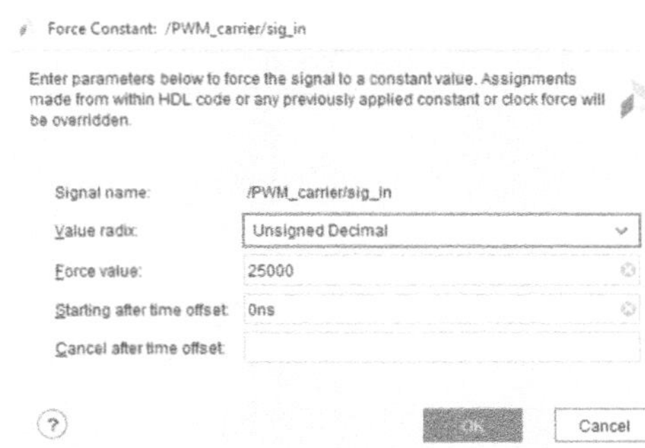

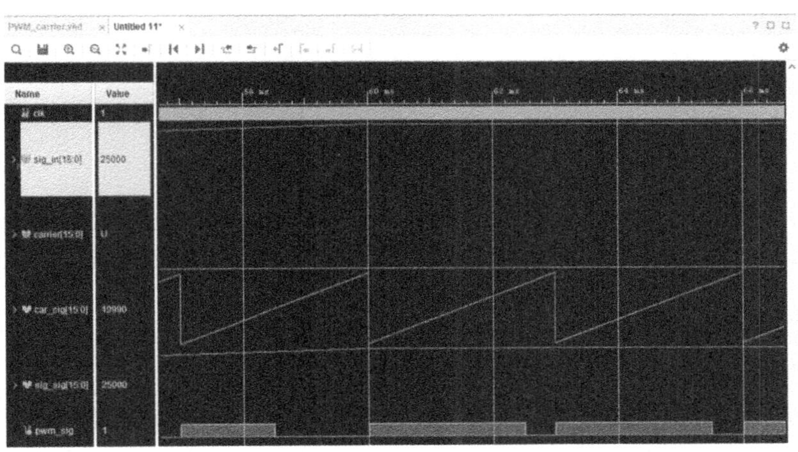

Figure 13.39 The signals on the waveforms window.

In the packaging setup window, and in the Review and Package tab, we click on the Package IP button, so the PWM_carrier IP is produced, and we can find it in IP Catalog as illustrated in Figures 13.48 and 13.49, respectively.

If we click on the PWM_carrier_v1_0 icon, we can see the Customize IP window as shown in Figure 13.50.

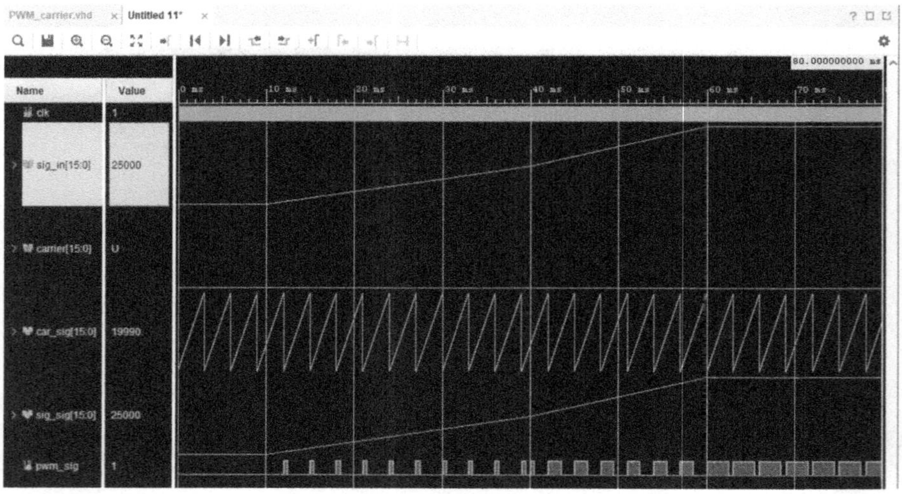

Figure 13.40 The total time simulation on waveforms window.

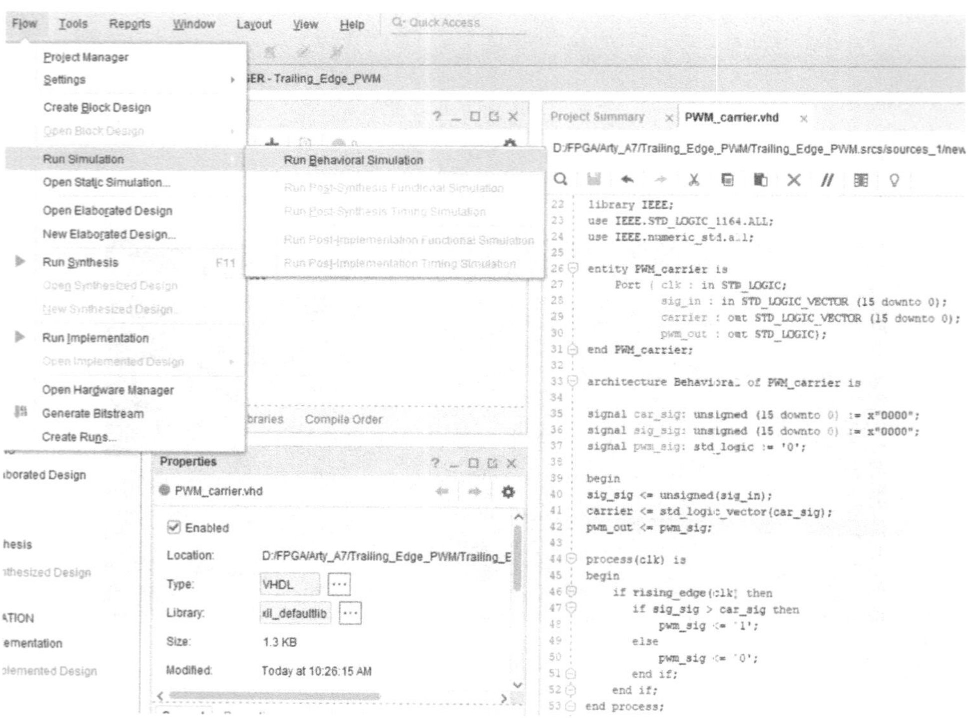

Figure 13.41 Running the complete codes behavioral simulation.

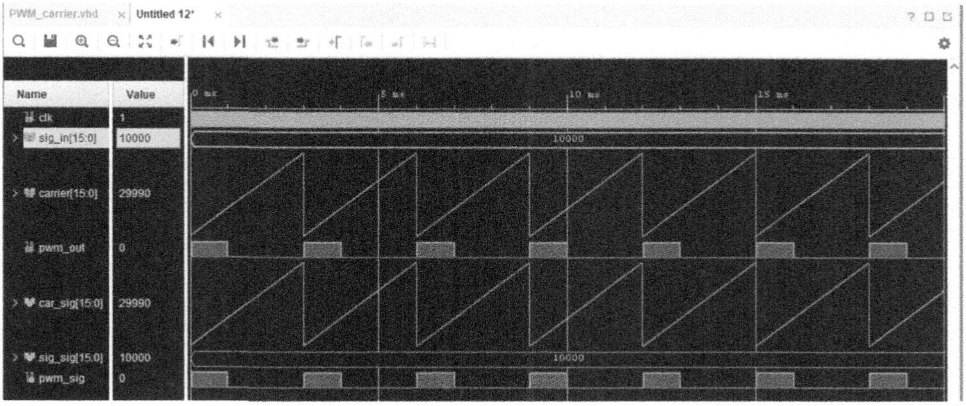

Figure 13.42 The complete signals on the waveforms window.

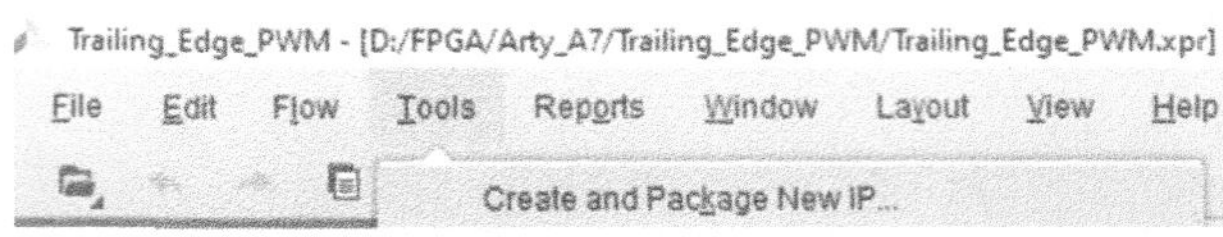

Figure 13.43 Selecting the Create and Package New IP....

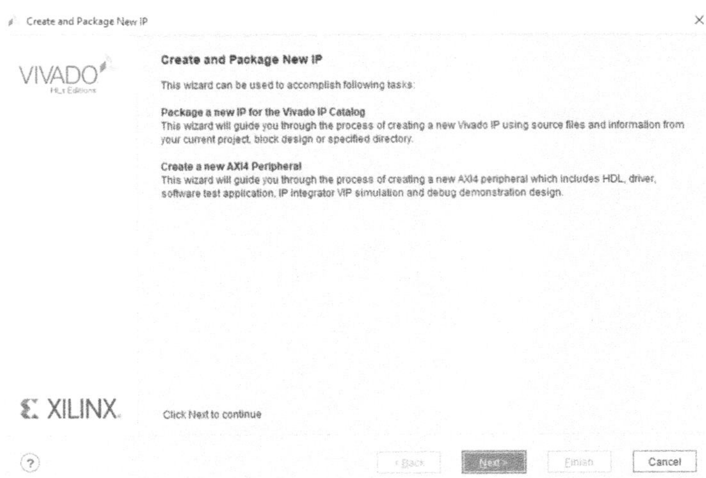

Figure 13.44 Clicking on the Next button.

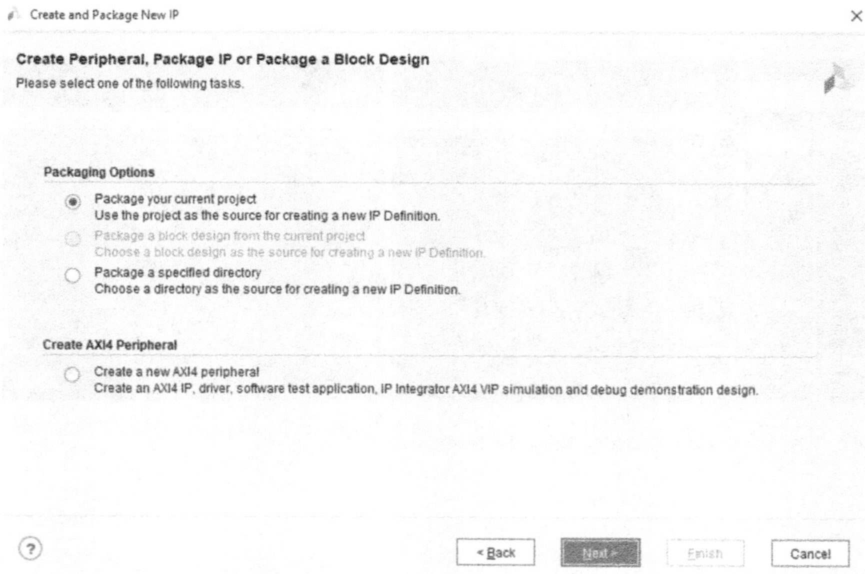

Figure 13.45 Selecting Package your current project.

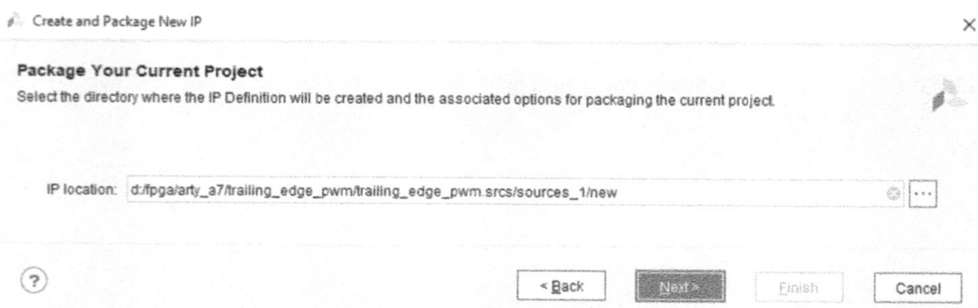

Figure 13.46 Clicking on the Next button.

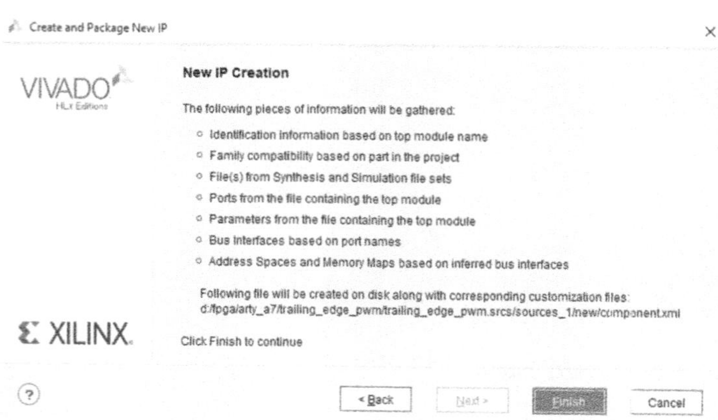

Figure 13.47 Clicking on the Finish button.

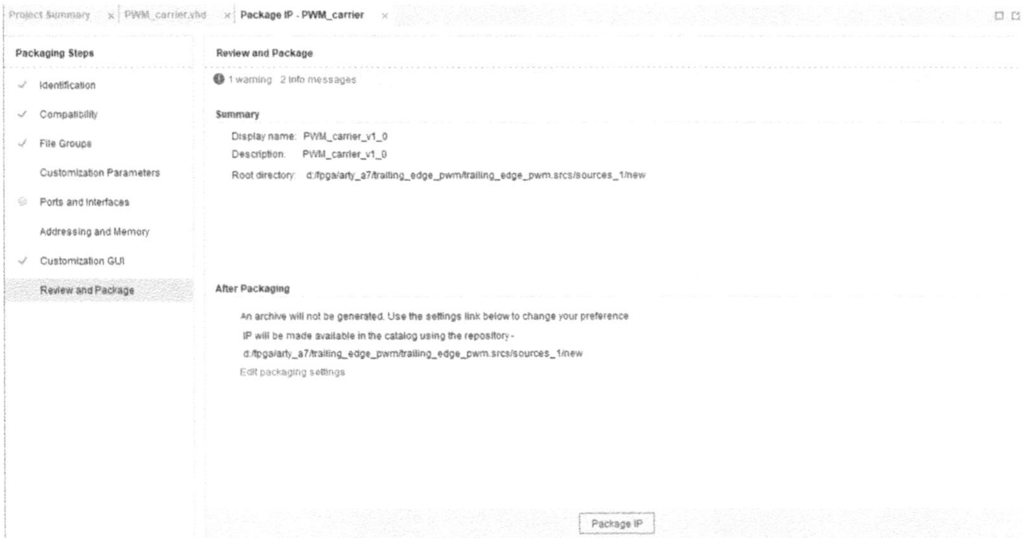

Figure 13.48 Clicking on the Package IP button.

Figure 13.49 The generated PWM_carrier IP.

13.5 The LUT-based Sine Wave in VHDL

In this section, we are going to create a sine wave based on a lookup table (LUT) in VHDL using Vivado. This technique is useful for functions like PWM. We will use Google Colab, which is a hosted Jupyter notebook service, to generate the LUT values for the sine waveform, and we use the following codes.

```
import numpy as np
from numpy import sin, pi
import matplotlib.pyplot as plt
x = np.linspace(0,1,500)
lut = sin(pi*x)*7500+7500
plt.plot(x,lut)
```

The generated sine half-wave signal is shown in Figure 13.51.

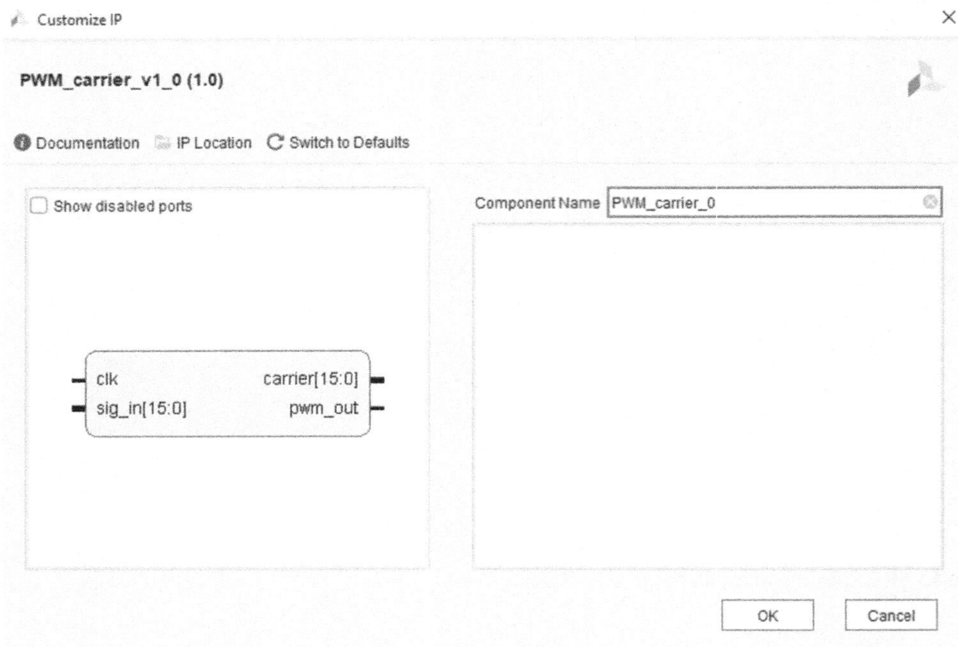

Figure 13.50 The Customize IP window of PWM_carrier.

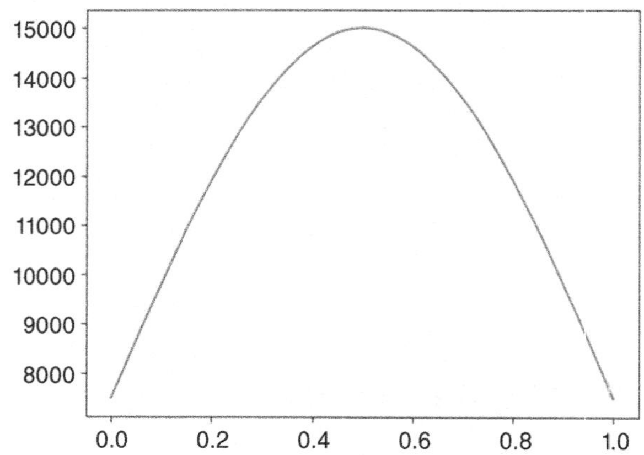

Figure 13.51 The generated sine half-wave signal.

The following codes are used for generating 500 LUT values with 20 values in a row.

```
for i,val in enumerate(lut):
    if i % 20:
       print(int(val),end=',')
    else:
       print(int(val),end=',')
       print(' ')
```

We create a VHDL file titled "sine_lut.vhd" and insert the following codes inside it.

```
library IEEE;
use IEEE.STD_LOGIC_1164.ALL;
use IEEE.NUMERIC_STD.ALL;

entity sine_lut is
     Port(
          clk: in std_logic;
          rst: in std_logic;
          sine: out std_logic_vector(15 downto 0));
end sine_lut;

architecture Behavioral of sine_lut is

type LUT_TYPE is array(0 to 499) of integer;
constant LUT: LUT_TYPE :=
(
7500,7547,7594,7641,7688,7736,7783,7830,7877,7924,7971,8018,8066,8113,8160,
8207,8254,8301,8348,8395,8441,
8488,8535,8582,8628,8675,8722,8768,8815,8861,8908,8954,9000,9047,9093,9139,
9185,9231,9277,9323,9368,
9414,9460,9505,9551,9596,9641,9687,9732,9777,9822,9866,9911,9956,10000,10045,
10089,10134,10178,10222,10266,
10310,10353,10397,10440,10484,10527,10570,10613,10656,10699,10741,10784,10826,
10869,10911,10953,10995,11036,11078,11119,
11161,11202,11243,11284,11324,11365,11405,11445,11485,11525,11565,11605,11644,
11683,11723,11761,11800,11839,11877,11916,
11954,11992,12029,12067,12104,12141,12178,12215,12252,12288,12324,12360,12396,
12432,12467,12503,12538,12573,12607,12642,
12676,12710,12744,12778,12811,12844,12877,12910,12943,12975,13007,13039,13071,
13102,13134,13165,13196,13226,13257,13287,
13317,13346,13376,13405,13434,13463,13491,13520,13548,13575,13603,13630,13657,
13684,13711,13737,13763,13789,13815,13840,
13865,13890,13915,13939,13963,13987,14010,14034,14057,14080,14102,14124,14146,
14168,14190,14211,14232,14252,14273,14293,
14313,14332,14352,14371,14390,14408,14426,14444,14462,14479,14497,14513,14530,
14546,14562,14578,14594,14609,14624,14638,
14653,14667,14680,14694,14707,14720,14733,14745,14757,14769,14780,14791,14802,
14813,14823,14833,14843,14852,14862,14871,
14879,14887,14895,14903,14910,14918,14924,14931,14937,14943,14949,14954,14959,
14964,14968,14972,14976,14980,14983,14986,
14989,14991,14993,14995,14996,14998,14999,14999,14999,14999,14999,14999,14998,
14996,14995,14993,14991,14989,14986,14983,
14980,14976,14972,14968,14964,14959,14954,14949,14943,14937,14931,14924,14918,
14910,14903,14895,14887,14879,14871,14862,
14852,14843,14833,14823,14813,14802,14791,14780,14769,14757,14745,14733,14720,
14707,14694,14680,14667,14653,14638,14624,
14609,14594,14578,14562,14546,14530,14513,14497,14479,14462,14444,14426,14408,
14390,14371,14352,14332,14313,14293,14273,
14252,14232,14211,14190,14168,14146,14124,14102,14080,14057,14034,14010,13987,
13963,13939,13915,13890,13865,13840,13815,
13789,13763,13737,13711,13684,13657,13630,13603,13575,13548,13520,13491,13463,
13434,13405,13376,13346,13317,13287,13257,
13226,13196,13165,13134,13102,13071,13039,13007,12975,12943,12910,12877,12844,
12811,12778,12744,12710,12676,12642,12607,
```

```
12573,12538,12503,12467,12432,12396,12360,12324,12288,12252,12215,12178,12141,
12104,12067,12029,11992,11954,11916,11877,
11839,11800,11761,11723,11683,11644,11605,11565,11525,11485,11445,11405,11365,
11324,11284,11243,11202,11161,11119,11078,
11036,10995,10953,10911,10869,10826,10784,10741,10699,10656,10613,10570,10527,
10484,10440,10397,10353,10310,10266,10222,
10178,10134,10089,10045,10000,9956,9911,9866,9822,9777,9732,9687,9641,9596,9551,
9505,9460,9414,9368,9323,
9277,9231,9185,9139,9093,9047,9000,8954,8908,8861,8815,8768,8722,8675,8628,8582,
8535,8488,8441,8395,
8348,8301,8254,8207,8160,8113,8066,8018,7971,7924,7877,7830,7783,7736,7688,7641,
7594,7547,7500
);

signal lut_sig: unsigned(15 downto 0);

begin

sine <= std_logic_vector(lut_sig);

process(clk,rst) is
variable cnt: integer range 0 to 500;
begin
    if rst = '0' then
       cnt := 0;
       lut_sig <= (others => '0');
    elsif rising_edge(clk) then
       lut_sig <= to_unsigned(LUT(cnt),16);
       if cnt > 498 then
          cnt := 0;
       else
          cnt := cnt + 1;
       end if;
    end if;
end process;

end Behavioral;
```

Then, from Flow menu, we select the Run Behavioral Simulation as depicted in Figure 13.52.

Then, we click on the Restart button in the opened waveforms window as illustrated in Figure 13.53.

We then right-click on the "rst" and select the Force Constant... option as demonstrated in Figure 13.54.

We enter the value of zero in the Force Constant window, as shown in Figure 13.55.

We right-click on the "clk" and select the Force Clock... option as depicted in Figure 13.56.

We set the Force Clock window for "clk" as shown in Figure 13.57.

Then, we run the simulation for 100 μs as illustrated in Figure 13.58.

After simulation for 100 μs, we right-click on the "rst" and set the Force value to 1, as shown in Figure 13.59.

We click on the Simulation for 100 μs button again and click on the Zoom Fit button. The resulted signal waveforms are depicted in Figure 13.60.

Next, we right-click on the lut_sig and select the Waveform Style as Analog as illustrated in Figure 13.61.

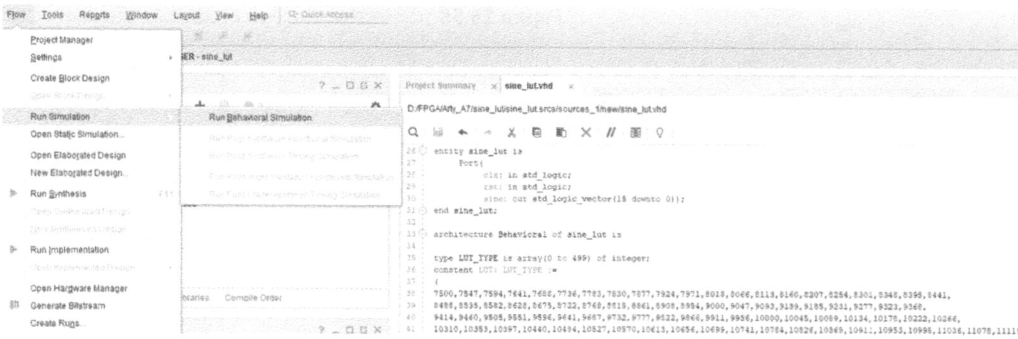

Figure 13.52 Selecting the Run Behavioral Simulation.

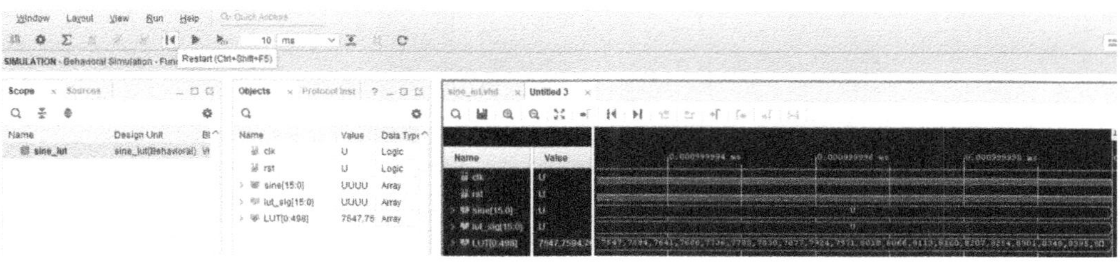

Figure 13.53 Clicking on the Restart button.

Figure 13.54 Selecting the Force Constant... option for "rst".

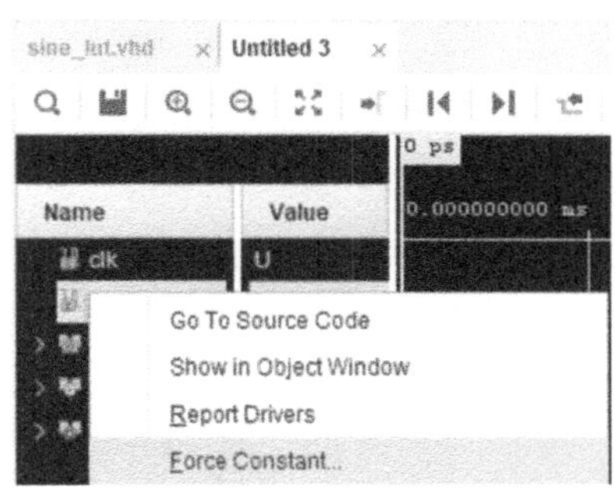

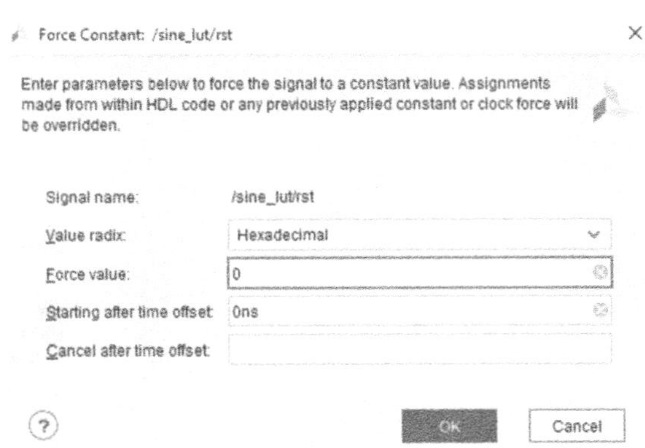

Figure 13.55 Entering the value of zero in the Force Constant window.

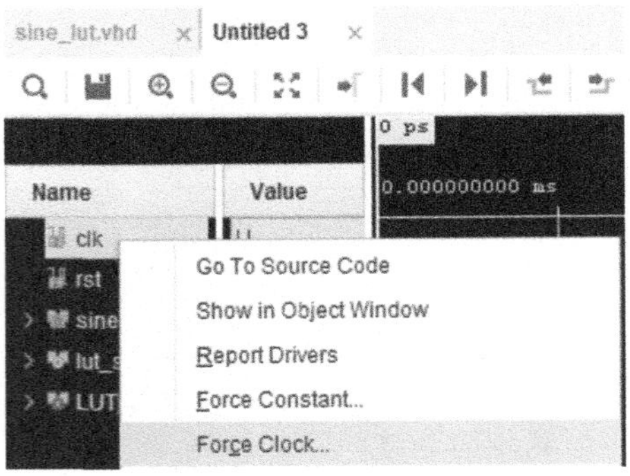

Figure 13.56 Selecting the Force Clock... option.

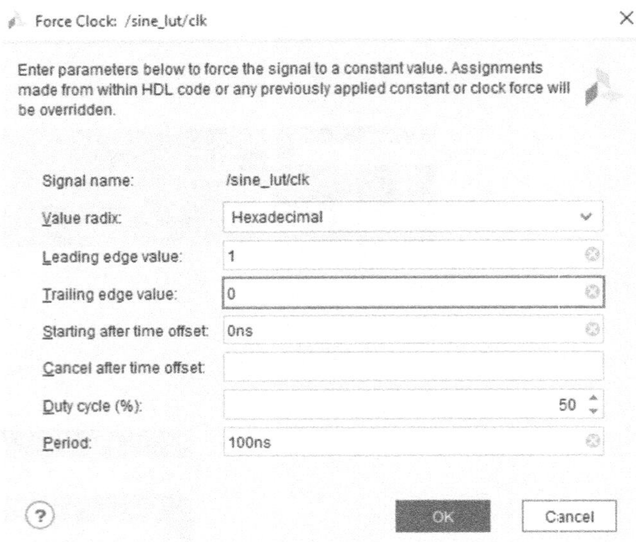

Figure 13.57 Setting the Force Clock window for "clk".

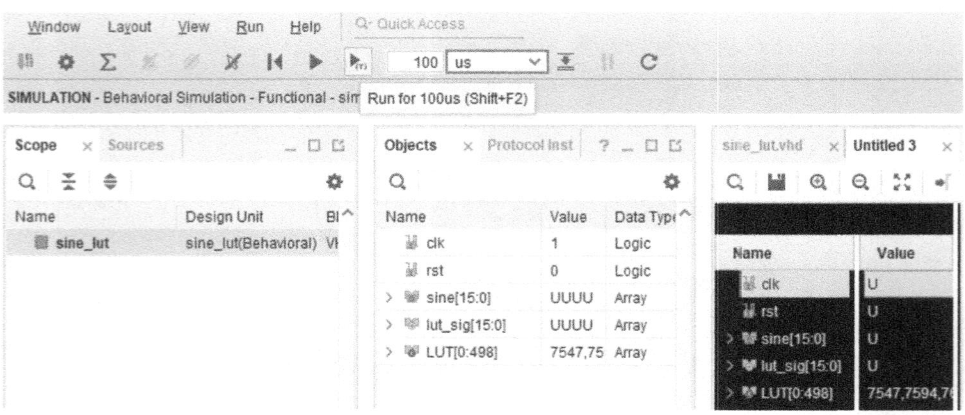

Figure 13.58 Running the simulation for 100 μs.

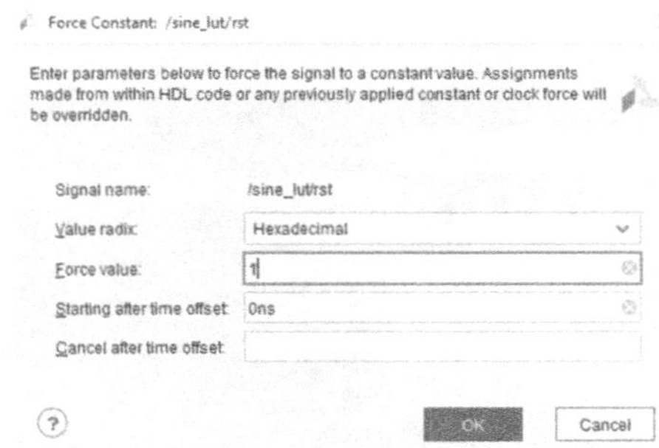

Figure 13.59 Changing the Force value of "rst" to 1.

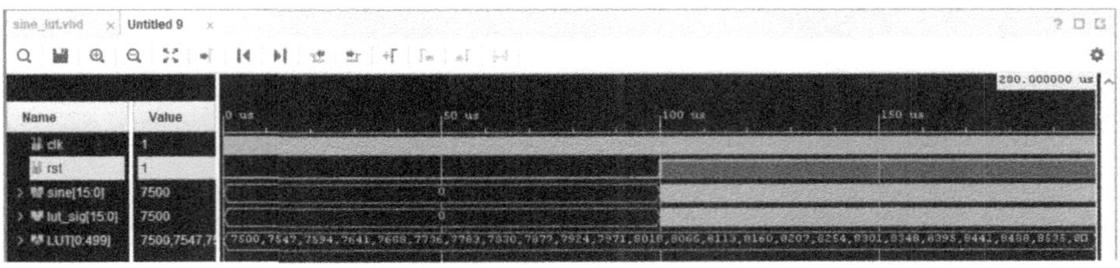

Figure 13.60 The signals on the waveforms window.

Figure 13.61 Selecting the Waveform Style as Analog.

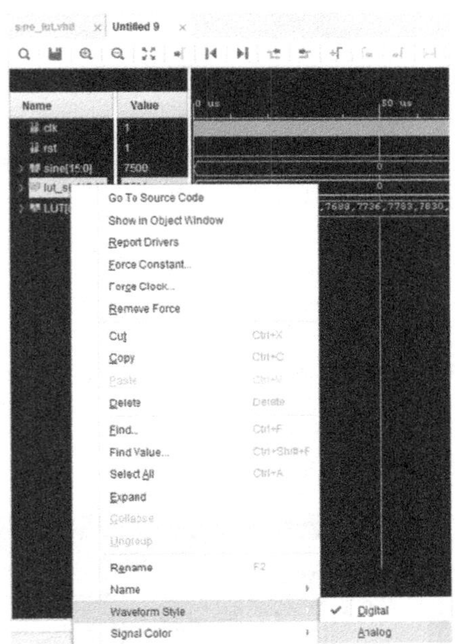

The resulting waveform of "lut_sig" is shown in Figure 13.62.

To perform a 180-degree rotation during the negative half-cycle, we define a signal "state" and modify the previous codes accordingly. We then run the behavioral simulation as depicted in Figure 13.63.

We restart the simulation, set the "rst" value to zero, force the clock "clk", and run the simulation for 100 μs. Then, as before, we set the "rst" value to 1 and continue to run the simulation for multiple 100 μs. This results in the "sine" waveform being produced as expected, as shown in Figure 13.64.

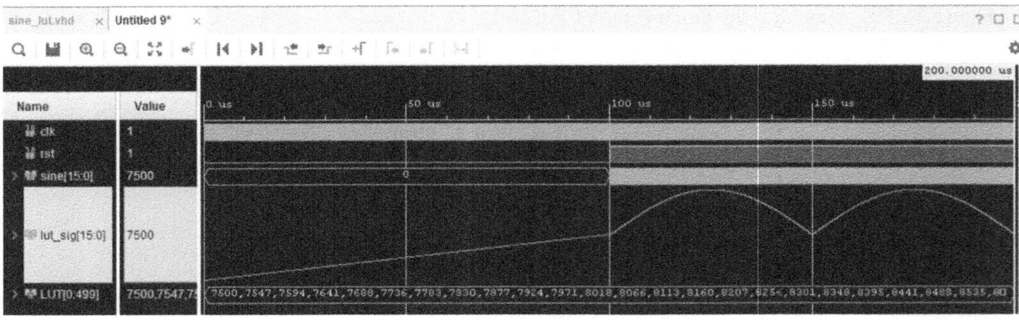

Figure 13.62 The resulting waveform of "lut_sig".

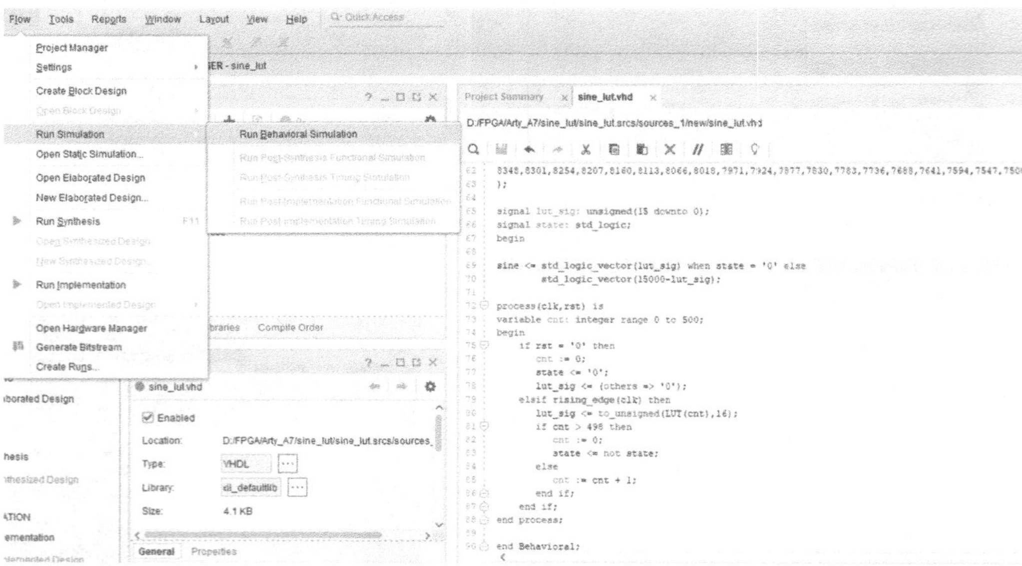

Figure 13.63 Modifying the codes and running the behavioral simulation.

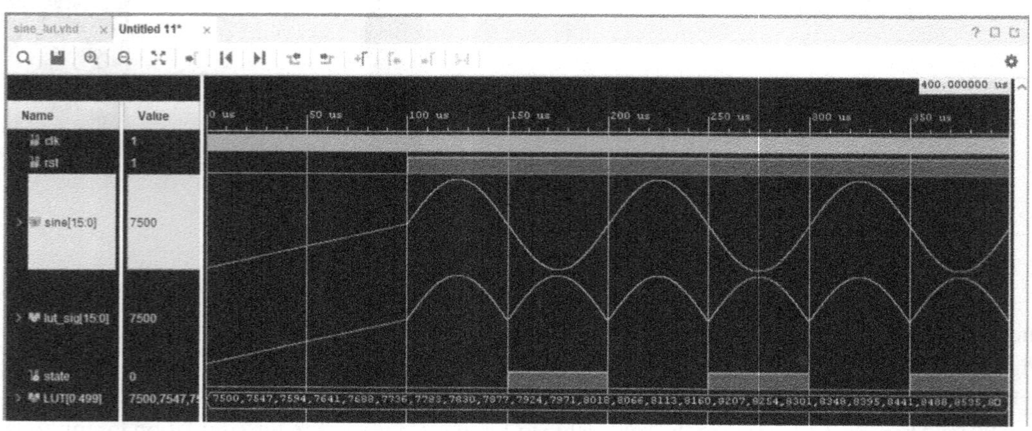

Figure 13.64 The produced "sine" waveform.

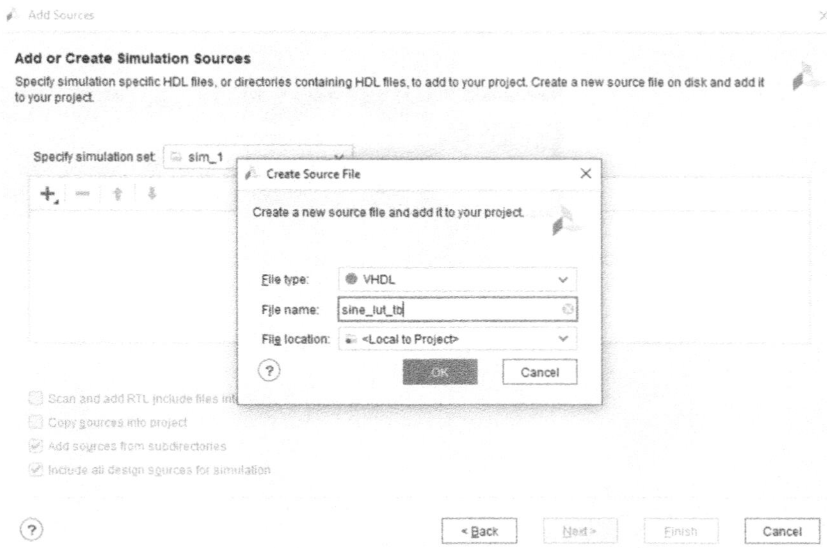

Figure 13.65 Creating a test bench source file for simulation.

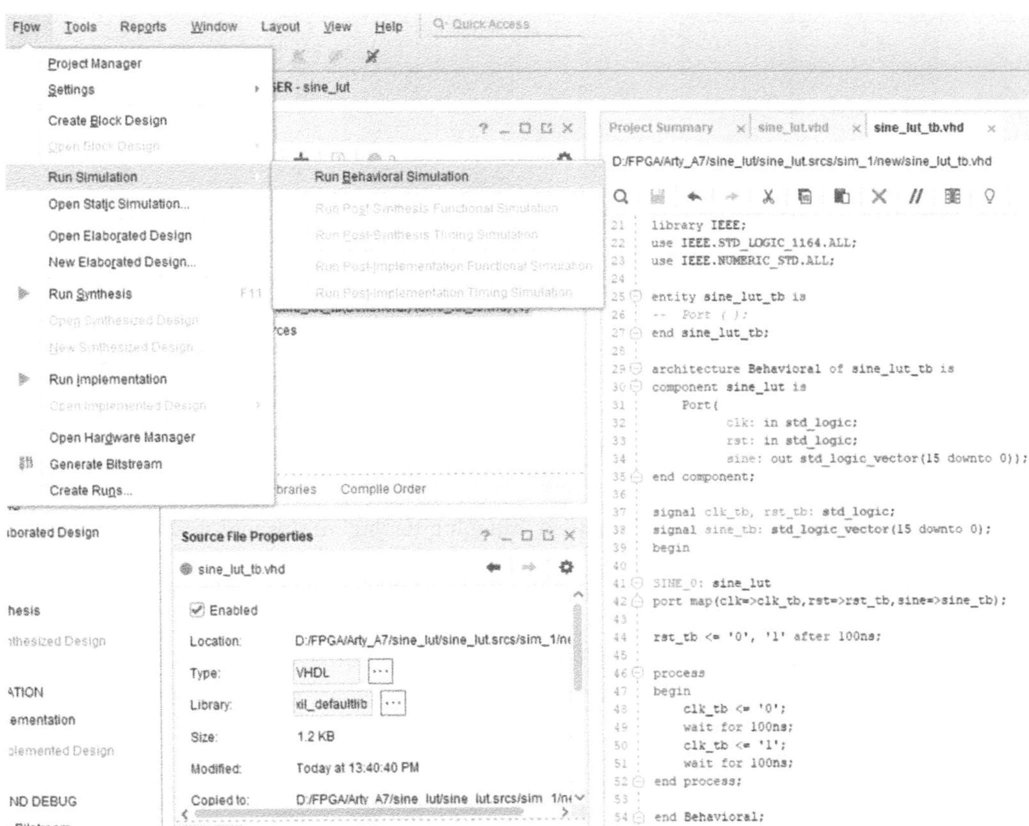

Figure 13.66 Selecting the Run Behavioral Simulation.

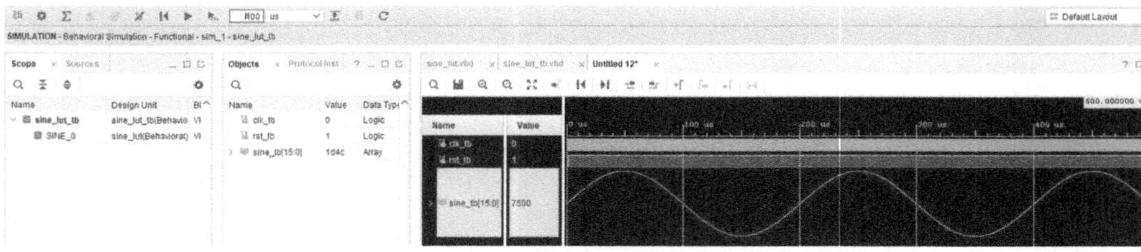

Figure 13.67 The resulted waveforms window for our test bench.

Figure 13.68 Selecting the Add Module to Block Design.

Then, we create a test bench source file titled "sine_lut_tb.vhd" in simulation sources folder as illustrated in Figure 13.65.

We write the codes inside the file "sine_lut_tb.vhd", and select Run Behavioral Simulation from Flow menu, as shown in Figure 13.66.

We restart the simulation, click the Run for 100 μs button multiple times, and select the waveform style of "sine_tb" as Analog. The resulting waveforms window for our test bench is depicted in Figure 13.67.

If we create a block design titled "test_IP", we can add our sine_lut.vhd as an IP core in the Diagram window. To do this, we right-click on the file name and select Add Module to Block Design, as illustrated in Figure 13.68.

The "sine_lut" module is added in the Diagram window as shown in Figure 13.69.

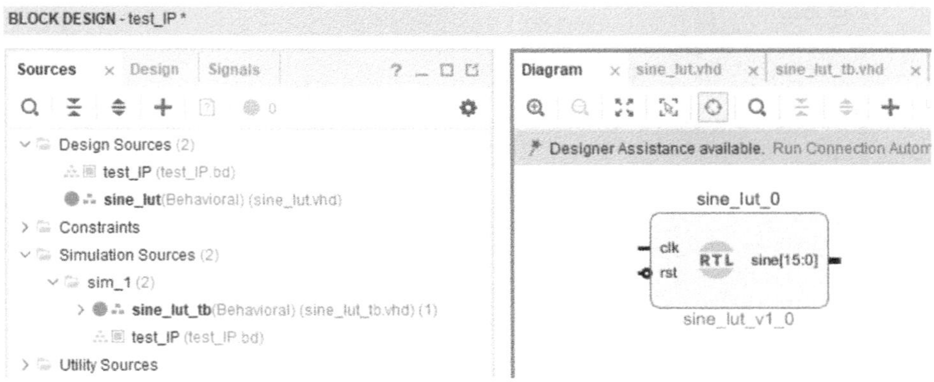

Figure 13.69 The added "sine_lut" module in the Diagram window.

13.6 The Symmetrical PWM in VHDL

In this section, we are going to implement a symmetrical PWM carrier for PWM generation, and compare it with a sine wave to generate two-level PWM signal, which can be used later in power electronics converters. We previously implemented a trial PWM carrier in Section 13.3. Now, we will make some changes in the trial PWM to be a symmetrical up–down counter. The codes for the file PWM_CAR.vhd are as follows.

```
library IEEE;
use IEEE.STD_LOGIC_1164.ALL;
use IEEE.numeric_std.all;

entity PWM_CAR is
    Port ( clk : in STD_LOGIC;
           rst : in STD_LOGIC;
           carrier : out STD_LOGIC_VECTOR (15 downto 0)
          );
end PWM_CAR;

architecture Behavioral of PWM_CAR is

signal car_sig: unsigned (15 downto 0) := x"0000";
signal car_dir: std_logic;
begin
carrier <= std_logic_vector(car_sig);

process(clk) is
begin
   if rst = '0' then
   -- Initial start up
      car_dir <= '1';
      car_sig <= (others => '0');
   elsif rising_edge(clk) then
    -- PWM direction flag
      if car_sig >= (15000-2) then
```

```
            car_dir <= '0';
            car_sig <= to_unsigned(15000,16);
        elsif car_sig <= 2 then
            car_dir <= '1';
            car_sig <= (others => '0');
        end if;
        -- Actual PWM counter
        if car_dir = '1' then
            car_sig <= car_sig + 2;
        else
            car_sig <= car_sig - 2;
        end if;
    end if;
end process;

end Behavioral;
```

Then we can run the behavioral simulation as shown in Figure 13.70.

We click on the Restart button in the opened window, then right-click on the "clk" signal, select the Force Clock option, and apply the settings as depicted in Figure 13.71.

Also, we right-click on the "rst" and force a constant value of zero, as illustrated in Figure 13.72.

We then click the Run for 100 μs button once, and then change the Force value of "rst" to 1, as demonstrated in Figure 13.73.

We click on the Run for 100 μs button multiple times and then click on the Zoom Fit button to view the entire waveforms window, as shown in Figure 13.74. Now, we can right-click on the "carrier" and select the Waveform Style as Analog, as depicted in Figure 13.75.

We can see in Figure 13.76 that a symmetrical triangular waveform is produced for "carrier".

It is also better to right-click on the "carrier" and select the Radix as Unsigned Decimal as illustrated in Figure 13.77.

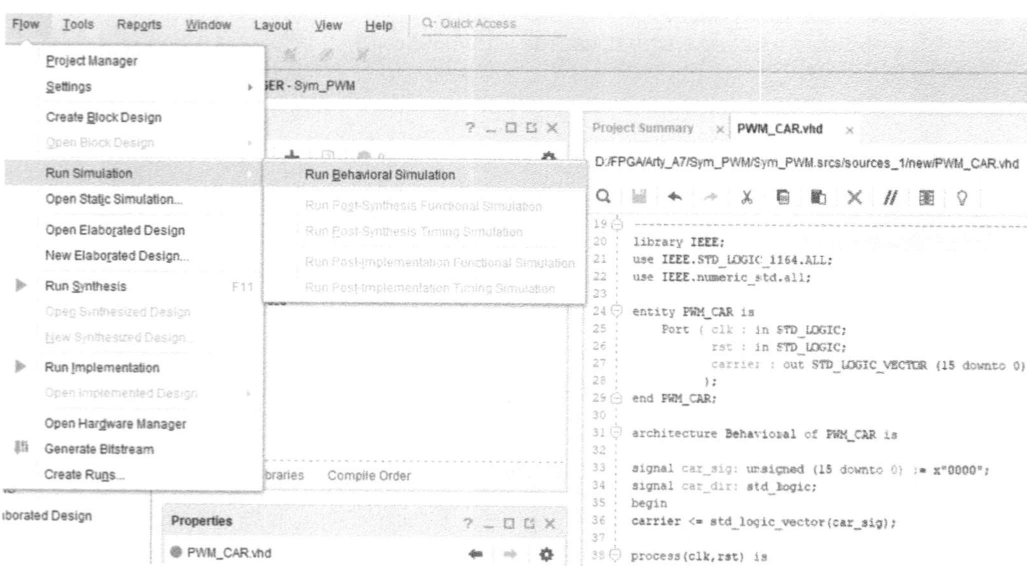

Figure 13.70 Running the behavioral simulation for PWM_CAR.vhd.

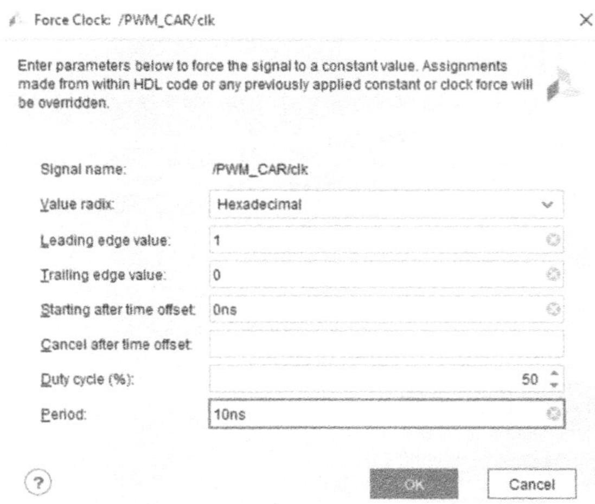

Figure 13.71 Selecting the Force Clock option for "clk" and configuring the settings.

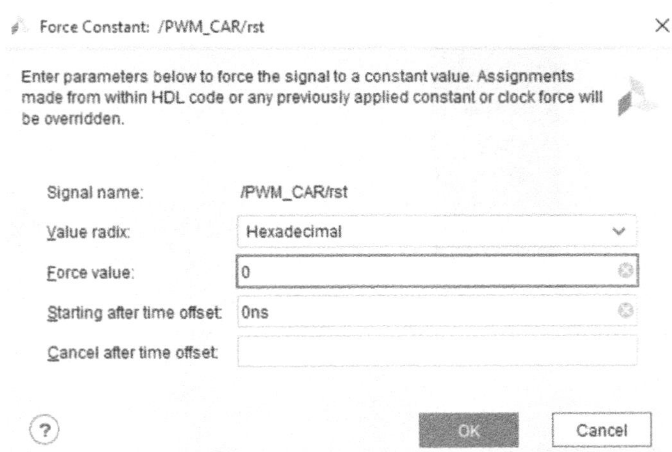

Figure 13.72 Forcing a constant value of zero for "rst".

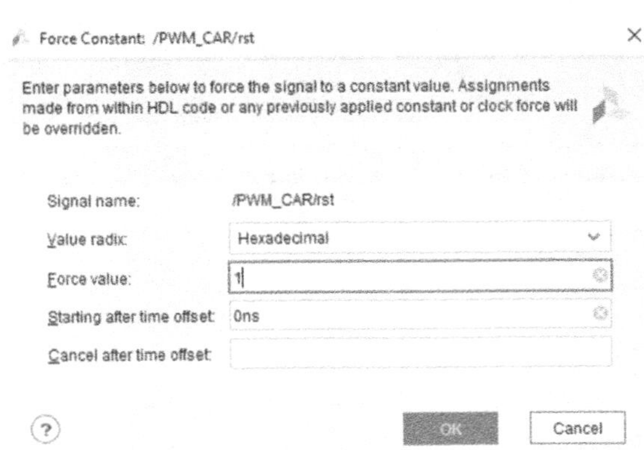

Figure 13.73 Changing the Force value of "rst" to 1.

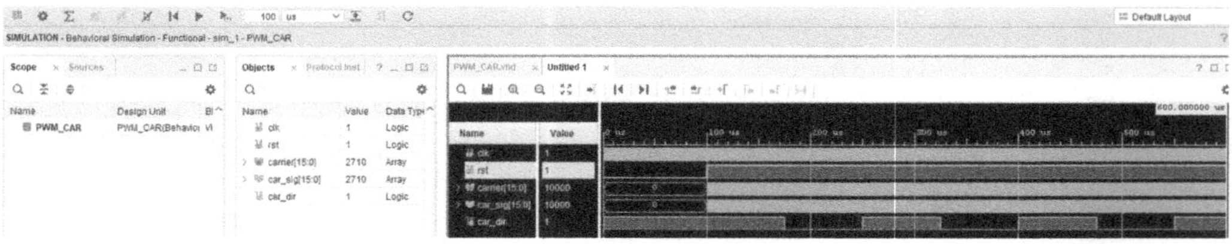

Figure 13.74 The signals on the waveforms window.

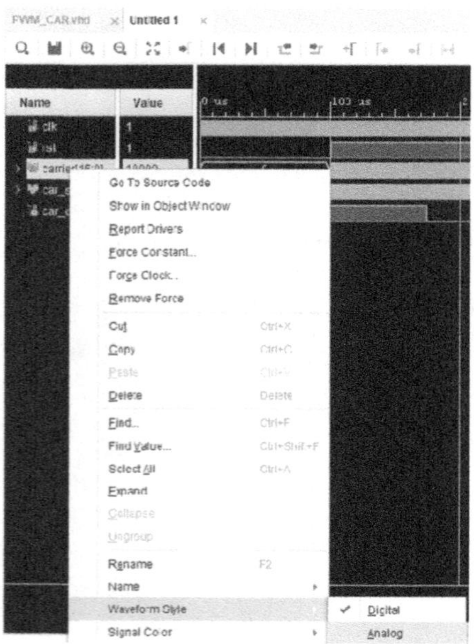

Figure 13.75 Selecting the "carrier" waveform style as analog.

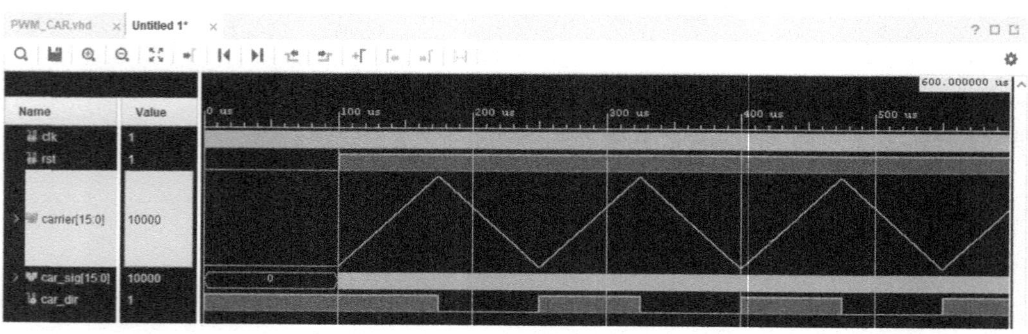

Figure 13.76 The "carrier" symmetrical triangular waveform.

Then, we can modify the codes for PWM_CAR.vhd by adding a generic pwm_step and variable pwm_var instead of "2" by the following modified codes.

```
library IEEE;
use IEEE.STD_LOGIC_1164.ALL;
use IEEE.numeric_std.all;

entity PWM_CAR is
```

Figure 13.77 Selecting the Radix as Unsigned Decimal for "carrier".

```
    generic(
        pwm_step: integer :=2
        );
    Port ( clk : in STD_LOGIC;
           rst : in STD_LOGIC;
           carrier : out STD_LOGIC_VECTOR (15 downto 0)
          );
end PWM_CAR;

architecture Behavioral of PWM_CAR is

signal car_sig: unsigned (15 downto 0) := x"0000";
signal car_dir: std_logic;
begin
carrier <= std_logic_vector(car_sig);

process(clk,rst) is
variable step_var: unsigned(15 downto 0);
begin
   if rst = '0' then
   -- Initial start up
      car_dir <= '1';
      car_sig <= (others => '0');
   elsif rising_edge(clk) then
    -- PWM direction flag
      step_var := to_unsigned(pwm_step,16);
      if car_sig >= (15000-step_var) then
```

```
            car_dir <= '0';
            car_sig <= to_unsigned(15000,16);
        elsif car_sig <= step_var then
            car_dir <= '1';
            car_sig <= (others => '0');
        end if;
        -- Actual PWM counter
        if car_dir = '1' then
            car_sig <= car_sig + step_var;
        else
            car_sig <= car_sig - step_var;
        end if;
    end if;
end process;
```

Then, we create a block design named "top" to open the Diagram window, as shown in Figure 13.78.

We right-click on the file PWM_CAR.vhd, then select the Add Module to Block Design option, and the module PWM_CAR is added to the Diagram window as depicted in Figures 13.79 and 13.80, respectively.

We right-click on the Design Sources folder and select the Add Sources... button as illustrated in Figure 13.81. We then click on the Add Files, and from the project folder directory of previous section, we select the file "sine_lut.vhd" as shown in Figure 13.82.

In the opened Add Sources window, we click on the Finish button as depicted in Figure 13.83.

The "sine_lut" is added to the Design Sources folder, and then we right-click on it and select the Add Module to Block Design option as illustrated in Figure 13.84. So, the module lut_sine is also added to the Diagram window.

Then, we right-click on the Design Sources folder and select the Add Sources option to create a VHDL file named "compare.vhd" as shown in Figure 13.85.

In the opened Add Sources window, we click on the Finish button as shown in Figure 13.86.

We add the codes inside the file "compare.vhd", and set it as top, so that we can simulate it as depicted in Figure 13.87.

Also, under the Project Manager, we select the Settings icon, then choose the Simulation tab, and select the simulation top module name as "compare" as illustrated in Figure 13.88. Next, we can run the behavioral simulation.

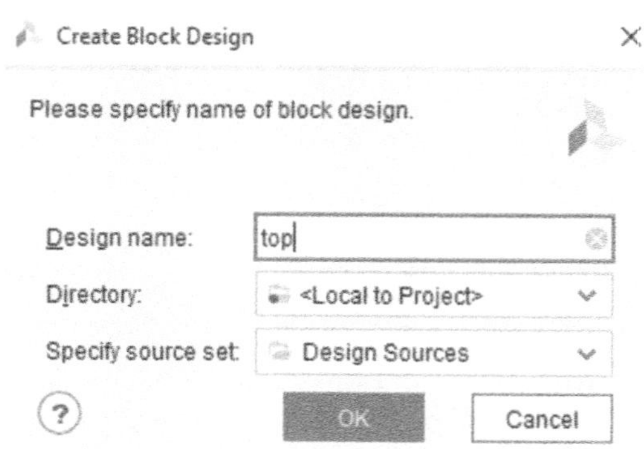

Figure 13.78 Creating a block design.

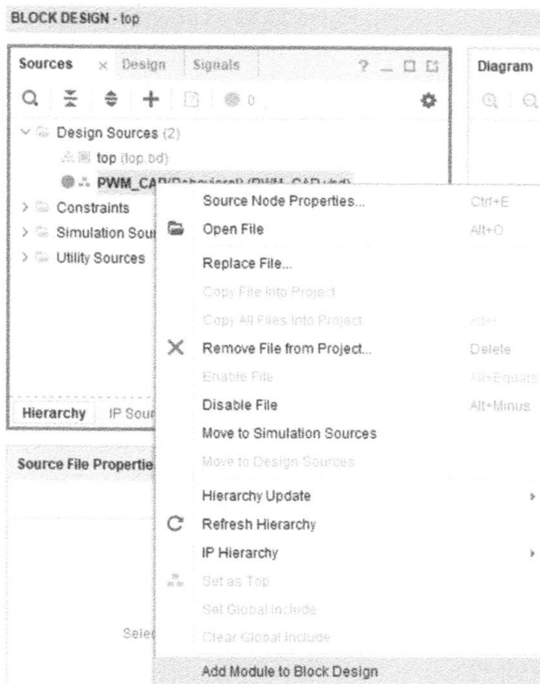

Figure 13.79 Selecting the Add Module to Block Design option.

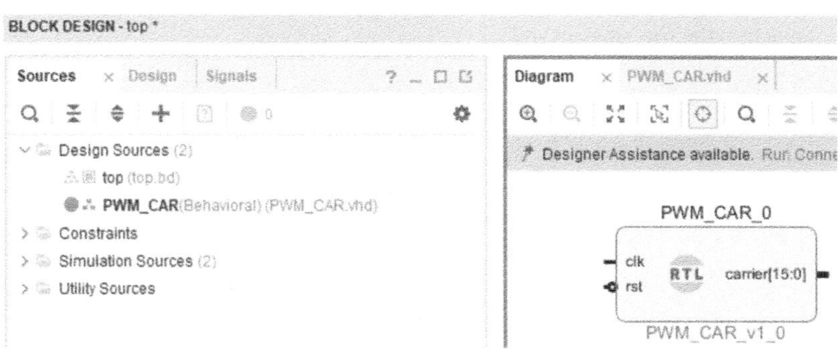

Figure 13.80 The added module PWM_CAR to Diagram window.

In the waveforms window, we click on the Restart button, and force clock for "clk", and force constant value of zero for "rst", "car", and "sig". Next, we run for 100 μs. After this, we change the force value of "rst" to 1, and run the simulation for another 100 μs. Now, we change the force value of "car" to 5h (=5) and run for another 100 μs. Finally, we change the force value of "sig" to 10 h (=16) for another 100 μs. We can see all the signals on the waveforms window in Figure 13.89.

After successful simulation, we can add the module "compare" to Diagram window as depicted in Figure 13.90.

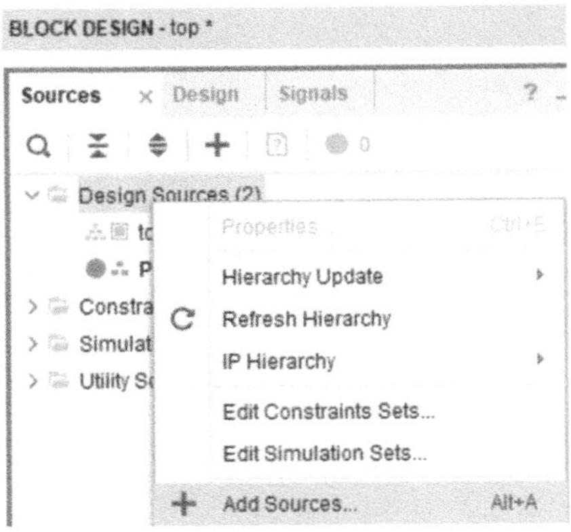

Figure 13.81 Clicking on the Add Sources... button.

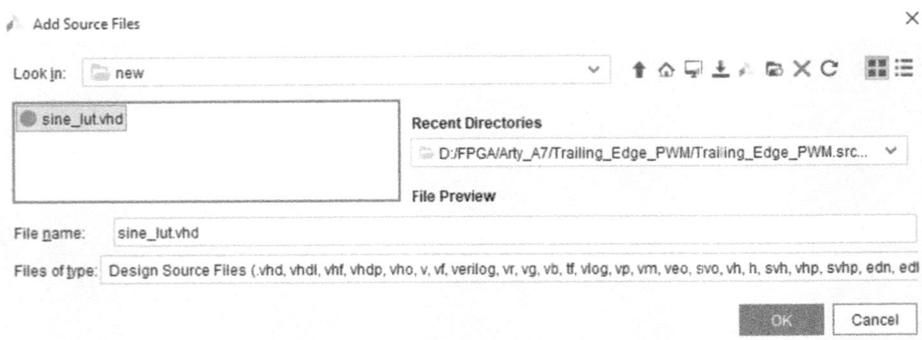

Figure 13.82 Selecting the file "sine_lut.vhd".

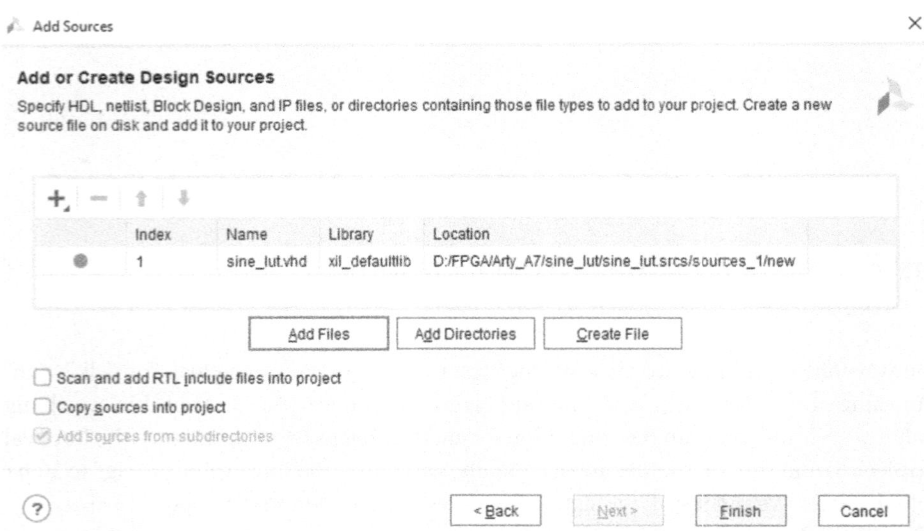

Figure 13.83 Clicking on the Finish button.

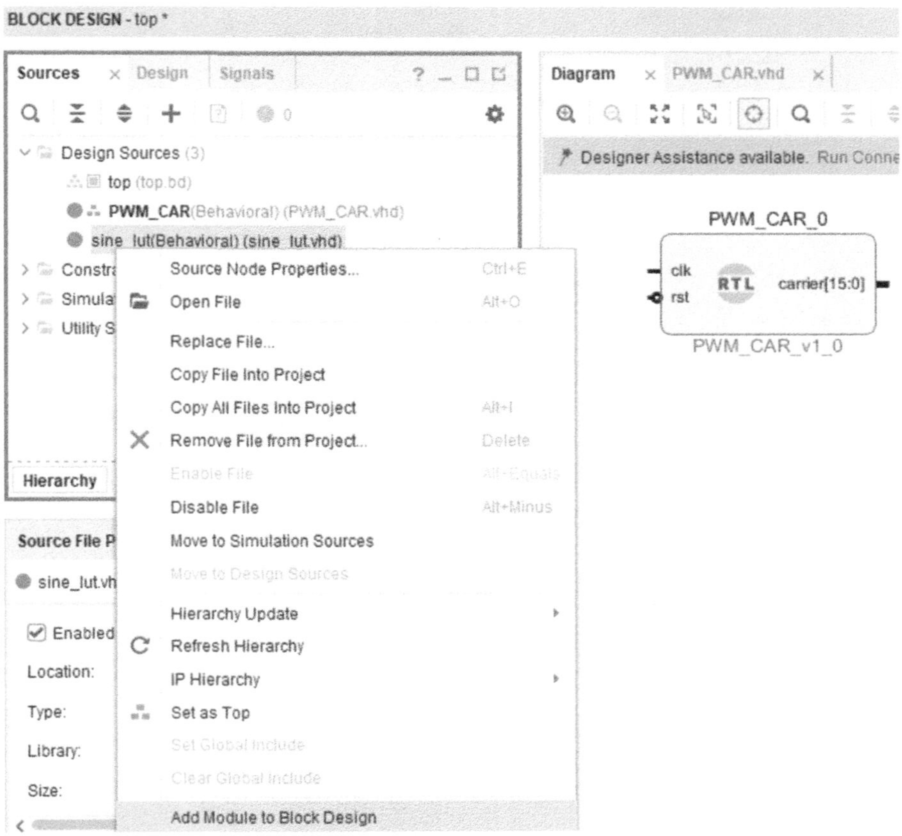

Figure 13.84 Selecting the Add Module to Block Design option.

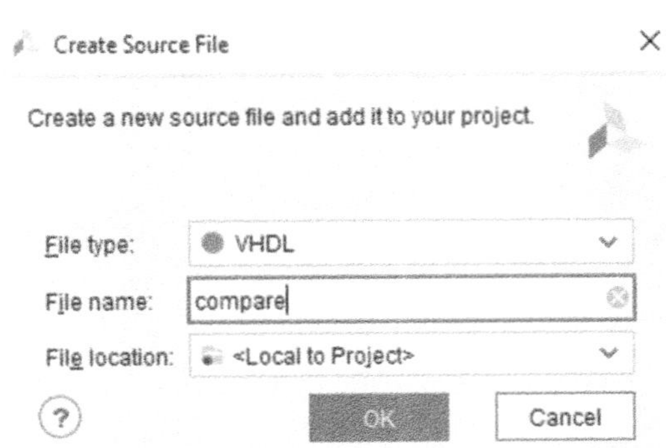

Figure 13.85 Creating a VHDL file.

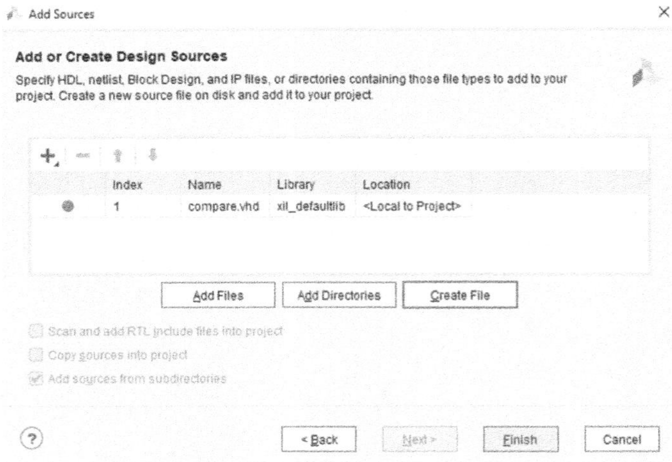

Figure 13.86 Clicking on the Finish button.

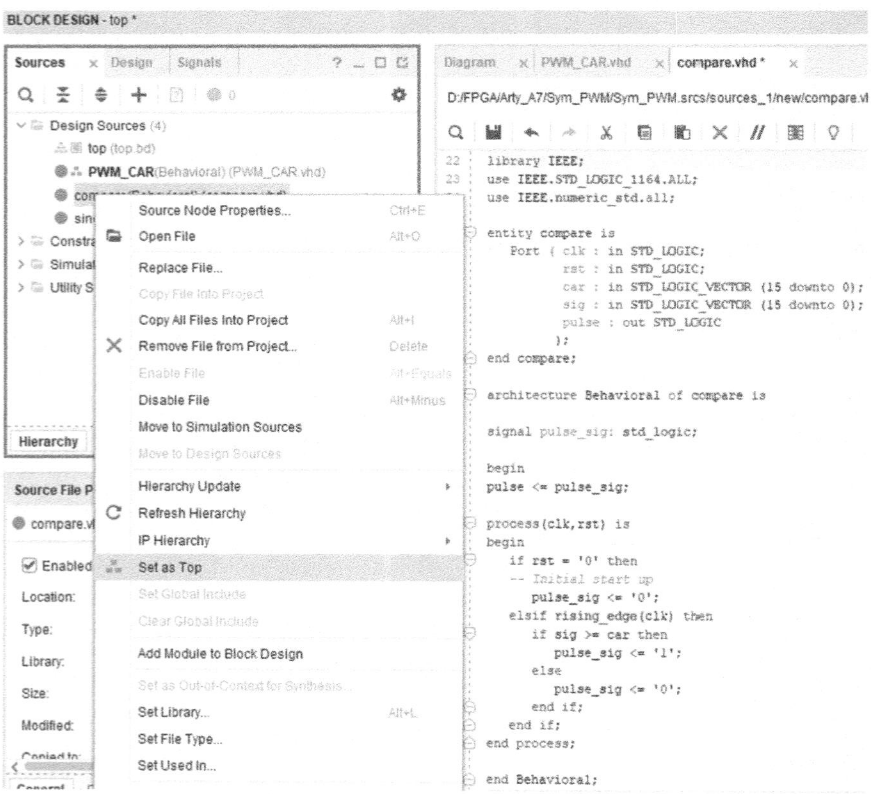

Figure 13.87 The codes of "compare.vhd".

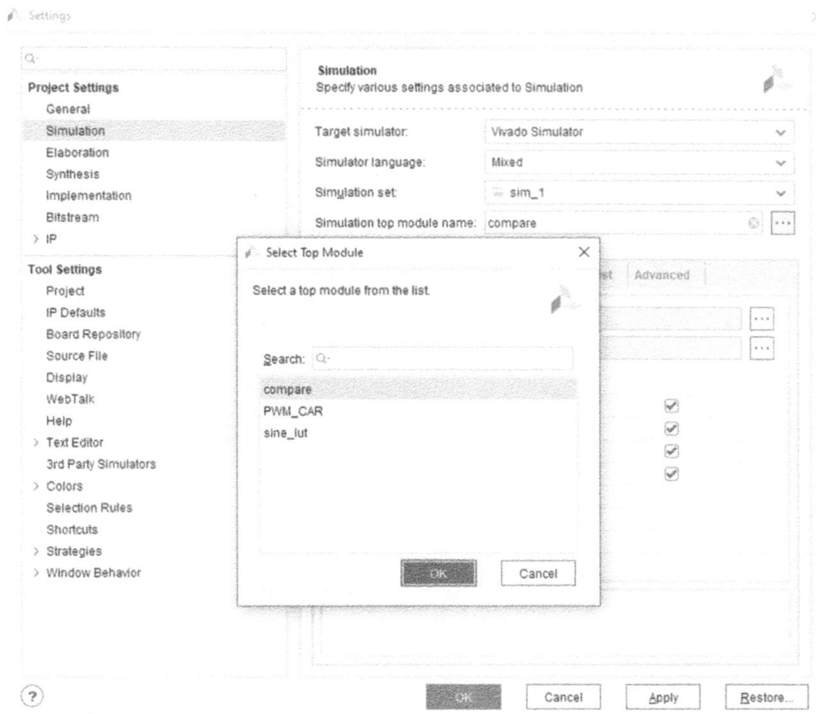

Figure 13.88 Selecting the simulation top module name as "compare".

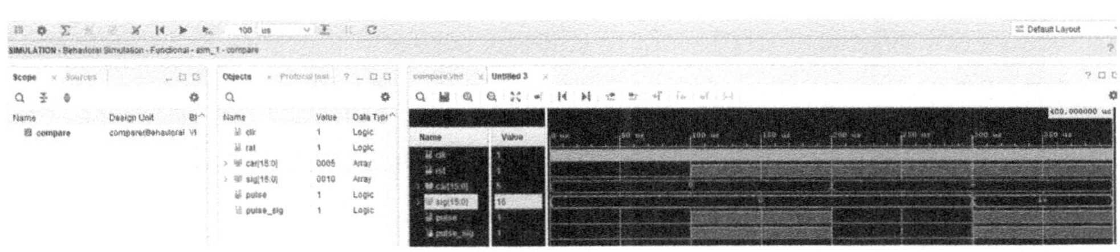

Figure 13.89 Simulation of all the signals on the waveforms window.

In the Diagram window, we right-click and select the Add IP... option as illustrated in Figure 13.91.

In the opened window, we search for clock and select the Simulation Clock Generator as shown in Figure 13.92.

If we double-click on the Simulation Clock Generator module, we can see the re-customize IP window, and we leave the values unchanged as depicted in Figure 13.93.

We connect the clock and reset signals of all modules together and link them to the simulation clock generator. Then, we select carrier, sine, and pulse ports, and press Ctrl+T to produce external

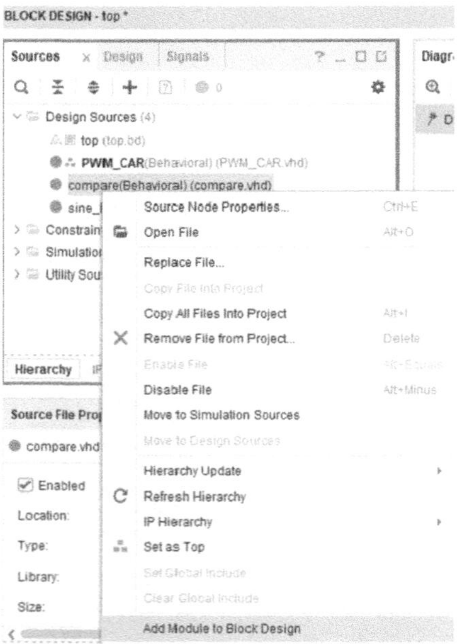

Figure 13.90 Clicking on the Add Module to Block Design option.

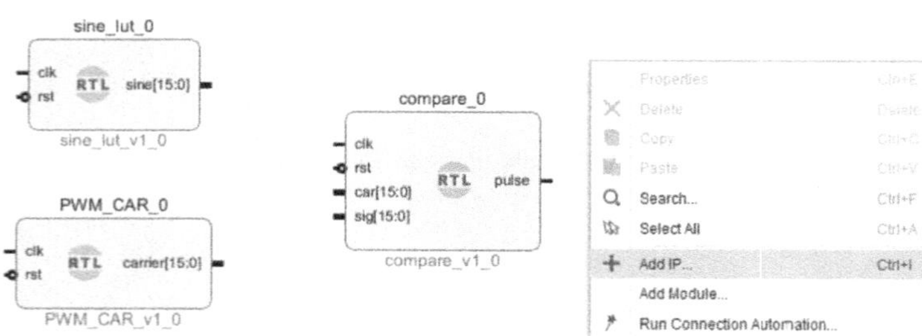

Figure 13.91 Selecting the Add IP... option.

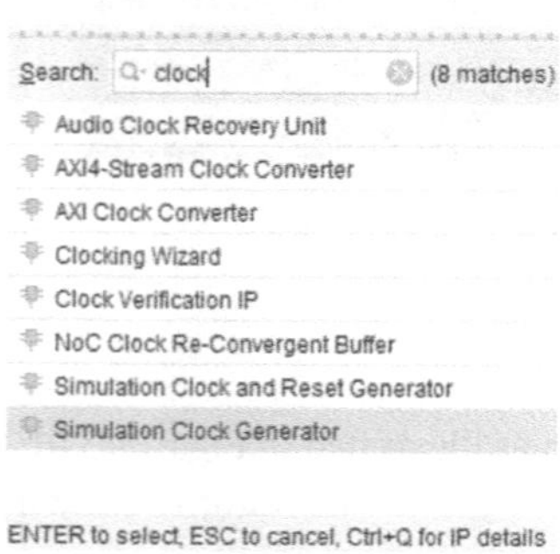

Figure 13.92 Selecting the Simulation Clock Generator.

ports for them. Next, we right-click in the Diagram window and select the Regenerate Layout option as illustrated in Figure 13.94.

We also double-click on the module PWM_CAR and change the PWM Step to 10 as demonstrated in Figure 13.95.

Then, we right-click on the "top" block diagram and select the Create HDL Wrapper... option, as shown in Figure 13.96.

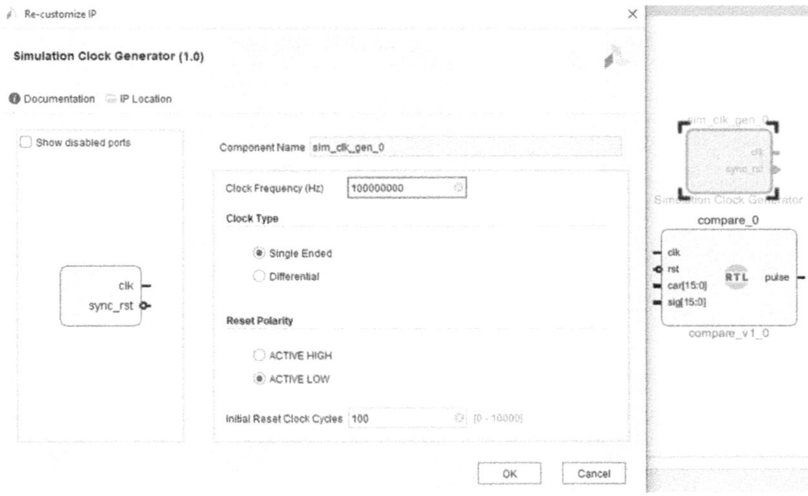

Figure 13.93 The re-customize IP window for Simulation Clock Generator.

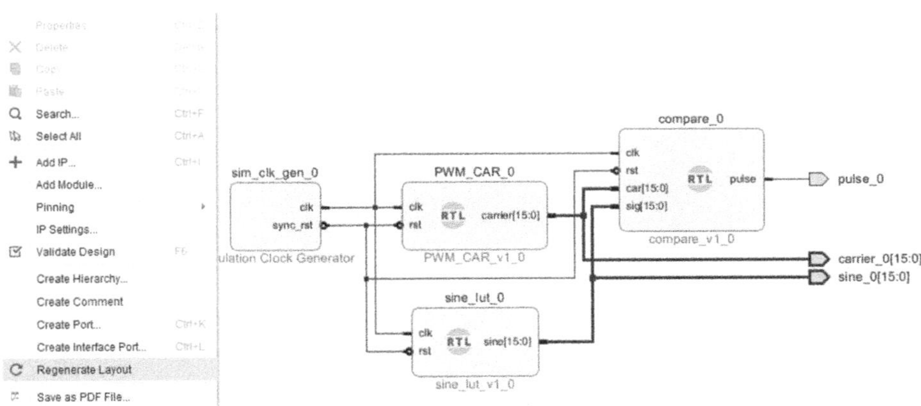

Figure 13.94 Connecting the modules and regenerating layout.

Figure 13.95 Double-clicking on the module PWM_CAR and changing the PWM Step to 10.

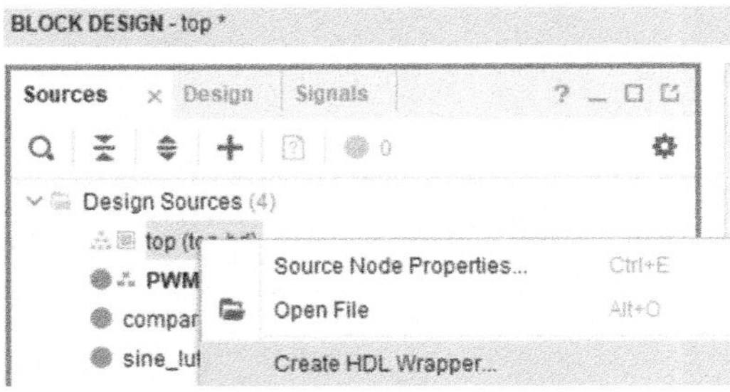

Figure 13.96 Selecting the Create HDL Wrapper... option.

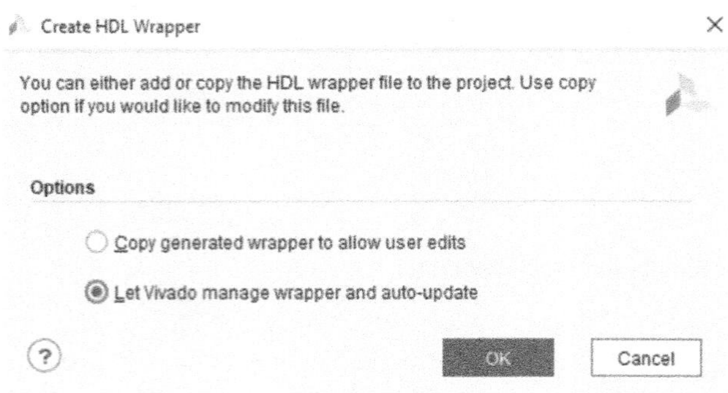

Figure 13.97 Letting Vivado to manage wrapper and auto-update.

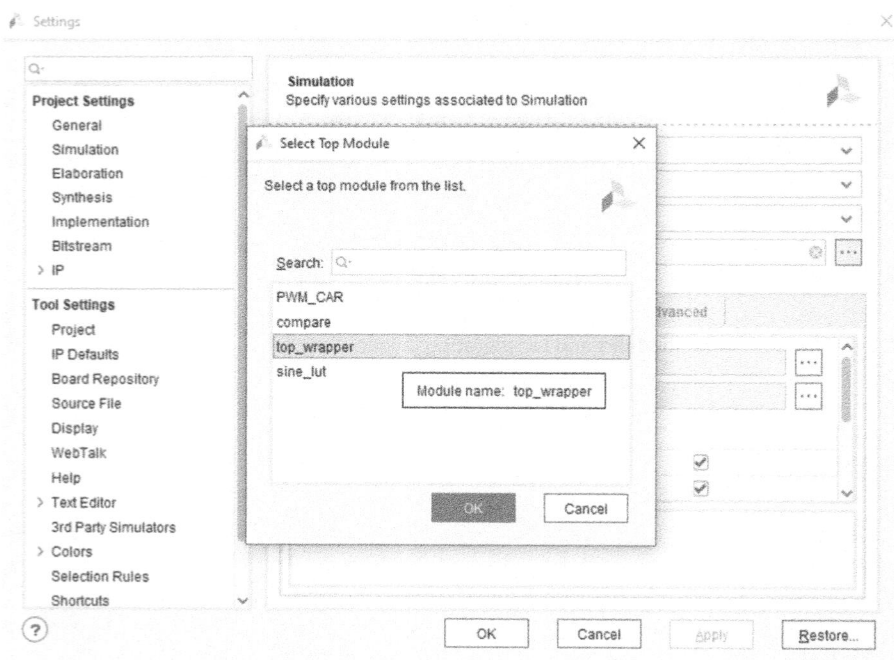

Figure 13.98 Selecting the simulation top module name as "top_wrapper".

In the opened window, we let Vivado manage wrapper and auto-update, and then click on the OK button, as depicted in Figure 13.97.

In the Project Manager and Settings window, we choose the Simulation tab and select the simulation top module name as "top_wrapper", as illustrated in Figure 13.98.

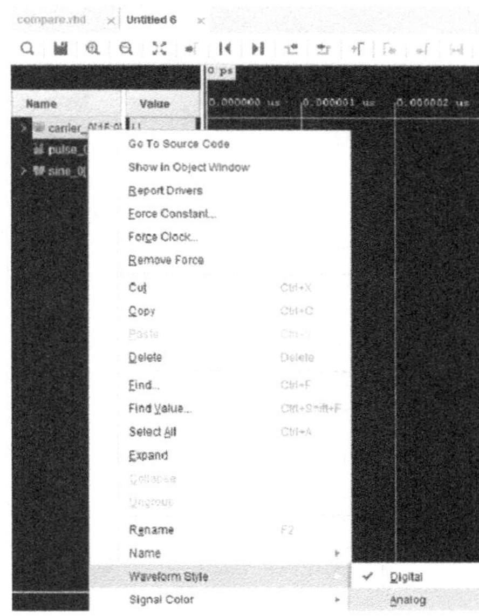

Figure 13.99 Selecting the Waveform Style as Analog for "carrier_0".

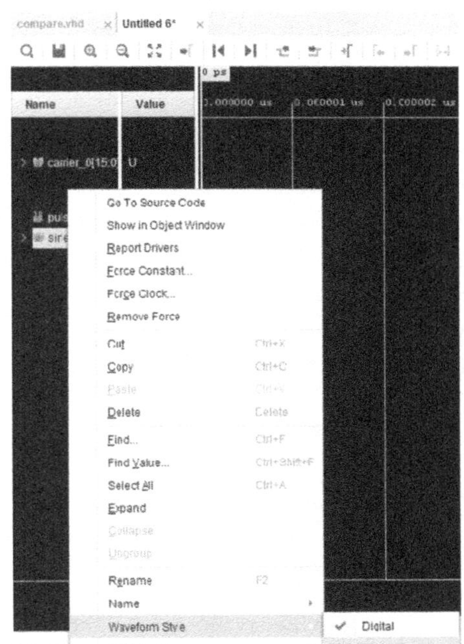

Figure 13.100 Selecting the Waveform Style as Analog for "sine_0".

In the waveforms window, we restart the simulation, and then we right-click on the carrier_0 and sine_0 ports, and select their waveform styles as Analog, as shown in Figures 13.99 and 13.100, respectively.

Finally, we run the simulation for 100 μs, and click on the Zoom Fit button to view the signals on the waveforms window, as depicted in Figure 13.101.

As you can see, the frequency of sine wave is higher than that of the PWM carrier. However, in practice, the frequency of the sine wave should be much lower than the PWM carrier. So, we need to reduce the frequency of the sine wave. As you know, we had only 500 values in the LUT of sine

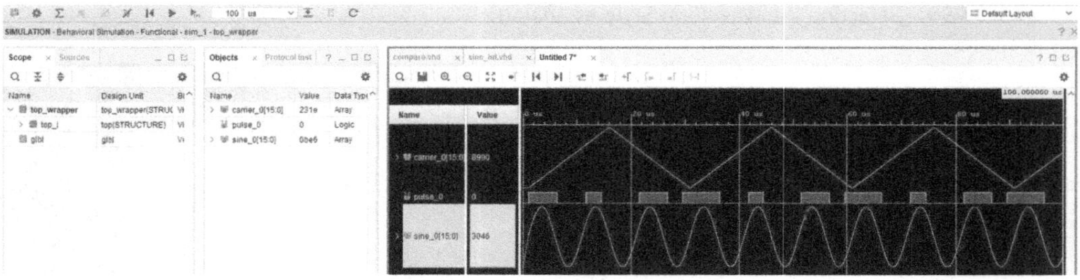

Figure 13.101 Viewing the signals on the waveforms window.

wave, so we need to increase this – for example, to 10000 values. Therefore, we use the following codes in Google Colab to produce 10000 LUT values for a half cycle of sine wave.

```
import numpy as np
from numpy import sin, pi
import matplotlib.pyplot as plt

x = np.linspace(0,1,10000)
lut = sin(pi*x)*7500+7500
for i,val in enumerate(lut):
    if i % 20:
       print(int(val),end=',')
    else:
       print(int(val),end=',')
       print('')
```

So, we remove the file "top_wrapper" from project, as illustrated in Figure 13.102.

We also modify the "sine_lut.vhd" file, as the following codes.

```
library IEEE;
use IEEE.STD_LOGIC_1164.ALL;
use IEEE.NUMERIC_STD.ALL;

entity sine_lut is
     Port(
          clk: in std_logic;
          rst: in std_logic;
          sine: out std_logic_vector(15 downto 0));
end sine_lut;

architecture Behavioral of sine_lut is

type LUT_TYPE is array(0 to 9999) of integer;
constant LUT: LUT_TYPE :=
(
-- We do copy and paste 10000 LUT values here
);

signal lut_sig: unsigned(15 downto 0);
signal state: std_logic := '0';
begin
```

```
sine <= std_logic_vector(lut_sig) when state = '0' else
        std_logic_vector(15000-lut_sig);

process(clk,rst) is
variable cnt: integer range 0 to 10000;
begin
    if rst = '0' then
       cnt := 0;
       state <= '0';
       lut_sig <= (others => '0');
    elsif rising_edge(clk) then
       lut_sig <= to_unsigned(LUT(cnt),16);
       if cnt > 9998 then
          cnt := 0;
          state <= not state;
       else
          cnt := cnt + 1;
       end if;
    end if;
end process;
end Behavioral;
```

Now, with modified codes for the file "sine_lut.vhd", if we simulate the top_wrapper again, we can get much better signals on the waveforms window, as shown in Figure 13.103.

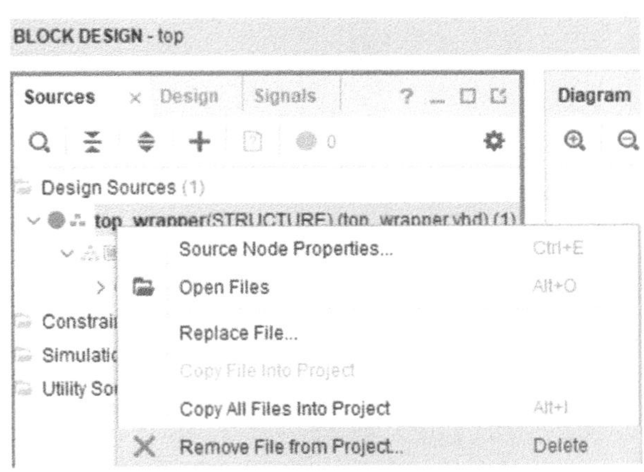

Figure 13.102 Removing the file "top_wrapper" from project.

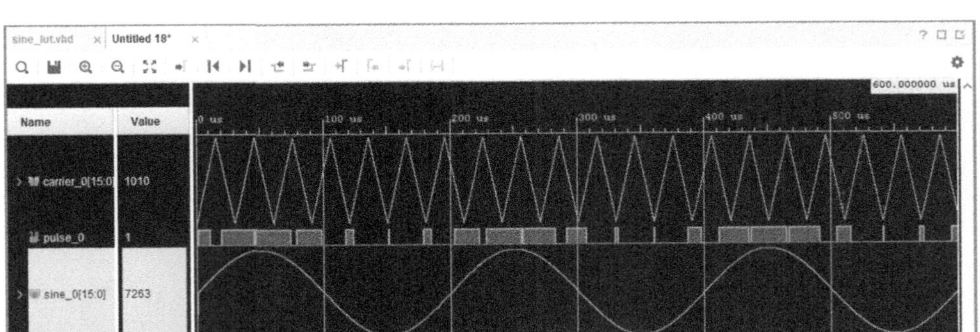

Figure 13.103 The sine wave PWM signals on the waveforms window.

14

Generating VHDL Code for Vivado Using Simulink HDL Coder

14.1 Introduction

In the evolving landscape of digital design, the need for efficient, reliable, and rapid development methodologies is more critical than ever. This chapter delves into one of the most powerful integrations available to engineers: the combination of Simulink HDL Coder and Xilinx Vivado. These tools represent a convergence of high-level modeling and hardware synthesis, enabling designers to transition seamlessly from abstract system representations to actual hardware implementations. Simulink provides a dynamic platform for simulating complex systems, allowing engineers to visualize and refine their designs through intuitive block diagrams. Meanwhile, HDL Coder empowers users to automatically generate synthesizable VHDL code directly from Simulink models, streamlining the path to FPGA and ASIC deployment. As designers increasingly turn to high-level abstractions, understanding the nuances of VHDL generation becomes essential for leveraging the full capabilities of modern design environments. This chapter will guide you through the critical processes involved in harnessing Simulink HDL Coder to create VHDL code tailored for implementation in Xilinx Vivado. From setting up your Simulink model to configuring the HDL Coder settings and navigating the intricacies of the generated code, you will gain insights into best practices that enhance design efficiency and performance. By the end of this chapter, you will not only appreciate the synergy between these powerful tools but also acquire practical skills that elevate your digital design workflow to new heights.

14.2 The Half Adder Circuit Design

In this section, we are going to design a half adder circuit in Simulink. In the command window of MATLAB, we type the following command.

```
>> simulink
```

After opening the Simulink environment, we use the following code to open the HDL Coder library browser.

```
>> hdllib
```

Then, using the HDL Coder library browser, we create a new model titled Half_Adder, as shown in Figure 14.1. The Half_Adder_HDL subsystem model is shown in Figure 14.2.

All the elements, except the rounding function and uniform random number generator, are available under the Commonly Used Blocks in the HDL Coder tab. In order to add the rounding

Fundamentals of VHDL for FPGA Programming Using Vivado, First Edition. Majid Pakdel.

Companion Website: www.wiley.com/go/pakdelVHDL

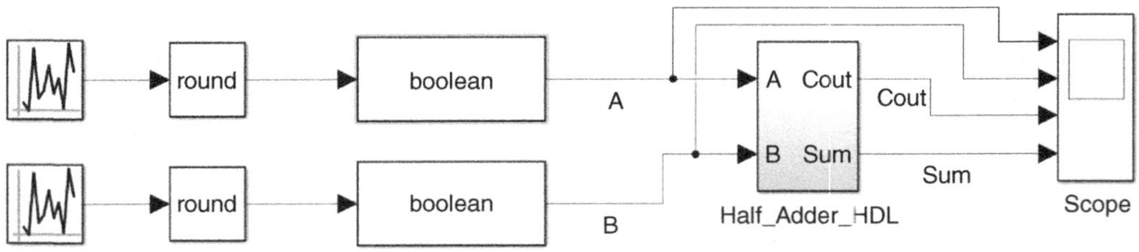

Figure 14.1 The Half_Adder model in Simulink.

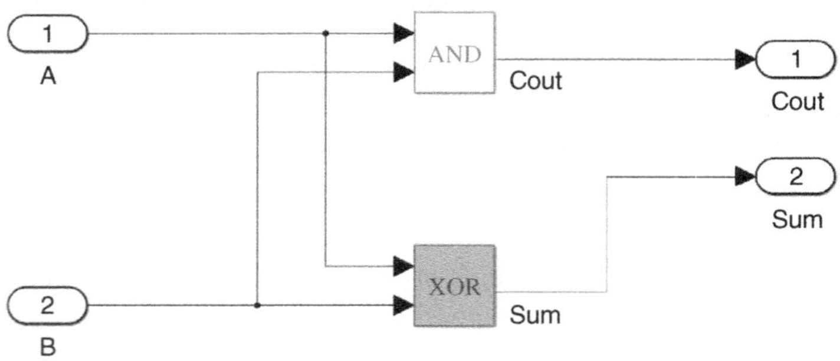

Figure 14.2 The Half_Adder_HDL subsystem model.

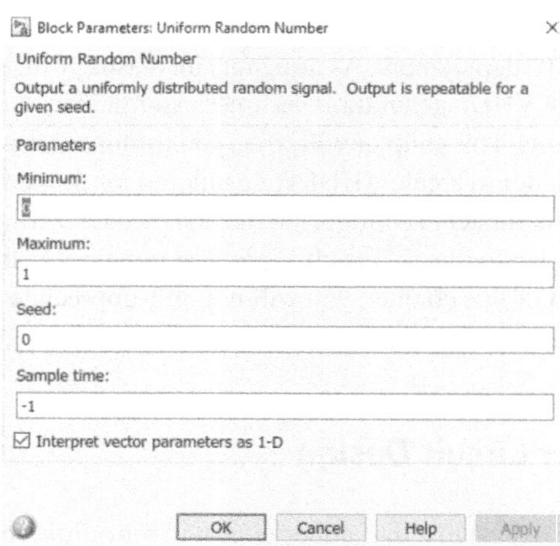

Figure 14.3 The upper uniform random number settings.

function and uniform random number, we need to run the following code in the MATLAB command window.

```
>> hdllib('off')
```

The configuration windows for the upper and lower settings of the uniform random number are shown in Figures 14.3 and 14.4, respectively.

The rounding function parameters setting window is illustrated in Figure 14.5.

Also, the convert parameters setting window is shown in Figure 14.6.

Then, we save the Simulink design in a desired folder (Half_Adder) and name it as Half_Adder.slx. Type the following code in the MATLAB command window.

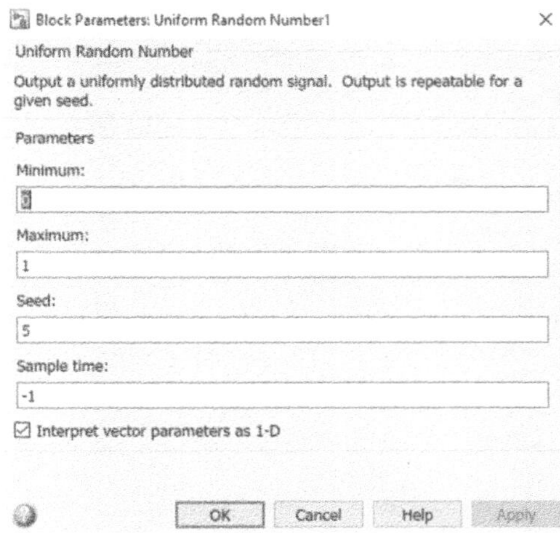

Figure 14.4 The lower uniform random number settings.

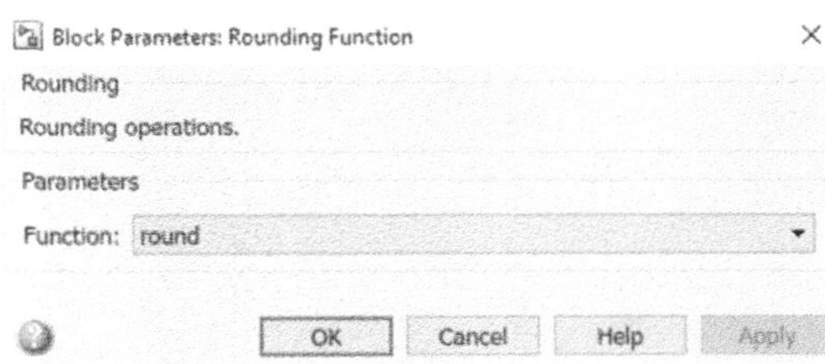

Figure 14.5 The rounding function parameters setting.

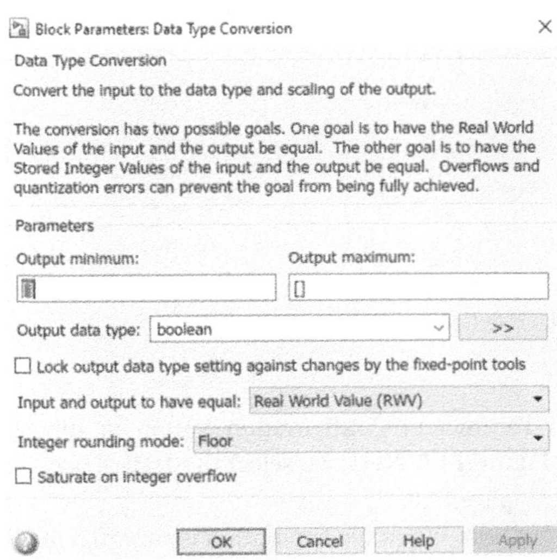

Figure 14.6 The convert parameters setting.

```
>> cd('H:\matlab_fpga\Half_Adder')
```

Now, we can type the following code in the MATLAB command window as well.

```
>> hdlsetup('Half_Adder')
```

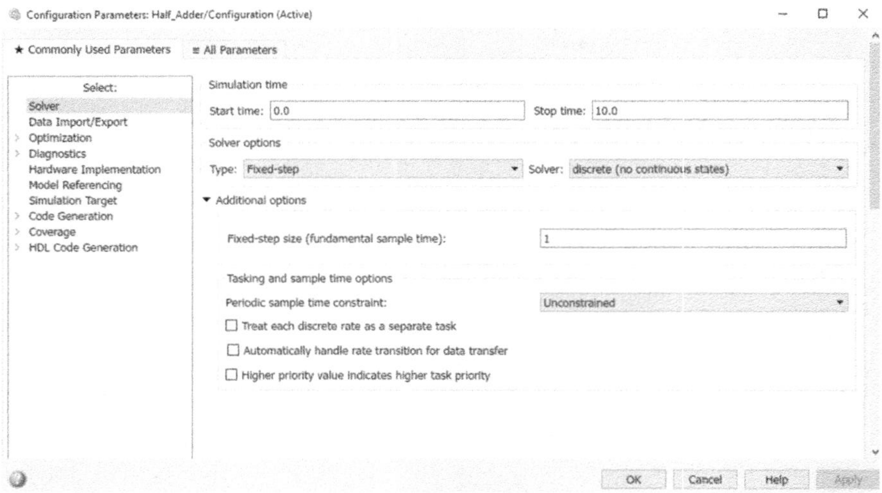

Figure 14.7 Setting the Solver parameters.

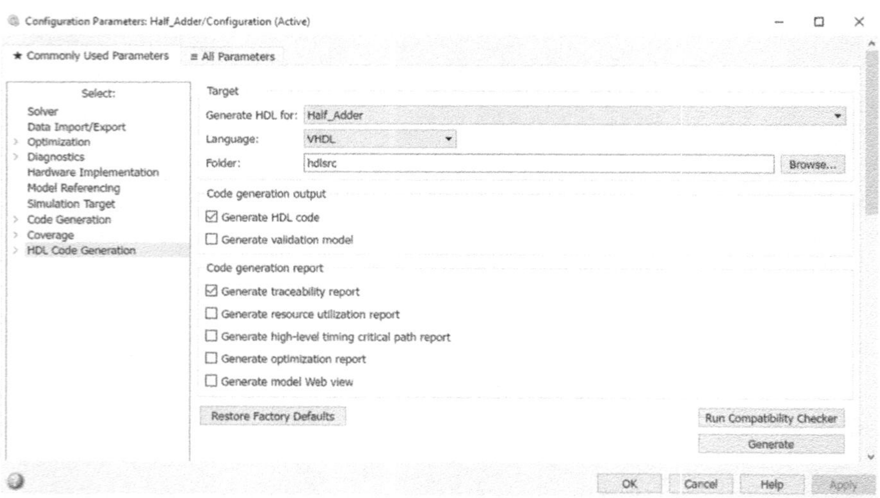

Figure 14.8 Setting the HDL Code Generation parameters.

We click on the Model Parameters Configuration button in the menu and set the Solver parameters as depicted in Figure 14.7. Next, we select the HDL Code Generation tab and set the parameters as illustrated in Figure 14.8.

Now, we can run the simulation. When the simulation is running, the Simulink model becomes red, and the port data types are displayed. If the data types are not displayed, we can right-click on the Simulink design, select Other Displays > Signals & Ports, and then choose Ports Data Types, as demonstrated in Figure 14.9.

The waveforms on the scope are shown in Figure 14.10.

Now, we write the following code in the MATLAB command window.

```
>> checkhdl('Half_Adder/Half_Adder_HDL')
```

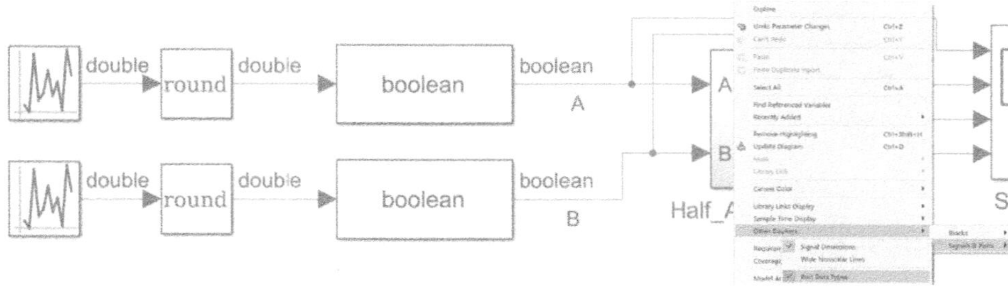

Figure 14.9 Displaying the Port Data Types.

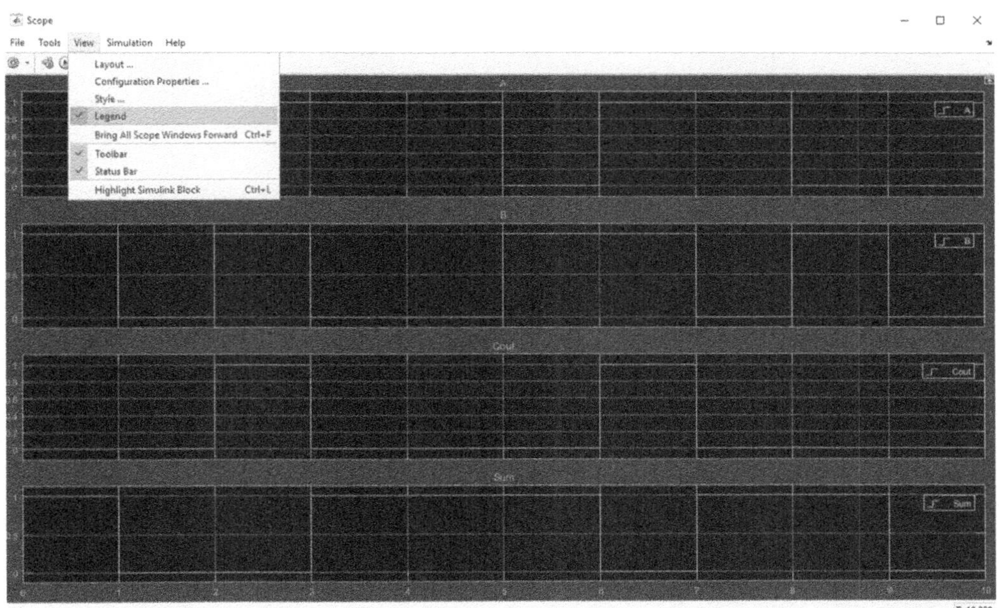

Figure 14.10 The waveforms on the scope.

A web browser HDL check report pops up, as depicted in Figure 14.11.

Finally, we type the following code in the MATLAB command window.

```
>> makehdl('Half_Adder/Half_Adder_HDL')
```

The Code Generation Report window is displayed, and the VHDL file (Half_Adder_HDL.vhd) is generated, as illustrated in Figure 14.12.

Now, we open the Vivado and create a new project titled Half_Adder_HDL. Click Next, and then click on Add files to the design sources section. However, before we click on the Add Files button, it is better to copy and paste the generated VHDL file (Half_Adder_HDL.vhd) into the Vivado-generated project folder (Half_Adder_HDL), and finally, we click on the Finish button. Then, we do Run Synthesis, and once it is successfully completed, we can click the Cancel button, as shown in Figure 14.13.

In the RTL analysis of the Project Manager section, we select the schematic. The Elaborate Design window will pop up, as depicted in Figure 14.14. Click the OK button.

Now, we can view the schematic, as illustrated in Figure 14.15.

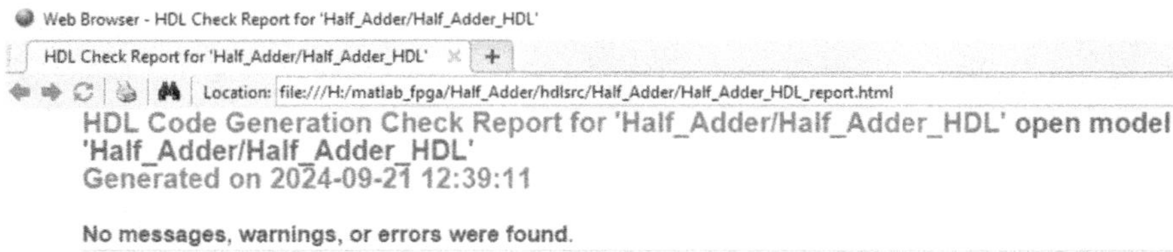

Figure 14.11 The Web Browser – HDL Check Report.

Figure 14.12 The Code Generation Report window.

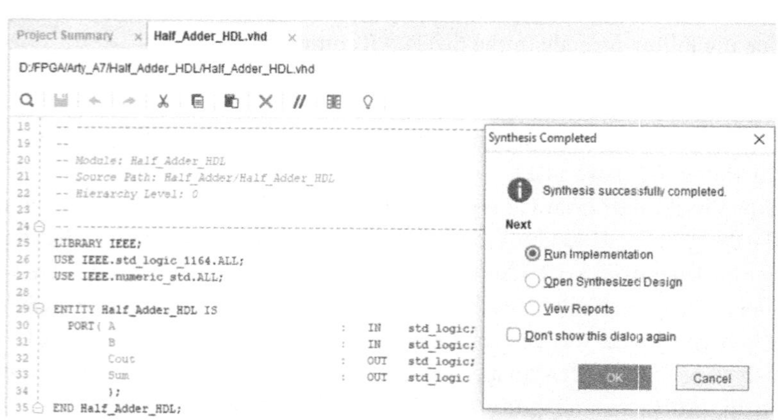

Figure 14.13 Successfully completed synthesis analysis.

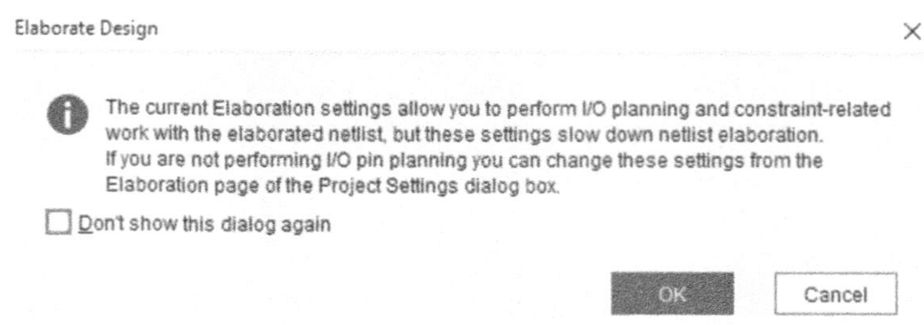

Figure 14.14 The Elaborate Design window.

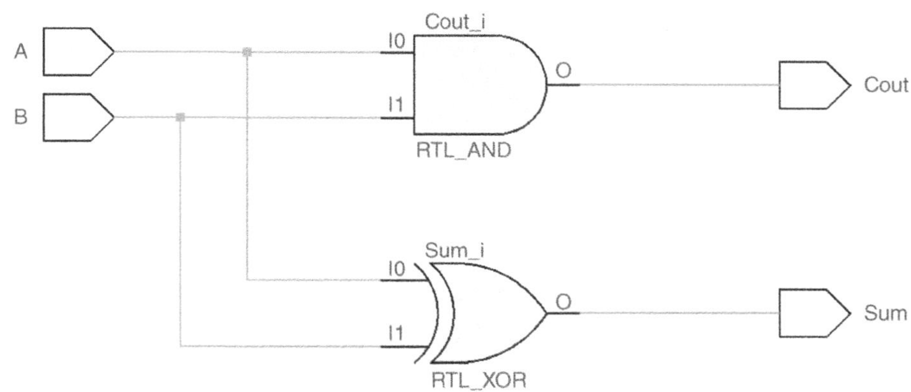

Figure 14.15 The schematic for the Half_Adder_HDL project.

In the IP Integrator section, we click on the Create Block Design option to create a block diagram titled "top". Then, right-click on the "Half_Adder_HDL.vhd" file and select Add Module to Block Design to insert it as a module into the diagram window, as shown in Figure 14.16.

14.3 Sequential Circuit and State Flow Chart Modeling

In this section, we are going to design a sequential circuit (a Johnson counter) and model a pulse generator using a state flow chart in Simulink. First, open the MATLAB command window and type the following command.

```
>> simulink
```

Also, we change the directory to where we want to save the Simulink project by the following code.

```
>> cd('H:\matlab_fpga\Johnson_Counter')
```

After opening the Simulink environment, we use the following code to open the HDL Coder library browser.

```
>> hdllib
```

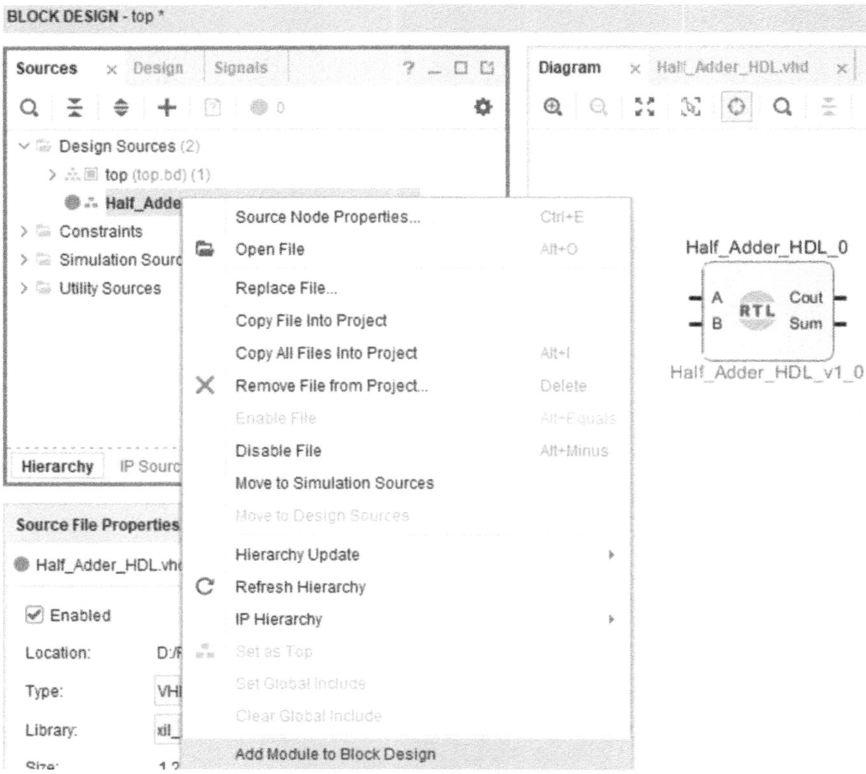

Figure 14.16 Selecting the Add Module to Block Design option.

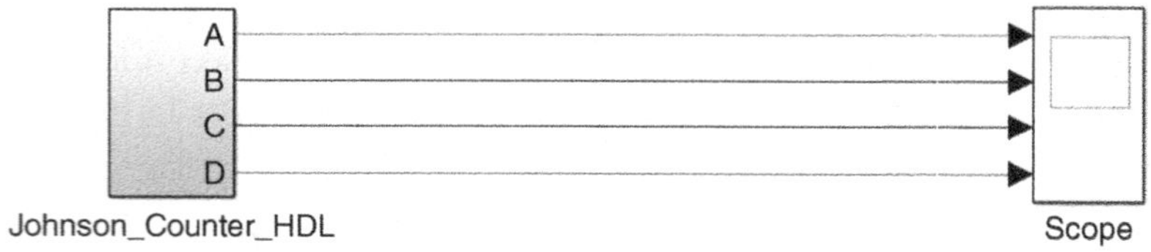

Figure 14.17 The Johnson_Counter model in Simulink.

Then, using the HDL Coder library browser, we create a new model titled Johnson_Counter, as shown in Figure 14.17. Also, the Johnson_Counter_HDL subsystem model is depicted in Figure 14.18.

We know that a 4-bit Johnson counter consists of four D flip-flops connected together, and its circuit diagram is demonstrated in Figure 14.19. The truth table of 4-bit Johnson counter is shown in Table 14.1.

Also, the Johnson_Counter subsystem model is illustrated in Figure 14.20.

The block parameters for delay (Z^{-1}) are shown in Figure 14.21.

Also, the block parameters for the NOT gate are depicted in Figure 14.22.

We use the chart in the stateflow library browser to create a finite state machine (FSM) model for clock pulse generation. We have two states (S0 and S1). To transition from S0 to S1, we can use

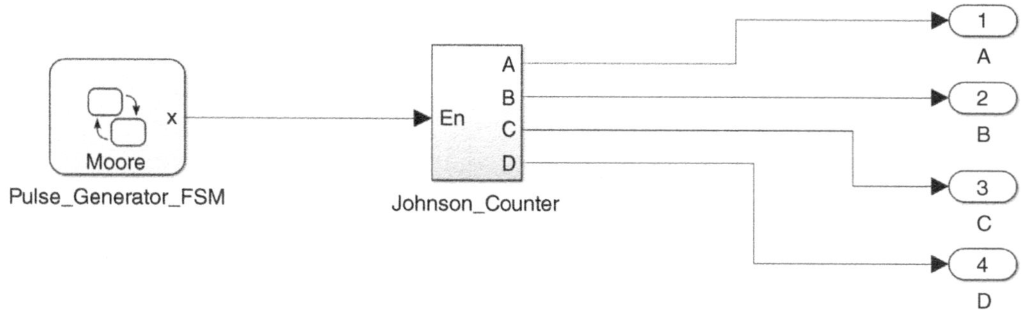

Figure 14.18 The Johnson_Counter_HDL subsystem model.

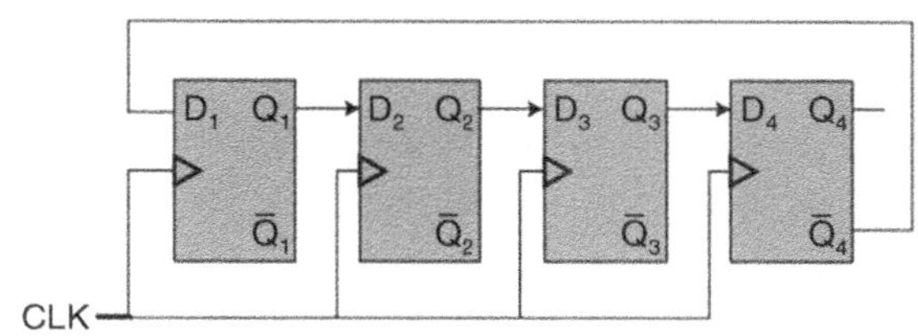

Figure 14.19 A 4-bit Johnson counter.

Table 14.1 The truth table of 4-bit Johnson counter.

CLK	Q_1	Q_2	Q_3	Q_4
0	0	0	0	0
1	1	0	0	0
2	1	1	0	0
3	1	1	1	0
4	1	1	1	1
5	0	1	1	1
6	0	0	1	1
7	0	0	0	1

a condition such as [after(1,sec)]. However, for HDL Coder, it is better to use a condition like [after(10,tick)]. In this case, if we set a fixed step size of 0.1 in the solver configuration parameters, the delay of 1s is produced as well, as illustrated in Figure 14.23.

We select the Model Explorer (Ctrl+H), as shown in Figure 14.24.

Then, we set the parameters of Model Explorer window, as depicted in Figure 14.25.

Now, we can type the following code in the MATLAB command window as well.

```
>> hdlsetup('Half_Adder')
```

Also, we set the fixed step size as 0.1s in the solver configuration parameters, as illustrated in Figure 14.26.

If we run the simulation, the generated pulses at the output of state flow chart (x) are shown in Figure 14.27.

Also, the Johnson counter waveforms of subsystem Jounson_Counter_HDL are depicted in Figure 14.28.

However, since the system clock frequency of the Arty A7 development board is 100 MHz, our fixed step size will be $1/10^8 = 10^{-8}$. Therefore, the solver configuration parameters window is demonstrated again in Figure 14.29.

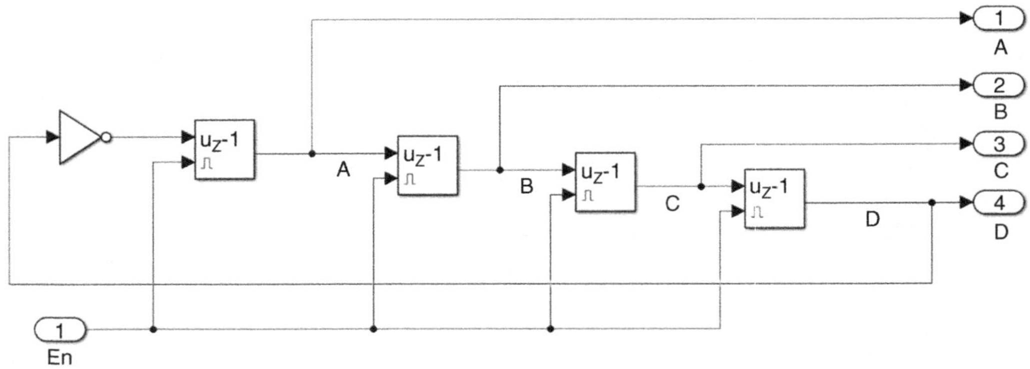

Figure 14.20 The Johnson_Counter subsystem model.

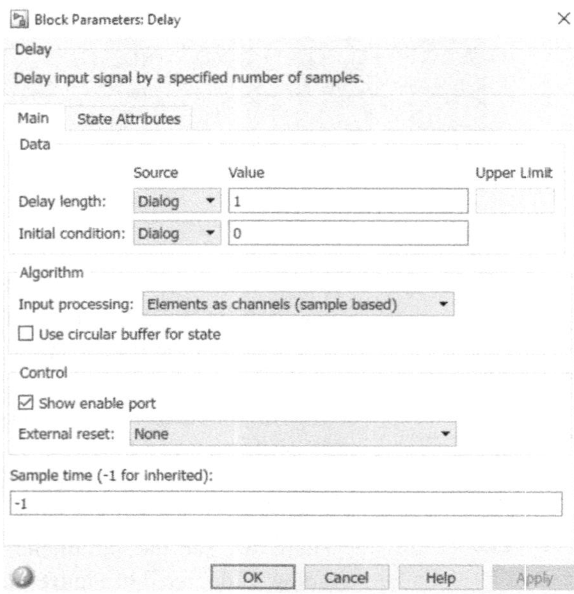

Figure 14.21 The block parameters for delay (Z^{-1}).

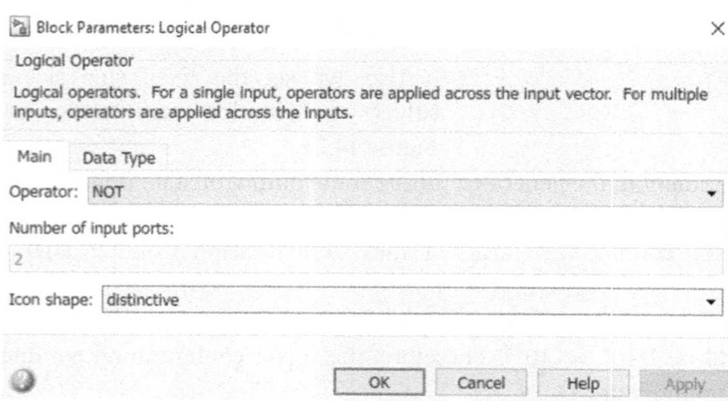

Figure 14.22 Block parameters for the NOT gate.

Figure 14.23 The state flow chart for clock pulse generator.

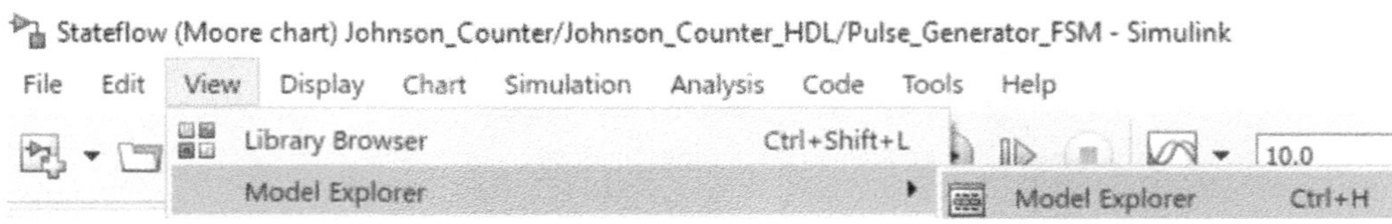

Figure 14.24 Selecting the Model Explorer (Ctrl+H).

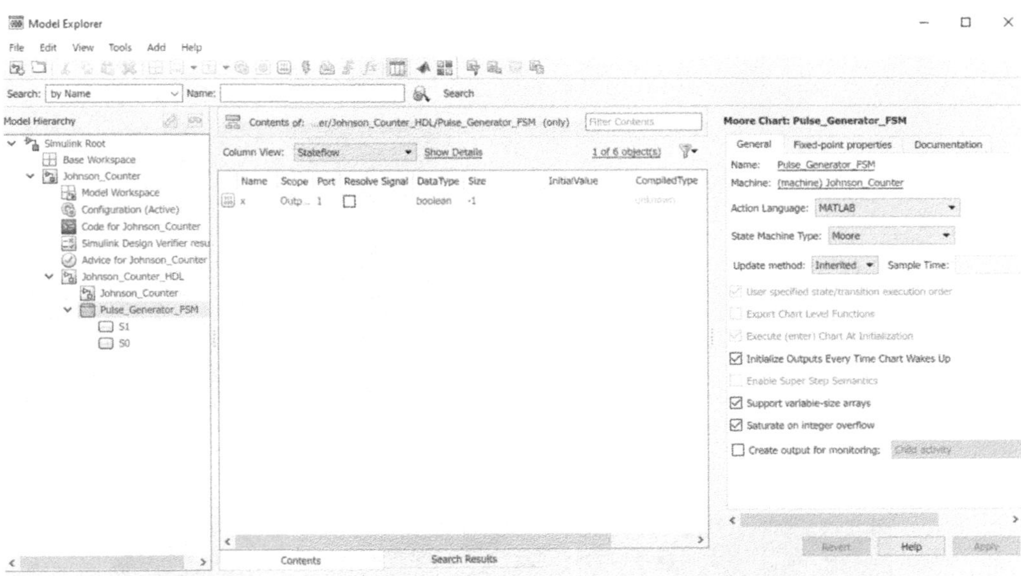

Figure 14.25 Setting the parameters of Model Explorer window.

Also, we set the HDL Code Generation tab parameters, as illustrated in Figure 14.30. Now, we write the following code in the MATLAB command window.

```
>> checkhdl('Johnson_Counter/Johnson_Counter_HDL')
```

A Web Browser – HDL Check Report is displayed, as shown in Figure 14.31.

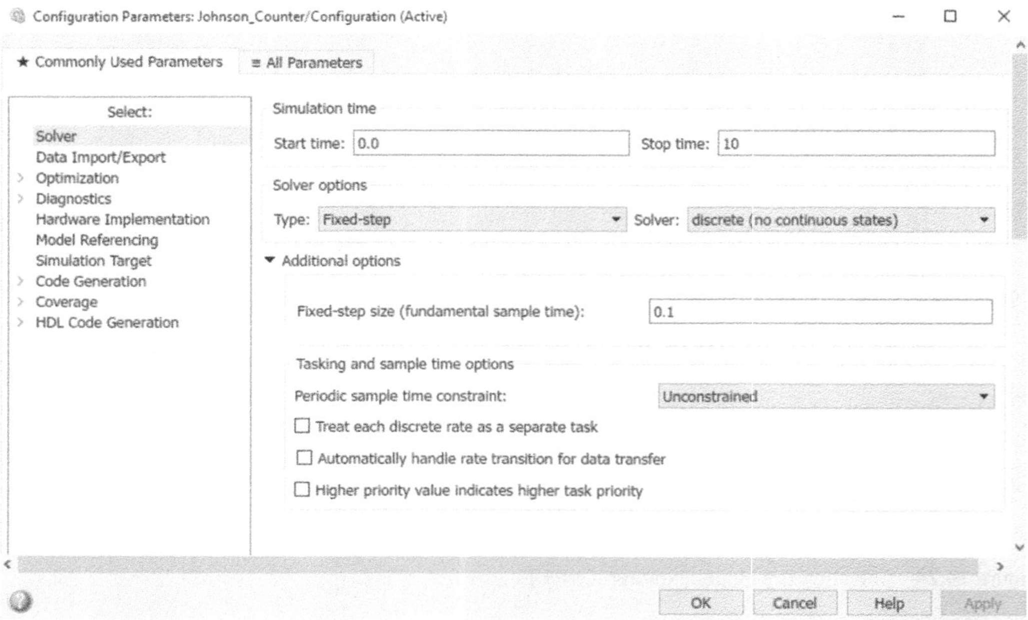

Figure 14.26 Setting the fixed step size as 0.1s.

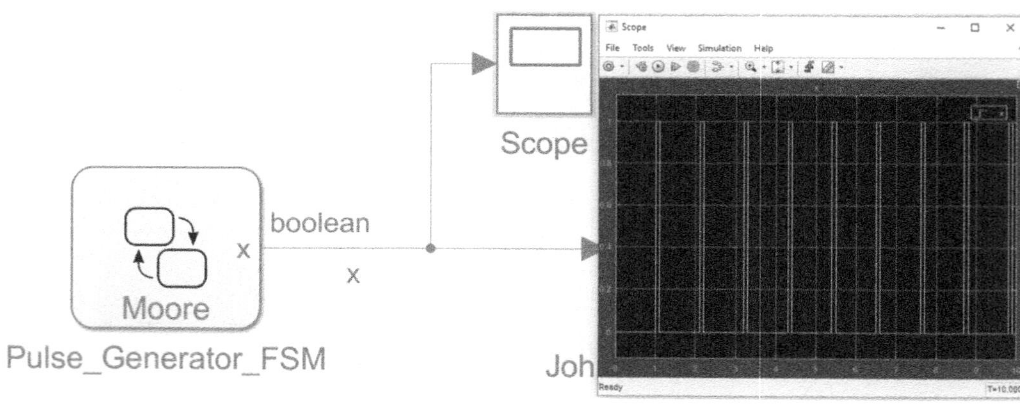

Figure 14.27 The generated pulses at the output of state flow chart.

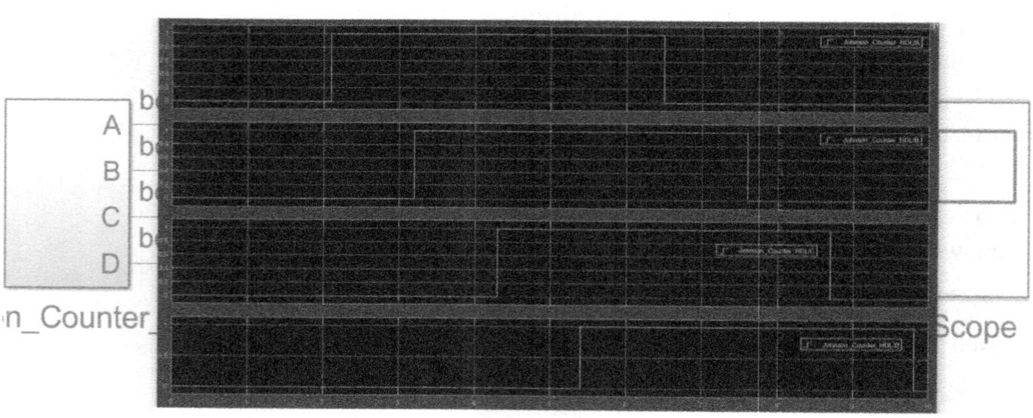

Figure 14.28 The waveforms of subsystem Jounson_Counter_HDL.

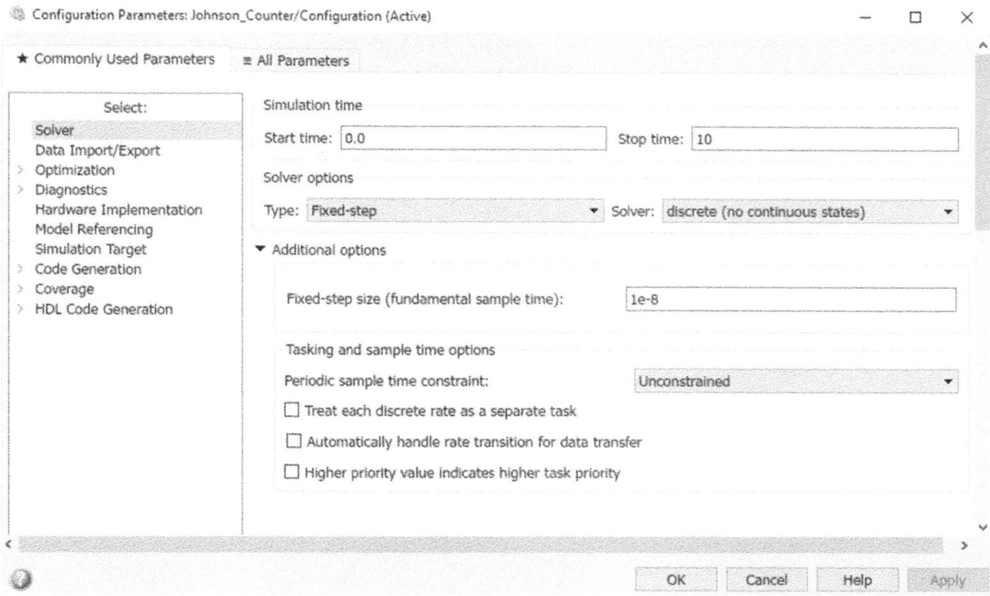

Figure 14.29 Setting the fixed-step size to 10^{-8} s.

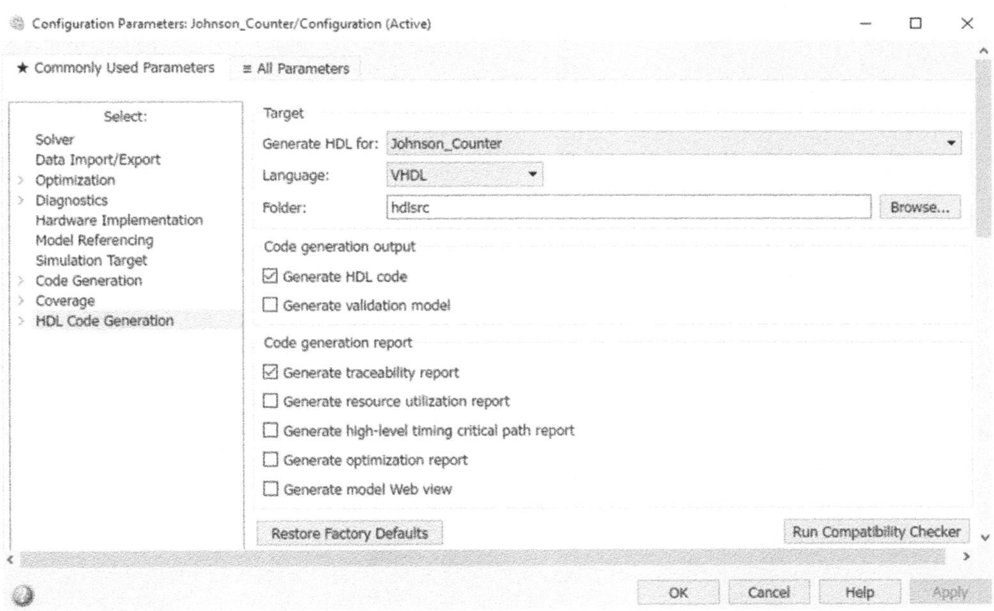

Figure 14.30 Setting the HDL Code Generation tab parameters.

Finally, we type the following code in the MATLAB command window.

```
>> makehdl('Johnson_Counter/Johnson_Counter_HDL')
```

The Code Generation Report window pops up, and four VHDL files are generated, as depicted in Figure 14.32.

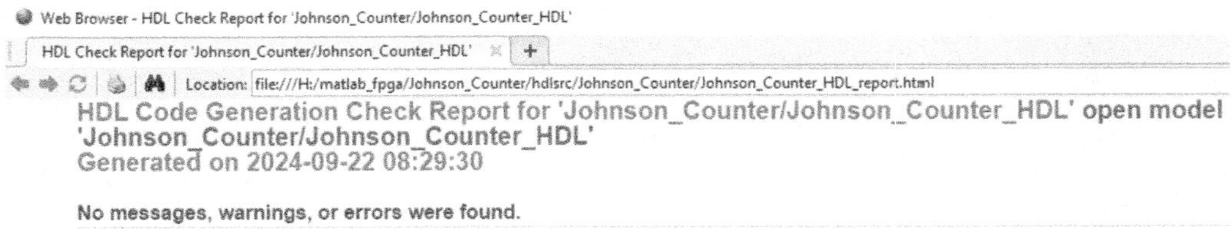

Figure 14.31 The Web Browser – HDL Check Report.

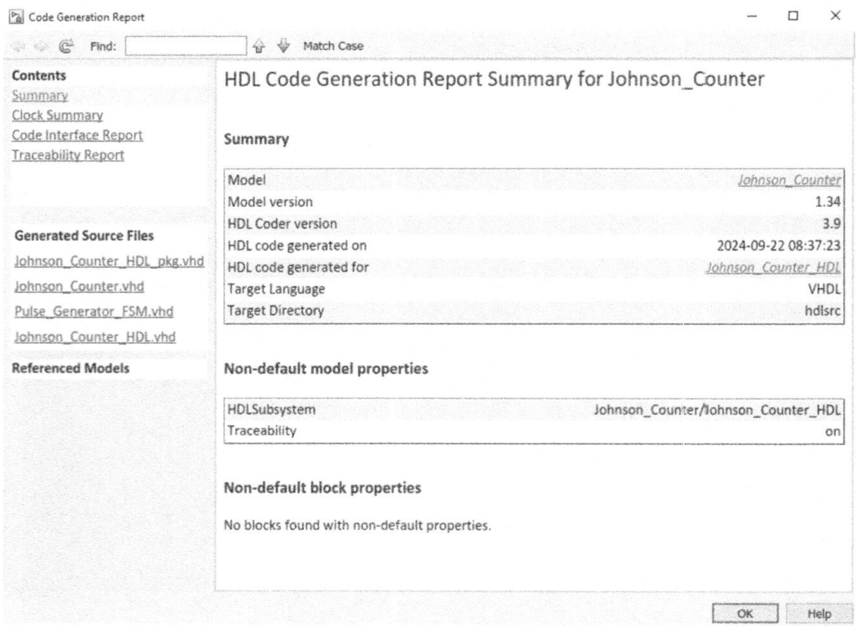

Figure 14.32 The Code Generation Report window.

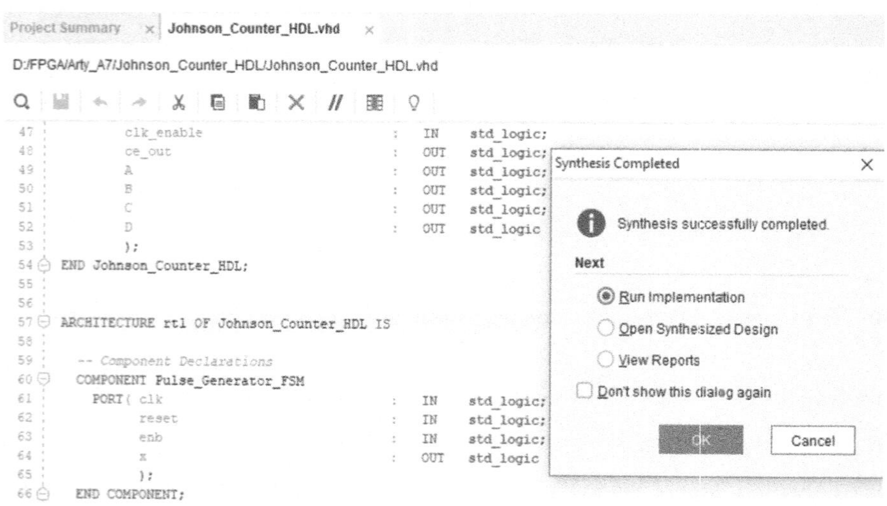

Figure 14.33 Successful completion of synthesis analysis.

Now, we open Vivado and create a new project titled Johnson_Counter_HDL. We click Next, and then select the Add files to the design sources. Before clicking the Add Files button, it is better to copy and paste the four generated VHDL files (shown in Figure 14.32) into the Vivado-generated folder (Johnson_Counter_HDL). Finally, we click the Finish button. Next, we run synthesis analysis, and it is successfully completed, so we click on the Cancel button, as shown in Figure 14.33.

In the RTL analysis on the Project Manager tab, we select the schematic. The Elaborate Design window pops up, and then we click on OK button. Now, we can see the schematic, as depicted in Figure 14.34.

Also, in the IP Integrator section, we click on Create Block Design option to create block diagram titled "top", and do right-click on "Johnson_Counter _HDL.vhd" file and select Add Module to Block Design option to add it as a module in the diagram window, as illustrated in Figure 14.35.

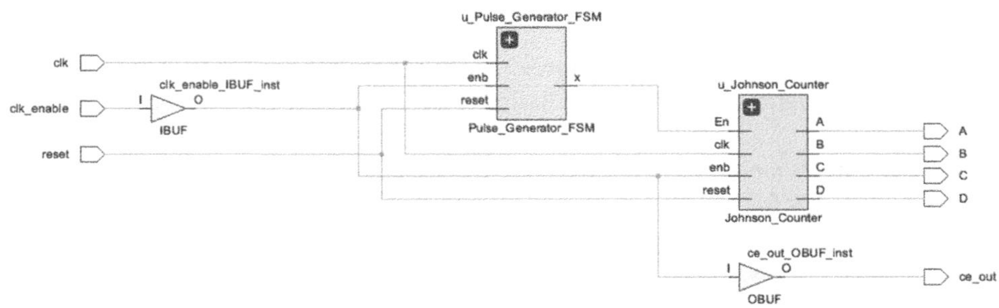

Figure 14.34 The schematic for the Johnson_Counter_HDL project.

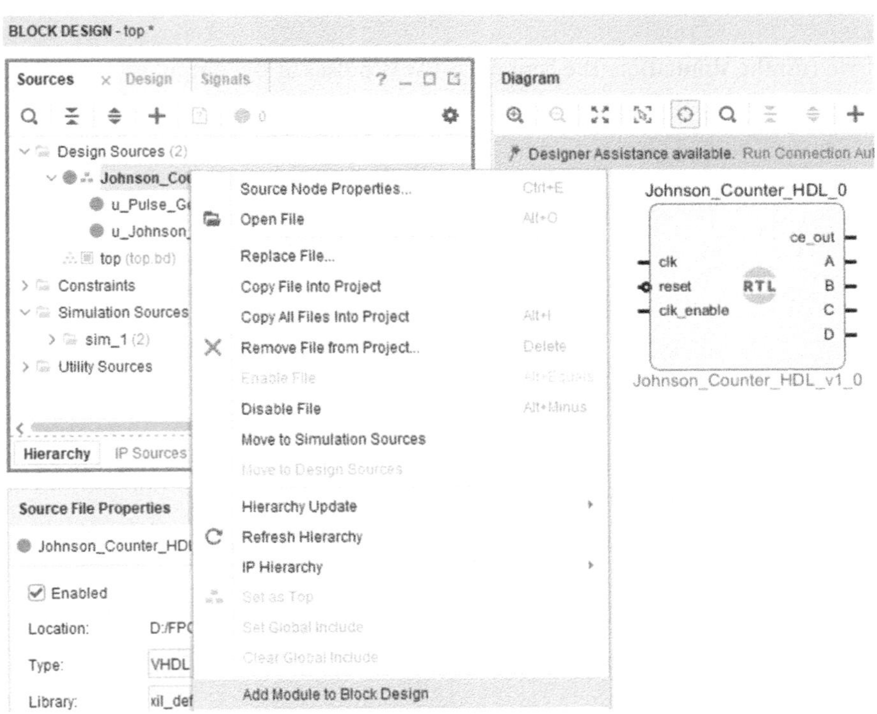

Figure 14.35 Selecting the Add Module to Block Design option.

14.4 The Basic PWM Design

In this section, we are going to design a basic PWM circuit and model it in Simulink. First of all, in the command window of MATLAB, we write the following command to open the Simulink.

```
>> simulink
```

Also, we change the directory to where we want to save the Simulink project by the following code.

```
>> cd('H:\matlab_fpga\Basic_PWM_Design')
```

After opening the Simulink environment, we use the following code to open the HDL Coder library browser.

```
>> hdllib
```

Then, using the HDL Coder library browser, we create a new model titled Basic_PWM_Design, as shown in Figure 14.36. Also, the Basic_PWM_Design subsystem model is depicted in Figure 14.37.

The block parameters for 4-bit counter free-running are illustrated in Figure 14.38.

The block parameters for constant main and signal attributes tabs are shown in Figures 14.39 and 14.40, respectively.

Now, we can type the following code in the MATLAB command window as well.

```
>> hdlsetup('Basic_PWM_Design')
```

We set the fixed step size as 1s in the solver configuration parameters, as depicted in Figure 14.41. Also, we set the HDL Code Generation tab parameters, as illustrated in Figure 14.42.

Now, if we run the simulation, the generated PWM pulses at the output of subsystem Basic_PWM_Design are shown in Figure 14.43.

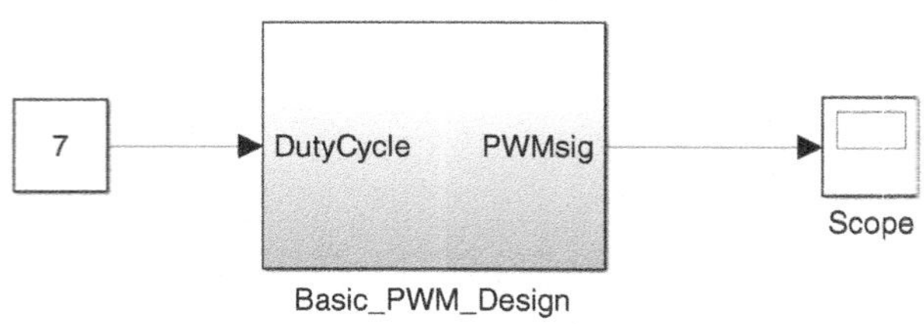

Figure 14.36 The Basic_PWM_Design model in Simulink.

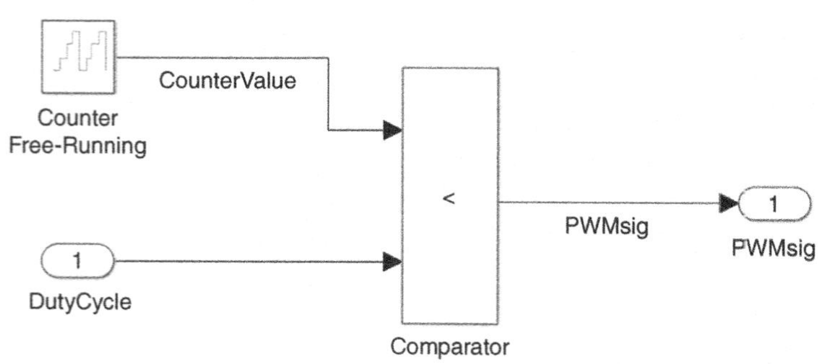

Figure 14.37 The Basic_PWM_Design subsystem model.

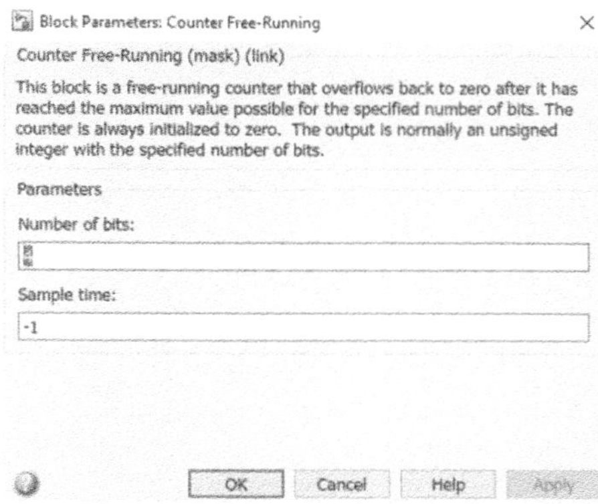

Figure 14.38 The block parameters for counter free-running.

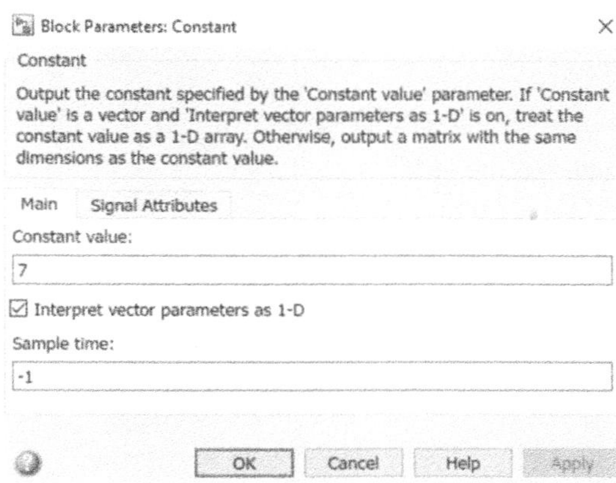

Figure 14.39 The block parameters for constant main tab.

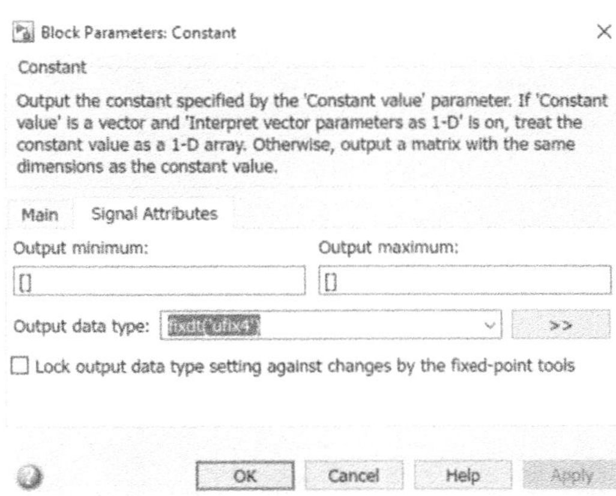

Figure 14.40 The block parameters for constant signal attributes tab.

The output waveform of 4-bit free-running counter and the PWM signal inside the subsystem Basic_PWM_Design model is depicted in Figure 14.44.

Before checking and making the HDL coder commands, it is better to delete the scope inside the subsystem Basic_PWM_Design model. Now, we write the following code in the MATLAB command window.

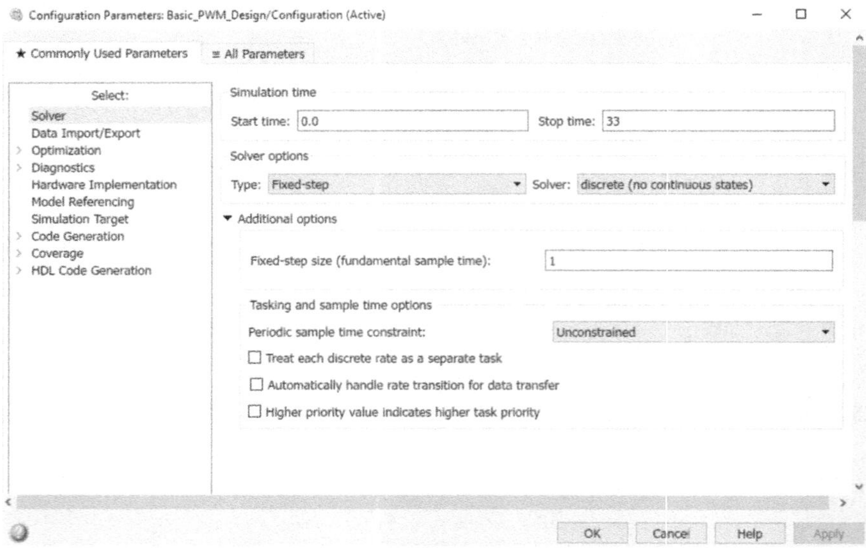

Figure 14.41 Setting the fixed-step size to 1s.

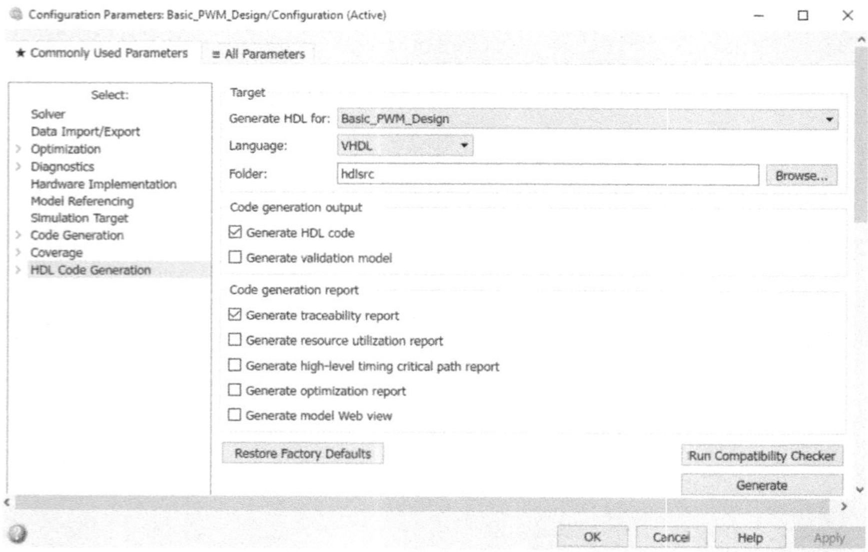

Figure 14.42 Setting the HDL Code Generation tab parameters.

```
>> makehdl('Basic_PWM_Design/Basic_PWM_Design')
```

A Web Browser – HDL Check Report is displayed, as illustrated in Figure 14.45.

Finally, we type the following code in the MATLAB command window.

```
>> makehdl('Basic_PWM_Design/Basic_PWM_Design')
```

The Code Generation Report window pops up, and the VHDL file (Basic_PWM_Design.vhd) is generated as shown in Figure 14.46.

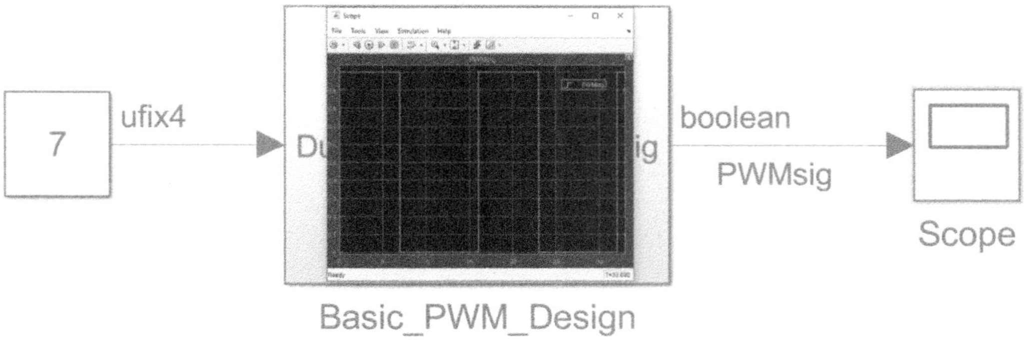

Figure 14.43 Generated PWM pulses at the output of subsystem Basic_PWM_Design.

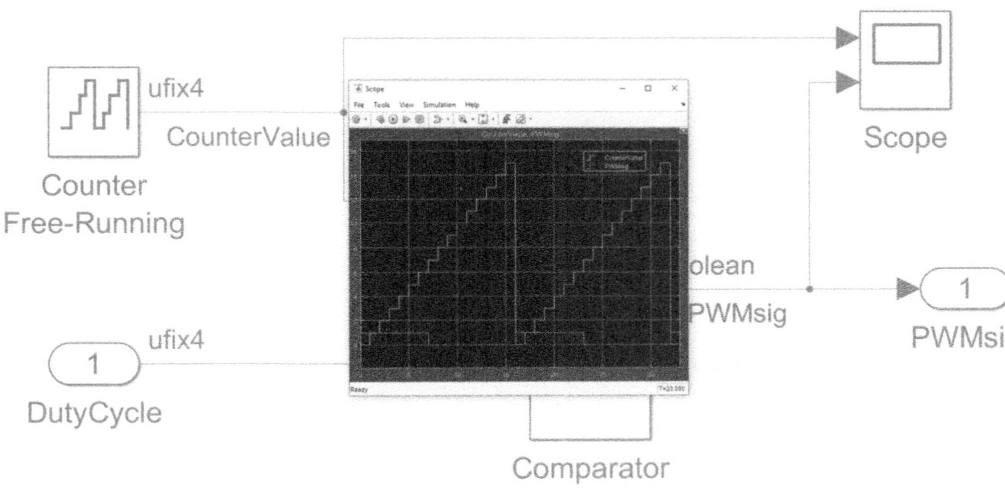

Figure 14.44 Output waveform of 4-bit free-running counter and the PWM signal.

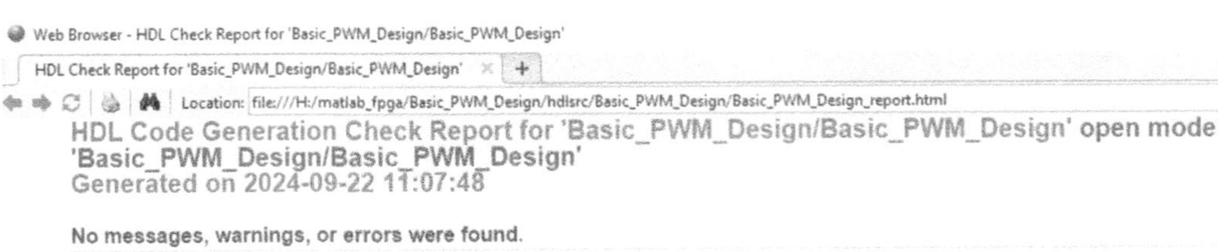

Figure 14.45 The Web Browser – HDL Check Report.

Now, we open Vivado and create a new project titled Basic_PWM. We click Next, and then select Add files to the design sources. Before clicking the Add Files button, it is better to copy and paste the generated VHDL file in Figure 14.46 into the Vivado-generated folder (Basic_PWM). Finally, we click the Finish button. Next, we run synthesis analysis, which is successfully completed, so we click the Cancel button, as depicted in Figure 14.47.

As you can see, we have added Count_Val as an output port, and its value is assigned by the unsigned signal CounterValue, as demonstrated in Figure 14.47. In the RTL analysis on the project

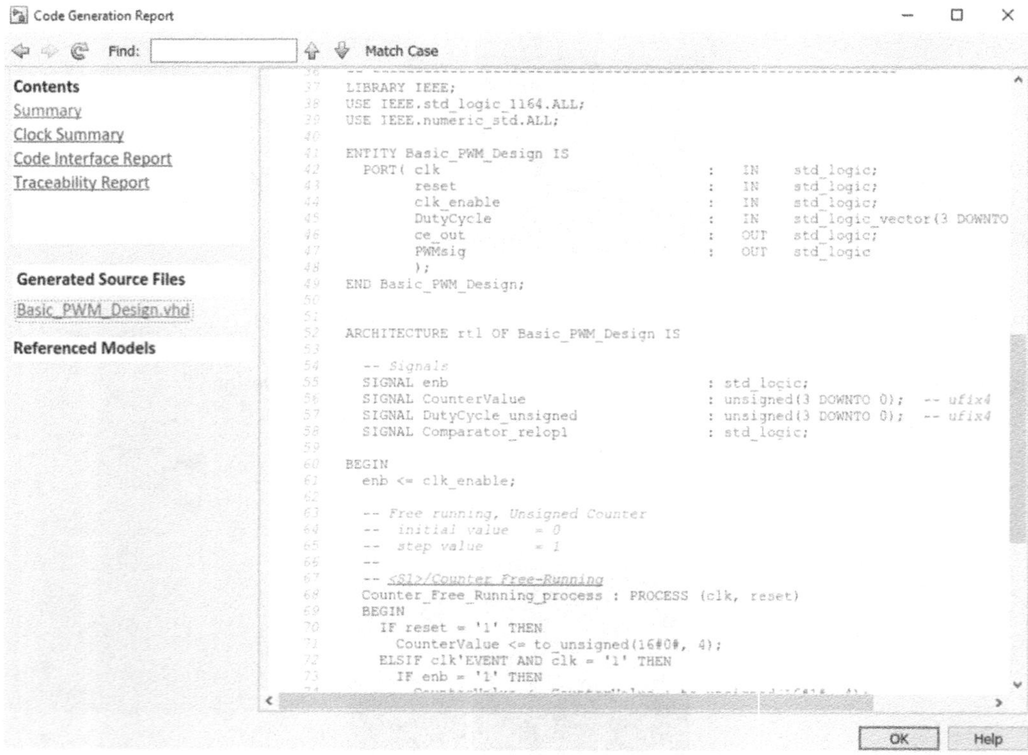

Figure 14.46 The Code Generation Report window.

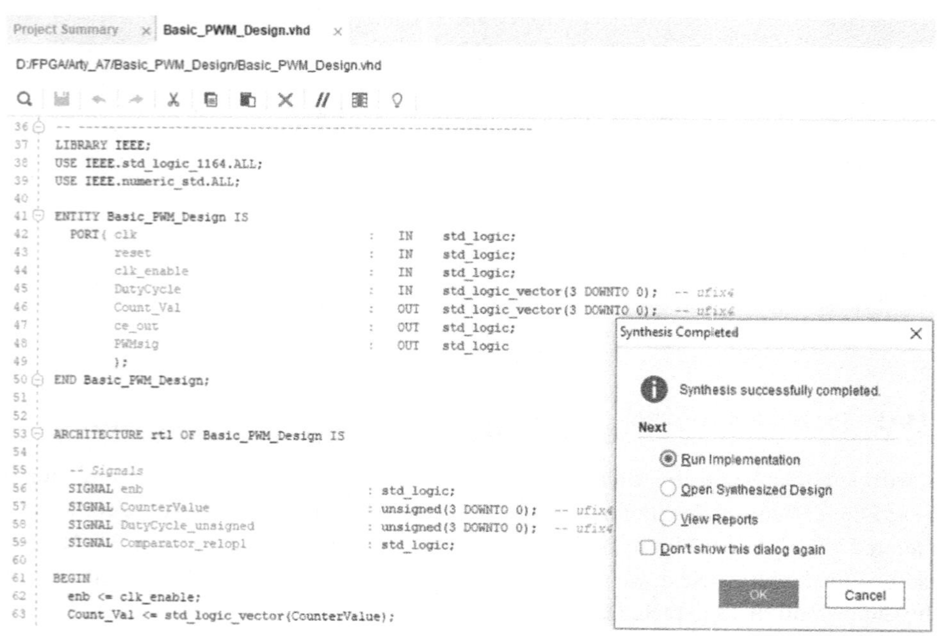

Figure 14.47 Successful completion of synthesis analysis.

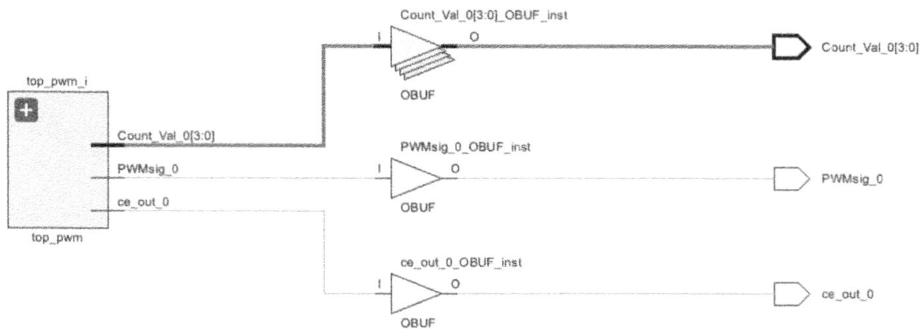

Figure 14.48 The schematic for the Basic_PWM project.

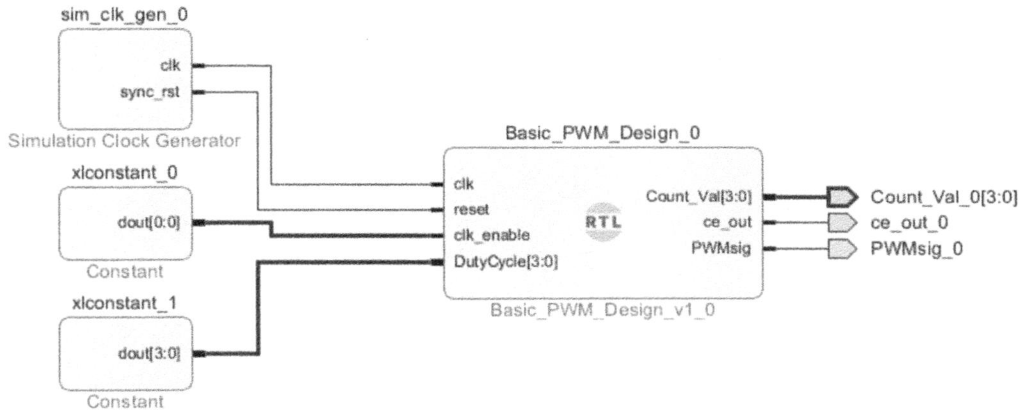

Figure 14.49 The block design in the diagram window.

Figure 14.50 The re-customize IP window for simulation clock generator.

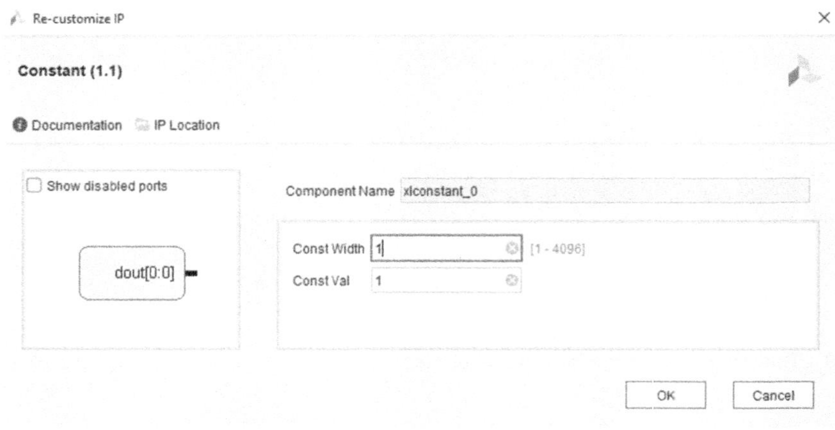

Figure 14.51 The re-customize IP parameters window for constant (clk_enable).

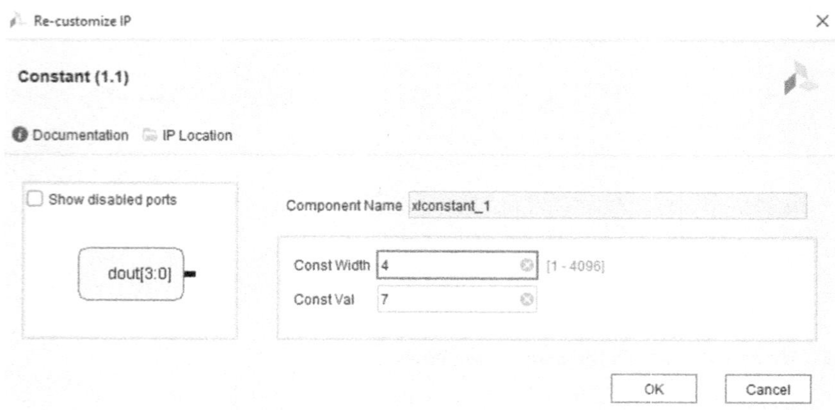

Figure 14.52 The re-customize IP parameters window for constant (DutyCycle).

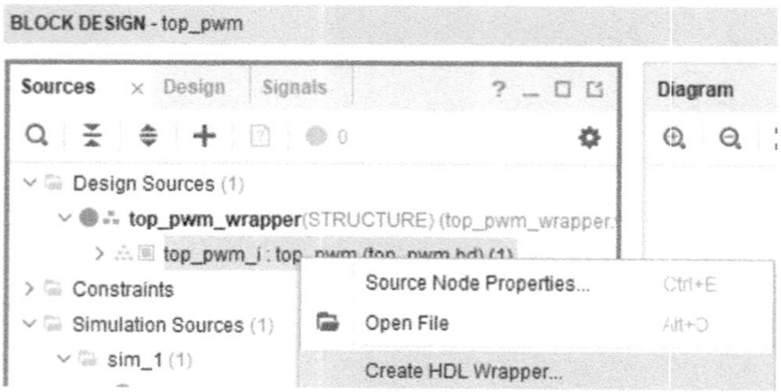

Figure 14.53 Selecting the create HDL wrapper....

manager subsection, we select the schematic. The Elaborate Design window pops up, and then we click on the OK button. Now, we can see the schematic, as depicted in Figure 14.48.

Also, in the IP Integrator section, we click on the Create Block Design option to create a block diagram titled "top_pwm". We then right-click on the "Basic_PWM_Design.vhd" file and select

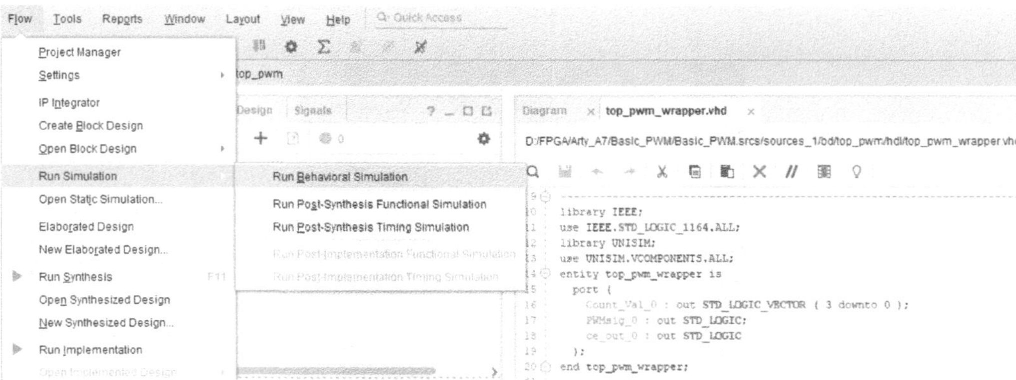

Figure 14.54 Selecting the Run Behavioral Simulation option.

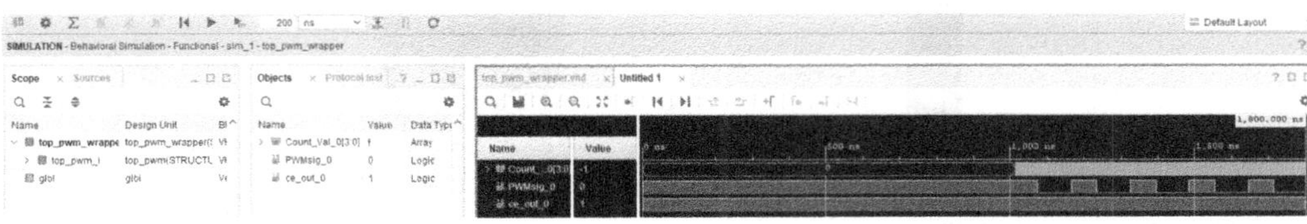

Figure 14.55 The generated PWM signal after 100 cycles.

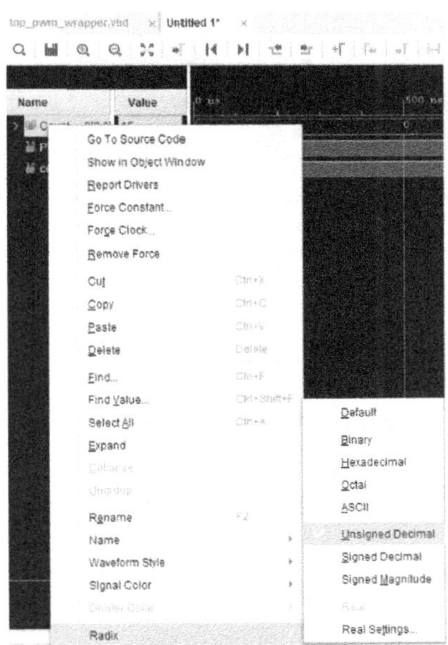

Figure 14.56 Selecting the radix as unsigned decimal.

Add Module to Block Design option to add it as a module in the diagram window. We also add a simulation clock generator, a constant for clk_enable, and a 4-bit constant for DutyCycle. Then, we select the Count_Val, ce_out, and PWMsig, and press Ctrl+T to create output ports for them, as illustrated in Figure 14.49.

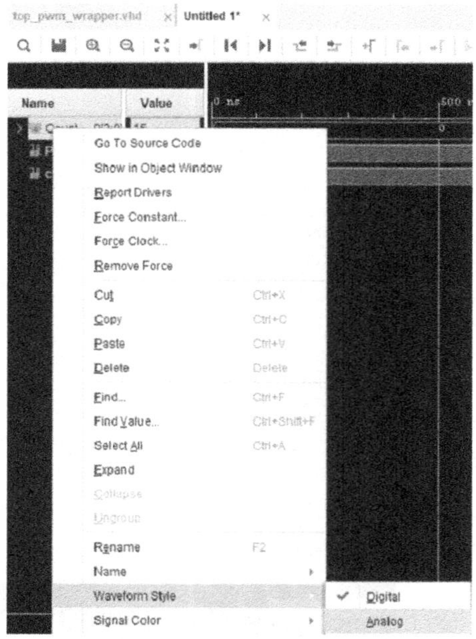

Figure 14.57 Selecting the waveform style as analog.

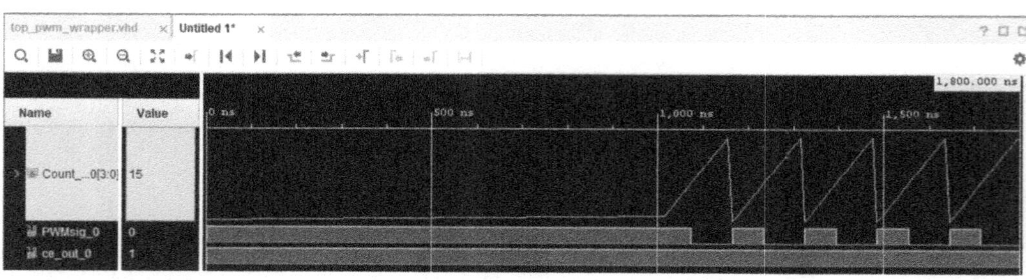

Figure 14.58 The triangular waveform of Count_Val.

The re-customize IP parameters window for simulation clock generator is shown in Figure 14.50. Also, the re-customize IP parameters windows for constants are depicted in Figures 14.51 and 14.52, respectively.

Then, we right-click on top_pwm.bd, and select the create HDL wrapper..., as illustrated in Figure 14.53.

Now, from the Flow menu, we select the Run Behavioral Simulation option, as shown in Figure 14.54.

First, we restart the simulation and run the simulation for 200 ns. After 100 cycles (1000 ns), the PWM signal is produced as depicted in Figure 14.55, since the initial reset clock cycle has been set to 100 in Figure 14.50.

We right-click on Count_Val and select the radix as unsigned decimal, as illustrated in Figure 14.56.

Also, we right-click on Count_Val and select the waveform style as analog, as shown in Figure 14.57.

Now, we can view the triangular waveform of Count_Val, as depicted in Figure 14.58.

15

Application of Flowcharts in Simulink for VHDL Code Generation

15.1 Flowchart of Multiplier

In this section, we are going to implement a multiplier using a flowchart in Simulink. The algorithms based on state flowcharts in HDL Coder for VHDL programming using Vivado are discussed in detail. The flowchart of the multiplier ($P = A \times B$) is shown in Figure 15.1. We assume that $B \neq 0$ and $P = 0$ initially.

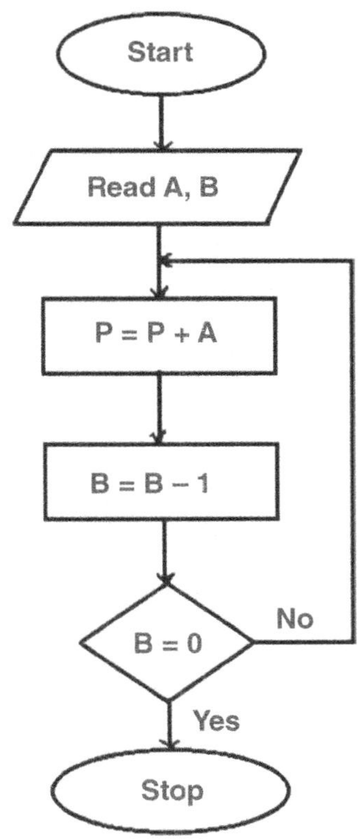

Figure 15.1 The flowchart of the multiplier (P = A × B).

Fundamentals of VHDL for FPGA Programming Using Vivado, First Edition. Majid Pakdel.

Companion Website: www.wiley.com/go/pakdelVHDL

We will use the fixed-point numeric type to define the parameters of the flowchart in a MATLB m-file. For example, if we write and run the following code, we will have the results as follows.

```
clear all; close all; clc
% Define 5-bit unsigned number
nt = numerictype(0,5,0);
a = fi(3,nt)
a.bin

a =

     3

          DataTypeMode: Fixed-point: binary point scaling
            Signedness: Unsigned
            WordLength: 5
        FractionLength: 0
ans =
00011
```

When starting the flowchart in the Stateflow environment of Simulink, we begin with the initial state (S0) by initializing all outputs to zero or predefined values using the constants defined in the MATLAB m-file. Then, we implement the flowchart step-by-step, recognizing each step as a separate state in Stateflow. If we encounter a complex operation, we divide it into multiple steps or states for clarity and manageability. Constants defined in the m-file are set as parameters, while intermediate variables are defined as locals in the Stateflow Model Explorer. We assign inputs to local variables and similarly define inputs and outputs in the Model Explorer. When performing any arithmetic operation within a state, make sure to convert it into fixed-point format using the fi command. For example, we have the following code.

```
P = fi(P+A,0,14,0); % Perform P = P + A and save it as 14-bit unsigned number
B = fi(B-a1,0,7,0); % Perform B = B - 1 and save it as 7-bit unsigned number
```

We can also consider using MATLAB functions or Simulink functions to simplify and compact the flowchart. The designed Stateflow chart in Simulink is shown in Figure 15.2.

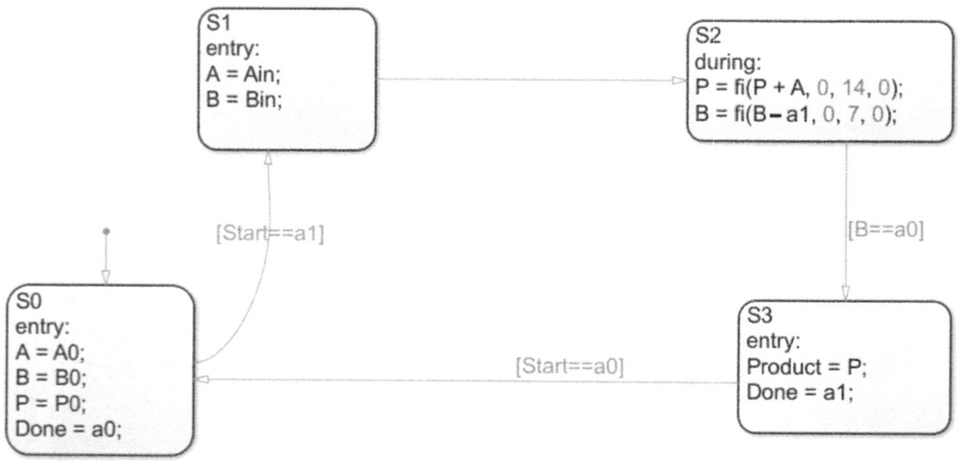

Figure 15.2 The designed Stateflow chart in Simulink.

Parameters (constants) are already defined inside the m-file (FC_Parameters.m) as the following codes.

```
clear all; close all; clc
N = 7; %nbits
nt1 = numerictype(0,1,0);
nt2 = numerictype(0,N,0);
nt3 = numerictype(0,2*N,0);

% 1-bit constants
a0 = fi(0,nt1);
a1 = fi(1,nt1);

% Inputs
Ain = fi(50,nt2);
Bin = fi(12,nt2);

% Initial values
A0 = fi(0,nt2);
B0 = fi(0,nt2);
P0 = fi(0,nt3);
```

To use these constants inside the Stateflow, we should run the m-file. The A0, B0, P0, a0, and a1 are fixed-point parameters defined inside the m-file. The A, B, and P are local or intermediate variables. Ain, Bin, and Start are the inputs, while Product and Done are the outputs. From View → Model Explorer → Model Explorer, we open the Model Explorer window as depicted in Figure 15.3.

To add parameters, local variables, inputs, and outputs for the Stateflow design in Figure 15.2, in the Model Explorer window, click on the Add in the menu bar and select the Data option, as illustrated in Figure 15.4.

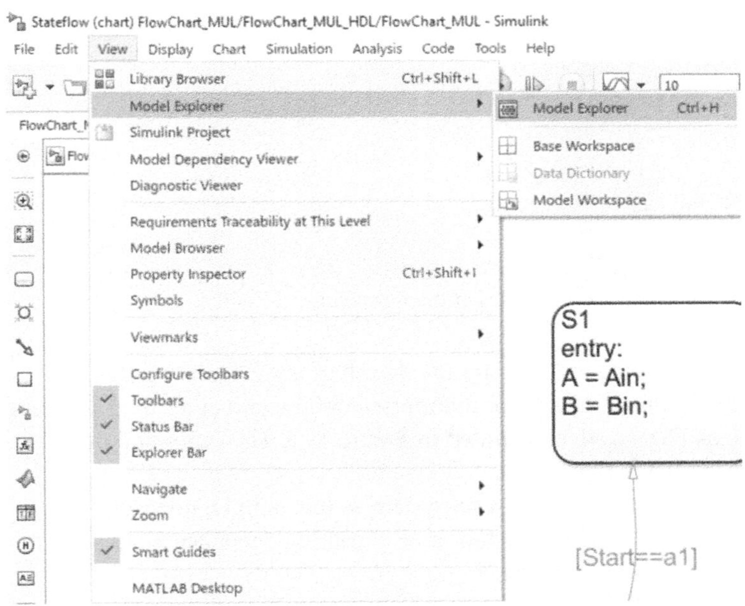

Figure 15.3 Selecting the Model Explorer option.

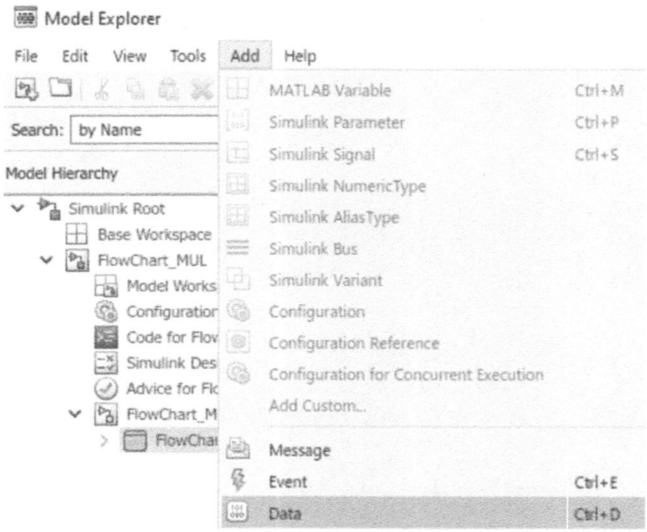

Figure 15.4 Selecting the Data option.

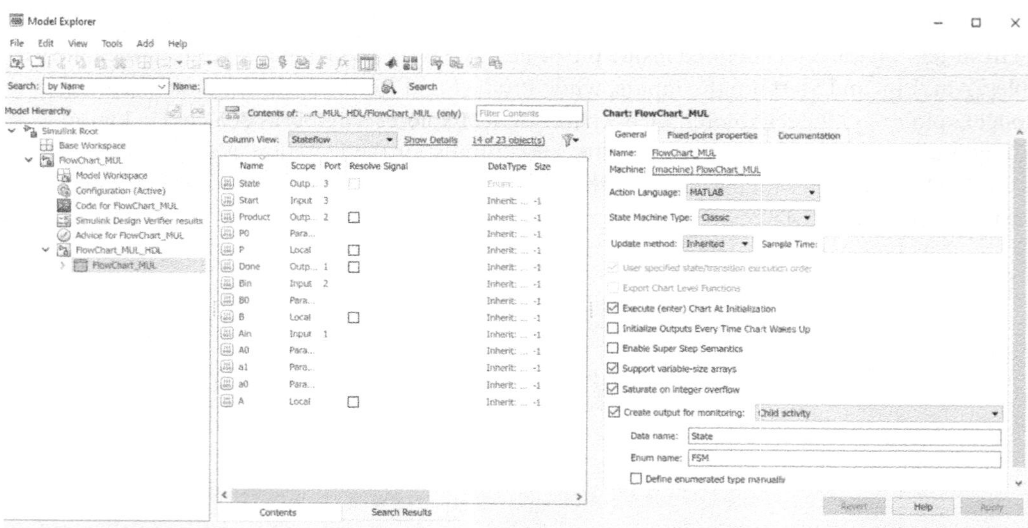

Figure 15.5 The complete added variables of the flowchart.

The complete list of added variables for the flowchart is shown in Figure 15.5.

We also enable the Create output for monitoring option, and enter the Data name as State and the Enum name as FSM, as demonstrated in Figure 15.5. The input and output variables of the flowchart are illustrated in Figure 15.6.

Next, we place the flowchart inside a subsystem as shown in Figure 15.7.

The block parameters for the Constant and Repeating Sequence Stair blocks are depicted in Figures 15.8, 15.9, and 15.10, respectively.

Before running the simulation, we set the Solver tab of Configuration Parameters window as illustrated in Figure 15.11.

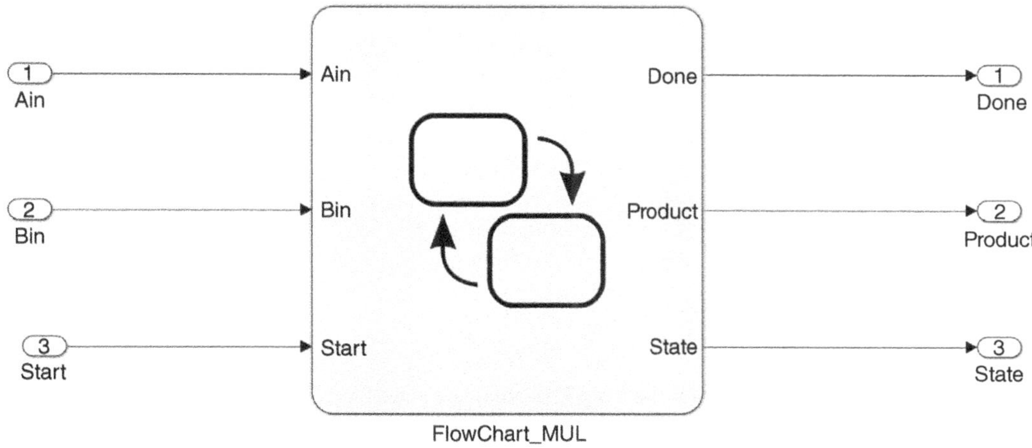

Figure 15.6 The flowchart's input and output variables.

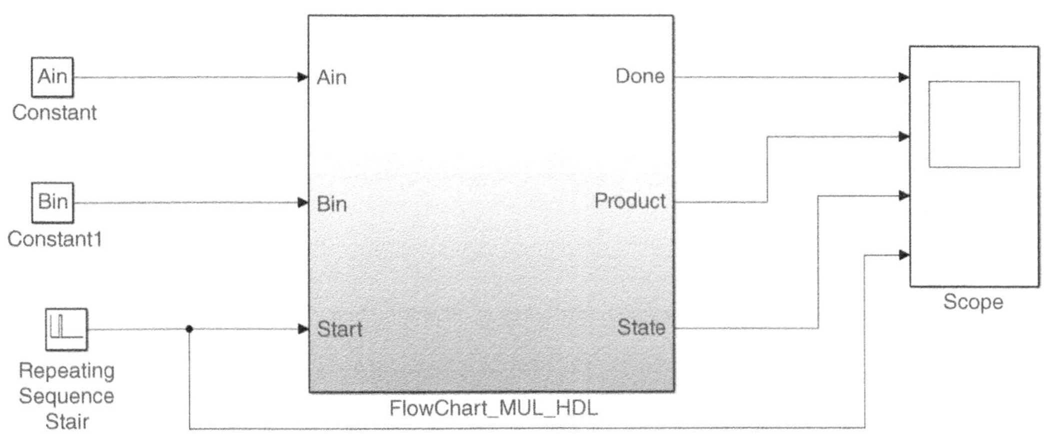

Figure 15.7 Putting the flowchart inside a subsystem.

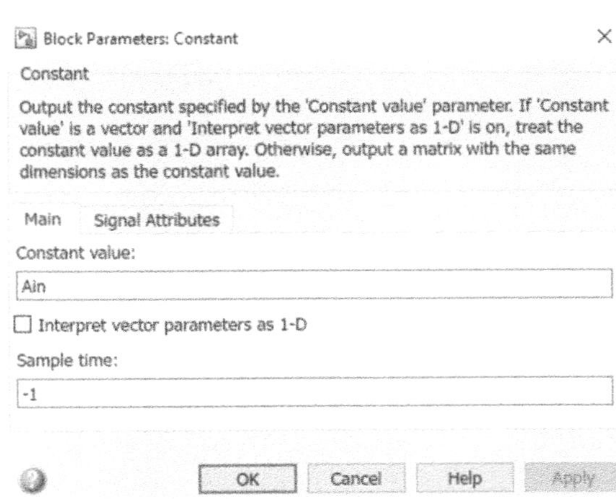

Figure 15.8 The block parameters for Constant.

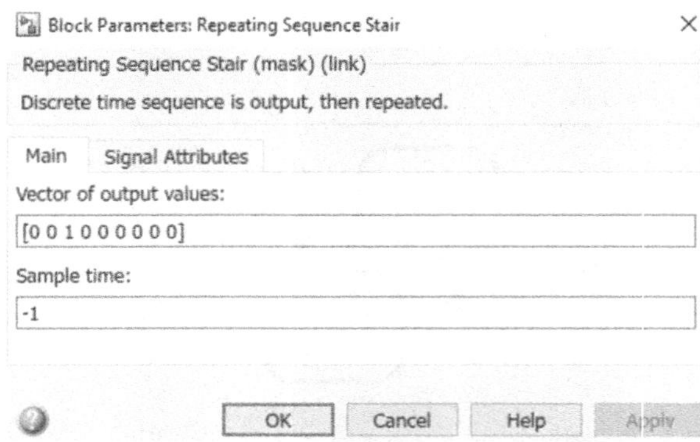

Figure 15.9 The block parameters for the main Repeating Sequence Stair.

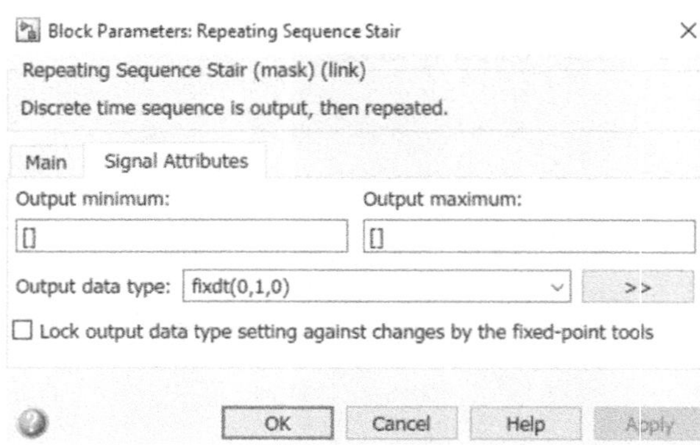

Figure 15.10 The block parameters for signal attributes of Repeating Sequence Stair.

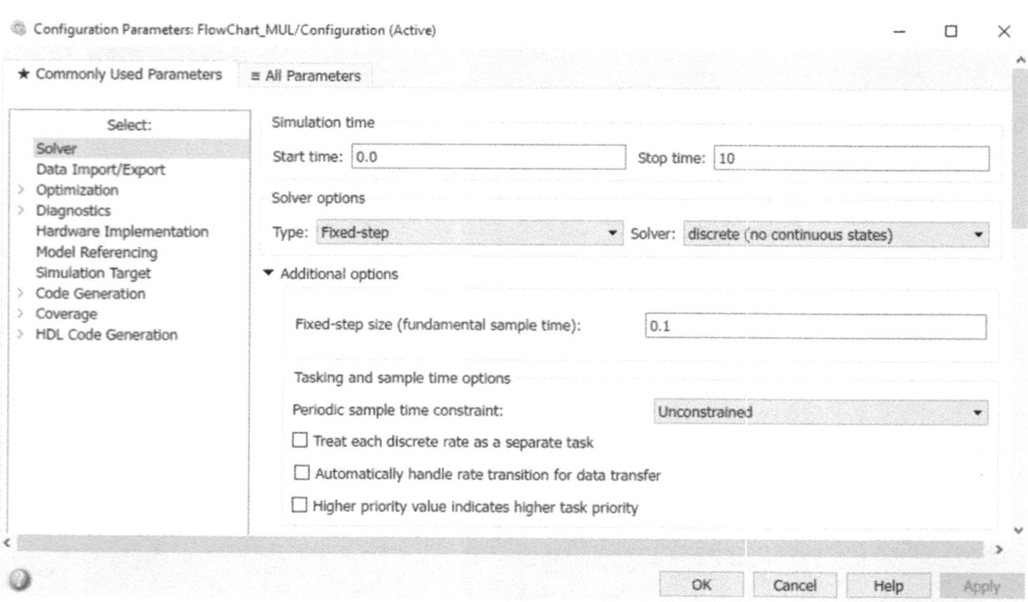

Figure 15.11 Setting the Solver tab of Configuration Parameters window.

Now, we can run the simulation. The output waveforms on the Scope block are shown in Figure 15.12.

Next, we generate the VHDL code for the subsystem "FlowChart_MUL_HDL" in Figure 15.7 using HDL Coder. To do that, we write the following code in the MATLAB command window.

```
>> hdlsetup('FlowChart_MUL')
```

Here, FlowChart_MUL is the name of the Simulink file (FlowChart_MUL.slx). Then, we configure the Solver tab of the Configuration Parameters window, as depicted in Figure 15.13.

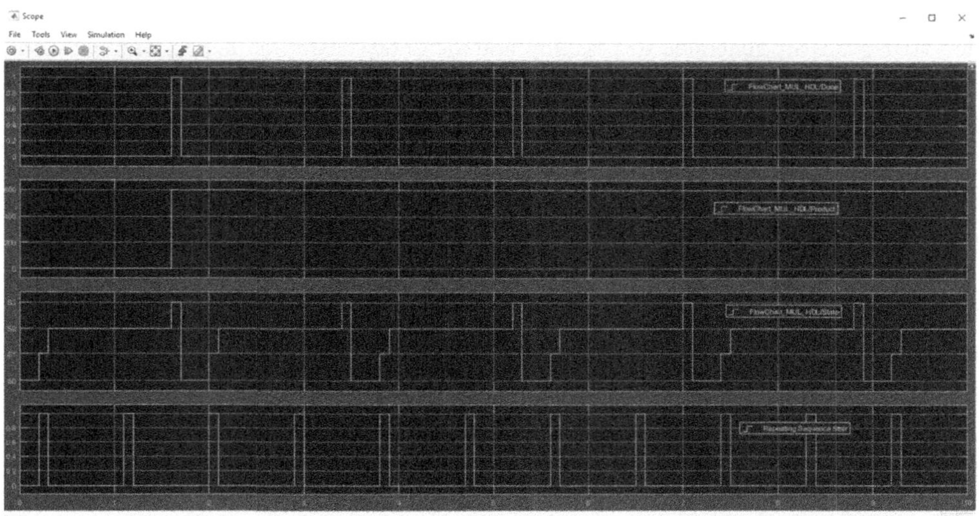

Figure 15.12 The waveforms on the Scope.

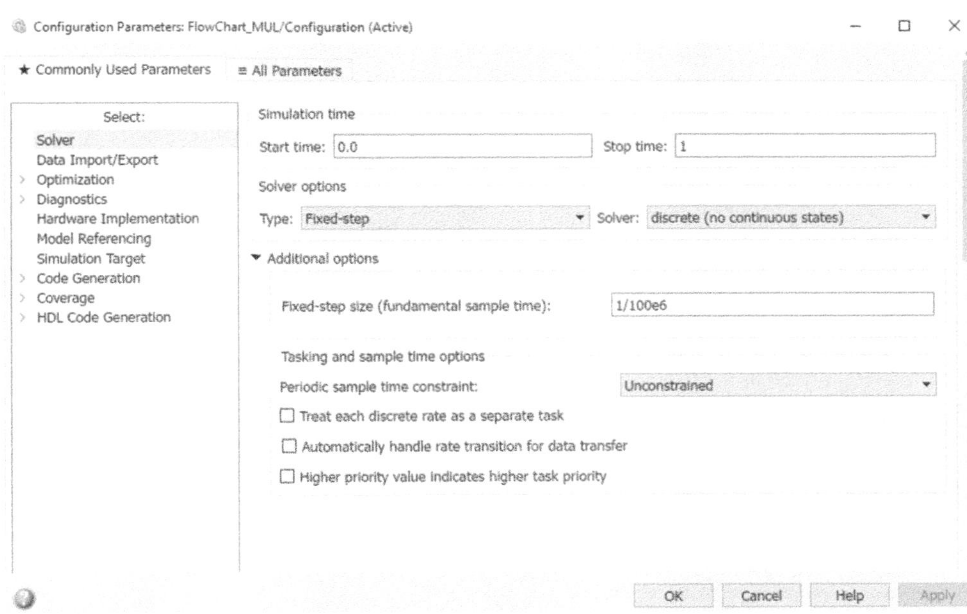

Figure 15.13 Setting the Solver tab of Configuration Parameters window.

Since the clock frequency of Xilinx FPGA development boards is usually 100 MHz, we set the fixed-step size as 1/100e6 as demonstrated in Figure 15.13. Also, we set the HDL Code Generation tab of Configuration Parameters window as illustrated in Figure 15.14.

Next, we type the following code in the MATLAB command window.

```
>> checkhdl('FlowChart_MUL/FlowChart_MUL_HDL')
```

This opens the HDL Check Report window in a web browser, as shown in Figure 15.15.

Now, we use the following code for HDL code generation in the MATLAB command window.

```
>> makehdl('FlowChart_MUL/FlowChart_MUL_HDL')
```

The VHDL codes are generated successfully, and the Code Generation Report window appears, as depicted in Figure 15.16.

Now, we run the Vivado and create a folder named "FlowChart." Inside this folder, we create a project titled "FlowChart" and click the Next button to proceed. In the Add Sources window, before adding the generated VHDL files, we first copy and paste the three generated VHDL files, as shown in Figure 15.16, into the FlowChart folder of the Vivado project. Now, by clicking the Add Files button, we can add them to the project. We then select our Xilinx development board

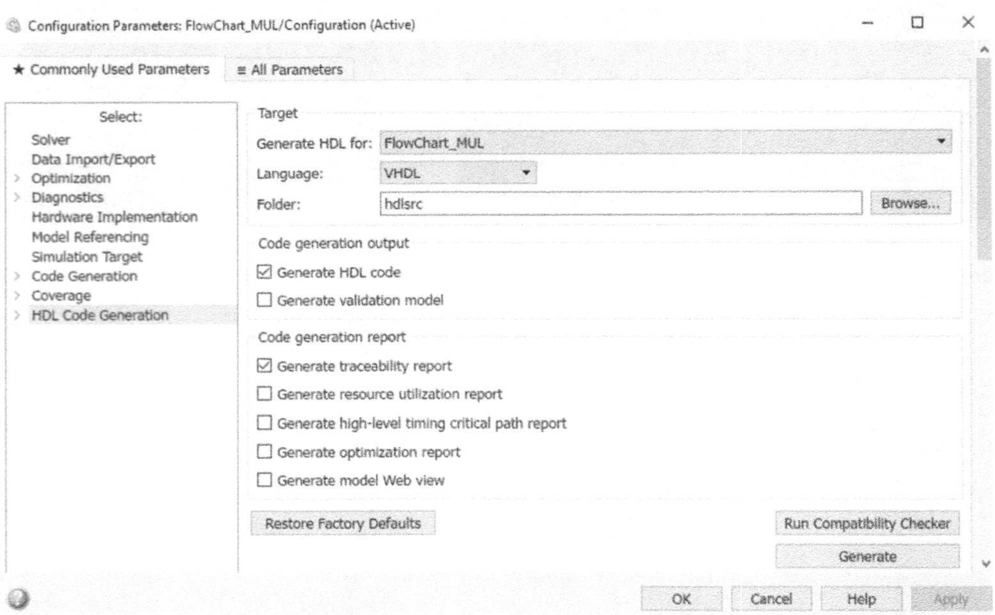

Figure 15.14 Setting the HDL Code Generation of Configuration Parameters window.

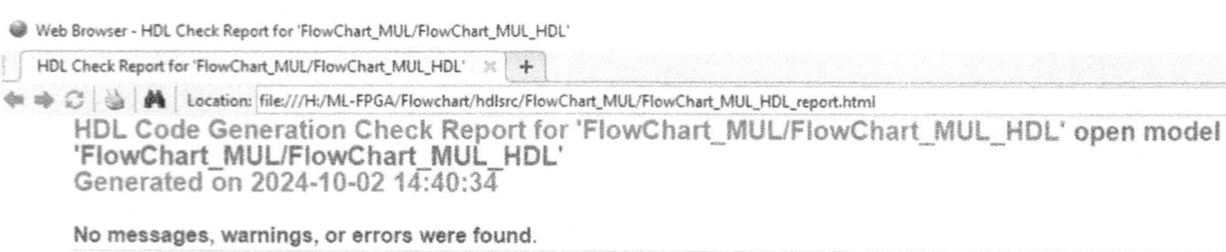

Figure 15.15 The HDL Check Report window.

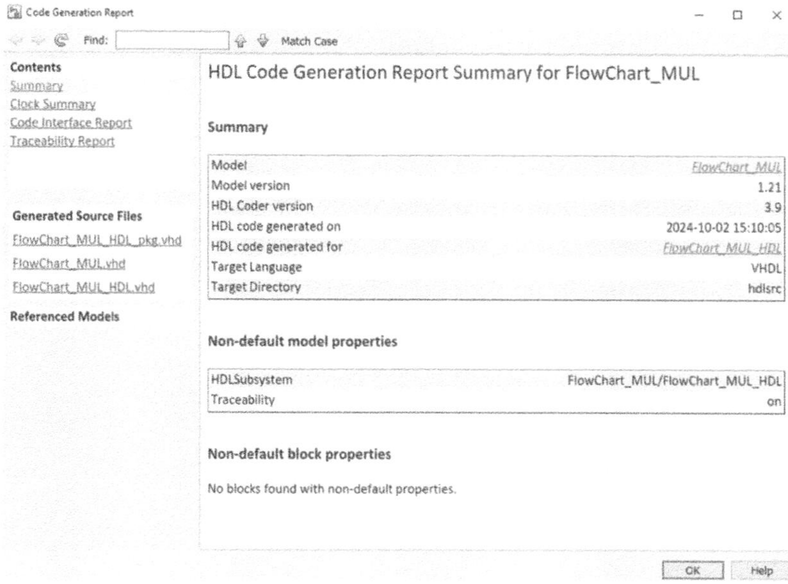

Figure 15.16 The Code Generation Report window.

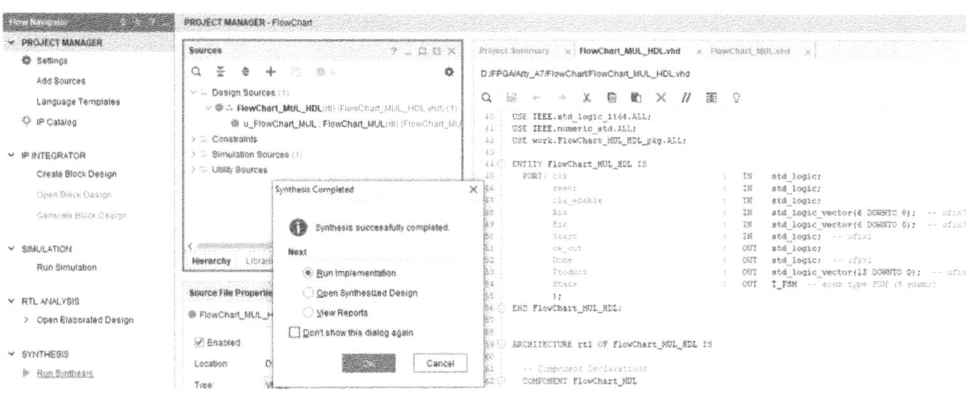

Figure 15.17 Running the synthesis.

and click on the Finish button to create the FlowChart project with added generated VHDL files. Finally, we run synthesis, as illustrated in Figure 15.17.

As demonstrated in Figure 15.17, the synthesis is successfully completed. We can now click the Cancel button. Then, we open the schematic in the RTL Analysis tab, as shown in Figure 15.18.

Now, we can simulate the project by selecting Flow → Run Simulation → Run Behavioral Simulation, as depicted in Figure 15.19.

In the opened simulation window, we click on the Restart button. Then, in the waveform window, we right-click on the "clk" and select "Force Clock..." option. Set its configuration as illustrated in Figure 15.20. Similarly, right-click on the "reset" and "clk_enable" signals, select "Force Constant..." option, and force its value to 1, as shown in Figure 15.21. We right-click on the "Ain", select "Force Constant..." option, and force its value to an unsigned decimal of 20, as illustrated in Figure 15.22. Similarly, we force the "Bin" constant value to unsigned decimal of 10.

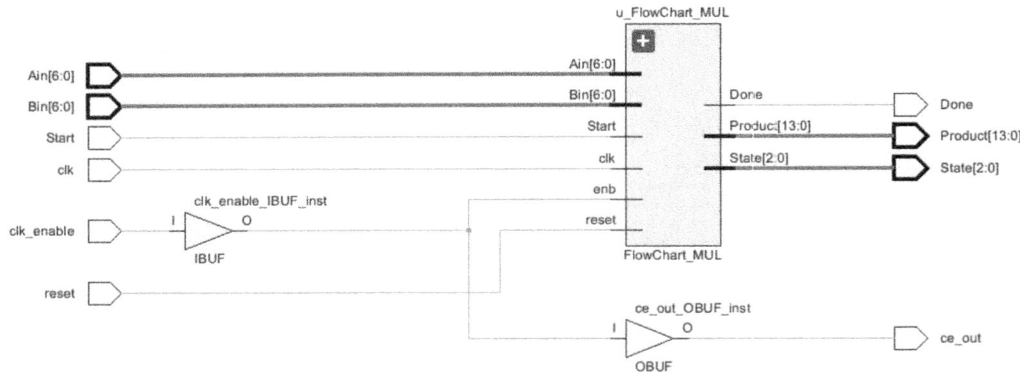

Figure 15.18 The schematic of "FlowChart_MUL_HDL."

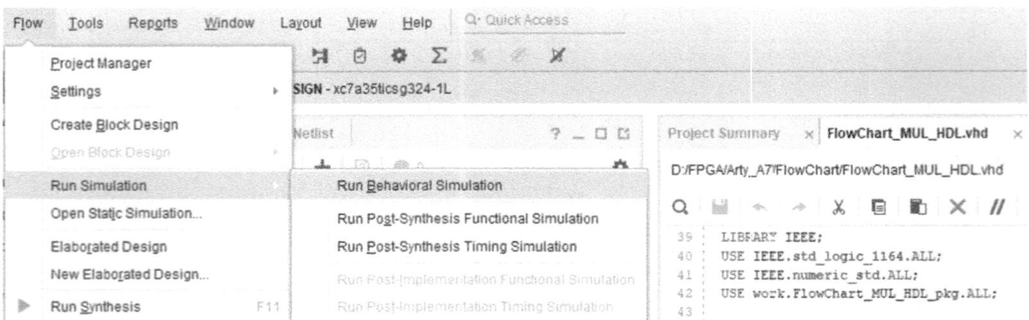

Figure 15.19 Running the simulation.

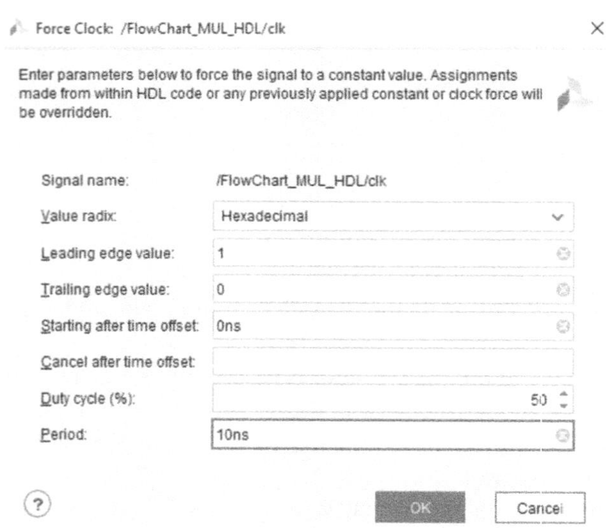

Figure 15.20 The Force Clock settings for "clk."

Next, we right-click on "Start", select "Force Clock..." option, and adjust its settings, as shown in Figure 15.23.

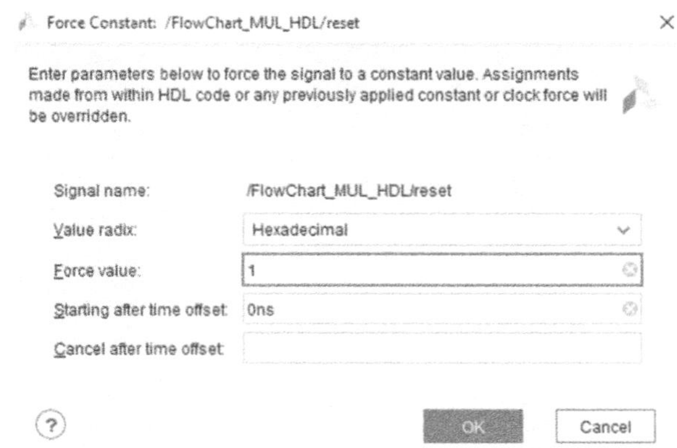

Figure 15.21 The Force Constant settings for "reset" and "clk_enable."

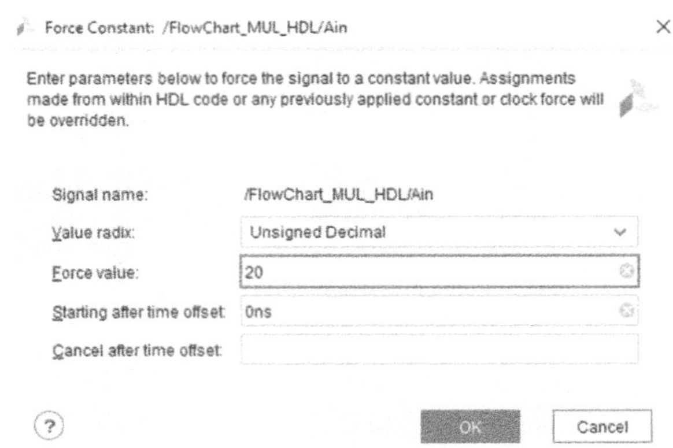

Figure 15.22 The Force Constant settings for "Ain."

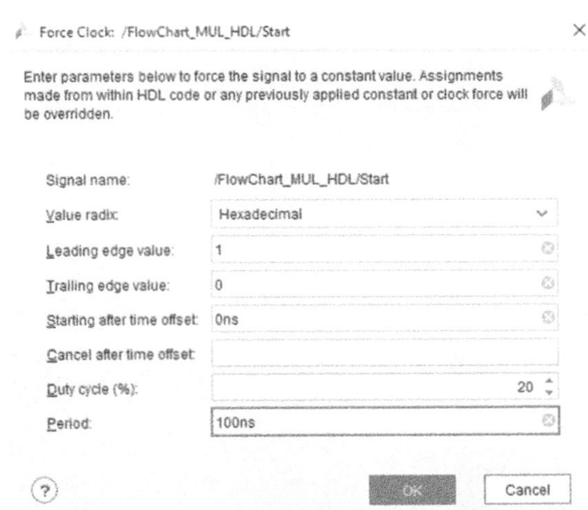

Figure 15.23 The force clock settings for "Start."

Now, click the Run for 20 ns button, as shown in Figure 15.24.

We right-click again on the "reset" and force its value to zero, as depicted in Figure 15.25.

Finally, click the Run for 20 ns button multiple times and click the Zoom Fit button in the waveforms window to view the simulated waveforms, as illustrated in Figure 15.26.

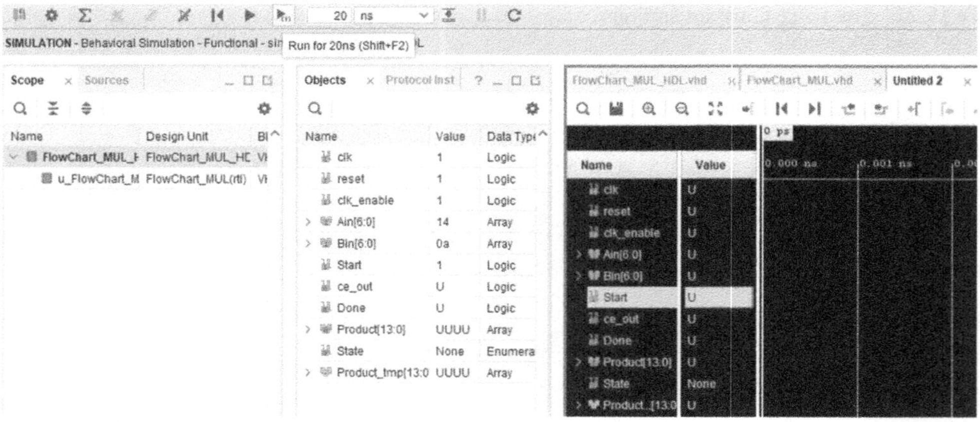

Figure 15.24 Clicking once on the Run for 20 ns button.

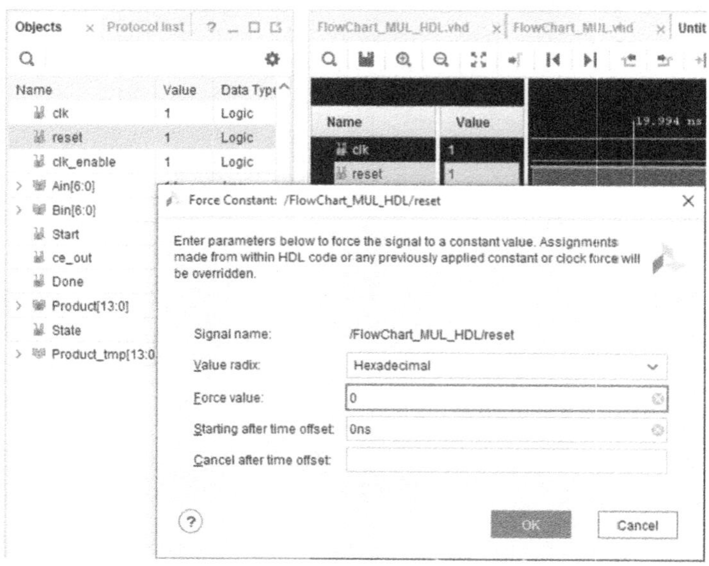

Figure 15.25 Forcing the "reset" value to zero.

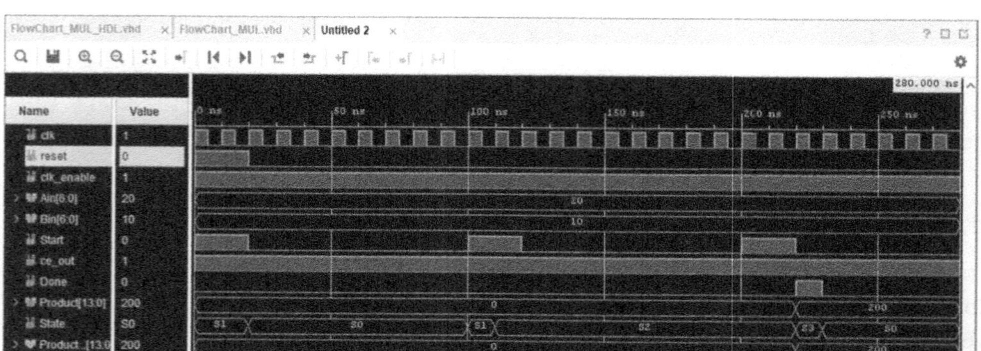

Figure 15.26 The simulated waveforms.

15.2 Flowchart of GCD Calculator

In this section, we will implement the GCD calculator algorithm directly from its flowchart using HDL Coder and simulate the design in the MATLAB Simulink environment to verify its functionality. Next, we will utilize the HDL Coder to generate synthesizable VHDL code from the designed GCD calculator implemented in Stateflow. Finally, we will synthesize and implement the generated VHDL code using Xilinx Vivado. The flowchart of the GCD calculator is shown in Figure 15.27.

To effectively implement the digital system design algorithm directly from the flowchart, we need to carefully analyze the flowchart and identify the inputs, outputs, constants, and intermediate variables. Also, we should initialize constants and inputs using a separate MATLAB script (m-file) using fixed-point numeric objects such as "numerictype" and "fi", as described in the previous section. When starting the flowchart in Stateflow within Simulink, begin with the initial state (S0) by initializing all outputs to zero or to predefined values using the constants previously defined in m-file. Then, implement the flowchart step-by-step, recognizing each step as a separate state in Stateflow chart. If you encounter a complex operation, divide it into multiple steps or states for clarity and manageability. Constants defined in the m-file should be added as parameters, and intermediate variables should be defined as local variables in the Model Explorer of Stateflow in Simulink. Inputs should be assigned to local variables. Similarly, we define inputs and outputs in the Model Explorer. When you perform any arithmetic operation at any state, make sure to convert it into fixed point using the "fi" function, as discussed in the previous section. Also, it is better to consider using MATLAB functions or Simulink functions to simplify and compact the flowchart within Stateflow. The Stateflow chart of the GCD calculator is depicted in Figure 15.28.

Parameters (constants) are already defined inside the m-file (FC_Parameters.m) as the following codes.

```
clear all; close all; clc
N = 7; %nbits
nt1 = numerictype(0,1,0);
nt2 = numerictype(0,N,0);
```

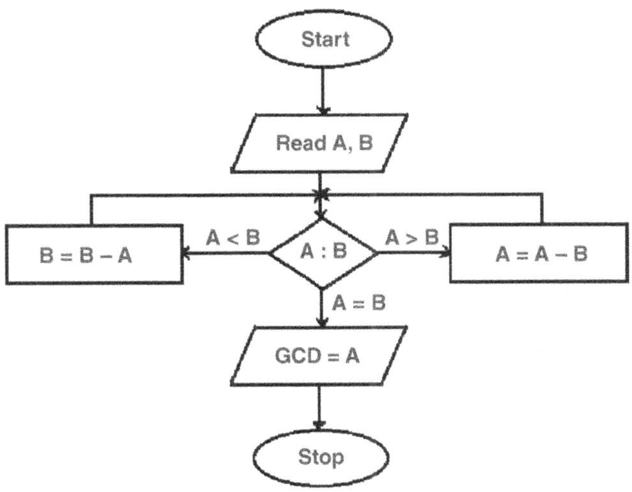

Figure 15.27 The flowchart of GCD calculator.

```
% 1-bit constants
a0 = fi(0,nt1);
a1 = fi(1,nt1);

% Inputs
Ain = fi(65,nt2);
Bin = fi(13,nt2);

% Initial values
A0 = fi(0,nt2);
B0 = fi(0,nt2);
```

To use these constants inside the Stateflow, we should run the m-file. The variables A0, B0, a0, and a1 are fixed-point parameters defined in the m-file. A and B are local or intermediate variables. Additionally, Ain, Bin, and Start are inputs, while GCD and Done are outputs. From View → Model Explorer → Model Explorer, we open the Model Explorer window in order to add parameters, local variables, inputs, and outputs for the Stateflow design shown in Figure 15.28. In the Model Explorer window, we click on Add in the menu bar, select the Data option, and begin adding the variables defined in the Stateflow chart. The complete list of added variables for the Stateflow chart is shown in Figure 15.29.

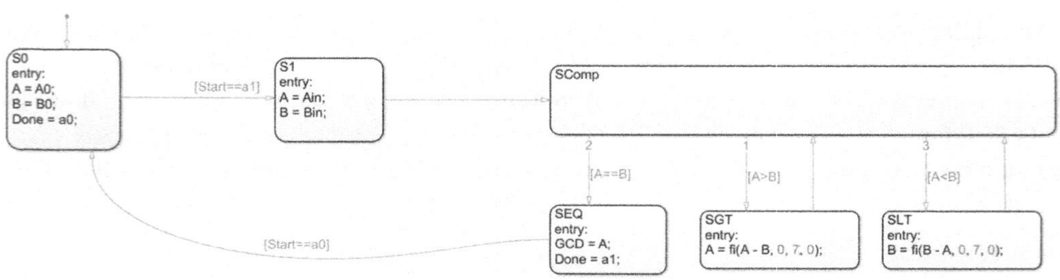

Figure 15.28 The Stateflow chart of the GCD calculator.

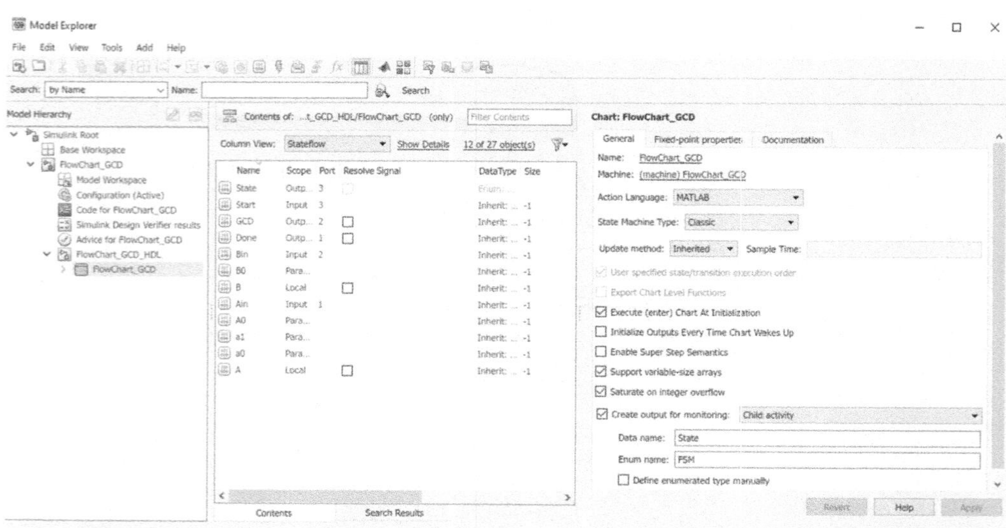

Figure 15.29 The complete list of added variables for the Stateflow chart.

We also tick on Create Output for Monitoring option, and enter the Data name as State and the Enum name as FSM, as demonstrated in Figure 15.29. The flowchart's input and output variables are illustrated in Figure 15.30.

Next, we place the flowchart inside a subsystem, as shown in Figure 15.31.

The block parameters for the Constant and Repeating Sequence Stair blocks in Figure 15.31 are similar to those described in the previous section and are depicted in Figures 15.8, 15.9, and 15.10, respectively. Before running the simulation, we set the Solver tab of Configuration Parameters window, as illustrated in Figure 15.11.

Now, we can run the simulation. The resulting scope waveforms are shown in Figure 15.32. If we type the following code in the MATLAB command window, we can access the library elements of HDL Coder.

```
>> hdllib
```

To return to the normal Simulink library, we should use the following code in the MATLAB command window.

```
>> hdllib('off')
```

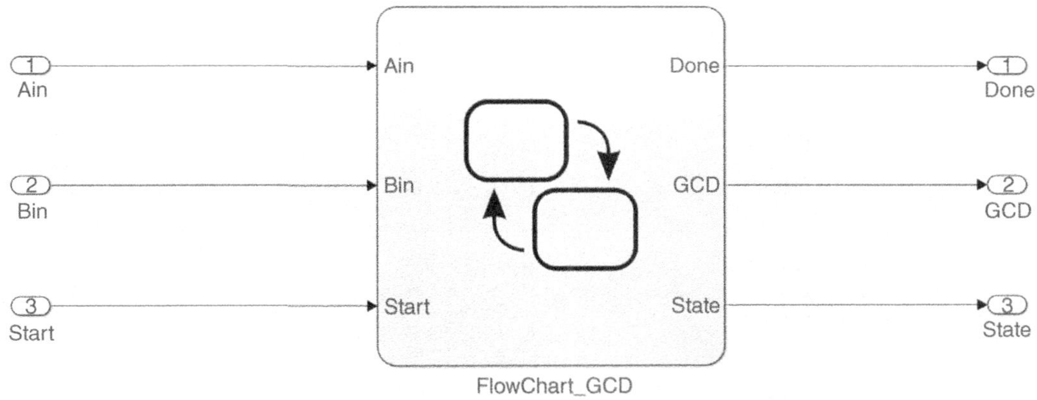

Figure 15.30 The flowchart's input and output variables.

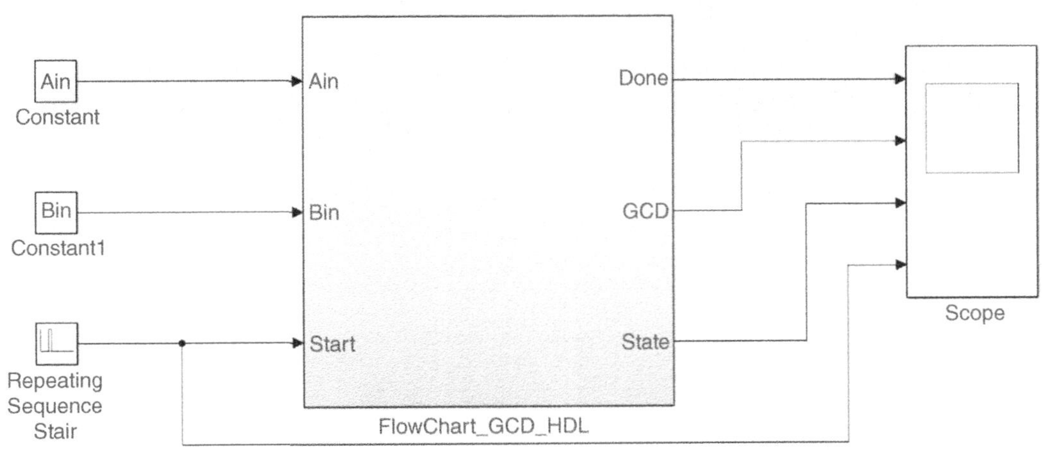

Figure 15.31 Placing the flowchart inside a subsystem.

Now we can generate the VHDL code for the subsystem "FlowChart_GCD_HDL" in Figure 15.31 using HDL Coder. To do this, we write the following code in MATLAB command window.

```
>> hdlsetup('FlowChart_GCD')
```

Here, FlowChart_GCD is the name of the Simulink file (flowchart_GCD.slx). Next, we configure the Solver and HDL Code Generation tabs in the Configuration Parameters window similarly to Figures 15.13 and 15.14, respectively. Then, we type the following code in the MATLAB command window.

```
>> checkhdl('FlowChart_GCD/FlowChart_GCD_HDL')
```

This opens the HDL Check Report window in a web browser, as shown in Figure 15.33.

Now, we use the following code for HDL code generation in the MATLAB command window.

```
>> makehdl('FlowChart_GCD/FlowChart_GCD_HDL')
```

The VHDL codes are generated successfully, and the Code Generation Report window is displayed as depicted in Figure 15.34.

Now, we run the Vivado and create a project named "FlowChart_GCD" and then click the Next button, as illustrated in Figure 15.35.

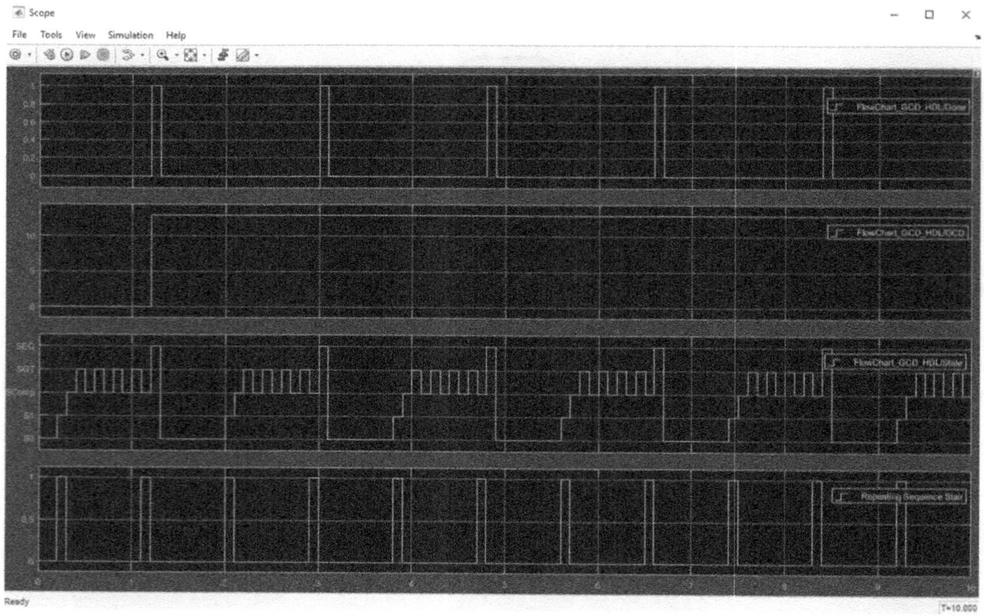

Figure 15.32 The waveforms of Scope.

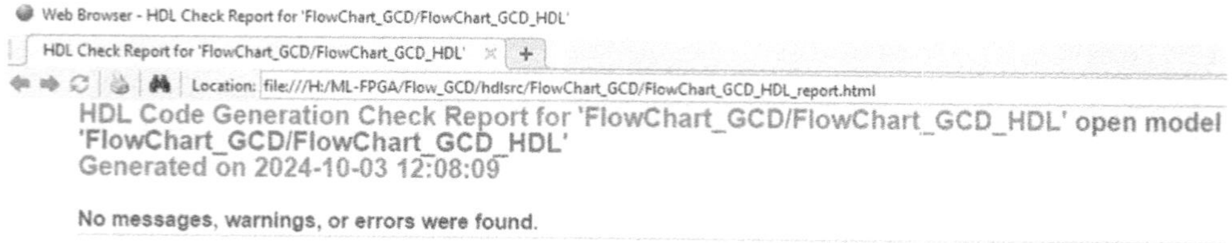

Figure 15.33 The HDL Check Report window.

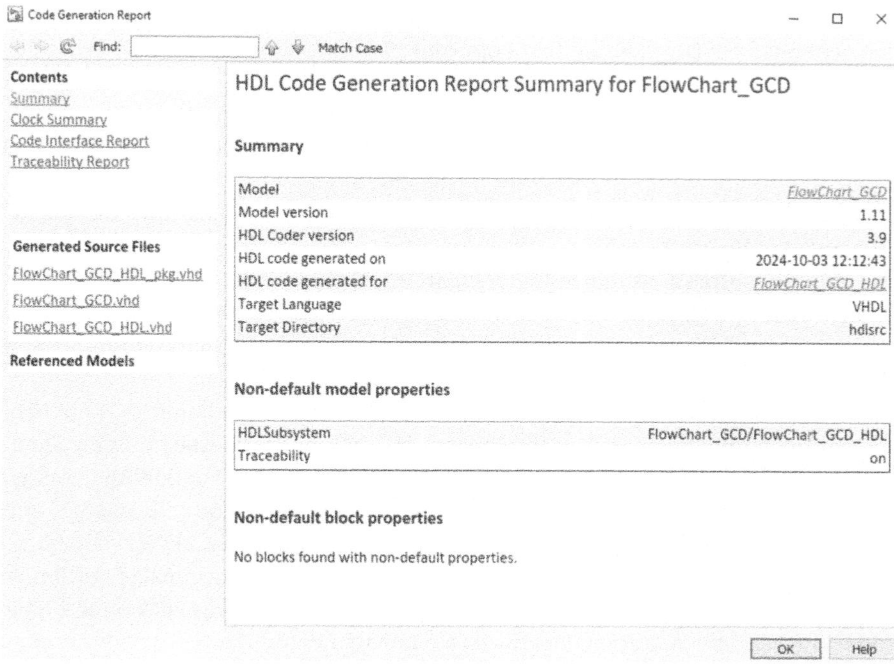

Figure 15.34 The Code Generation Report window.

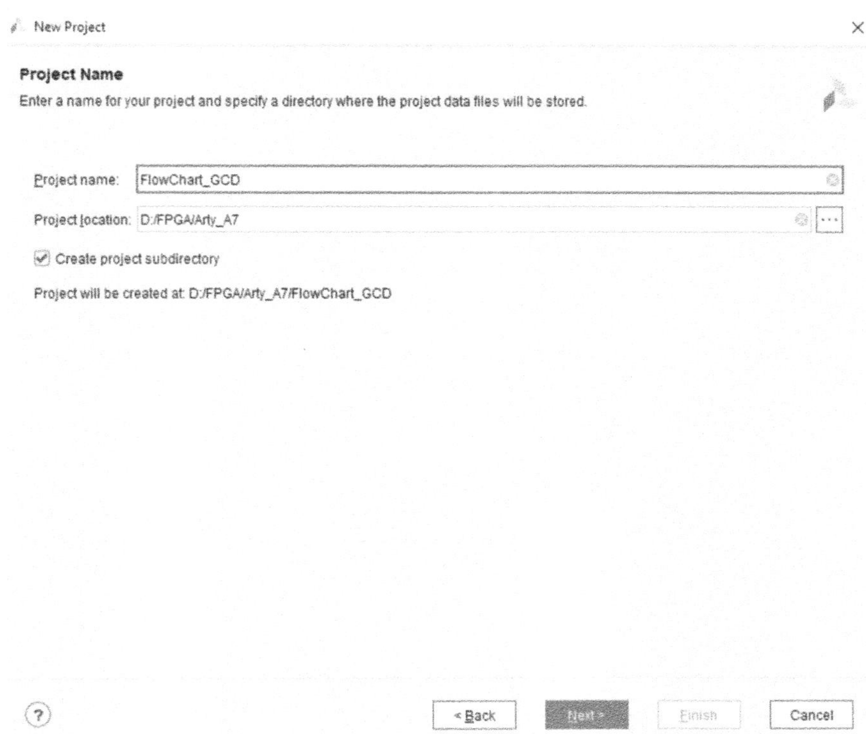

Figure 15.35 Creating a project named "FlowChart_GCD."

We select the RTL Project type and click the Next button. In the Add Sources window, before adding the generated VHDL files, we first copy and paste the three generated VHDL files from Figure 15.34 into the FlowChart_GCD folder within the Vivado project directory. Now, by clicking the Add Files button, we add them to the project, as shown in Figure 15.36.

Next, we select our Xilinx development board and click on the Next button, as depicted in Figure 15.37.

Finally, we click the Finish button to create the FlowChart_GCD project with the added VHDL files, as illustrated in Figure 15.38.

Then, we run the synthesis, as illustrated in Figure 15.39.

As demonstrated in Figure 15.39, the synthesis is successfully completed, and we can click on the Cancel button. Then, we open the schematic in the RTL Analysis tab, as shown in Figure 15.40.

Now, we can simulate the project by selecting Flow → Run Simulation → Run Behavioral Simulation. In the simulation window that opens, we click on the Restart button. Then in the waveform window, we right-click on "clk" and select "Force Clock..." option and configure the settings, as illustrated in Figure 15.20. Also, we right-click on "reset" and "clk_enable" and select "Force Constant..." option and force its value to 1, as shown in Figure 15.21. We right-click on the "Ain", select "Force Constant..." option, and force its value to unsigned decimal of 40. Similarly, we force the "Bin" constant value to unsigned decimal of 25. Finally, we right-click on the "Start" and select "Force Clock..." option and set its settings, as shown in Figure 15.23. Then, we click once on

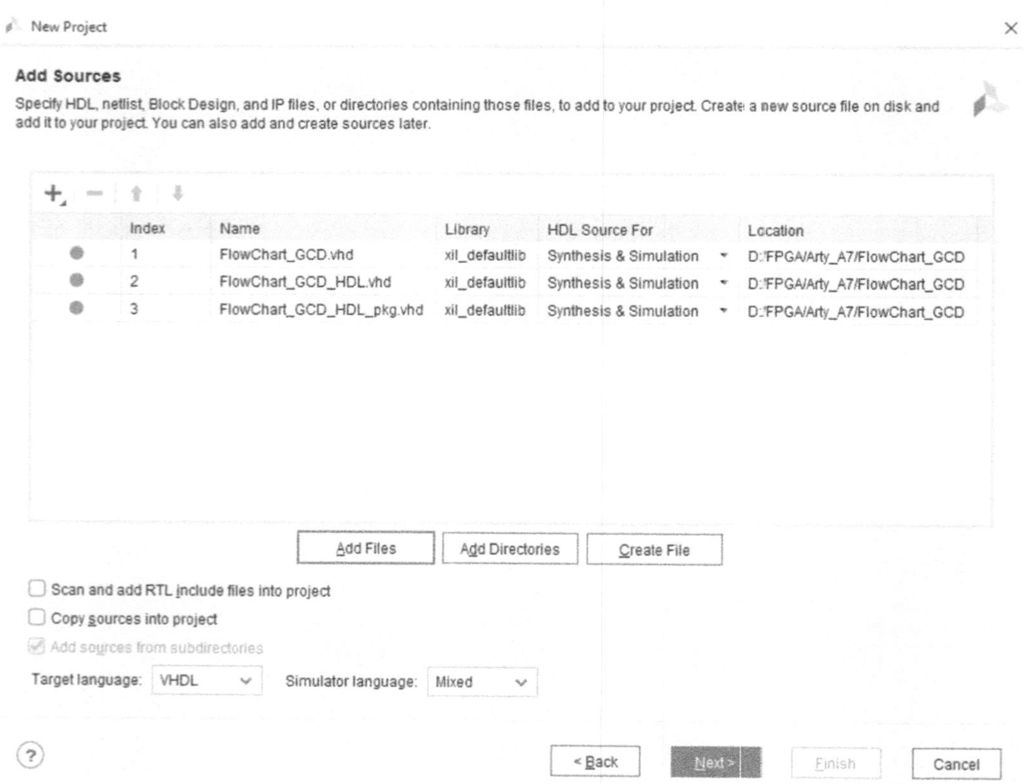

Figure 15.36 Adding generated VHDL files to the Sources folder.

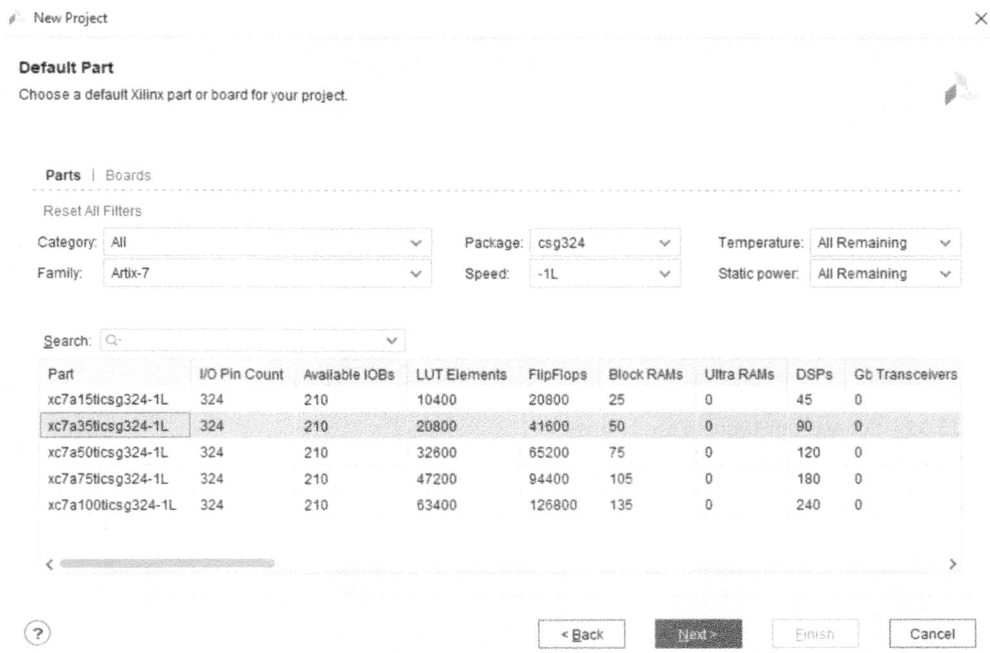

Figure 15.37 Selecting our Xilinx development board.

Figure 15.38 Clicking on the Finish button.

the Run for 20 ns button, as shown in Figure 15.24. We right-click again on the "reset" and force its value to zero as depicted in Figure 15.25. Then, we click on the Run for 20 ns button multiple times and click on the Zoom Fit button in waveforms window to get the simulated waveforms, as illustrated in Figure 15.41.

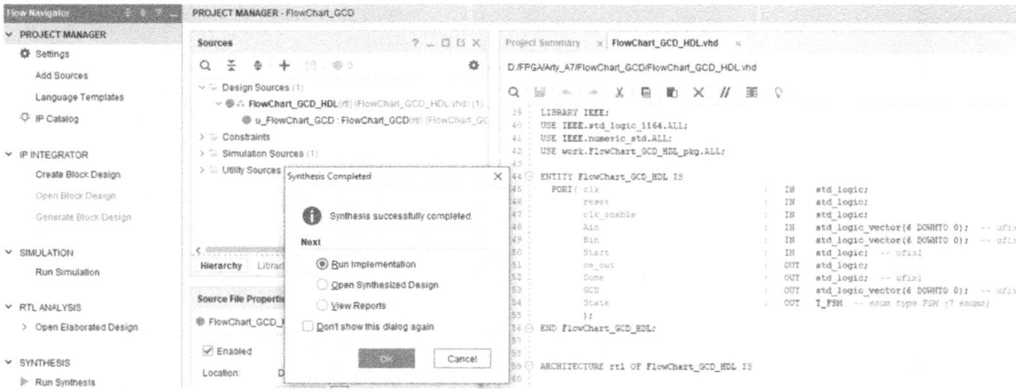

Figure 15.39 Successfully running the synthesis.

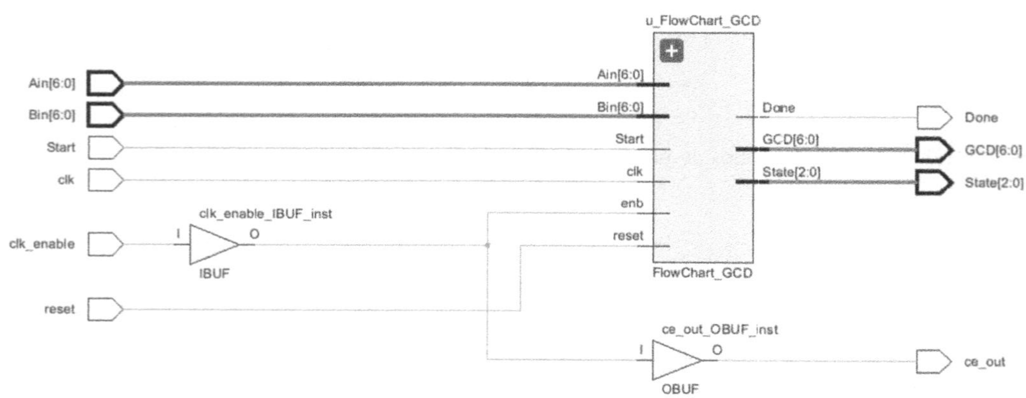

Figure 15.40 The schematic of "FlowChart_GCD_HDL."

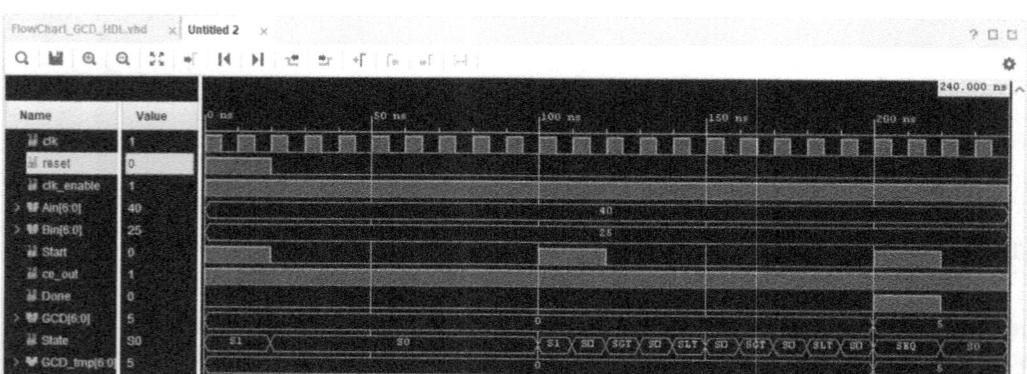

Figure 15.41 The simulated waveforms.

15.3 Flowchart of Booth's Multiplication

In this section, we are going to implement Booth's multiplication algorithm directly from its flowchart. We will simulate the design in the MATLAB Simulink environment to verify its functionality. Then, we will utilize the HDL Coder to generate synthesizable VHDL code from the designed Booth's multiplication algorithm implemented in the Stateflow, and finally, we will synthesize and implement the generated VHDL code using Xilinx Vivado. The flowchart for Booth's multiplication algorithm is shown in Figure 15.42. In order to implement the digital system design algorithm directly from the flowchart, we must carefully analyze it and identify the inputs, outputs, constants, and intermediate variables. Constants and inputs should be initialized using a separate MATLAB script (m-file), where all values are defined using fixed-point numeric objects such as "numerictype" and "fi", as described in the previous sections. We usually begin with the initial state (S0) by initializing all outputs to zero or predefined values, using the constants previously defined in the m-file.

Then, we implement the flowchart step by step, recognizing each step as a separate state in the Stateflow of Simulink. If we encounter a complex operation, it is better to divide it into multiple steps or states for clarity and manageability. Constants defined in the m-file are declared as parameters, while intermediate variables are defined as local variables in the Model Explorer of Stateflow. Inputs are assigned to local variables and similarly, inputs and outputs are defined in the Model Explorer as well. When we perform any arithmetic operation at any state, we should make sure that the values are converted to fixed point using "fi", as discussed in previous sections. It is also better to consider using MATLAB functions or Simulink functions to simplify and compact the flowchart design in the Stateflow of Simulink. The Stateflow diagram for Booth's multiplication algorithm is depicted in Figure 15.43. You should note that

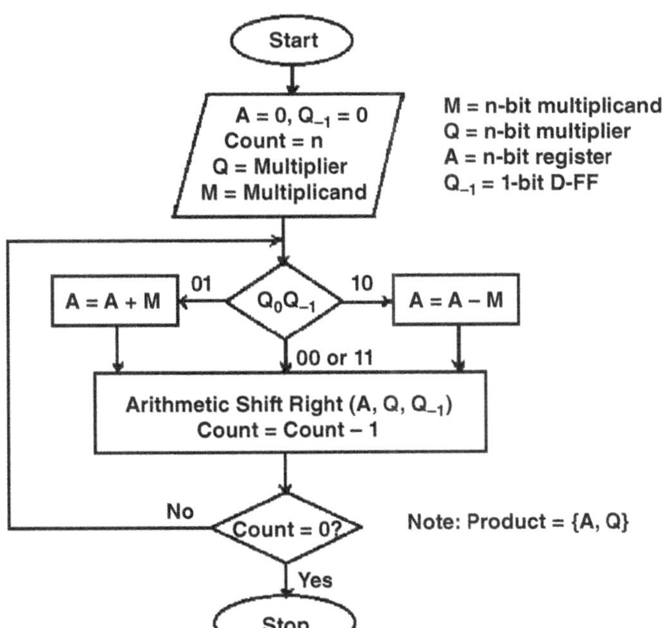

Figure 15.42 The flowchart of Booth's multiplication algorithm.

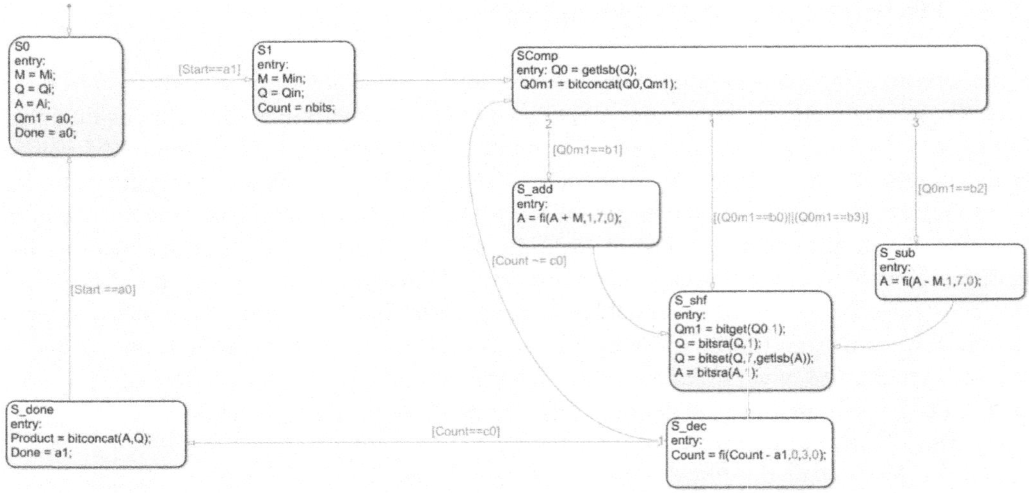

Figure 15.43 The Stateflow chart of Booth's multiplication algorithm.

in Figure 15.43, Q_{-1} is represented as Qm1. To extract the least significant bit (LSB) of Q, we can use getlsb(Q) or bitget(Q,1). The command for performing a 1-bit right shift of variable A is bitsra(A,1). Also, in order to set the most significant bit (MSB) of Q (the seventh bit in this case) with the LSB of A, we use the command bitset(Q,7,getlsb(A)). Moreover, to concatenate the variables A and Q, we use the command bitconcat(A,Q), as demonstrated in Figure 15.43.

Parameters (constants) are already defined inside the m-file (FC_Parameters.m) as the following codes.

```
clear all; close all; clc
N = 7; %nbits
n = ceil(log2(N));
ntc = numerictype(0,n,0);
nbits = fi(N,ntc);
c0 = fi(0,ntc);

nt = numerictype(1,N,0);

% Inputs
Min = fi(36,nt);
Qin = fi(24,nt);

% Initial conditions
Mi = fi(0,nt);
Qi = fi(0,nt);
Ai = fi(0,nt);

% 1-bit constants
nt1 = numerictype(0,1,0);
a0 = fi(0,nt1);
a1 = fi(1,nt1);
```

```
% 2-bit constants
nt2 = numerictype(0,2,0);
b0 = fi(0,nt2);
b1 = fi(1,nt2);
b2 = fi(2,nt2);
b3 = fi(3,nt2);
```

To use these constants inside the Stateflow, we first need to run the m-file. Mi, Qi, Ai, nbits, c0, a0, a1, b0, b1, b2, and b3 are fixed-point parameters defined within the m-file. M, Q, and A are local or intermediate variables. Also, Min, Qin, and Start are inputs, while State, Qm1, Q0m1, Product, Count, and Done are outputs. From View → Model Explorer → Model Explorer, we open the Model Explorer window, in order to add parameters, local variables, inputs, and outputs for the Stateflow design in Figure 15.43. In the Model Explorer window, we click on the Add in the menu bar and select the Data option, and begin adding the variables as defined in Stateflow chart. The complete list of added variables of the Stateflow chart is shown in Figure 15.44.

We also check the Create output for monitoring option and enter the Data name as State and the Enum name as FSM, as demonstrated in Figure 15.44. The flowchart's input and output variables are illustrated in Figure 15.45.

Next, we place the flowchart within a subsystem, as shown in Figure 15.46.

The block parameters for Constant and Repeating Sequence Stair in Figure 15.31 are similar to those described in the previous sections, as depicted in Figures 15.8, 15.9, and 15.10, respectively. Before running the simulation, we set the Solver tab of Configuration Parameters window, as illustrated in Figure 15.11. Now, we can run the simulation. The resulting waveforms of scope are shown in Figure 15.47.

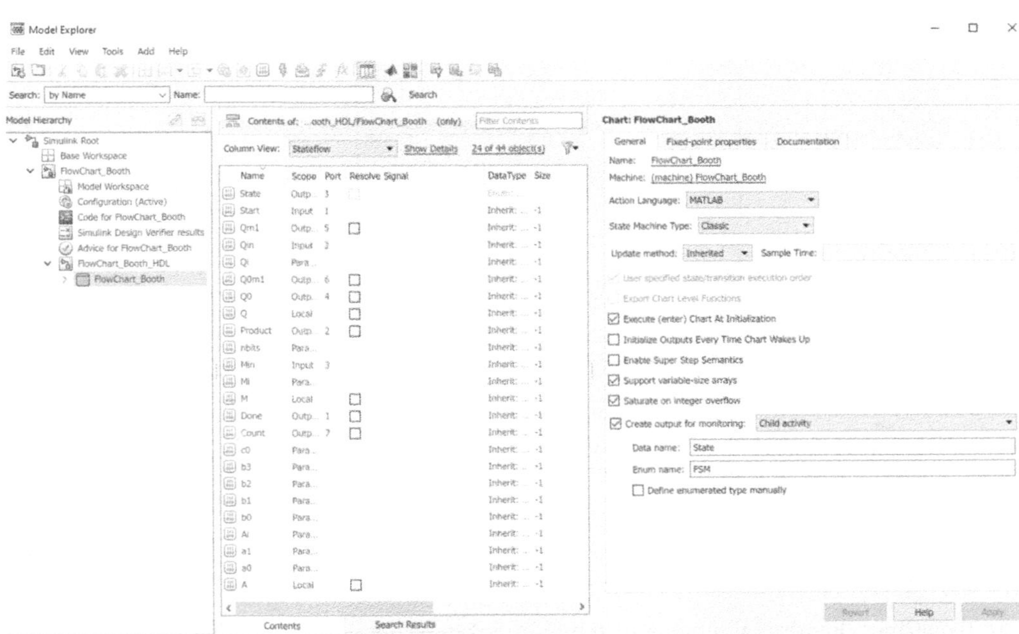

Figure 15.44 The complete list of added variables for the flowchart.

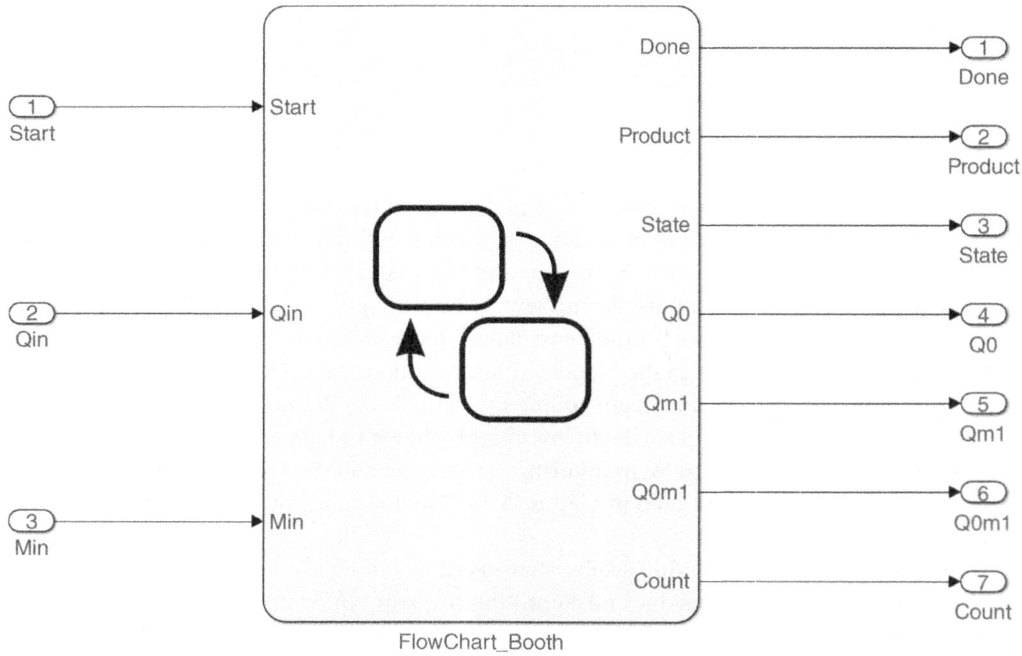

Figure 15.45 The flowchart's input and output variables.

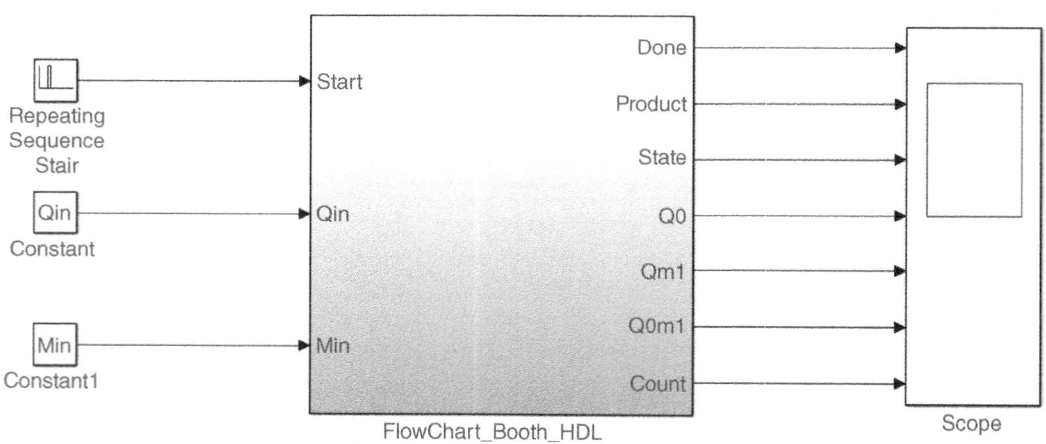

Figure 15.46 Placing the flowchart inside a subsystem.

Now, we can generate the VHDL code for the subsystem "FlowChart_GCD_HDL" in Figure 15.46 using HDL Coder. To do this, we write the following code in the MATLAB command window.

```
>> hdlsetup('FlowChart_Booth')
```

The FlowChart_Booth is the name of the Simulink file (flowchart_Booth.slx). Then, we configure the Solver and HDL Code Generation tabs of the Configuration Parameters window as shown in Figures 15.13 and 15.14, respectively. Then, we type the following code in the MATLAB command window.

```
>> checkhdl('FlowChart_Booth/FlowChart_Booth_HDL')
```

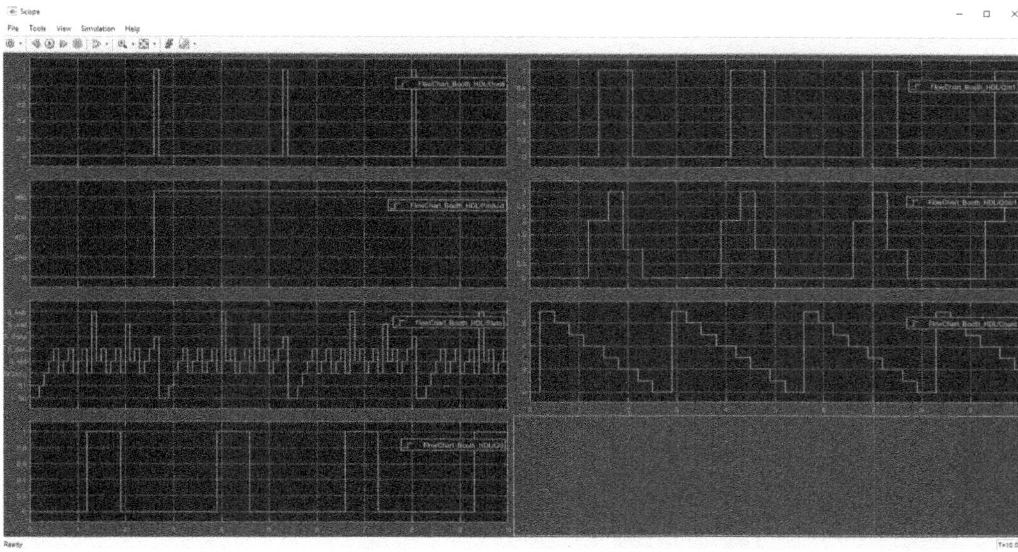

Figure 15.47 The waveforms of scope.

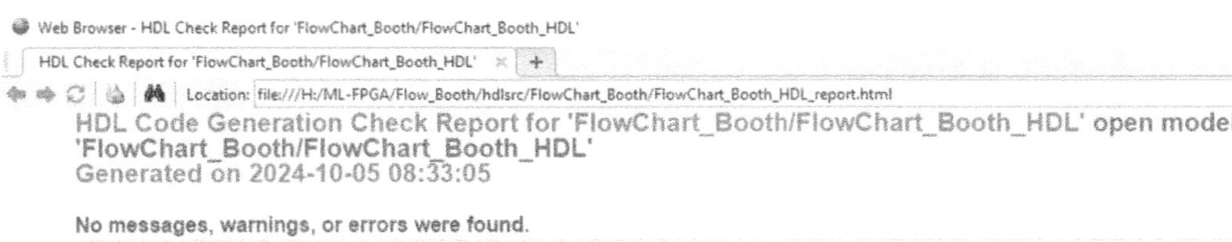

Figure 15.48 The HDL Check Report window.

This opens the HDL Check Report Window in a web browser, as shown in Figure 15.48.

Now, we use the following code for HDL Code Generation in the MATLAB command window.

```
>> makehdl('FlowChart_Booth/FlowChart_Booth_HDL')
```

The VHDL codes are generated successfully, and the Code Generation Report window is displayed, as depicted in Figure 15.49.

We can select the "MATLAB function" option from the left toolbar of the stateflow chart and incorporate the MATLAB function titled "ASR" into our design, as illustrated in Figure 15.50.

The MATLAB function titled "ASR" codes are as follows.

```
function [Ao,Qo] = ASR(Ax,Qx)
Qo = bitsra(Qx,1);
Qo = bitset(Qo,7,getlsb(Ax));
Ao = bitsra(Ax,1);
```

We can generate the VHDL code for this subsystem using HDL Coder as well. To do this, we write the following code in the MATLAB command window.

```
>> hdlsetup('FlowChart_Booth2')
```

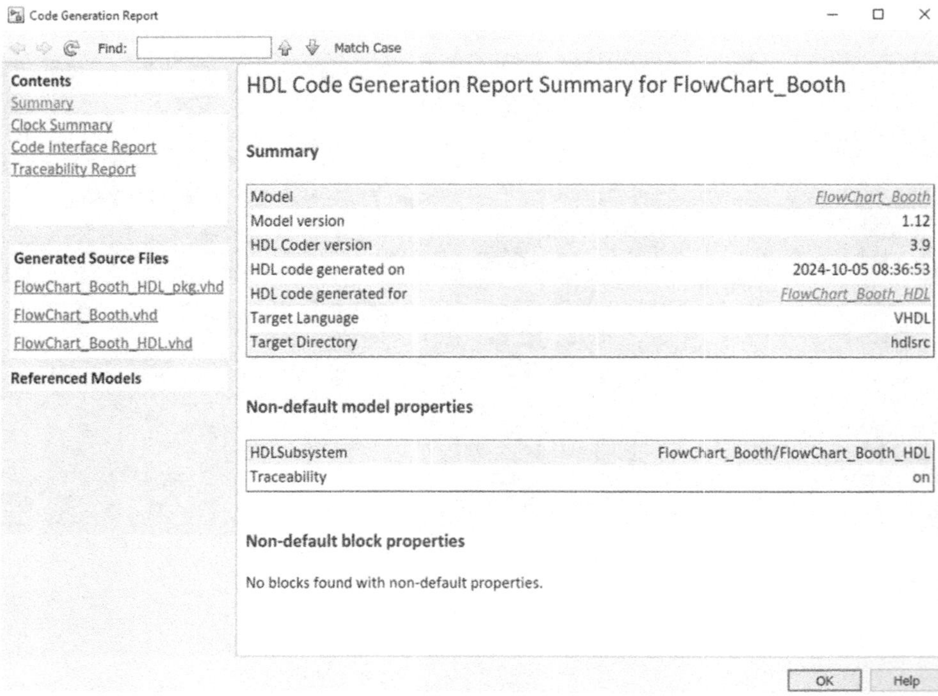

Figure 15.49 The Code Generation Report window.

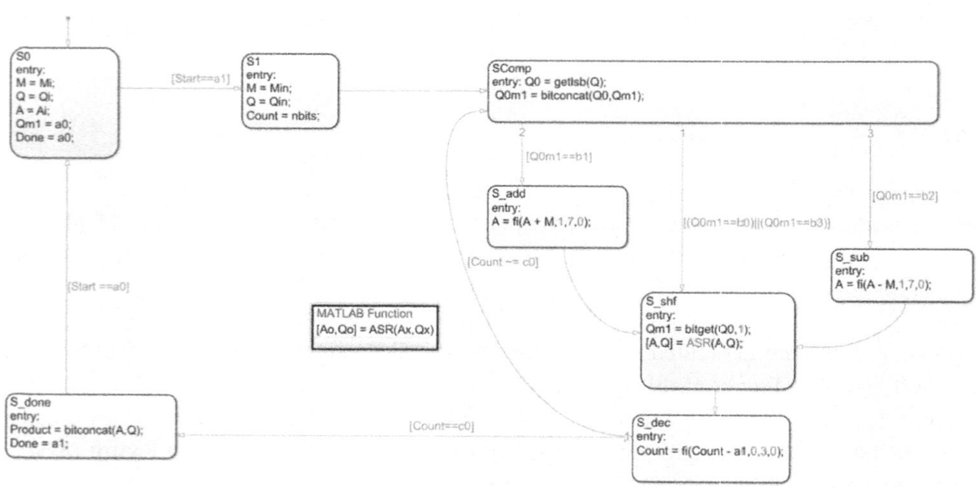

Figure 15.50 Incorporating the MATLAB function titled "ASR" into our design.

The FlowChart_Booth2 is the name of new Simulink file (flowchart_Booth2.slx). Then, we configure the Solver and HDL Code Generation tabs of the Configuration Parameters window similar to Figures 15.13 and 15.14, respectively. Then, we type the following code in the MATLAB command window.

```
>> checkhdl('FlowChart_Booth2/FlowChart_Booth_HDL')
```

This opens the HDL Check Report window in a web browser, as shown in Figure 15.51.

Now, we use the following code for HDL code generation in the MATLAB command window.

```
>> makehdl('FlowChart_Booth2/FlowChart_Booth_HDL')
```

The VHDL codes are generated successfully, and the Code Generation Report window is displayed as depicted in Figure 15.52.

Now, we run the Vivado and create a project named "FlowChart_Booth2" and then click the Next button. We select the project type as RTL Project and click the Next button. In the Add Sources window, before adding the generated VHDL files, we copy and paste the three generated VHDL files in Figure 15.52 to the folder FlowChart_Booth2 of Vivado project. Now, by clicking on the Add Files button, we can add them to the project, as shown in Figure 15.53.

Next, we select our Xilinx development board and click on the Next button. Finally, we click on the Finish button to create the FlowChart_Booth2 project with added generated VHDL files. Then we run the synthesis as illustrated in Figure 15.54.

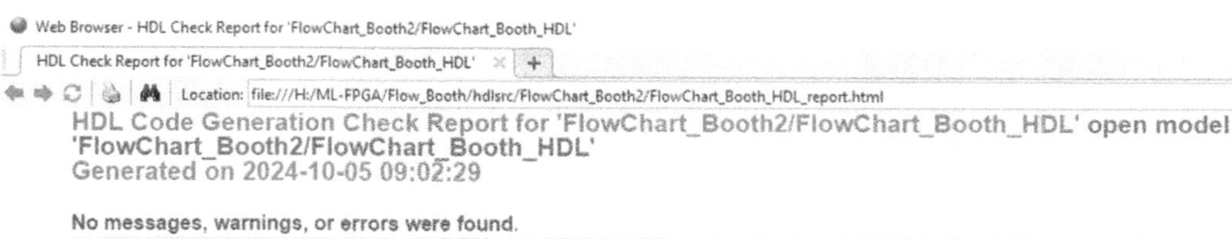

Figure 15.51 The HDL Check Report window.

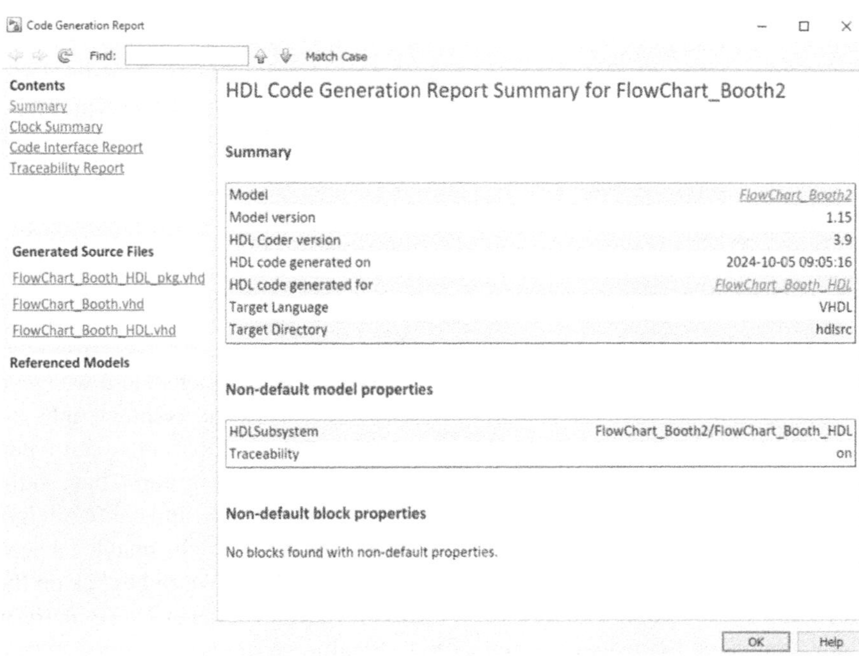

Figure 15.52 The Code Generation Report window.

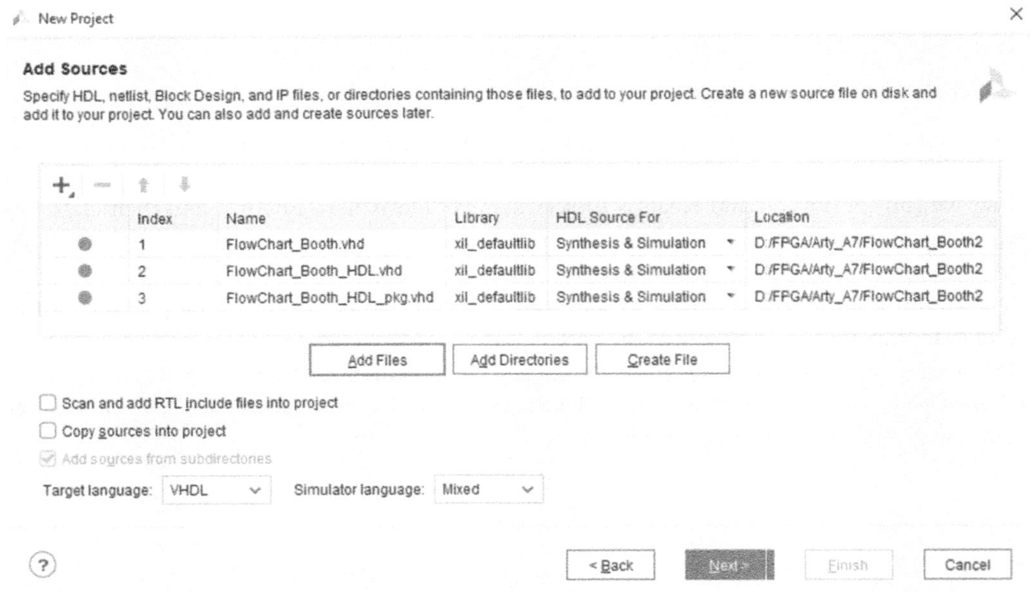

Figure 15.53 Adding generated VHDL files to the Sources folder.

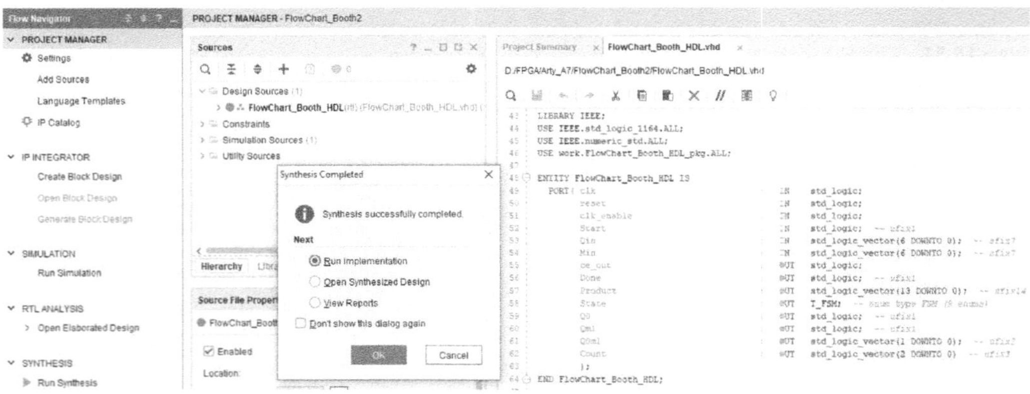

Figure 15.54 Successfully running the synthesis.

As demonstrated in Figure 15.54, the synthesis is successfully completed, and we can click on the Cancel button. Then, we open the schematic in the RTL Analysis tab, as shown in Figure 15.55.

Now, we can simulate the project by selecting Flow → Run Simulation → Run Behavioral Simulation. In the opened simulation window, we click on the Restart button. Then, in the waveform window, we right-click on the "clk", select "Force Clock..." option, and configure its settings as illustrated in Figure 15.20. Also, we right-click on the "reset" and "clk_enable", select "Force Constant..." option, and force its value to 1, as shown in Figure 15.21. We right-click on the "Qin", select "Force Constant..." option, and force its value to unsigned decimal of 15. Similarly, we force the "Min" constant value to unsigned decimal of 20. Finally, we right-click on the "Start", select "Force Clock..." option, and do its settings as shown in Figure 15.23. Then, we click once on the

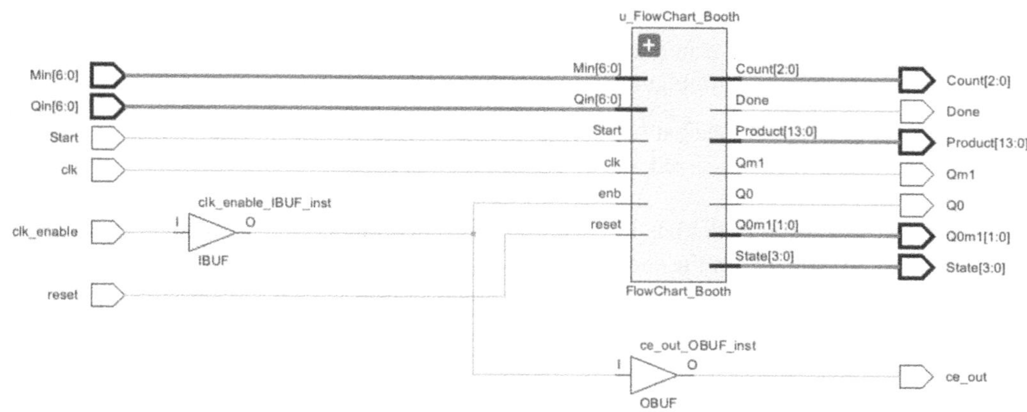

Figure 15.55 The schematic of "FlowChart_Booth_HDL."

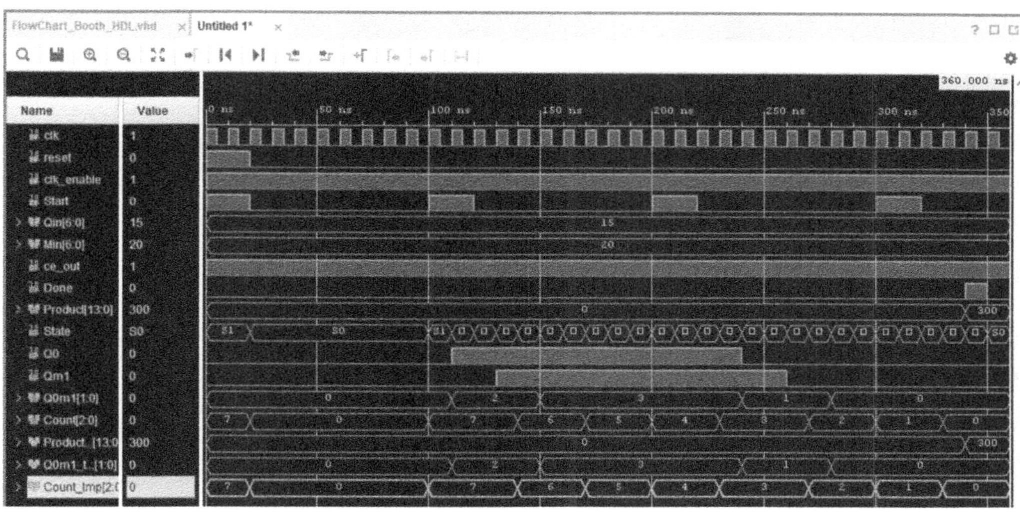

Figure 15.56 The simulated waveforms.

Run for 20 ns button as shown in Figure 15.24. We right-click again on the "reset" and force its value to zero as depicted in Figure 15.25. Then, we click on the Run for 20 ns button multiple times and click on the Zoom Fit button in waveforms window. We also select Q0m1, Count, Product_tmp, Q0m1_tmp, and Count_tmp and right-click and choose Radix → Unsigned Decimal, then we get the simulated waveforms, as depicted in Figure 15.56.

15.4 Flowchart of an Industrial Control System

In this section, we are going to implement an industrial control system directly from its flowchart. We will simulate the design in the MATLAB Simulink environment to verify its functionality. Then, we will utilize the HDL Coder to generate the synthesizable VHDL code from the designed industrial control system implemented in the Stateflow environment of Simulink. Finally, we will synthesize and implement the generated VHDL code using Xilinx Vivado. The sequence of operations for the system illustrated in Figure 15.57 is as follows:

- The box is initially positioned at LS1 (LS1 is closed).
- When the start button is pressed, the conveyor motor activates, moving the box to position A (LS1 opens).
- The conveyor transports the box to position A and then halts (the position is confirmed by eight off-to-on pulses from the encoder to the up counter).
- Following a 10-s delay, the conveyor resumes movement, carrying the box to LS2 where it stops (LS2 closes).
- An emergency stop push button is available to halt the process at any moment.
- If the process is interrupted using the emergency stop push button, both the timer and counter will automatically reset.

The flowchart of the industrial control system (Figure 15.57) is shown in Figure 15.58.

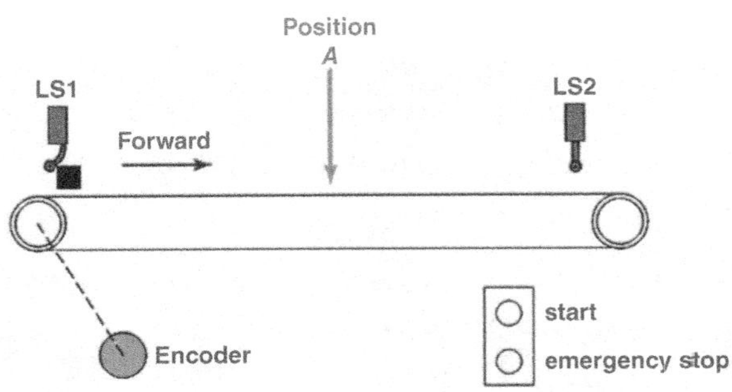

Figure 15.57 An industrial control system.

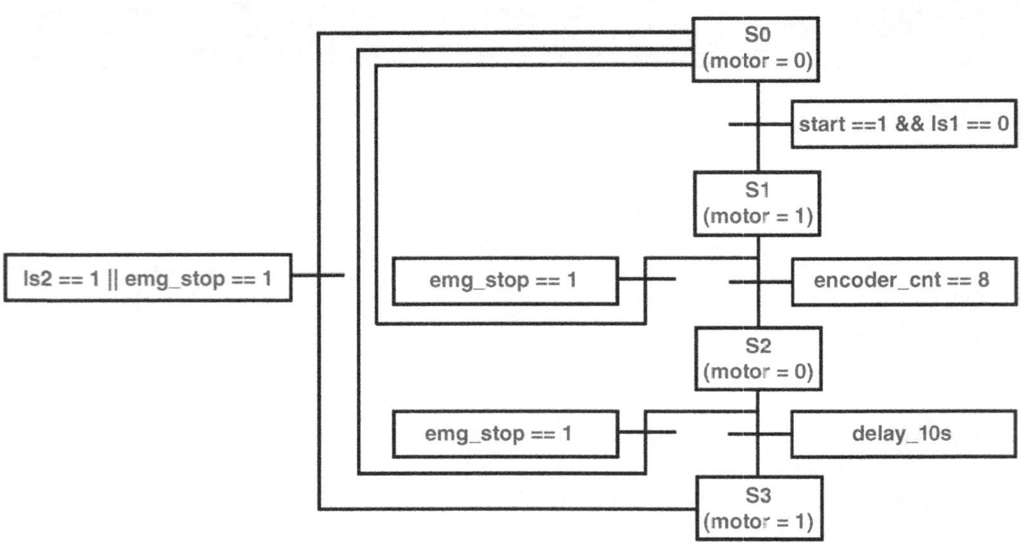

Figure 15.58 The flowchart of the industrial control system (Figure 15.57).

The state flowchart of the industrial control system is depicted in Figure 15.59. The parameters (constants) are already defined inside the m-file (FC_Parameters.m) as the following codes.

```
close all; clear all; clc
a0 = fi(0,0,1,0);
a1 = fi(1,0,1,0);
en_cnt0 = fi(0,0,4,0);
cnt_val = fi(8,0,4,0);
```

To use these constants inside the Stateflow, we first need to run the m-file. The parameters cnt_val, en_cnt0, a0, and a1 are fixed-point parameters defined inside the m-file. The encoder_cnt is a local or intermediate variable. Also, start, ls1, ls2, emg_stop, and encoder are inputs, and State and motor are outputs. From View → Model Explorer → Model Explorer, we open the Model Explorer window, in order to add parameters, local variables, inputs, and outputs for the Stateflow design in Figure 15.59. In the Model Explorer window, we click on the Add menu in the menu bar and select the Data option and begin to add the variables defined in the Stateflow chart. The complete list of added variables of the Stateflow chart is shown in Figure 15.60.

We also check the Create output for monitoring option and enter the Data name as State and the Enum name as FSM, as demonstrated in Figure 15.60. The flowchart's input and output variables are illustrated in Figure 15.61.

Also, we put the flowchart inside a subsystem as shown in Figure 15.62.

The block parameters for Repeating Sequence Stair of "start" in Figure 15.62 are similar to those described in the previous sections, as depicted in Figures 15.9 and 15.10 respectively. Also, the block parameters for Repeating Sequence Stair of "ls1" in Figure 15.62 are exact inverse of the Repeating Sequence Stair of "start." The block parameters for Repeating Sequence Stair of "ls2" is depicted in Figure 15.63, and in the Signal Attributes tab, the output data type is fixdt(0,1,0).

Finally, the block parameters for Repeating Sequence Stair of "encoder" is depicted in Figure 15.64, and in the Signal Attributes tab, the output data type is fixdt(0,1,0) as well.

Figure 15.59 The state flowchart of industrial control system.

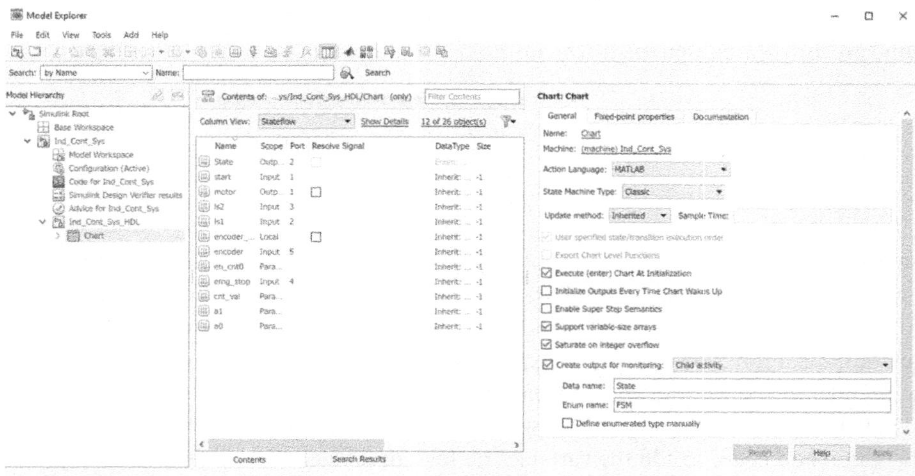

Figure 15.60 The complete list of added variables for the flowchart.

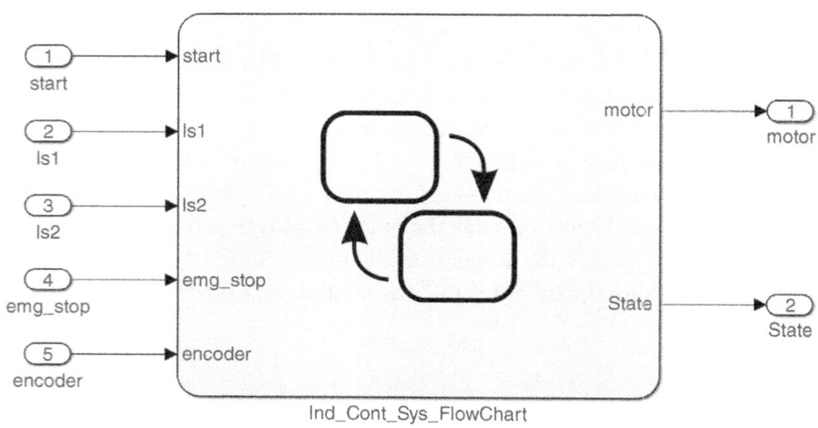

Figure 15.61 The flowchart's input and output variables.

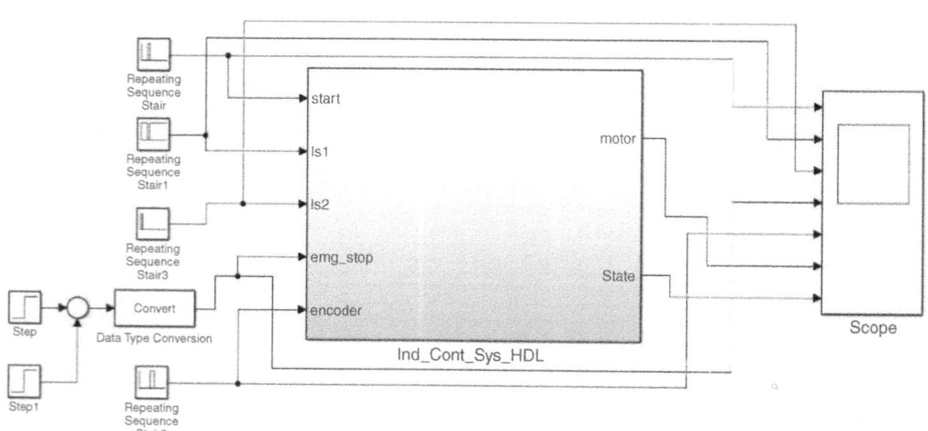

Figure 15.62 Placing the flowchart inside a subsystem.

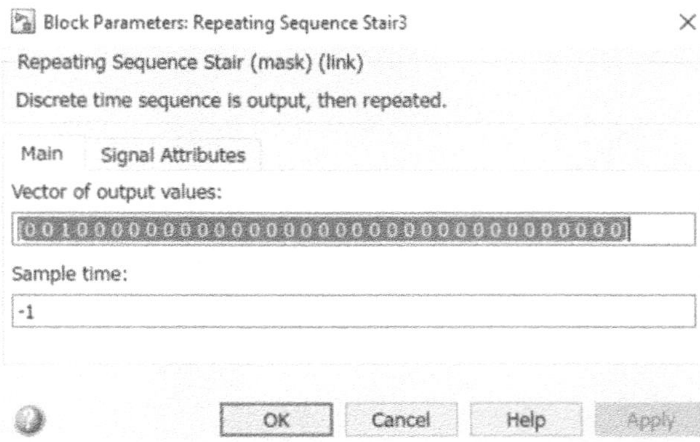

Figure 15.63 The block parameters for Repeating Sequence Stair of "ls2."

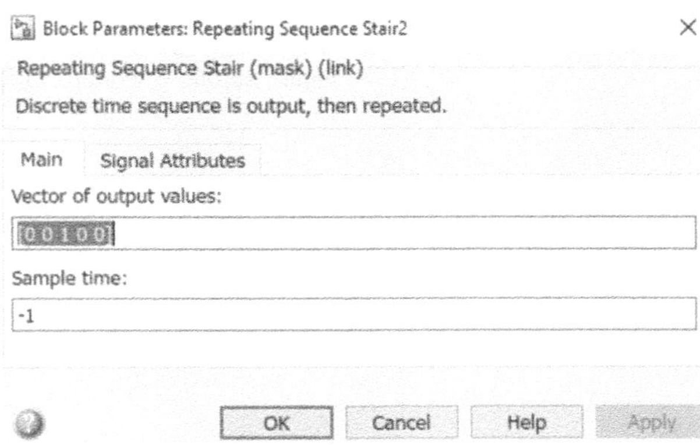

Figure 15.64 The block parameters for Repeating Sequence Stair of "encoder."

Also, the step source (Step, Step1) and data type conversion parameters are illustrated in Figures 15.65, 15.66, and 15.67, respectively.

Before running the simulation, we set the Solver tab of Configuration Parameters window as illustrated in Figure 15.68.

We should note that since the fixed-step size is 0.1 s, for getting a delay of 10 s in Figure 15.59, we have used the "[after(100,tick)]" in transition from S2 to S3. Now, we can run the simulation, and the waveforms of scope are shown in Figure 15.69.

Now, we can generate the VHDL code for the subsystem "Ind_Cont_Sys_HDL" in Figure 15.46 using HDL Coder. To do that, we write the following code in the MATLAB command window.

```
>> hdlsetup('Ind_Cont_Sys')
```

The "Ind_Cont_Sys" is the name of the Simulink file (Ind_Cont_Sys.slx). Now, if we use 10^{-8} s (1/100e6) as fixed-step size in Configuration Parameters window, to achieve a delay of 10 s, we should use [after(1e9,tick)] (we cannot use [after(10,s)] since HDL Coder is unable to handle it and gives an error in code generation). However, for simulation purposes we will use the fixed-step size of 0.1 s. Also, we set the HDL Code Generation tab of Configuration Parameters window similar to Figure 15.14. Then we type the following code in the MATLAB command window.

```
>> checkhdl('Ind_Cont_Sys/Ind_Cont_Sys_HDL')
```

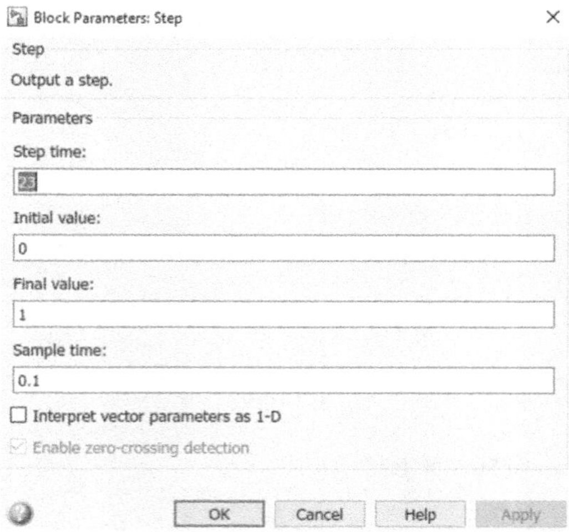

Figure 15.65 The Step block parameters.

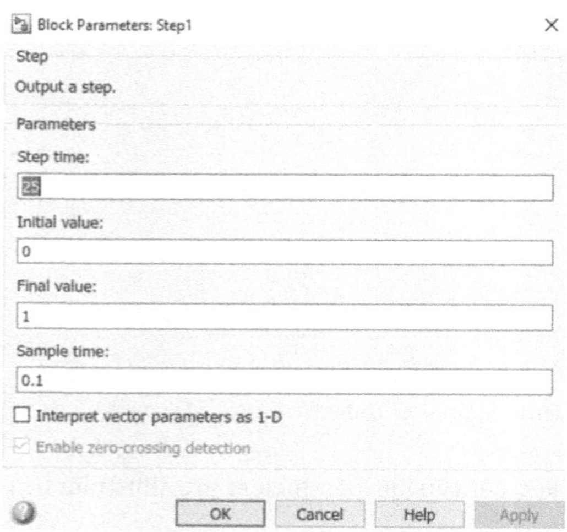

Figure 15.66 The Step1 block parameters.

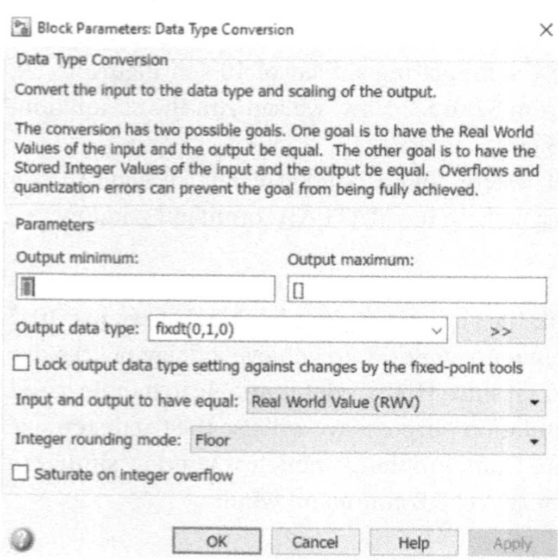

Figure 15.67 The data type conversion parameters.

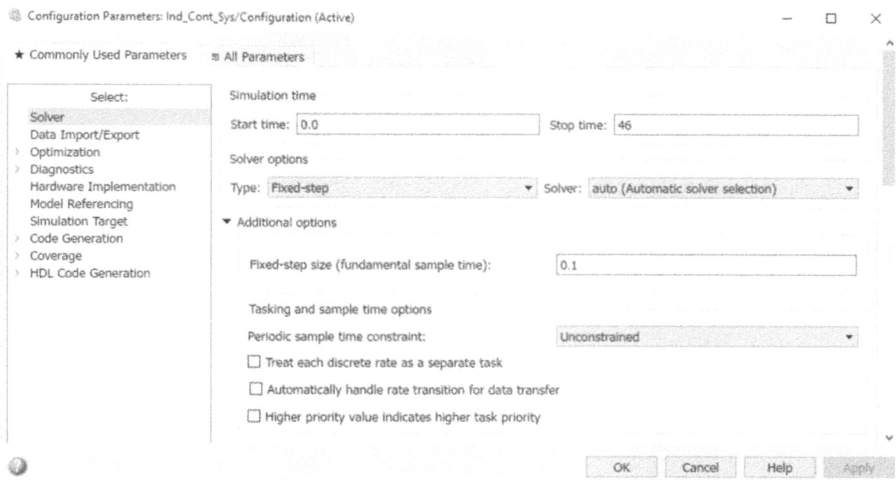

Figure 15.68 The Solver tab of Configuration Parameters window settings.

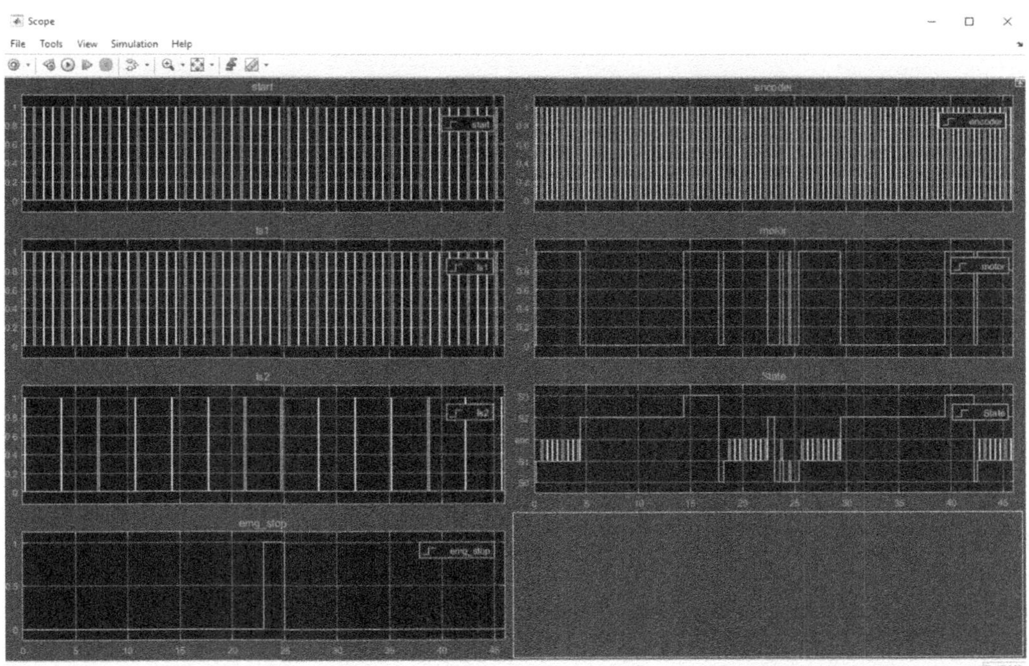

Figure 15.69 The waveforms of scope.

This opens the HDL Check Report window in a web browser, as shown in Figure 15.70.

Now, we use the following code for HDL code generation in the MATLAB command window.

```
>> makehdl('Ind_Cont_Sys/Ind_Cont_Sys_HDL')
```

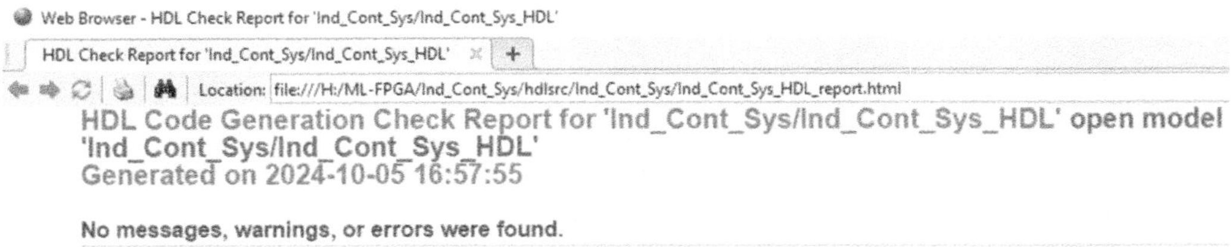

Figure 15.70 The HDL Check Report window.

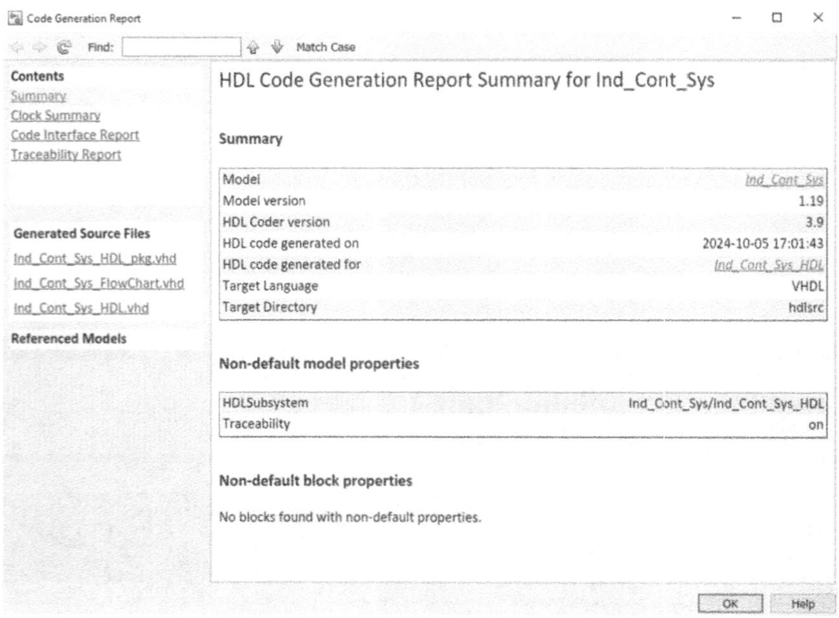

Figure 15.71 The Code Generation Report window.

The VHDL codes are generated successfully, and the Code Generation Report window is displayed as depicted in Figure 15.71. Now, we run the Vivado and create a project named "Ind_Cont_Sys" and then click the Next button. Then we select project type as RTL project and click the Next button. In the Add Sources window, before adding the generated VHDL files, we copy and paste the three generated VHDL files in Figure 15.71 to the folder Ind_Cont_Sys of Vivado project. Now, by clicking on the Add Files button, we can add them to the project as shown in Figure 15.72.

Next, we select our Xilinx development board and click on the Next button. Finally, we click on the Finish button to create the Ind_Cont_Sys project with added generated VHDL files. Then, we run the synthesis as illustrated in Figure 15.73.

As demonstrated in Figure 15.73, the synthesis is successfully completed, and we can click on the Cancel button. Then, we open the schematic in the RTL Analysis tab as shown in Figure 15.74.

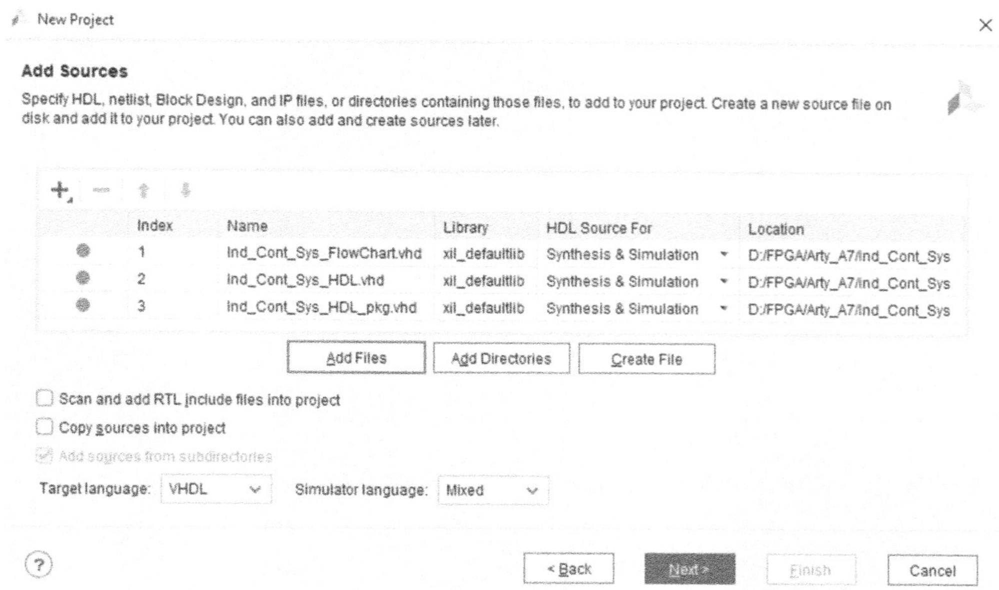

Figure 15.72 Adding generated VHDL files to the Sources folder.

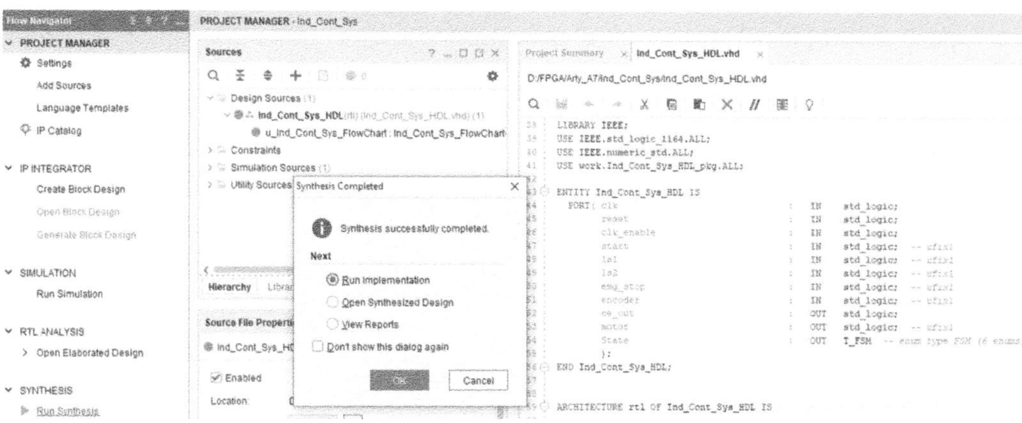

Figure 15.73 Successfully running the synthesis.

Now, we can simulate the project by selecting Flow → Run Simulation → Run Behavioral Simulation. In the opened simulation window, we click on the Restart button. Then in the waveform window, we right-click on the "clk" and select "Force Clock..." option and configure its settings, as illustrated in Figure 15.20. Also, we right-click on the "reset" and "clk_enable", select "Force Constant..." option, and force its value to 1, as shown in Figure 15.21. We right-click on the "start", select "Force Clock..." option, and set its parameters as shown in Figure 15.75.

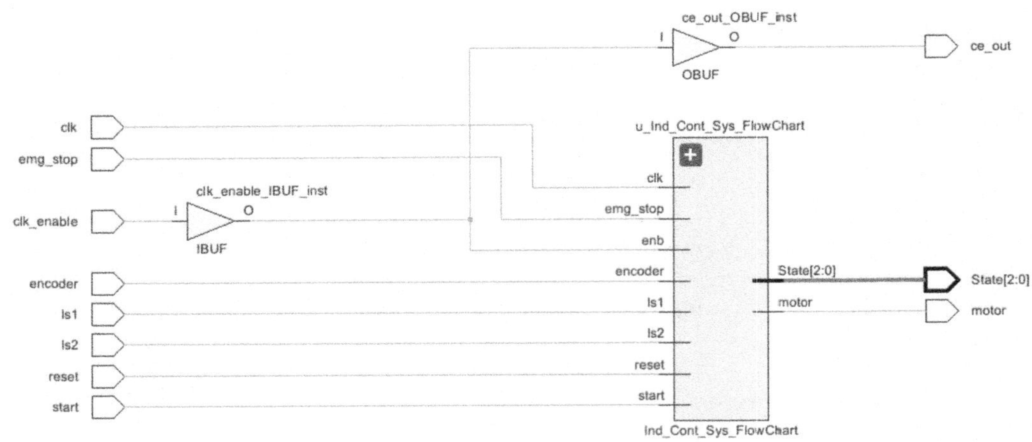

Figure 15.74 The schematic of "Ind_Cont_Sys_HDL."

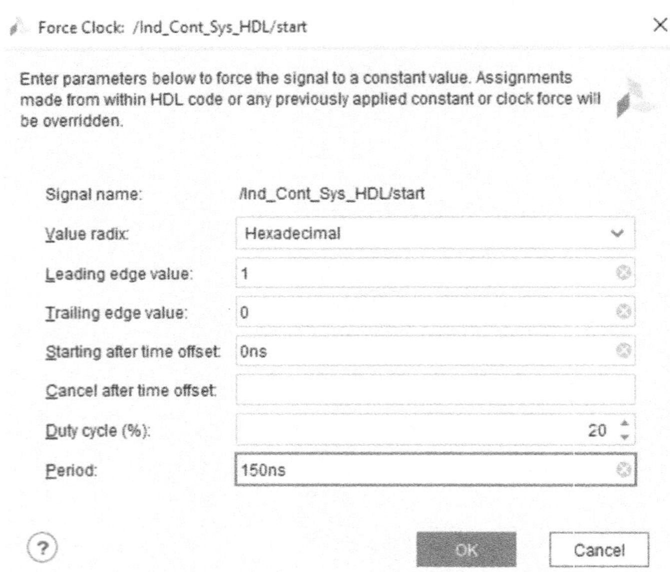

Figure 15.75 The force clock settings for "start."

Similarly, we do force clock the "ls1" parameters as depicted in Figure 15.76. Also, we do force clock the "ls2" parameters as depicted in Figure 15.77.

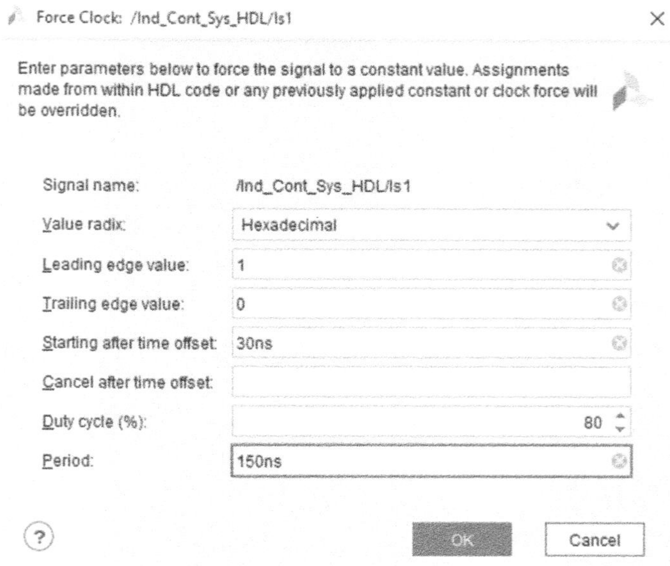

Figure 15.76 The force clock settings for "ls1."

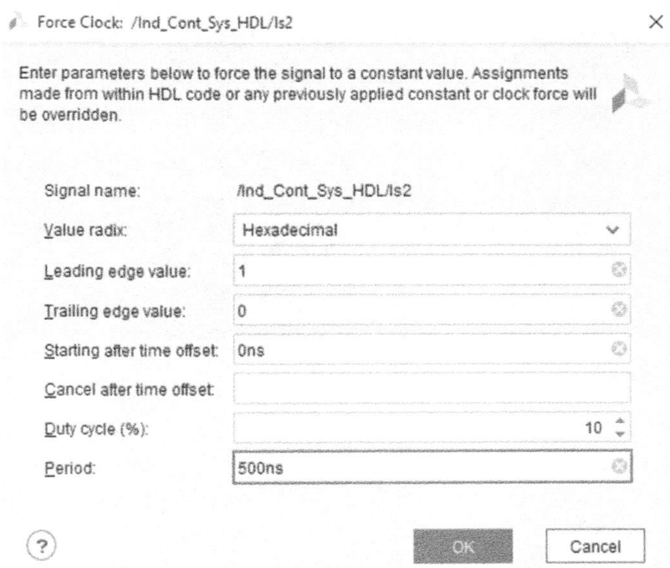

Figure 15.77 The force clock settings for "ls2."

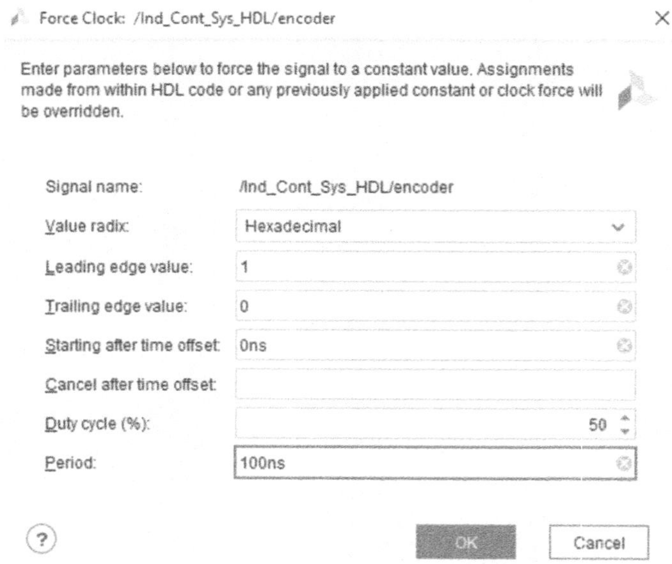

Figure 15.78 The force clock settings for "encoder."

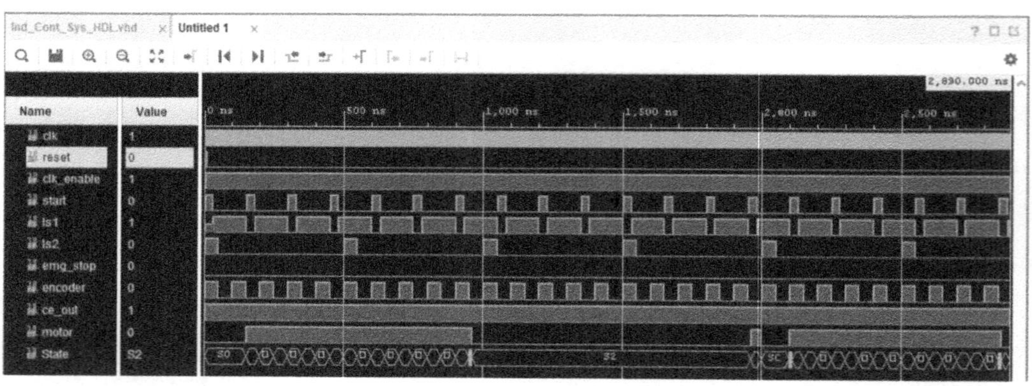

Figure 15.79 The simulated waveforms.

We right-click on the "emg_stop," select "Force Constant..." option, and force its value to zero. Finally, we right-click on the "encoder", select "Force Clock..." option, and configure its settings, as illustrated in Figure 15.78.

Then, we click once on the Run for 10 ns button. We right-click again on the "reset" and force its value to zero. Then, we click on the Run for 10 ns button multiple times and click on the Zoom Fit button in waveforms window to get the simulated waveforms, as illustrated in Figure 15.79.

16

Machine Learning Regression Model on FPGA

16.1 Introduction

In this chapter, we are going to implement a machine-learning regression model on FPGA using Simulink and Vivado. First of all, we download a dataset from Kaggle that is suitable for the implementation of linear regression. Using MATLAB, we train the linear regression model based on the least squares method. We use the admission dataset to predict the chances of admission to a university based on various input parameters. Then, we implement our trained machine learning regression model and generate synthesizable VHDL code using HDL Coder. Finally, using Vivado, we run synthesis analysis and simulation to verify the designed VHDL project.

16.2 The Kaggle Dataset

In this chapter, we will use the admission dataset from Kaggle to predict the admission chance in a university based on various input parameters. The weblink for downloading the admission dataset is as follows.

https://www.kaggle.com/datasets/ks2001/admission-dataset

The overview of the admission dataset in Excel software is shown in Figure 16.1.

16.3 Machine Learning Regression Model for Admission Dataset

We use the following MATLAB m-file (RegModel.m) to implement the machine learning regression model for the admission dataset.

```
clear all; close all; clc
cd('H:\ML-FPGA\Regression');
% Load Data from excel file
d = xlsread('adm_data.xlsx');

% Histogram Plots
histogram(d(:,2));
grid on;
xlabel('Score')
```

Fundamentals of VHDL for FPGA Programming Using Vivado, First Edition. Majid Pakdel.

Companion Website: www.wiley.com/go/pakdelVHDL

	A	B	C	D	E	F	G	H	I
1	Serial No.	GRE Score	TOEFL Score	University Rating	SOP	LOR	CGPA	Research	Chance of Admit
2	1	337	118	4	4.5	4.5	9.65	1	0.92
3	2	324	107	4	4	4.5	8.87	1	0.76
4	3	316	104	3	3	3.5	8	1	0.72
5	4	322	110	3	3.5	2.5	8.67	1	0.8
6	5	314	103	2	2	3	8.21	0	0.65
7	6	330	115	5	4.5	3	9.34	1	0.9
8	7	321	109	3	3	4	8.2	1	0.75
9	8	308	101	2	3	4	7.9	0	0.68
10	9	302	102	1	2	1.5	8	0	0.5
11	10	323	108	3	3.5	3	8.6	0	0.45

Figure 16.1 The overview of admission dataset in Excel.

```
ylabel('Frequency')
title('GRE Score')

figure
histogram(d(:,3));
grid on;
xlabel('Score')
ylabel('Frequency')
title('TOFEL Score')

figure
c1 = cellstr(string(d(:,4)));
histogram(categorical(c1),'BarWidth',0.3);
grid on;
xlabel('Rating')
ylabel('Frequency')
title('University Rating')

figure
c2 = cellstr(string(d(:,5)));
histogram(categorical(c2),'BarWidth',0.3);
grid on;
xlabel('SOP')
ylabel('Frequency')
title('SOP')

figure
c3 = cellstr(string(d(:,6)));
histogram(categorical(c3),'BarWidth',0.3);
grid on;
xlabel('LOR')
ylabel('Frequency')
title('LOR')

figure
histogram(d(:,7));
grid on;
xlabel('CGPA')
ylabel('Frequency')
title('CGPA')
```

```
figure
c4 = cellstr(string(d(:,8)));
histogram(categorical(c4),'BarWidth',0.3);
grid on;
xlabel('Research')
ylabel('Frequency')
title('Research')

figure
histogram(d(:,9));
grid on;
xlabel('Chance')
ylabel('Frequency')
title('Chances of Admission')

Xfeatures = d(:,2:end-1);
Y = d(:,end);
X = [ones(size(Y,1),1) Xfeatures];
AllData = [X Y];

cv = cvpartition(size(AllData,1),'HoldOut',0.2);
Index = cv.test;

Test_Data = AllData(Index,:);
Train_Data = AllData(~Index,:);

XTrain = Train_Data(:,1:end-1);
YTrain = Train_Data(:,end);

XTest = Test_Data(:,1:end-1);
YTest = Test_Data(:,end);

% Linear Regression
Xt = XTrain';
Th = (pinv(Xt*XTrain))*Xt*YTrain;

ypr = (Th')*XTest(1:end,:)';
ypr = ypr';
% Calculate MSE
disp('MSE Linear Regression')
MSE = immse(YTest, ypr)

NTe = length(ypr);
n = 0:1:(NTe-1);
% Plot graph
figure
scatter(n,YTest);
grid on;
hold on
plot(n, ypr);
legend('Testing Data','Predicted Data')
ylabel('Chance of Admit')
title('Linear Regression')
```

```
% Teta (Learned Parameters)
t0 = Th(1);
t1 = Th(2);
t2 = Th(3);
t3 = Th(4);
t4 = Th(5);
t5 = Th(6);
t6 = Th(7);
t7 = Th(8);

% Test Data
x0 = Convert_to_Fixed_Point(XTest(:,1));
x1 = Convert_to_Fixed_Point(XTest(:,2));
x2 = Convert_to_Fixed_Point(XTest(:,3));
x3 = Convert_to_Fixed_Point(XTest(:,4));
x4 = Convert_to_Fixed_Point(XTest(:,5));
x5 = Convert_to_Fixed_Point(XTest(:,6));
x6 = Convert_to_Fixed_Point(XTest(:,7));
x7 = Convert_to_Fixed_Point(XTest(:,8));
```

The MATLAB function (Convert_to_Fixed_Point.m) is defined by the following codes.

```
function xf = Convert_to_Fixed_Point(xin,hstep)
% Converting double type array into fixed type array
% hstep is optional parameter
Xmax = max(xin);
Xmin = min(xin);
IP_max = fix(Xmax);
IP_min = fix(Xmin);
fractional_parts = abs(xin - floor(xin));
non_zero_fractional_parts = fractional_parts(fractional_parts > 0);
FP_min = min(non_zero_fractional_parts);
if ((IP_max > 0) && (IP_min < 0))
    WL_int = ceil(log2((abs(IP_max) + abs(IP_min)) + 1));
elseif ((IP_max > 0) && (IP_min >= 0))
    IP_min = 0;
    WL_int = ceil(log2((abs(IP_max) + abs(IP_min)) + 1));
else
    IP_max = 0;
    WL_int = ceil(log2((abs(IP_max) + abs(IP_min)) + 1));
end

if nargin < 2
    hfactor = 10;
    if isempty(FP_min)
        FL = 0;
    else
        FL = ceil(log2(1/abs(FP_min/hfactor)));
    end
else
    FL = ceil(log2(1/abs(hstep)));
end
WL = WL_int + FL + 1; % 1 extra bit is for sign
s = 1;
xf = fi(xin,s,WL,FL);
```

In linear regression, we assume that the output variable y (a scalar) can be described as a combination of the n input variables $x_1, x_2, \ldots, x_n$.

$$y = \theta_0 + \theta_1 x_1 + \theta_2 x_2 + \cdots + \theta_n x_n \tag{16.1}$$

where we have the following expressions.

1. The coefficients $\theta_0, \theta_1, \ldots, \theta_n$ are the parameters of the model.
2. The θ_0 is the intercept (offset or bias) term.

The linear regression model described in Equation (16.1) can also be expressed in vector notation. Given the parameter vector $\theta = [\theta_0\ \theta_1\ \ldots\ \theta_n]^T$ and the input vector x with an extra element 1 in its first position, Equation (16.1) can be rewritten as follows.

$$y = [\theta_0\ \theta_1 \ldots \theta_n]\begin{bmatrix} 1 \\ x_1 \\ \ldots \\ x_n \end{bmatrix} = \theta^T x \tag{16.2}$$

In order to find θ, we use the following equation:

$$\hat{\theta} = \arg_{\theta}^{\min} \frac{1}{n}\sum_{i=1}^{n} L(\hat{y}(x_i;\ \theta), y_i) \tag{16.3}$$

The cost function ($J(\theta)$) is defined as follows:

$$J(\theta) = \frac{1}{n}\sum_{i=1}^{n} L(\hat{y}(x_i;\ \theta), y_i) \tag{16.4}$$

Also, the squared error loss function is given by:

$$L(\hat{y}(x_i;\ \theta), y_i) = (\hat{y}(x_i;\ \theta) - y)^2 \tag{16.5}$$

Using linear algebra, the solution to this problem is to determine $\hat{\theta}$ using the following equation:

$$\hat{\theta} = (X^T X)^{-1} X^T y \tag{16.6}$$

In summary, the linear regression method is described as follows.

Learning of linear regression is done with squared error loss as follows:

Input: Training data $\tau = \{x_i, y_i\}_{i=1}^{N}$

Output: Learned parameter vector $\hat{\theta}$ (computed by solving Equation (16.6))

Prediction with linear regression is done as follows:

Input: Learned parameter vector $\hat{\theta}$ and test input x^*

Output: Prediction of $\hat{y}(x^*)$ using the following equation:

$$\hat{y}(x^*) = \hat{\theta}^T x^* \tag{16.7}$$

When we run the MATLAB m-file RegModel.m, we can view the histogram plots of GRE Score, TOFEL Score, University Rating, SOP, LOR, CGPA, Research, and Chances of Admission. Also, the linear regression plot of testing and predicted data for Chance of Admit is depicted in Figure 16.2.

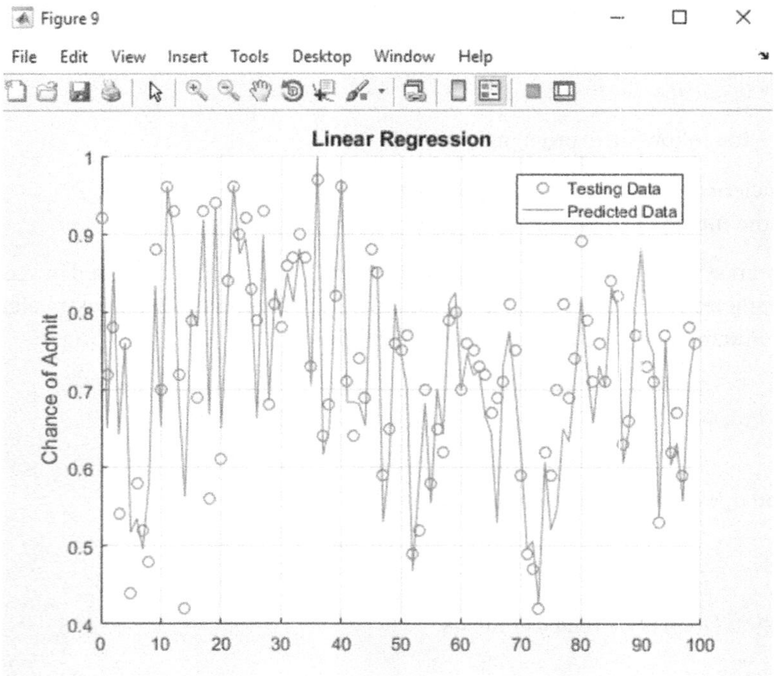

Figure 16.2 The linear regression plot of testing and predicted data.

Moreover, the following expression appears in the command window:

```
MSE Linear Regression
MSE =
    0.0027
```

So, we see that the mean squared error is 0.0031, and the final learned parameters of θ are obtained using the following command:

```
>> Th
Th =
   -1.2459
    0.0017
    0.0028
    0.0068
    0.0002
    0.0191
    0.1215
    0.0247
```

16.4 The Regression Model in Simulink

Now, we open the Simulink by using the following command:

```
>> simulink
```

The Simulink design (ML_Regression.slx) is illustrated in Figure 16.3.

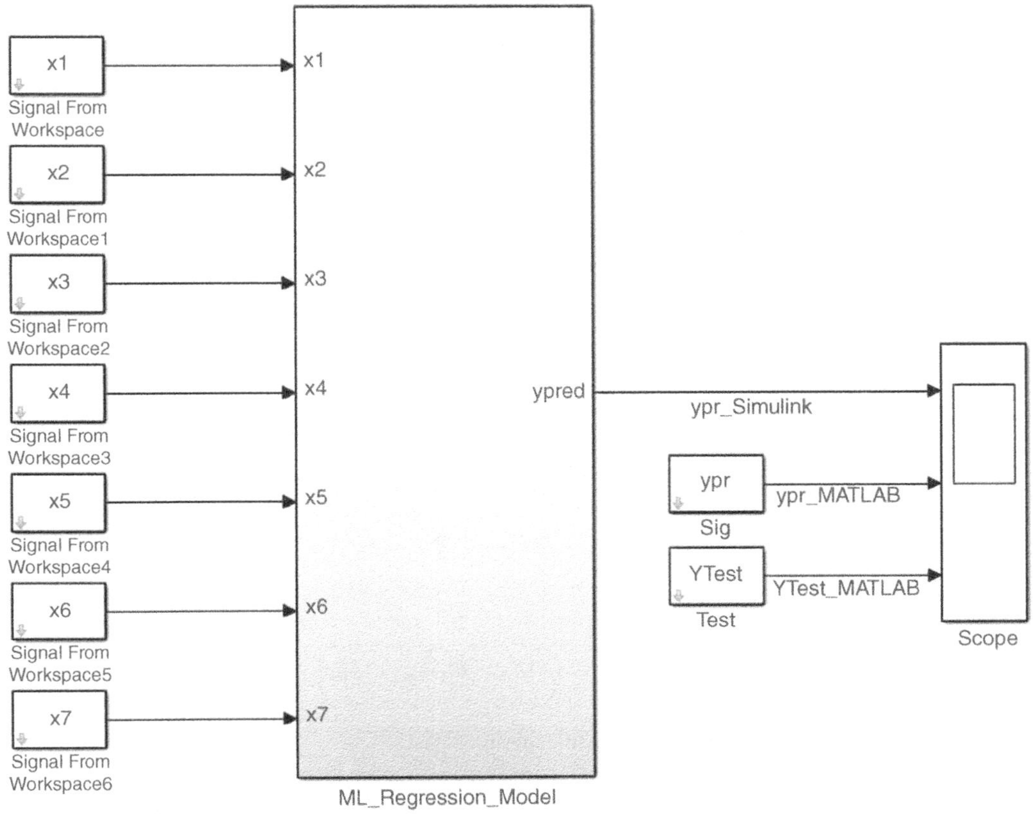

Figure 16.3 The Simulink design for regression model.

The subsystem "ML_Regression_Model" design is shown in Figure 16.4.

The Signal From Workspace block parameters for x1 is depicted in Figure 16.5.

We should note that x0 = XTest(:,1) = ones(100,1). So, we have used only constant t0 in Figure 16.4. The block parameters for constant t0 are illustrated in Figures 16.6 and 16.7, respectively.

Also, the block parameters for the gain t1 are shown in Figures 16.8 and 16.9, respectively.

Furthermore, the Signal Attributes tab of Sum in Figure 16.4 is depicted in Figure 16.10.

Now, we can run the simulation, and to view the port data types, we right-click on the Simulink design and select Other Displays → Signals & Ports → Port Data Types. We set the Solver tab of the Configuration Parameters window, as illustrated in Figure 16.11. Then, we can run the simulation, and in the Scope window, we click on View → Legend and also View → Style... and select the appropriate marker for each line to get the Simulink prediction, MATLAB prediction, and YTest waveforms, as shown in Figure 16.12.

Now, we can generate the VHDL code for the subsystem "ML_Regression_Model" in Figure 16.3 using HDL Coder. To do that, we write the following code in MATLAB command window.

```
>> hdlsetup('ML_Regression')
```

The ML_Regression is the name of Simulink design. Then, we go to the Configuration Parameters window and set the Solver tab values, as demonstrated in Figure 16.11. Since size(XTest) = 100×8, we have set the stop time to 100 in Figure 16.11. Next, we go to the HDL Code Generation tab and set the parameters, as depicted in Figure 16.13.

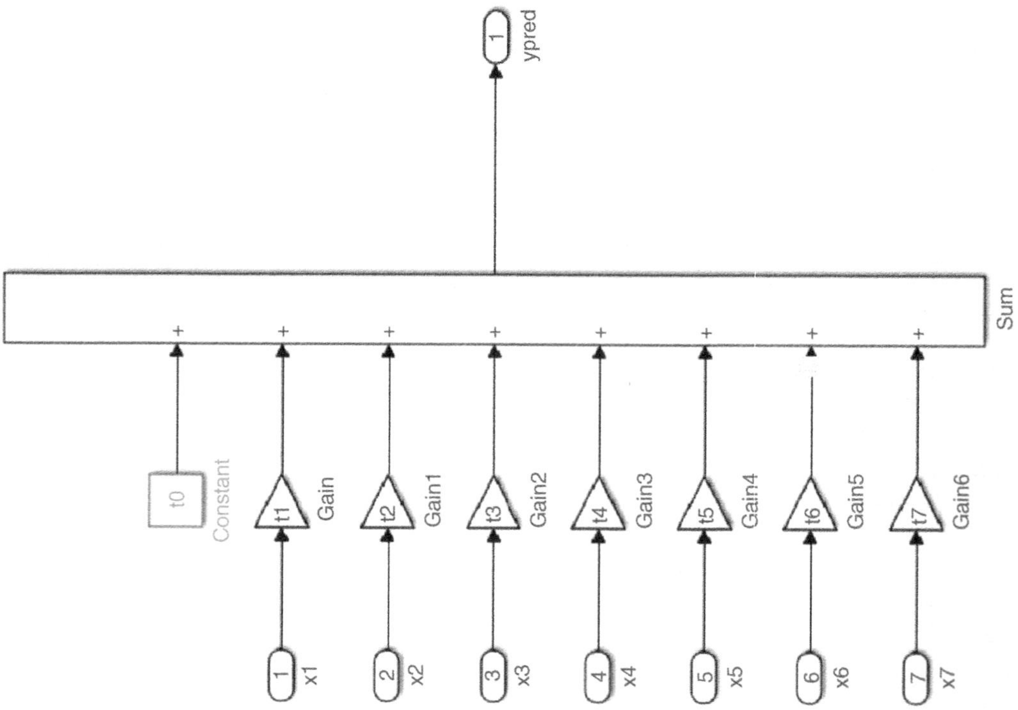

Figure 16.4 The subsystem "ML_Regression_Model" design.

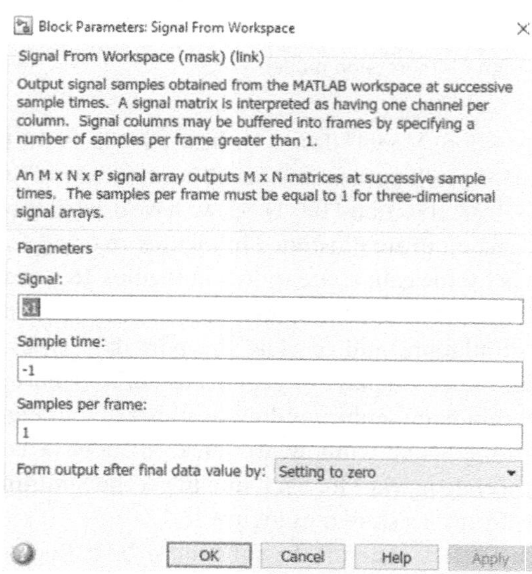

Figure 16.5 The Signal From Workspace block parameters for x1.

Then, we type the following code in the MATLAB command window.

```
>> checkhdl('ML_Regression/ML_Regression_Model')
```

The HDL check report window as a web browser pops up, as shown in Figure 16.14.

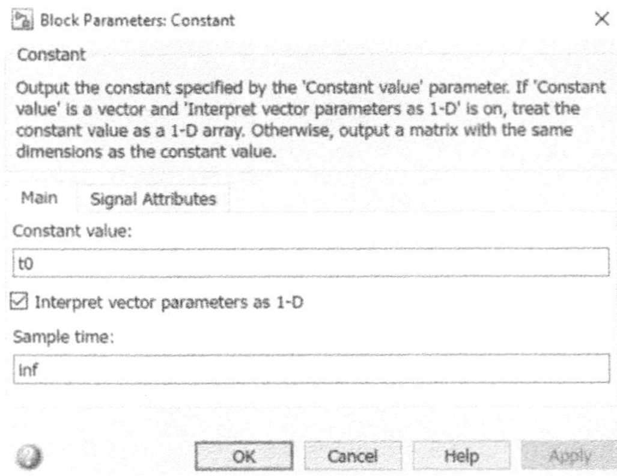

Figure 16.6 Block Parameters (Constant) for Main tab.

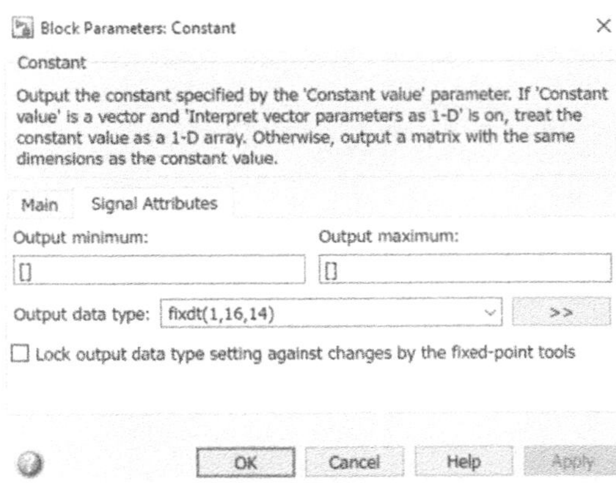

Figure 16.7 Block Parameters (Constant) for Signal Attributes tab.

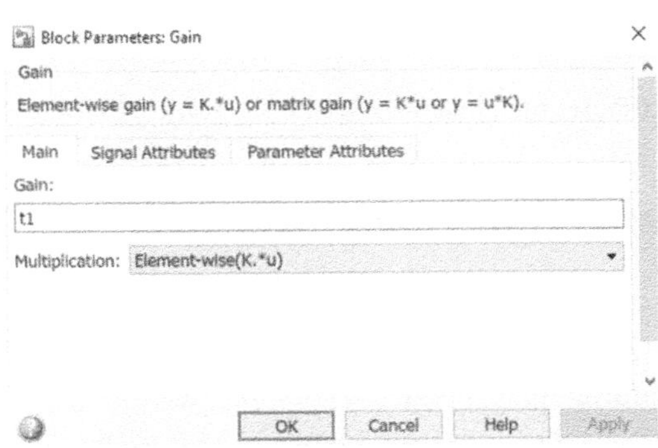

Figure 16.8 Block Parameters (Gain) for Main tab.

Now, we use the following code for HDL Code Generation in the MATLAB command window.

```
>> makehdl('ML_Regression/ML_Regression_Model')
```

The VHDL codes are generated successfully, and the Code Generation Report window is displayed, as depicted in Figure 16.15.

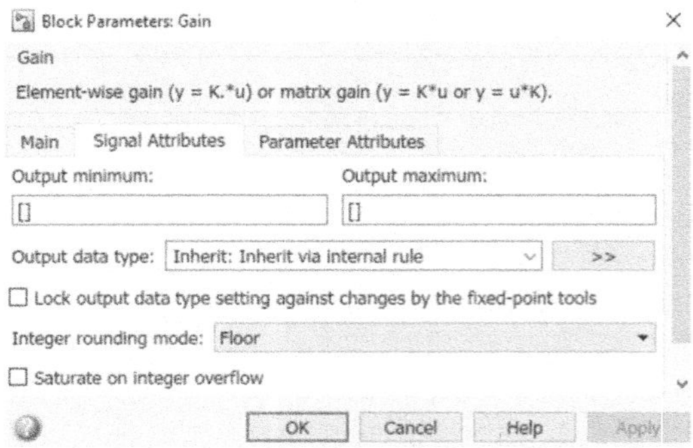

Figure 16.9 Block Parameters (Gain) for Signal Attributes tab.

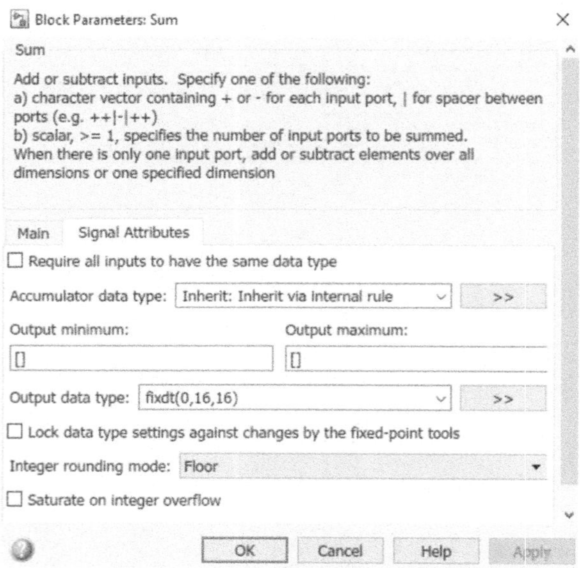

Figure 16.10 The Signal Attributes tab of Sum block.

16.5 Using Generated VHDL Code in Vivado

Now, we run the Vivado and create a project named "ML_Regression" and then click the Next button. Then, we select the project type as RTL project and click the Next button. In the Add Sources window, before adding the generated VHDL file, we first copy and paste the generated VHDL file (ML_Regression_Model.vhd) from Figure 16.15 into the ML_Regression folder of the Vivado project. Now, by clicking on the Add Files button, we can add it to the project, as shown in Figure 16.16.

Next, we select our Xilinx development board, click the Next button, and finally, click the Finish button to create the "ML_Regression" project with the added generated VHDL file. Then, we run the synthesis, as illustrated in Figure 16.17.

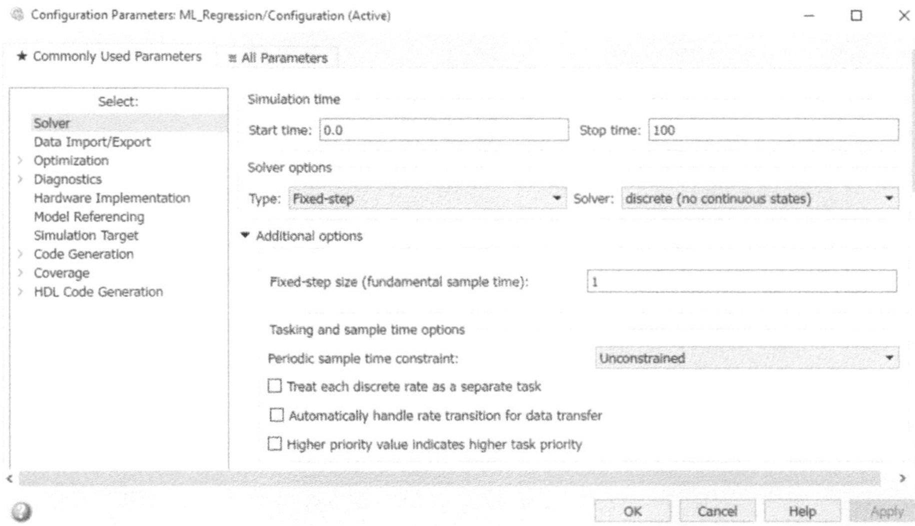

Figure 16.11 The Solver tab of Configuration Parameters window.

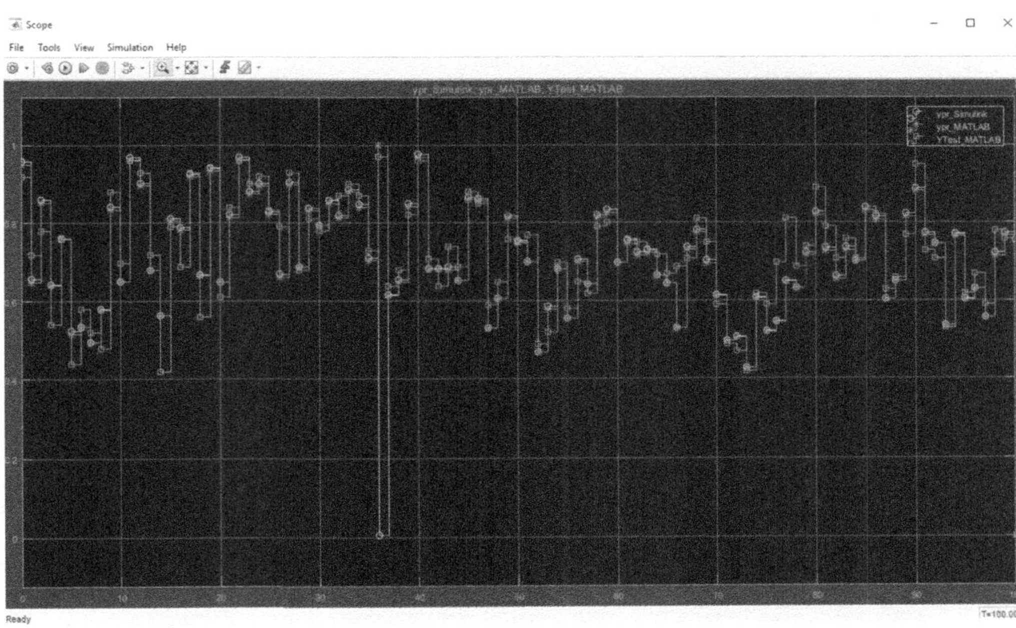

Figure 16.12 The Simulink prediction, MATLAB prediction, and YTest waveform.

As demonstrated in Figure 16.17, the synthesis is successfully completed, and we can click on the Cancel button. Then, we open the schematic in the RTL Analysis tab as shown in Figure 16.18.

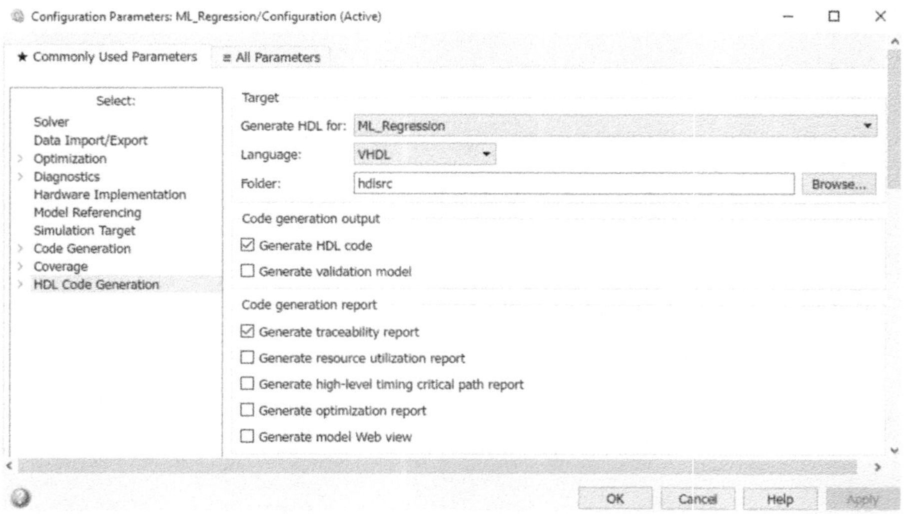

Figure 16.13 The HDL Code Generation tab of Configuration Parameters window.

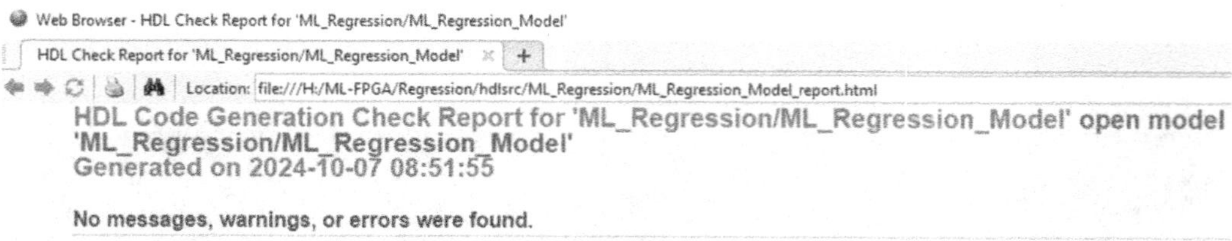

Figure 16.14 The HDL Check Report window.

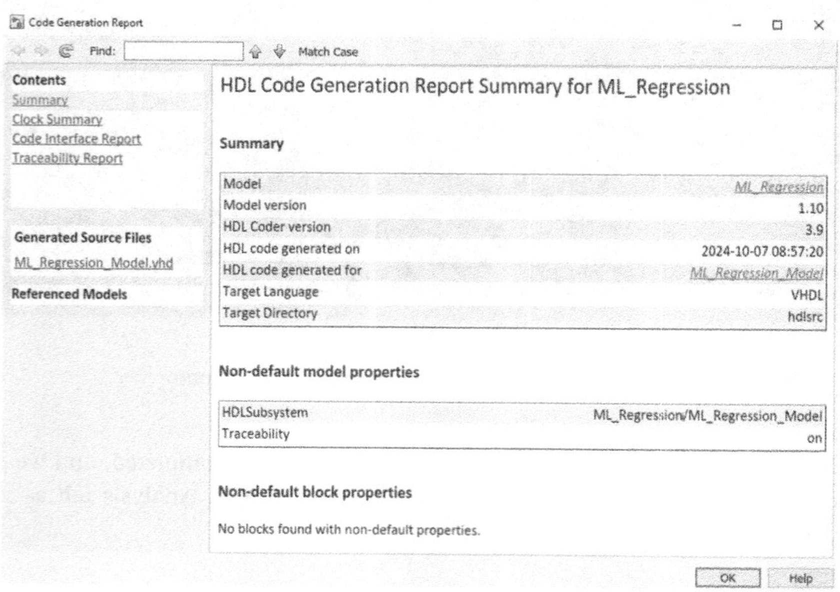

Figure 16.15 The Code Generation Report window.

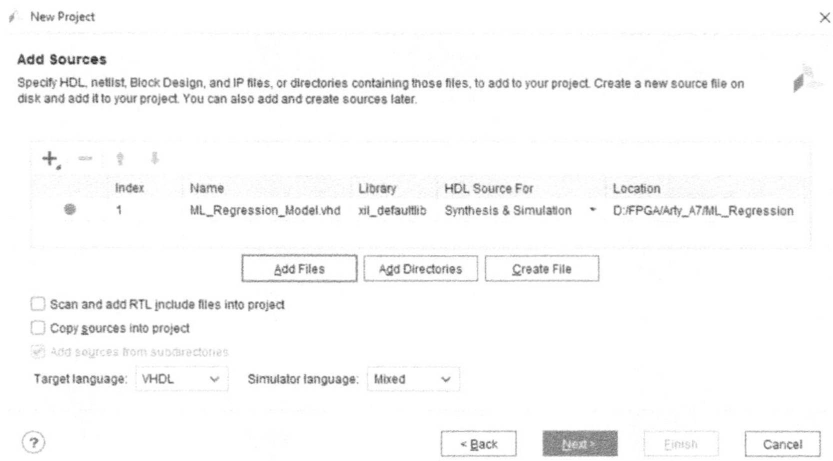

Figure 16.16 Adding generated VHDL file to the Sources folder.

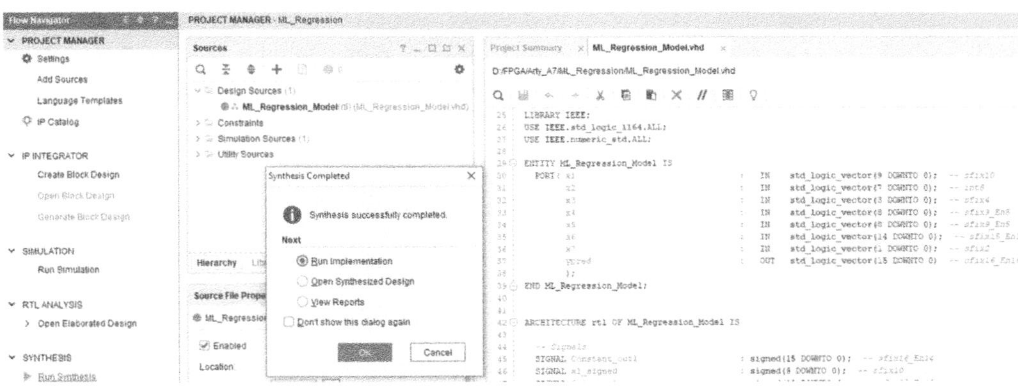

Figure 16.17 Successful synthesis execution.

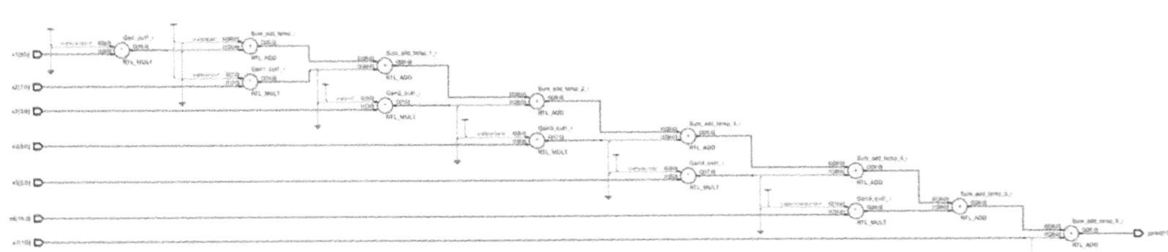

Figure 16.18 The schematic of "ML_Regression_Model."

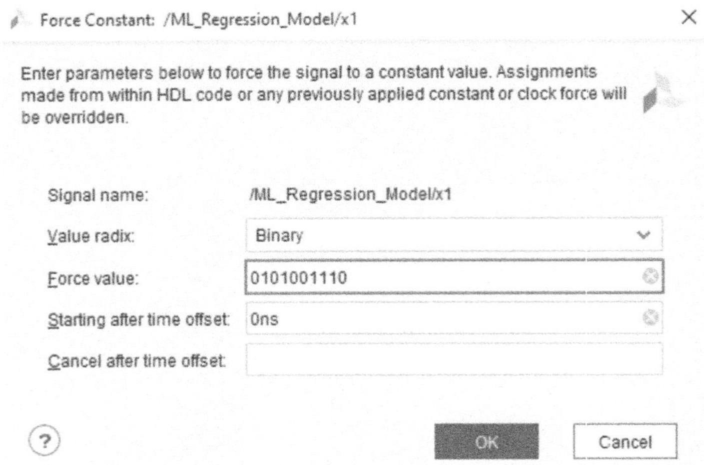

Figure 16.19 The Force Constant window for x1.

Now, we can simulate the project by selecting Flow → Run Simulation → Run Behavioral Simulation. In the opened simulation window, we click on the Restart button. We then write the following code in the MATLAB command window.

```
>> x1_20 = x1(20)
x1_20 =
   334
          DataTypeMode: Fixed-point: binary point scaling
            Signedness: Signed
            WordLength: 10
        FractionLength: 0
>> x1_20.bin
ans =
0101001110
```

Then, in the Waveform window, we right-click on "x1" and select the "Force Constant..." option, setting its parameter, as illustrated in Figure 16.19.

We repeat the same process for x2 = 01110111, x3 = 0101, x4 = 010010000, x5 = 010010000, x6 = 010010111101100, and x7 = 01. Then, we click the Run button for 10 ns once and click the Zoom Fit button in the waveform window to obtain the simulated waveforms, as illustrated in Figure 16.20. We also right-click on "ypred" and select Radix → Binary to view the "ypred" value as a binary number. We know that the "pred" data type is fixdt(0,16,16), meaning it is unsigned with a word length of 16 bits and a fraction length of 16 bits. Since there is no integer part, only the fractional part exists. Therefore, we have the following equation.

$$\text{pred}(20) = 1111000100001111 =$$

$$2^{-1} + 2^{-2} + 2^{-3} + 2^{-4} + 2^{-8} + 2^{-13} + 2^{-14} + 2^{-15} + 2^{-16} = 0.9416 \tag{16.8}$$

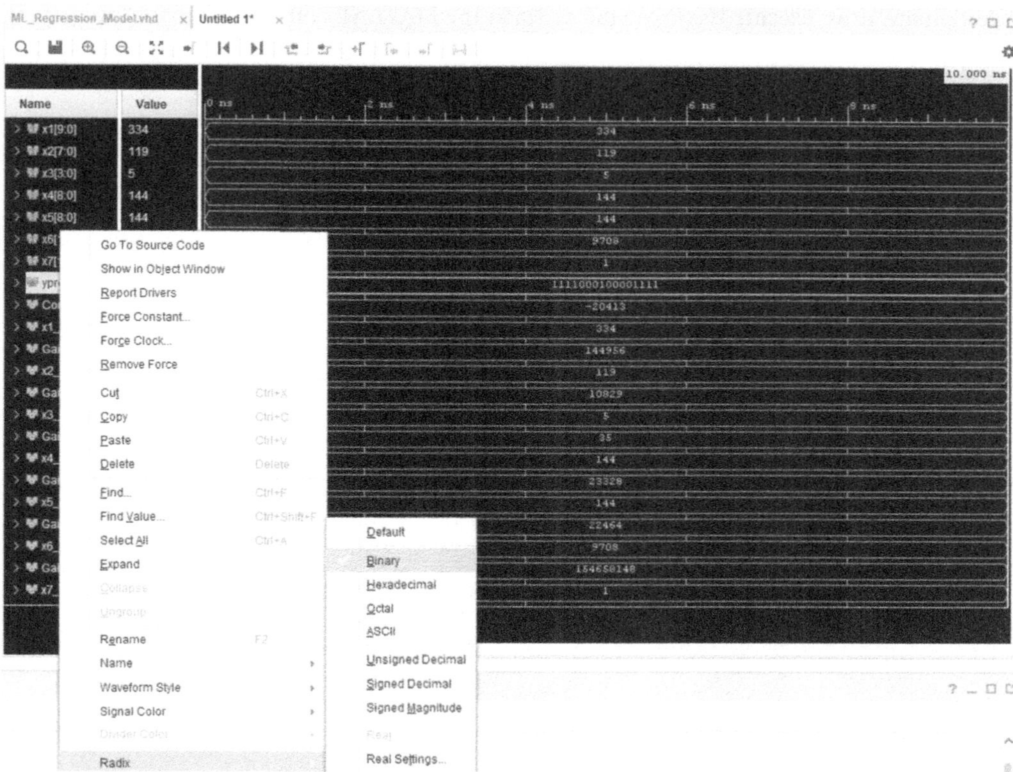

Figure 16.20 The simulated waveforms.

We can use the bin2float function to get the float value of a binary number, as shown in the following codes.

```
function out = bin2float(bin)

ind = find(bin == '.');
decpart = bin(1:ind-1);
floatpart = bin(ind+1:end);

out = bin2dec(decpart);
for n = 1:length(floatpart)
    if floatpart(n) == '1'
        out = out + 2^(-n);
    end
end
```

So, if we enter the following code in the MATLAB command window, we get the following result.

```
>> bin2float('0.1111000100001111')
ans =
    0.9416
```

Additionally, if we execute the following codes in the MATLAB command window, we get the following results.

```
>> ypr(20)
ans =
    0.9344
>> YTest(20)
ans =
    0.9400
```

Thus, we successfully achieve a good prediction using an FPGA as well.

17

Machine Learning Binary Classification Model on FPGA

17.1 Introduction

In this section, we are going to implement a machine learning binary classification model on FPGA using Simulink and Vivado. First, we download a dataset from Kaggle that is suitable for the implementation of binary classification. Using MATLAB, we train the binary classification model by minimizing the negative log-likelihood cost function. We use the customer satisfaction dataset to predict customer satisfaction based on various input parameters. Then, we implement the trained machine learning model of binary classification using HDL Coder and generate synthesizable VHDL code. Finally, using Vivado, we run synthesis analysis and simulation to verify the designed VHDL project.

17.2 The Kaggle Dataset

In this chapter, we will use the customer satisfaction dataset from Kaggle to predict customer satisfaction based on several input parameters. The weblink for downloading the dataset is as follows.

https://www.kaggle.com/datasets/raminhuseyn/airline-customer-satisfaction

The overview of part of customer satisfaction dataset in the Excel software is shown in Figure 17.1.

17.3 The Binary Classification (Logistic Regression)

In this section, we describe the binary classification or logistic regression, and our aim is to learn function $g(x)$ parameterized by θ, which predicts the binary class.

$$z = \theta_0 + \theta_1 x_1 + \theta_2 x_2 + \cdots + \theta_n x_n = \boldsymbol{\theta}^T \boldsymbol{x} \tag{17.1}$$

The z takes values on the entire real line, but we need a function that returns a value in the interval [0, 1]. The logistic function is defined by the following equation.

$$h(z) = \frac{1}{1+e^{-z}} = \frac{e^z}{1+e^z} \text{ or } g(x) = \frac{1}{1+e^{-\theta^T x}} = \frac{e^{\theta^T x}}{1+e^{\theta^T x}} \tag{17.2}$$

The following codes are used to plot $h(z)$ with respect to z in the interval [−10, 10], and the resulted waveform is depicted in Figure 17.2.

Fundamentals of VHDL for FPGA Programming Using Vivado, First Edition. Majid Pakdel.

Companion Website: www.wiley.com/go/pakdelVHDL

	A	B	C	D	E	F	G	H	I	J	K	L
1	satisfaction	Customer Type	Age	Type of Travel	Class	Flight Distance	Seat comfort	Departure/Arrival time convenient	Food and drink	Gate location	Inflight wifi service	Inflight entertainment
2	satisfied	Loyal Customer	65	Personal Trave	Eco	265	0	0	0	2	2	4
3	satisfied	Loyal Customer	47	Personal Trave	Business	2464	0	0	0	3	0	2
4	satisfied	Loyal Customer	15	Personal Trave	Eco	2138	0	0	0	3	2	0
5	satisfied	Loyal Customer	60	Personal Trave	Eco	623	0	0	0	3	3	4
6	satisfied	Loyal Customer	70	Personal Trave	Eco	354	0	0	0	3	4	3
7	satisfied	Loyal Customer	30	Personal Trave	Eco	1894	0	0	0	3	2	0
8	satisfied	Loyal Customer	66	Personal Trave	Eco	227	0	0	0	3	2	5
9	satisfied	Loyal Customer	10	Personal Trave	Eco	1812	0	0	0	3	2	0
10	satisfied	Loyal Customer	56	Personal Trave	Business	73	0	0	0	3	5	3

Figure 17.1 The overview of part of customer satisfaction dataset.

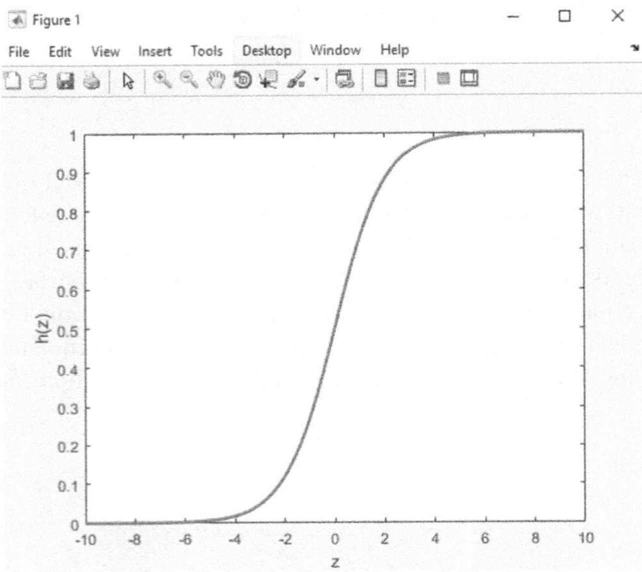

Figure 17.2 The plot of $h(z)$ with respect to z.

```
z = -10:0.1:10;
hz = zeros(1,length(z));
for i = 1:length(z)
   hz = 1./(1+exp(-z));
end
plot(z,hz,'LineWidth',2);
xlabel('z');
ylabel('h(z)')
```

The class prediction or binary classification is defined by the following equation.

$$\hat{y}(\boldsymbol{x},*) = \begin{cases} 1 \text{ if } g((\boldsymbol{x},*) > r \\ 0 \text{ elsewhere} \end{cases} \tag{17.3}$$

The decision threshold is $r = 0.5$.

The parameter vector $\boldsymbol{\theta}$ is found by the following equation.

$$\boldsymbol{\theta} = \min J(\theta) \tag{17.4}$$

The cost function (negative log-likelihood cost) in logistic regression is given by the following equation:

$$J(\theta) = -\frac{1}{n}\left[\sum_{i=1}^{n} y^{(i)} \ln\left(h_\theta\left(x^{(i)}\right)\right) + (1 - y^{(i)}) \ln\left(1 - h_\theta\left(x^{(i)}\right)\right)\right] \tag{17.5}$$

The gradient of the cost function is a vector of the same length as $\boldsymbol{\theta}$, given by the following equation:

$$\frac{\partial J(\theta)}{\partial \theta_j} = \frac{1}{n}\sum_{i=1}^{n}\left(h_\theta\left(x^{(i)}\right) - y^{(i)}\right)x_j^{(i)} \tag{17.6}$$

Also, the logistic regression function is defined by the following equation:

$$h_\theta\left(x^{(i)}\right) = g\left(\theta^T x\right) = \frac{1}{1 + e^{-\theta^T x}} \tag{17.7}$$

In summary, the logistic regression method is described as follows.

Learning of binary regression is done as follows.

Input: Training data $\tau = \{x_i, y_i\}_{i=1}^{N}$ (with output classes y $=$ $\{0,1\}$)

Output: Learned parameter vector $\boldsymbol{\theta}$ (computed numerically by Equation (17.4))

Prediction with binary logistic regression is done as follows.

Input: Learned parameter vector $\boldsymbol{\theta}$, and test input $\boldsymbol{x}^*$

Output: Prediction of $\hat{y}(\boldsymbol{x}^*)$ by the following equation.

Compute $g(\boldsymbol{x},*)$, by the following equation.

$$g(\boldsymbol{x},*) = \frac{e^{\theta T x^*}}{1 + e^{\theta^T x^*}} \tag{17.8}$$

If $g(\boldsymbol{x},*) > 0.5$, return $\hat{y}(\boldsymbol{x},*) = 1$; otherwise, return $\hat{y}(\boldsymbol{x},*) = 0$.

17.4 Machine Learning Logistic Regression Model for Dataset

We use the following MATLAB m-file (ClassModel.m) to implement the machine learning binary classification model for customer satisfaction dataset.

```
clear all; close all; clc
cd('H:\ML-FPGA\Classification');
t = readtable('airline_customer.csv');
% Indices of different types of variables
Text_Var = [1 2 4 5];
Num_Disc_Var = [7 8 9 10 11 12 13 14 15 16 17 18 19 20];
Num_Cont_Var = [3 6 21 22];
% Text Data
Text_Data = t(:,Text_Var);
Text_Data_Cell = table2cell(Text_Data);
Text_Var_str = string(Text_Data_Cell);
% Discrete Numeric Data
NumD_Data = t(:,Num_Disc_Var);
NumD_Arr = table2array(NumD_Data);
NumD_str = string(NumD_Arr);
NumD_cell = cellstr(NumD_str);
% Continuous Numeric Data
NumC_Data = t(:,Num_Cont_Var);
```

```
NumC_Arr = table2array(NumC_Data);
% Convert Textual Data to Numerical Data
Text_Count = ones(size(Text_Var_str));
for i = 1:length(Text_Count)
    if (Text_Var_str(i,1) == 'satisfied')
        Text_Count(i,1) = 1;
    else
        Text_Count(i,1) = 0;
    end
end

for i = 1:length(Text_Count)
    if (Text_Var_str(i,2) == 'Loyal Customer')
        Text_Count(i,2) = 1;
    else
        Text_Count(i,2) = 0;
    end
end

for i = 1:length(Text_Count)
    if (Text_Var_str(i,3) == 'Business travel')
        Text_Count(i,3) = 0;
    else
        Text_Count(i,3) = 1;
    end
end

for i = 1:length(Text_Count)
    if (Text_Var_str(i,4) == 'Business')
        Text_Count(i,4) = 1;
    elseif (Text_Var_str(i,4) == 'Eco')
        Text_Count(i,4) = 2;
    else
        Text_Count(i,4) = 3;
    end
end

% Histogram's Plots of features
figure
subplot(121)
histogram(categorical(Text_Data_Cell(:,1)),'BarWidth',0.3);
ylabel('Frequency');
title('satisfaction');
grid on;

subplot(122)
histogram(categorical(Text_Data_Cell(:,2)),'BarWidth',0.3);
ylabel('Frequency');
title('Customer Type');
grid on;

figure
subplot(211)
histogram(categorical(Text_Data_Cell(:,3)),'BarWidth',0.3);
```

```
ylabel('Frequency');
title('Type of Travel');
grid on;

subplot(212)
histogram(categorical(Text_Data_Cell(:,4)),'BarWidth',0.3);
ylabel('Frequency');
title('Class');
grid on;

figure
subplot(221)
histogram(categorical(NumD_cell(:,1)),'BarWidth',0.3);
ylabel('Frequency');
title('Seat comfort');
grid on;

subplot(222)
histogram(categorical(NumD_cell(:,2)),'BarWidth',0.3);
ylabel('Frequency');
title('Departure/Arrival time convenient');
grid on;

subplot(223)
histogram(categorical(NumD_cell(:,3)),'BarWidth',0.3);
ylabel('Frequency');
title('Food and drink');
grid on;

subplot(224)
histogram(categorical(NumD_cell(:,4)),'BarWidth',0.3);
ylabel('Frequency');
title('Gate location');
grid on;

figure
subplot(121)
histogram(categorical(NumD_cell(:,5)),'BarWidth',0.3);
ylabel('Frequency');
title('Inflight wifi service');
grid on;

subplot(122)
histogram(categorical(NumD_cell(:,6)),'BarWidth',0.3);
ylabel('Frequency');
title('Inflight entertainment');
grid on;

figure
subplot(211)
histogram(categorical(NumD_cell(:,7)),'BarWidth',0.3);
ylabel('Frequency');
title('Online support');
grid on;
```

```
subplot(212)
histogram(categorical(NumD_cell(:,8)),'BarWidth',0.3);
ylabel('Frequency');
title('Ease of Online booking');
grid on;

figure
subplot(131)
histogram(categorical(NumD_cell(:,9)),'BarWidth',0.3);
ylabel('Frequency');
title('On-board service');
grid on;

subplot(132)
histogram(categorical(NumD_cell(:,10)),'BarWidth',0.3);
ylabel('Frequency');
title('Leg room service');
grid on;

subplot(133)
histogram(categorical(NumD_cell(:,11)),'BarWidth',0.3);
ylabel('Frequency');
title('Baggage handling');
grid on;

figure
subplot(131)
histogram(categorical(NumD_cell(:,12)),'BarWidth',0.3);
ylabel('Frequency');
title('Checkin service');
grid on;

subplot(132)
histogram(categorical(NumD_cell(:,13)),'BarWidth',0.3);
ylabel('Frequency');
title('Cleanliness');
grid on;

subplot(133)
histogram(categorical(NumD_cell(:,14)),'BarWidth',0.3);
ylabel('Frequency');
title('Online boarding');
grid on;

figure
subplot(221)
histogram(NumC_Arr(:,1));
ylabel('Frequency');
title('Age');
grid on;

subplot(222)
histogram(NumC_Arr(:,2));
ylabel('Frequency');
```

```
title('Flight Distance');
grid on;

subplot(223)
histogram(NumC_Arr(:,3));
xlim([0 140]);
ylabel('Frequency');
title('Departure Delay in Minutes');
grid on;

subplot(224)
histogram(NumC_Arr(:,4));
xlim([0 120]);
ylabel('Frequency');
title('Arrival Delay in Minutes');
grid on;

% Text_Count = Text_Count(1:100,:);
% NumD_Arr = NumD_Arr(1:100,:);
% NumC_Arr = NumC_Arr(1:100,:);

All_Data = [Text_Count NumD_Arr NumC_Arr];
rnan = all(~isnan(All_Data),2);
All_Data = All_Data(rnan,:);

% Matrix presentation of entire data
Xfeatures = All_Data(:,2:end);
Yout = All_Data(:,1);
Xf = [ones(size(Yout,1),1) Xfeatures];

%Splitting data into training and testing
P = 0.8;
Pn = 1-P;
XTrain = Xf(1:round(P*length(Xf)),:);
YTrain = Yout(1:round(P*length(Xf)),:);

XTest = Xf(end-round(Pn*length(Xf)):end,:);
YTest = Yout(end-round(Pn*length(Xf)):end,:);

%Implementation of logistic regression for classification
[Rows, Columns] = size(Xfeatures);
Theta_initial = zeros(Columns + 1, 1);

%Initial value of cost function
[Ji, Gradi] = Cost_Function(Theta_initial, XTrain, YTrain);
options = optimoptions(@fminunc,'Display','iter','Algorithm','Quasi-
Newton','GradObj','on','MaxIter',500);
[Theta, cost] = fminunc(@(t)(Cost_Function(t, XTrain, YTrain)),Theta_
initial,options);

%Final value of cost function
[Jf, Gradf] = Cost_Function(Theta_initial, XTrain, YTrain);
```

```
%Sigmoid function evaluation
z = XTest * Theta;
hz = 1 ./ (1 + exp(-z));  %Sigmoid function implementation

%Output Prediction
ypred = round(hz);

%Accuracy Prediction
Accuracy = 100*mean(ypred == YTest);
fprintf('Accuracy of Log Reg Method = %0.4f \n',Accuracy)

%Plot Graphs
figure
Out_Ys = [YTest ypred];
plot(Out_Ys(:,2),'*r')
grid on;
hold on

plot(Out_Ys(:,1),'ob')
ylim([-0.2 1.2])
legend('Predicted','Actual')
title('Logistic Regression Method')

figure
C1 = categorical(Out_Ys(:,2),[0 1], {'Not Satisfied','Satisfied'});
histogram(C1,'BarWidth',0.3);
hold on;
grid on;

C2 = categorical(Out_Ys(:,1),[0 1], {'Not Satisfied','Satisfied'});
histogram(C2,'BarWidth',0.15);
legend('Predicted','Actual')
title('Logistic Regression Method')
ylabel('Frequency')

%Quantization
x0 = Convert_to_Fixed_Point(XTest(:,1));
x1 = Convert_to_Fixed_Point(XTest(:,2));
x2 = Convert_to_Fixed_Point(XTest(:,3));
x3 = Convert_to_Fixed_Point(XTest(:,4));
x4 = Convert_to_Fixed_Point(XTest(:,5));
x5 = Convert_to_Fixed_Point(XTest(:,6));
x6 = Convert_to_Fixed_Point(XTest(:,7));
x7 = Convert_to_Fixed_Point(XTest(:,8));
x8 = Convert_to_Fixed_Point(XTest(:,9));
x9 = Convert_to_Fixed_Point(XTest(:,10));
x10 = Convert_to_Fixed_Point(XTest(:,11));
x11 = Convert_to_Fixed_Point(XTest(:,12));
x12 = Convert_to_Fixed_Point(XTest(:,13));
x13 = Convert_to_Fixed_Point(XTest(:,14));
x14 = Convert_to_Fixed_Point(XTest(:,15));
```

```
x15 = Convert_to_Fixed_Point(XTest(:,16));
x16 = Convert_to_Fixed_Point(XTest(:,17));
x17 = Convert_to_Fixed_Point(XTest(:,18));
x18 = Convert_to_Fixed_Point(XTest(:,19));
x19 = Convert_to_Fixed_Point(XTest(:,20));
x20 = Convert_to_Fixed_Point(XTest(:,21));
x21 = Convert_to_Fixed_Point(XTest(:,22));

Thf = fi(Theta,1,16,12);
YTest = YTest(1:1000);
```

First of all, we plot the histogram of different columns in the customer satisfaction dataset. We have also dedicated 80% of the data for training and 20% for testing. The codes for the function "Cost_Function" is shown as follows.

```
function [J, Grad] = Cost_Function(Theta, X, y)
n = length(y);
z = X * Theta;
Grad = zeros(size(Theta));
h = 1 ./ (1 + exp(-z));
J = -(1/n)*sum((y.*log(h))+((1-y).*log(1-h)));
for j=1:size(Theta,1)
    Grad(j)=(1/n)*sum((h-y).*X(:,j));
end
end
```

Also, the same function "Convert_to_Fixed_Point" from the previous chapter is used here as well. The predicted and actual histograms for customer satisfaction are illustrated in Figure 17.3. Furthermore, the predicted and actual plots for customer satisfaction are shown in Figure 17.4.

The accuracy of logistic regression method is displayed in MATLAB command window as follows.

```
Accuracy of Log Reg Method = 94.2505
```

Also, the optimum values of θ are as follows:

```
>> Theta'
ans =
  Columns 1 through 13
   -5.9870    1.6159   -0.5531   -0.4730    0.2791   -0.2126   -0.2270    0.1238
-0.0781    0.7168    0.1137    0.1442    0.2946
  Columns 14 through 22
    0.2091    0.0706    0.2812    0.0460    0.1497   -0.0096   -0.0002    0.0035
-0.0088
```

The length of θ is as follows:

```
>> length(Theta)
ans =
    22
```

Since min(Theta) = −5.9870, and max(Theta) = 1.6159, we can display the integer part of Theta using 4 bits (1 bit for the sign). We use the command Thf = fi(Theta,1,16,12), where the word length is 16 bits and the fraction length is 12 bits. To evaluate each element of Theta to at least 32 significant digits, we use the command vpa(Theta), which provides variable-precision arithmetic for floating-point numbers. The fixed-point values of θ are shown as follows.

```
>> Thf'
ans =
  Columns 1 through 13
   -5.9871     1.6160    -0.5530    -0.4731     0.2791    -0.2126    -0.2271     0.1238
-0.0781     0.7168     0.1138     0.1443     0.2947
  Columns 14 through 22
    0.2090     0.0706     0.2813     0.0459     0.1497    -0.0095    -0.0002     0.0034
-0.0088
          DataTypeMode: Fixed-point: binary point scaling
            Signedness: Signed
            WordLength: 16
        FractionLength: 12
```

We can use the command vpa(Thf') to get the variable-precision arithmetic for floating-point numbers of Thf' in the MATLAB command window as follows.

```
>> vpa(Thf')

ans =

[ -5.987060546875, 1.615966796875, -0.552978515625, -0.47314453125, 0.279052734375,
-0.212646484375, -0.22705078125, 0.123779296875, -0.078125, 0.716796875, 0.1137695
3125, 0.144287109375, 0.294677734375, 0.208984375, 0.070556640625, 0.28125, 0.0458
984375, 0.149658203125, -0.009521484375, -0.000244140625, 0.00341796875, -0.008789
0625]
```

As you can see, we obtain a maximum of 12 digits in the fractional part.

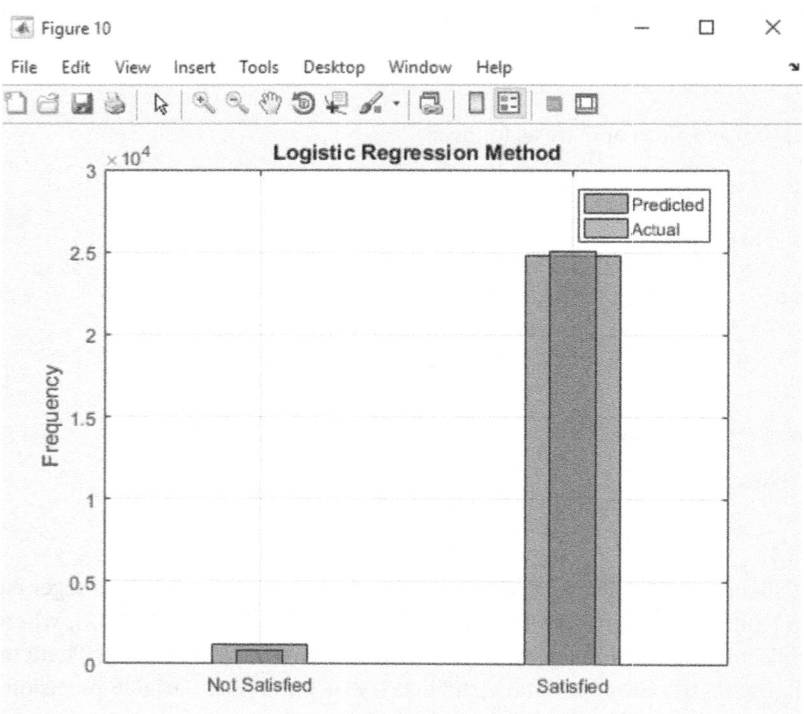

Figure 17.3 The predicted and actual histograms for customer satisfaction.

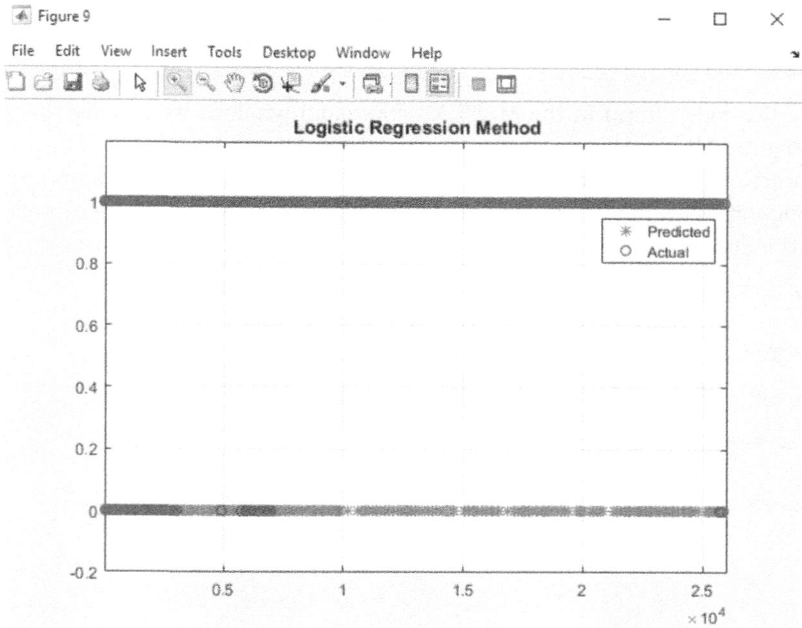

Figure 17.4 The predicted and actual plots for customer satisfaction.

17.5 The Logistic Regression Model in Simulink

Now, we open Simulink using the following command.

```
>> simulink
```

The Simulink design (Bin_Class.slx) is illustrated in Figure 17.5.

The subsystem "Bin_Class_Model" design is shown in Figure 17.4. The MATLAB function "WeightedSum" codes in Figure 17.6 are as follows.

```
function z = WeightedSum(x0,x1,x2,x3,x4,x5,x6,x7,x8,x9,x10,x11,x12,x13,...
    x14,x15,x16,x17,x18,x19,x20,x21,Theta)

T = Theta;
t0 = T(1); t1 = T(2); t2 = T(3); t3 = T(4); t4 = T(5);
t5 = T(6); t6 = T(7); t7 = T(8); t8 = T(9); t9 = T(10);
t10 = T(11); t11 = T(12); t12 = T(13); t13 = T(14); t14 = T(15);
t15 = T(16); t16 = T(17); t17 = T(18); t18 = T(19); t19 = T(20);
t20 = T(21); t21 = T(22);

WS0 = t0*x0 + t1*x1 + t2*x2 + t3*x3 + t4*x4;
WS1 = t5*x5 + t6*x6 + t7*x7 + t8*x8 + t9*x9;
WS2 = t10*x10 + t11*x11 + t12*x12 + t13*x13 + t14*x14;
WS3 = t15*x15 + t16*x16 + t17*x17 + t18*x18 + t19*x19;
WS4 = t20*x20 + t21*x21;
```

```
Wsum = WS0 + WS1 + WS2 + WS3 + WS4;

z = fi(Wsum,1,16,11);
```

If we write the code plot(z) in the MATLAB command window, we can see the plot of z as depicted in Figure 17.7.

We can observe that the z values are in the interval [−6, 6]. So, z = fi(Wsum,1,16,11) in the MATLAB function "WeightedSum" with 5 bits for integer part (1 bit for sign) and 11 bits for fractional part is suitable. The MATLAB function "Sigmoid" codes are as follows.

```
function hz = Sigmoid(z)
zd = double(z);
hz = 1/(1+exp(-zd));
```

Finally, the following codes are used inside the MATLAB function "Prediction".

```
function y = fcn(u)
if u > 0.5
    y = fi(1,0,1,0);
else
    y = fi(0,0,1,0);
end
```

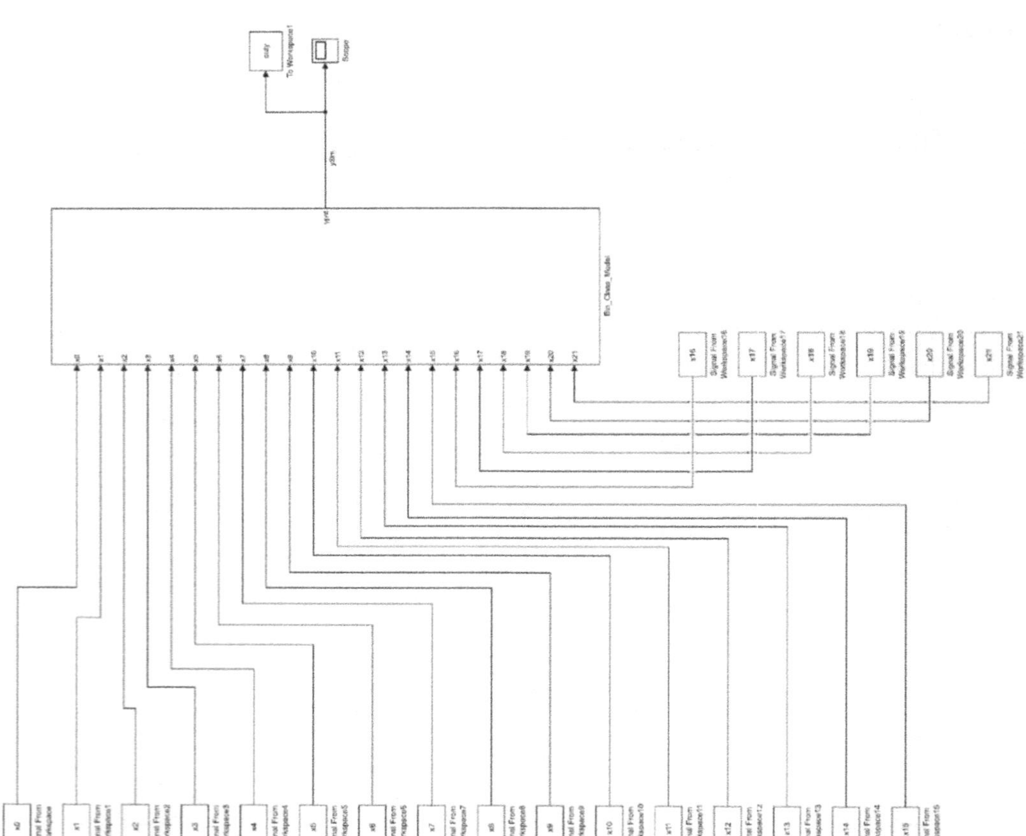

Figure 17.5 The Simulink model of binary classification.

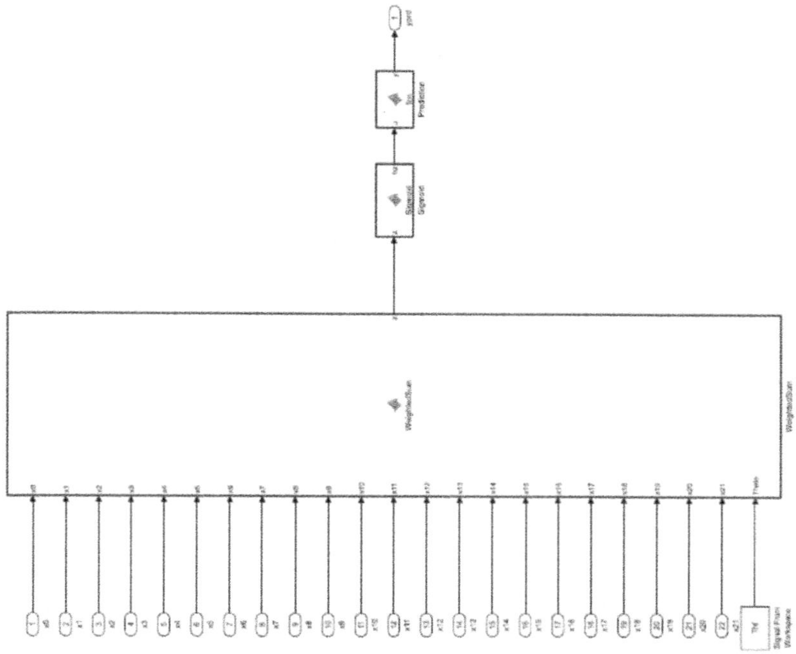

Figure 17.6 The subsystem "Bin_Class_Model" design.

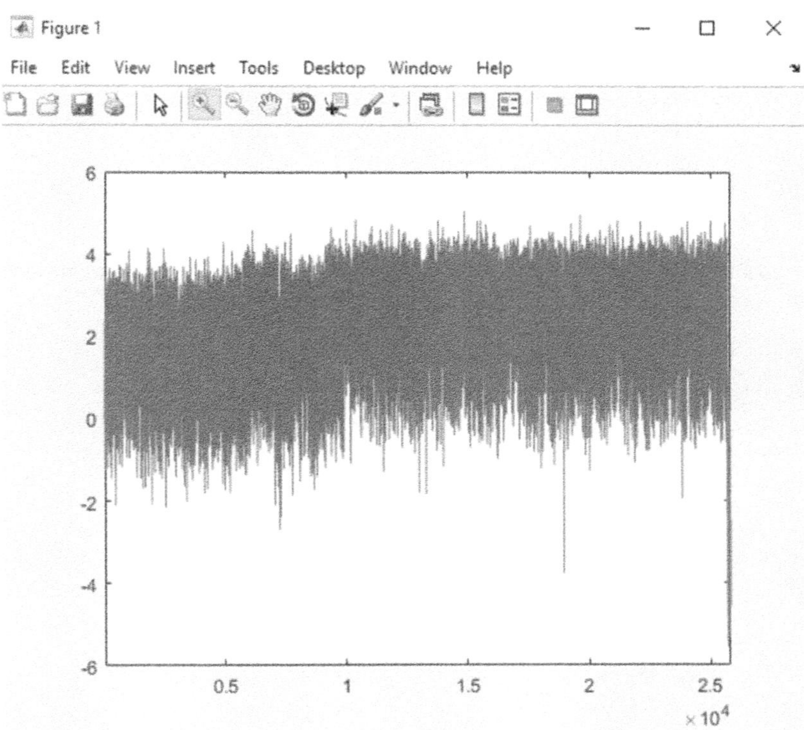

Figure 17.7 The plot of z.

The signal from workspace block parameters for "Thf" in Figure 17.6 is illustrated in Figure 17.8.

The signal from workspace block parameters for "x0" in Figure 17.5 is shown in Figure 17.9 (similar parameters are used for x1, ..., x21).

Also, the configuration parameters for Solver tab are depicted in Figure 17.10.

Now, if we run the simulation, we can view the prediction yprd (ySim) values on the scope, as illustrated in Figure 17.11.

The "to workspace" block parameters in Figure 17.5 are shown in Figure 17.12.

For calculating accuracy, we use the following m-file (Calculate_Accuracy.m).

```
close all; clc
y2 = outy;
y2(end) = [];
AccuracySim = 100*mean(y2 == YTest)
```

By running this m-file, we get the following result in the MATLAB command window.

```
AccuracySim =

   77.3000
```

The HDL Coder cannot generate codes for the MATLAB function "Sigmoid" and returns an error. Therefore, it is better to use a 1-D Lookup Table instead. The block parameters for the 1-D Lookup Table are depicted in Figure 17.13.

Next, we click the "Edit table and breakpoints..." button. In the opened window, from the Edit tab, we select the Add Row option to add the required rows. We use the following m-file to get sigmoid function values in the interval [−6:0.5:6].

```
z = -6:0.5:6;
hz = zeros(1,length(z));
for i = 1:length(z)
   hz = 1./(1+exp(-z));
end
plot(z,hz,'LineWidth',2);
xlabel('z');
ylabel('h(z)');
```

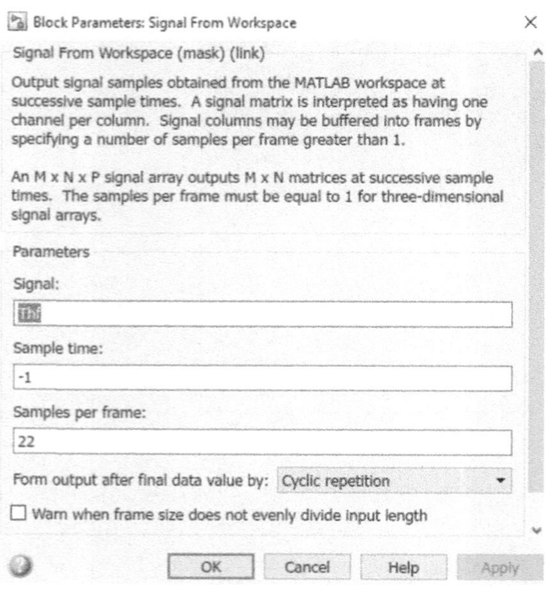

Figure 17.8 The signal from workspace block parameters for "Thf".

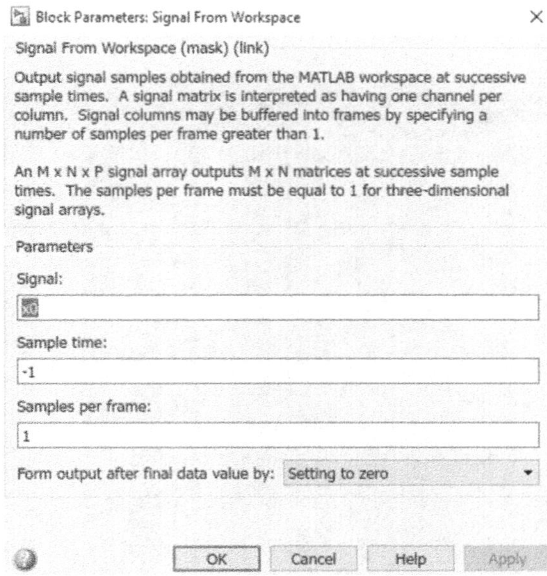

Figure 17.9 The signal from workspace block parameters for "x0".

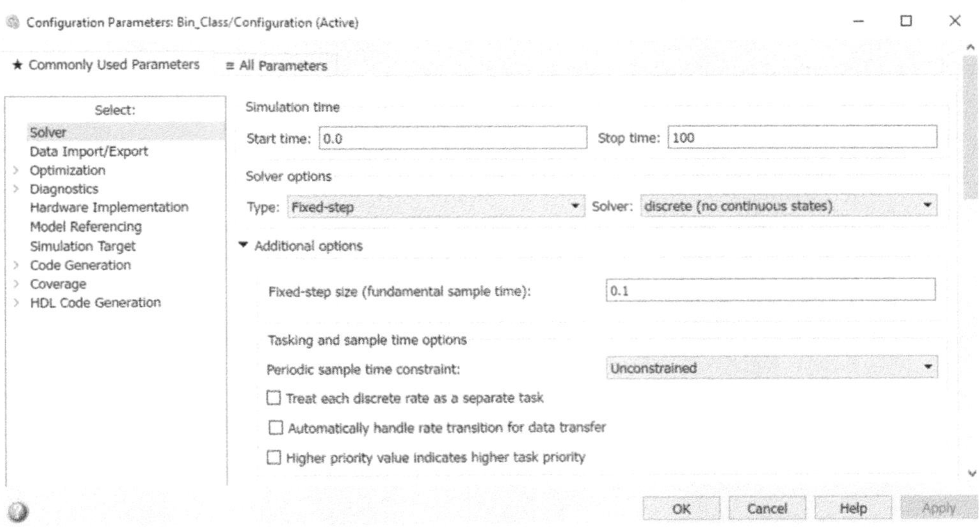

Figure 17.10 The configuration parameters for Solver tab.

After running the m-file, if we type "hz" in the MATLAB command window, we will get the following values.

```
>> hz

hz =

  Columns 1 through 13
    0.0025    0.0041    0.0067    0.0110    0.0180    0.0293    0.0474    0.0759
0.1192    0.1824    0.2689    0.3775    0.5000
  Columns 14 through 25
    0.6225    0.7311    0.8176    0.8808    0.9241    0.9526    0.9707    0.9820
0.9890    0.9933    0.9959    0.9975
```

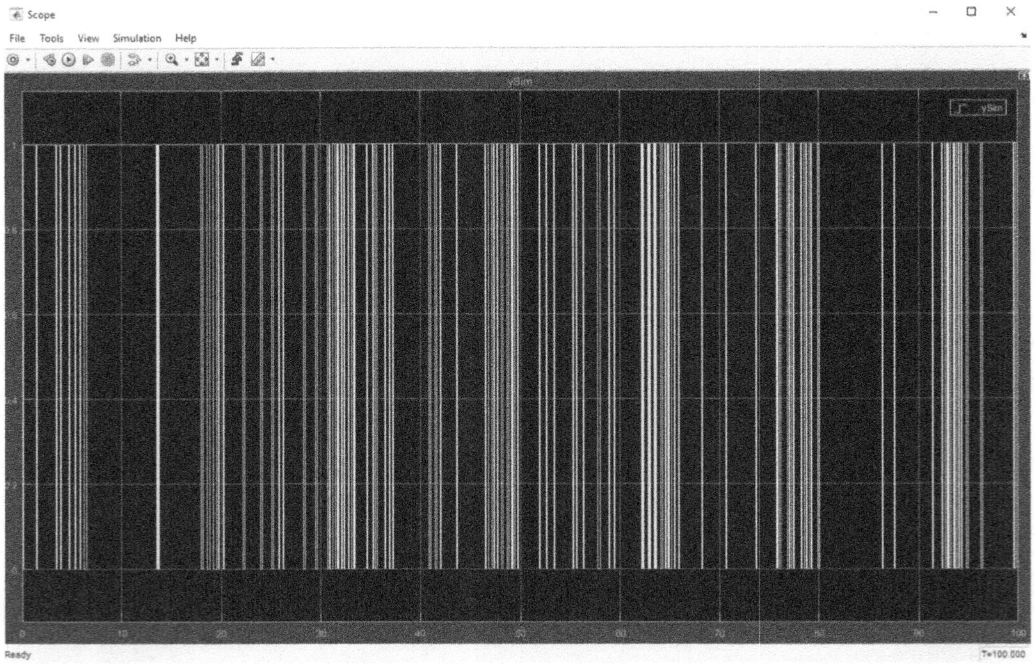

Figure 17.11 The prediction yprd (ySim) values on the scope.

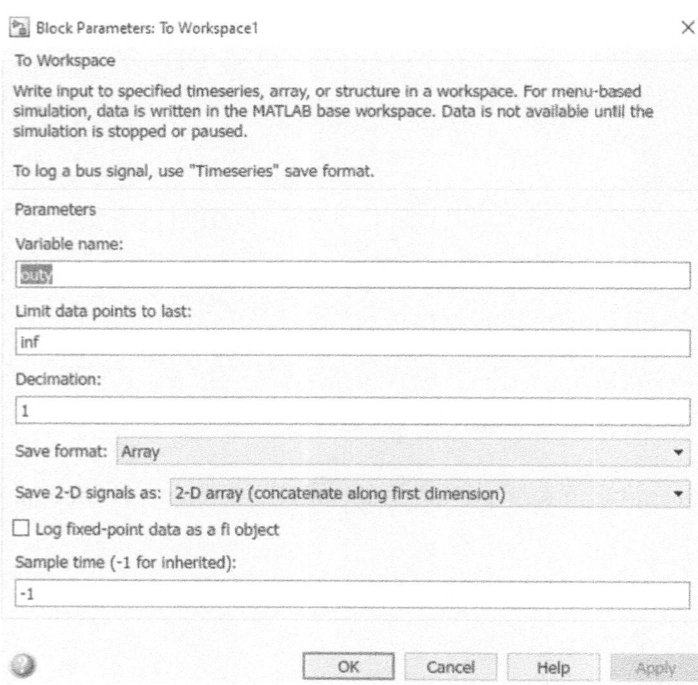

Figure 17.12 The “to workspace” block parameters for “outy”.

We can copy and paste these values in the Table data of Figure 17.13. By clicking on the linear plot in the Lookup Table editor window, we can see the linear plot of the sigmoid function in the interval [−6:0.5:6], as illustrated in Figure 17.14.

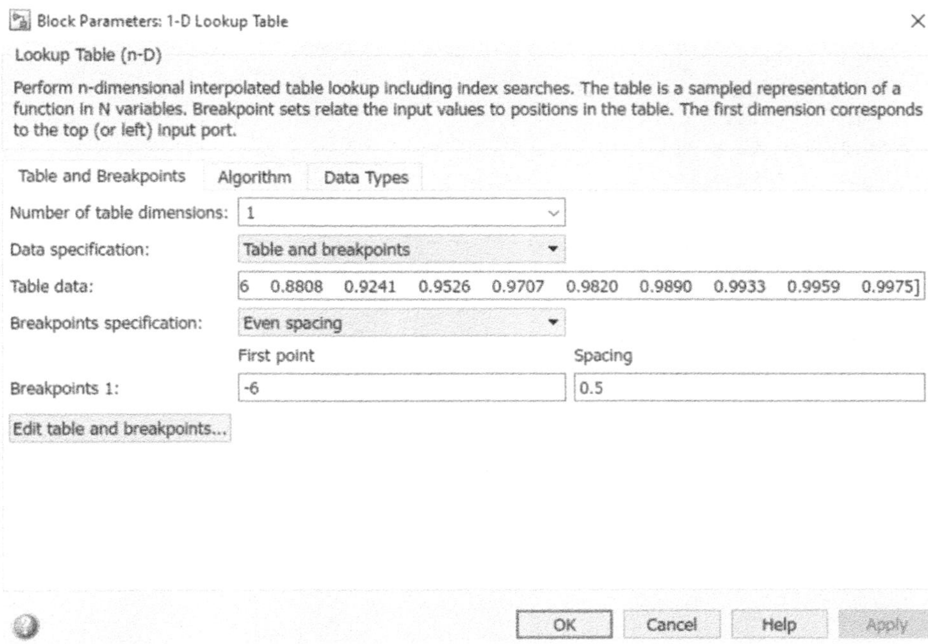

Figure 17.13 The block parameters for 1-D Lookup Table.

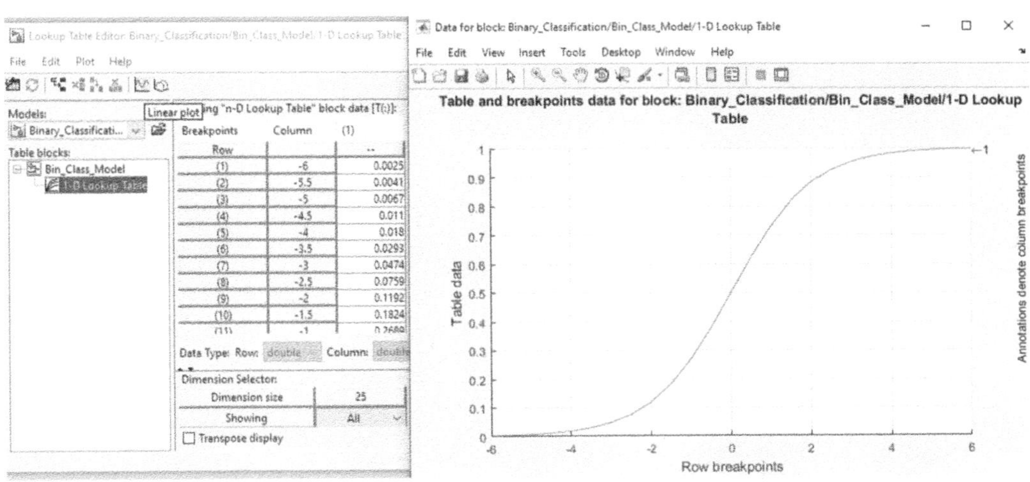

Figure 17.14 The linear plot of sigmoid function in the interval [−6:0.5:6].

Another solution is to use the piecewise linear approximation of the sigmoid function as a MATLAB function using the following codes.

```
function y = fcn(u)
u_in = fi(u,1,16,11);
if u_in < -3
    y = fi(0,1,16,11);
elseif (u_in >= -3 && u_in < -2)
    y = fi(0.072*u_in + 0.22,1,16,11);
elseif (u_in >= -2 && u_in < 2)
    y = fi(0.21*u_in + 0.5,1,16,11);
elseif (u_in >= 2 && u_in < 3)
    y = fi(0.072*u_in + 0.78,1,16,11);
else
    y = fi(1,1,16,11);
end
```

Since the HDL Coder cannot generate code for the signal from workspace block of "Thf", we use 22 constant blocks and a multiplexer instead, as shown in Figure 17.15.

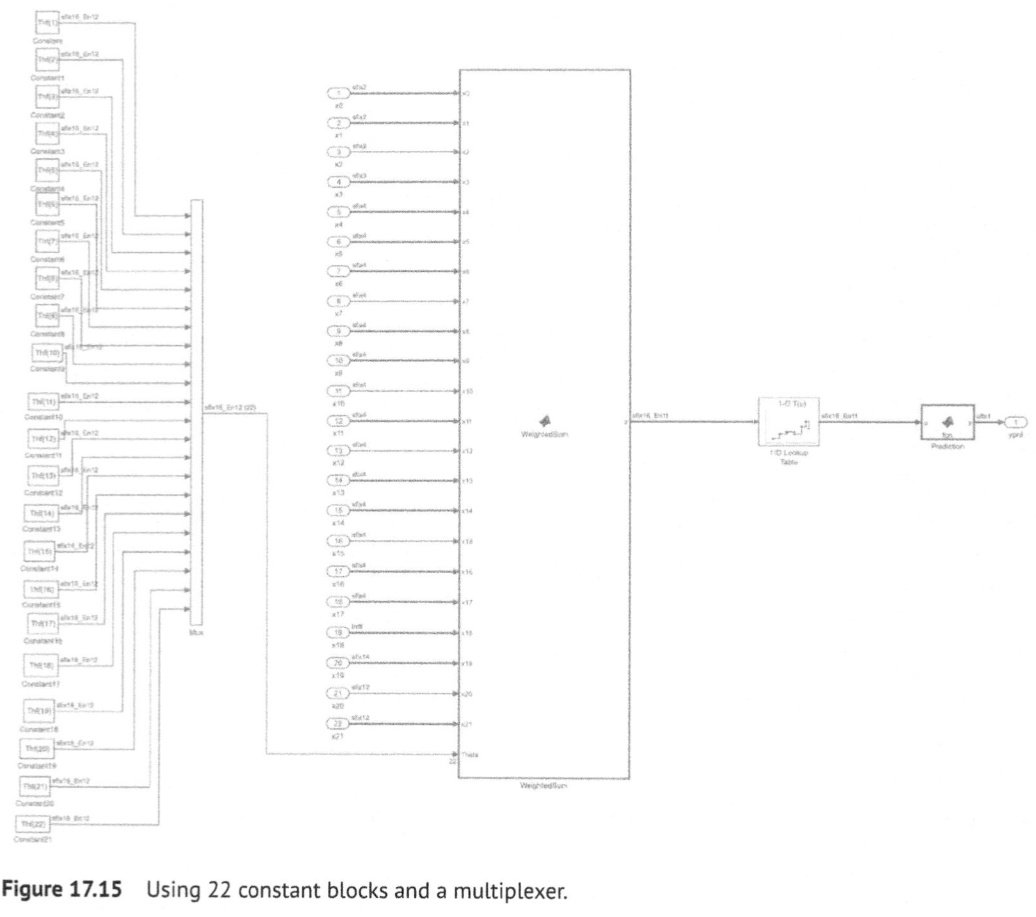

Figure 17.15 Using 22 constant blocks and a multiplexer.

The constant block parameters for the first element of "Thf" is depicted in Figure 17.16.

Next, we can generate the VHDL code for the subsystem "Bin_Class_Model" in Figure 17.15 using HDL Coder. To do this, we write the following code in the MATLAB command window.

```
>> hdlsetup('Binary_Classification')
```

Here, Binary_Classification is the name of the Simulink design. Then, we go to the Configuration Parameters window, set the Solver tab values as demonstrated in Figure 17.10, and configure the HDL Code Generation tab as depicted in Figure 17.17.

Then, we type the following code in the MATLAB command window.

```
>> checkhdl('Binary_Classification/Bin_Class_Model')
```

This opens the HDL Check Report window in a web browser, as shown in Figure 17.18.

Figure 17.16 The constant block parameters for Thf(1).

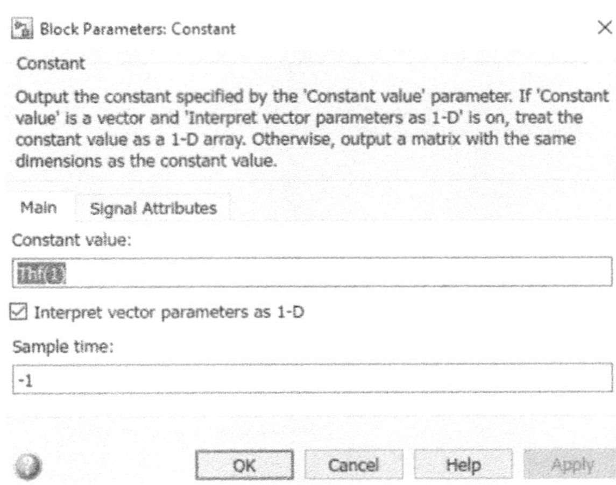

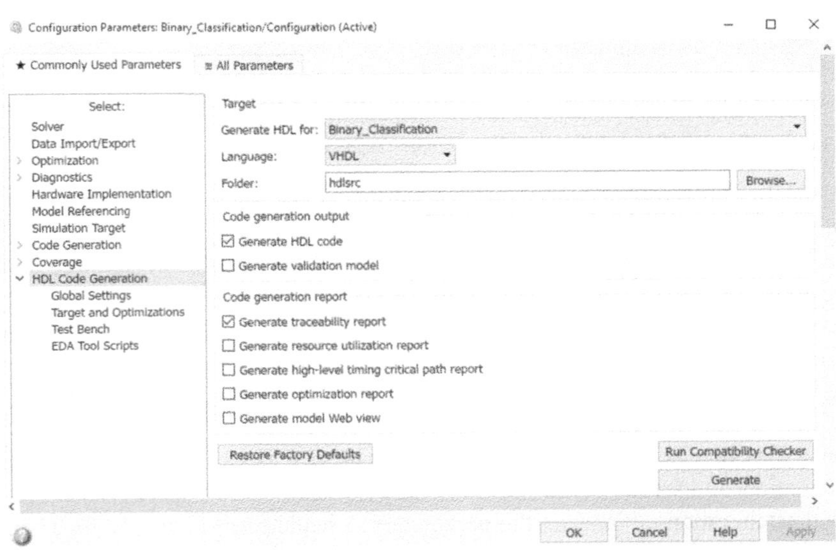

Figure 17.17 The HDL Code Generation tab of Configuration Parameters window.

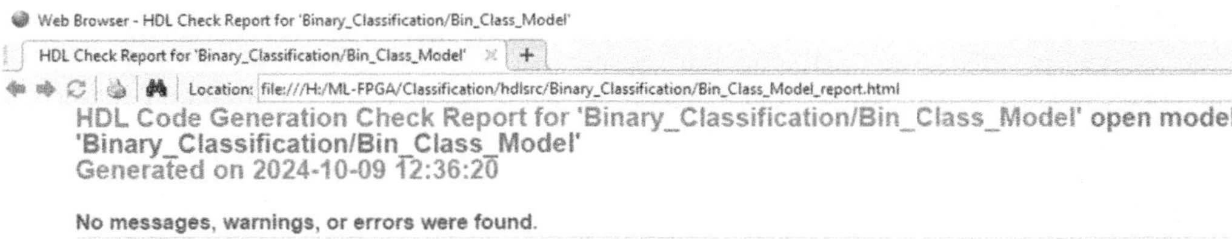

Figure 17.18 The HDL Check Report window.

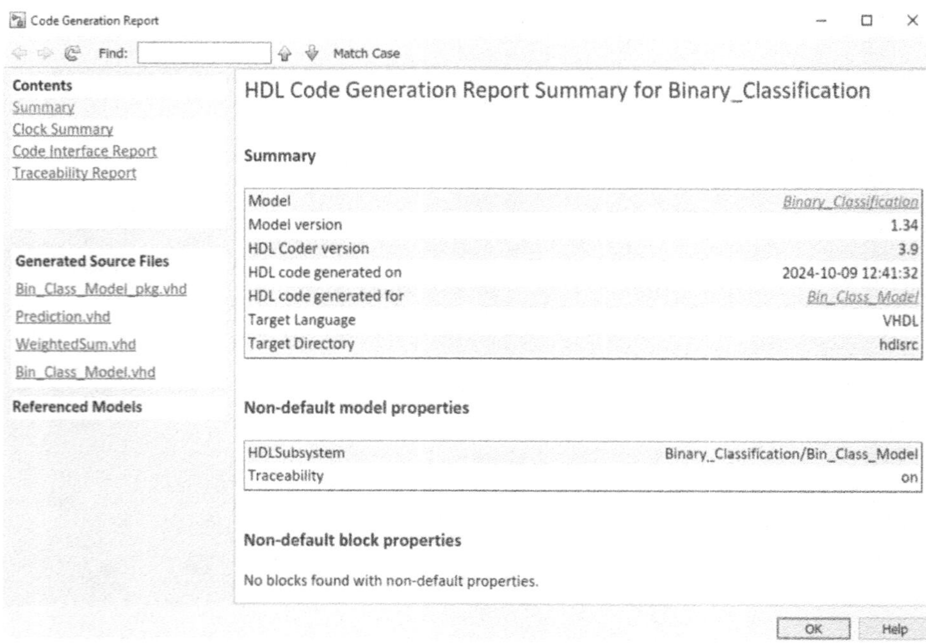

Figure 17.19 The Code Generation Report window.

Now, we use the following code for HDL code generation in the MATLAB command window.

```
>> makehdl('Binary_Classification/Bin_Class_Model')
```

The VHDL codes are generated successfully, and the Code Generation Report window is displayed, as depicted in Figure 17.19.

We can also use the piecewise linear approximation MATLAB function of sigmoid as mentioned earlier instead of 1-D Lookup Table, as illustrated in Figure 17.20.

Now, we can generate the VHDL code for the subsystem “Bin_Class_Model” in Figure 17.15 using HDL Coder. To do this, we write the following code in MATLAB command window.

```
>> hdlsetup('Binary_Classifier')
```

The Binary_Classifier is the name of the Simulink design. Then, we go to the Configuration Parameters window and set the Solver tab values as demonstrated in Figure 17.10. Next, we go to the HDL Code Generation tab and set the parameters as depicted in Figure 17.17. Then, we type the following code in the MATLAB command window.

```
>> checkhdl('Binary_Classifier/Bin_Class_Model')
```

This opens the HDL Check Report Window in a web browser, as shown in Figure 17.21.

Now, we use the following code for HDL code generation in the MATLAB command window.

```
>> makehdl('Binary_Classifier/Bin_Class_Model')
```

The VHDL codes are generated successfully, and the Code Generation Report window is displayed, as depicted in Figure 17.22.

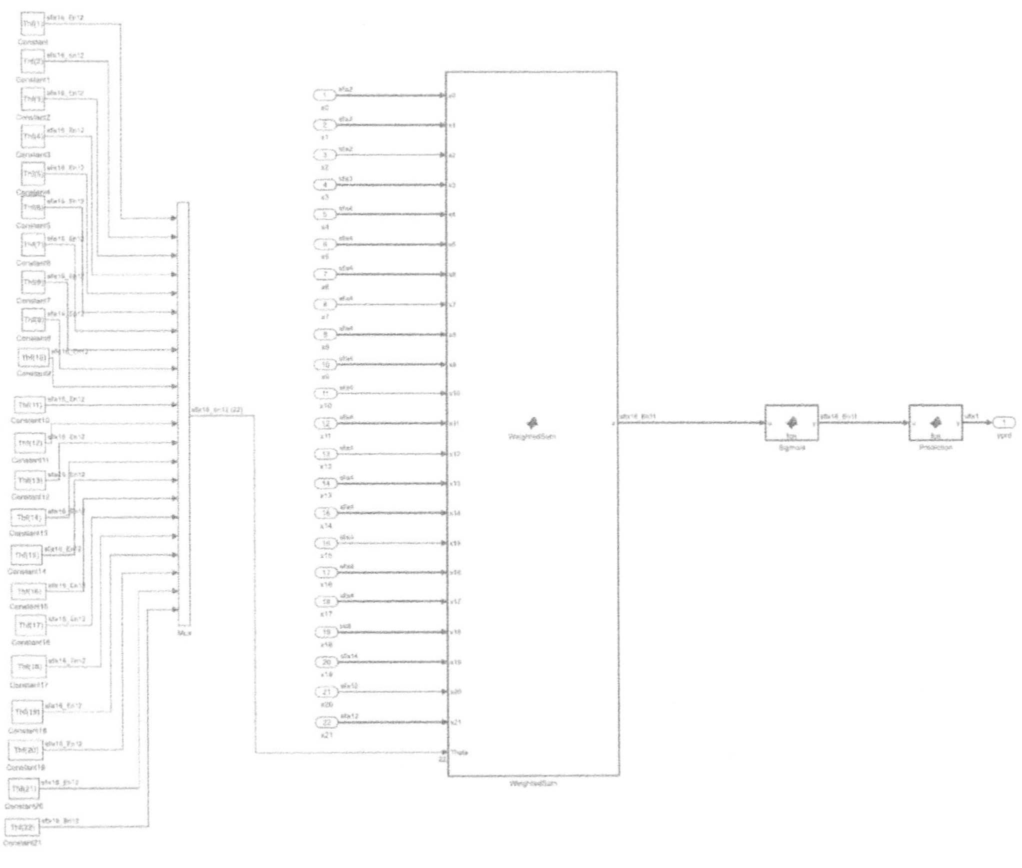

Figure 17.20 Using the piecewise linear approximation of sigmoid.

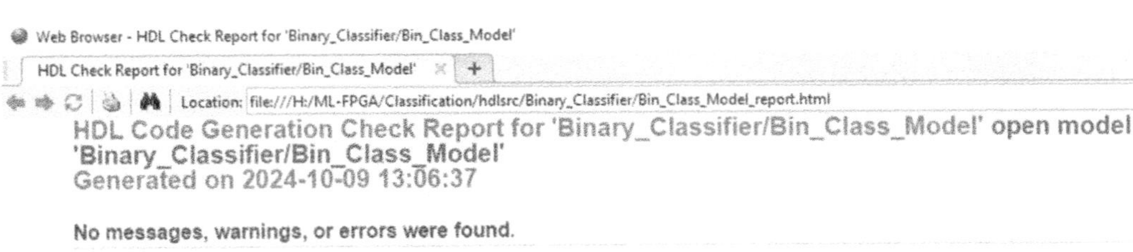

Figure 17.21 The HDL Check Report window.

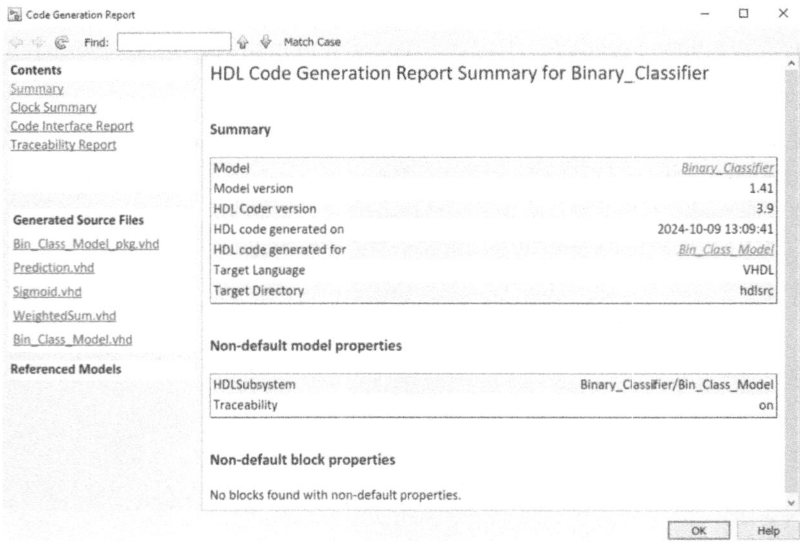

Figure 17.22 The Code Generation Report window.

17.6 Using Generated VHDL Code in Vivado

Now, we run Vivado and create a project named "Binary_Classification", then click the Next button. Next, we select RTL Project as the project type and click the Next button. In Add Sources window, before adding the generated VHDL files, we go ahead to copy and paste the generated VHDL files in Figure 17.19 into the Binary_Classification folder of the Vivado project. Now, by clicking on the Add Files button, we can add them to the project, as shown in Figure 17.23.

Next, we select our Xilinx development board and click the Next button. Finally, we click the Finish button to create the "Binary_Classification" project with added generated VHDL files. Then, we run synthesis, as illustrated in Figure 17.24. As demonstrated in Figure 17.24, the synthesis is successfully completed, and we can click the Cancel button. Then, we open the schematic in the RTL Analysis tab, as shown in Figure 17.25.

Now, we can simulate the project by selecting Flow → Run Simulation → Run Behavioral Simulation. In the opened simulation window, we click the Restart button. We write the following code in the MATLAB command window.

```
>> XTest(100,:)

ans =

  Columns 1 through 11

           1            1            0            1            2
2            2            2            5            4            4
  Columns 12 through 22

           4            4            4            4            3
4            4           44         3062            0            3
>> YTest(100)
ans =
```

```
        1

>> ypred(100)

ans =

        1
```

In the waveform window, we right-click on "x0" and select the "Force Constant..." option, then configure its settings, as illustrated in Figure 17.26.

We repeat the same process for x1 = 1, x2 = 0, x3 = 1, x4 = 2, x5 = 2, x6 = 2, x7 = 2, x8 = 5, x9 = 4, x10 = 4, x11 = 4, x12 = 4, x13 = 4, x14 = 4, x15 = 3, x16 = 4, x17 = 4, x18 = 44, x19 = 3062, x20 = 0, and x21 = 3. Next, we click the Run button for 10 ns once, and click the Zoom Fit button in the waveforms window to get the simulated waveforms, as illustrated in Figure 17.27.

By doing this, we achieve the same prediction using FPGA with the 1-D Lookup Table for the sigmoid function. Now, we create a new project named "Binary_Classifier" and then click the Next button. Then, we select project type as RTL Project and click the Next button. In Add Sources

Figure 17.23 Adding generated VHDL files to the Sources folder.

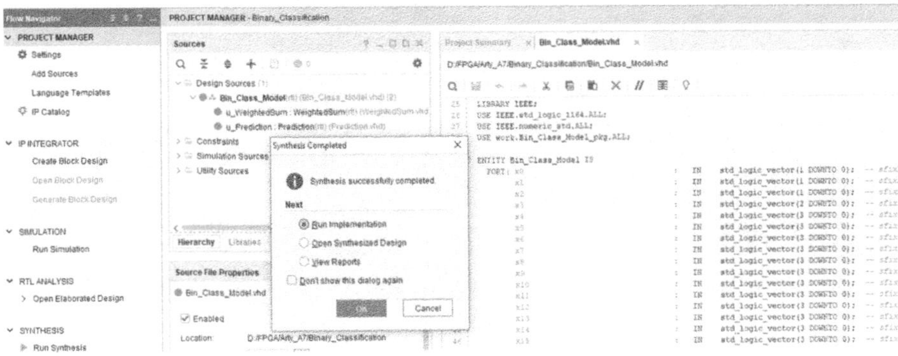

Figure 17.24 Successfully running the synthesis.

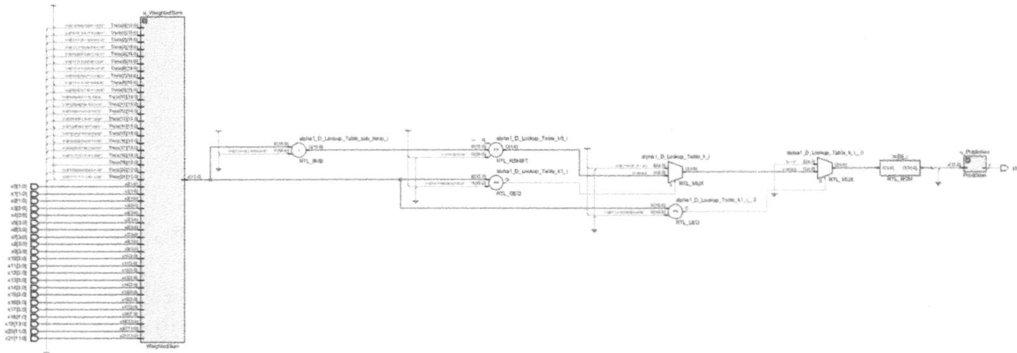

Figure 17.25 The schematic of "Bin_Class_Model".

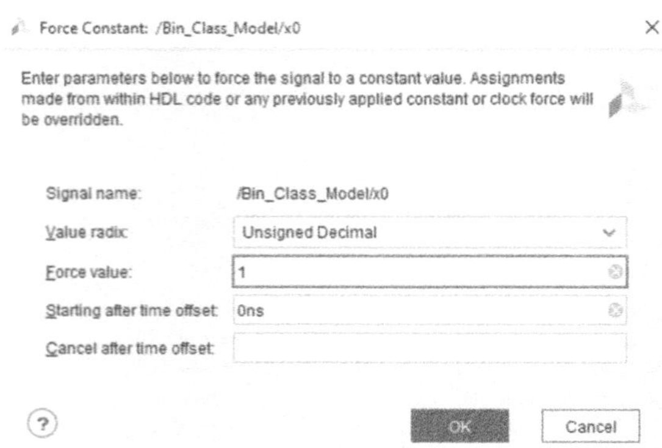

Figure 17.26 The Force Constant window for x0.

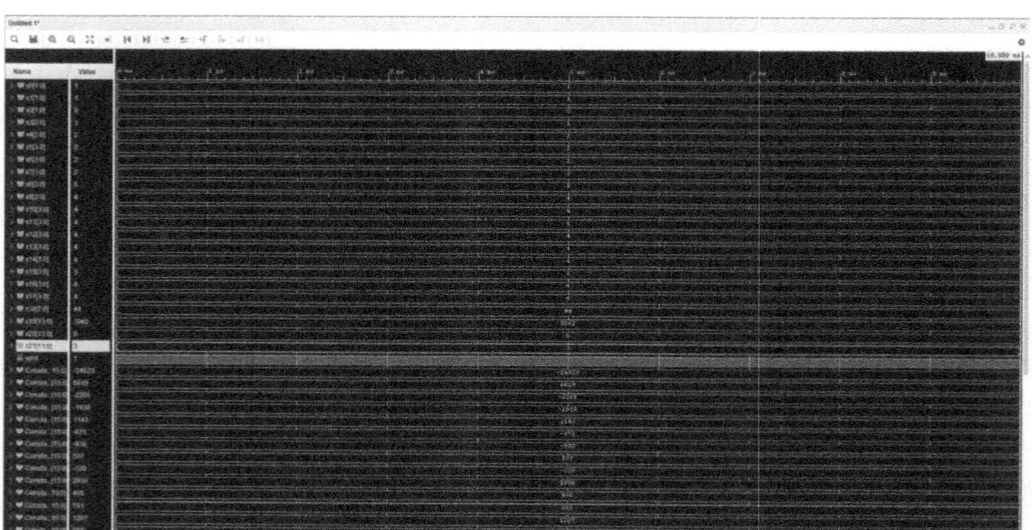

Figure 17.27 The simulated waveforms.

window, before adding the generated VHDL files, we copy and paste the generated VHDL files in Figure 17.22 into the folder Binary_Classifier of Vivado project. Now, by clicking on the Add Files button, we can add them to the project, as shown in Figure 17.28.

Next, we select our Xilinx development board and click the Next button. Finally, we click on the Finish button to create the "Binary_Classifier" project with added generated VHDL files. Then, we run synthesis as illustrated in Figure 17.29. As demonstrated in Figure 17.29, the synthesis is successfully completed, and we can click on the Cancel button. Then, we open the schematic in the RTL Analysis tab, as shown in Figure 17.30.

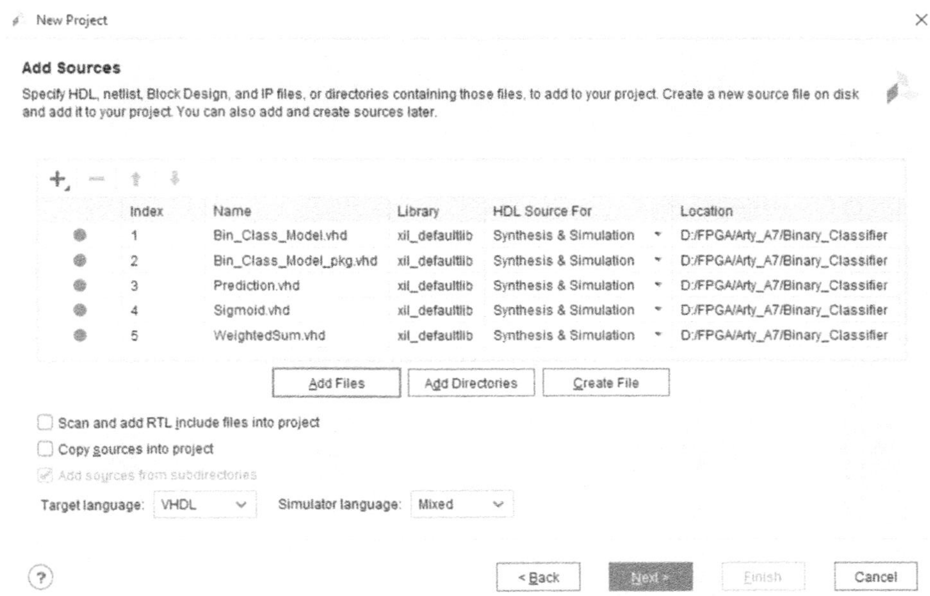

Figure 17.28 Adding generated VHDL files to the Sources folder.

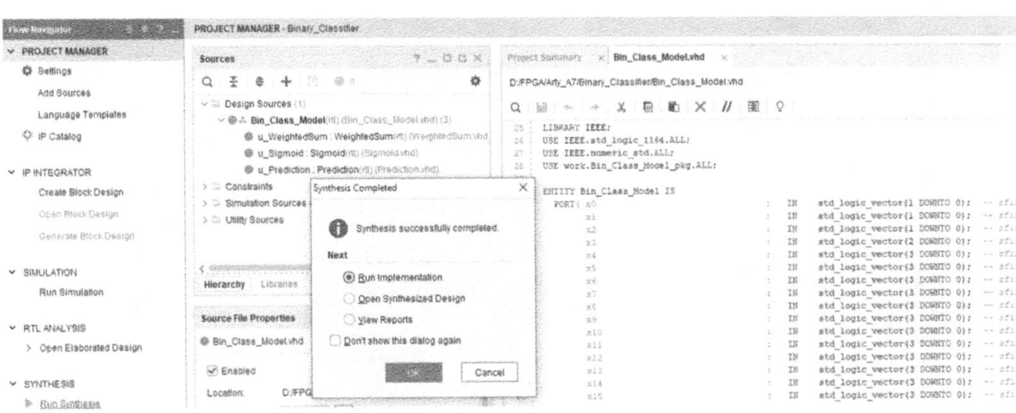

Figure 17.29 Successfully running the synthesis.

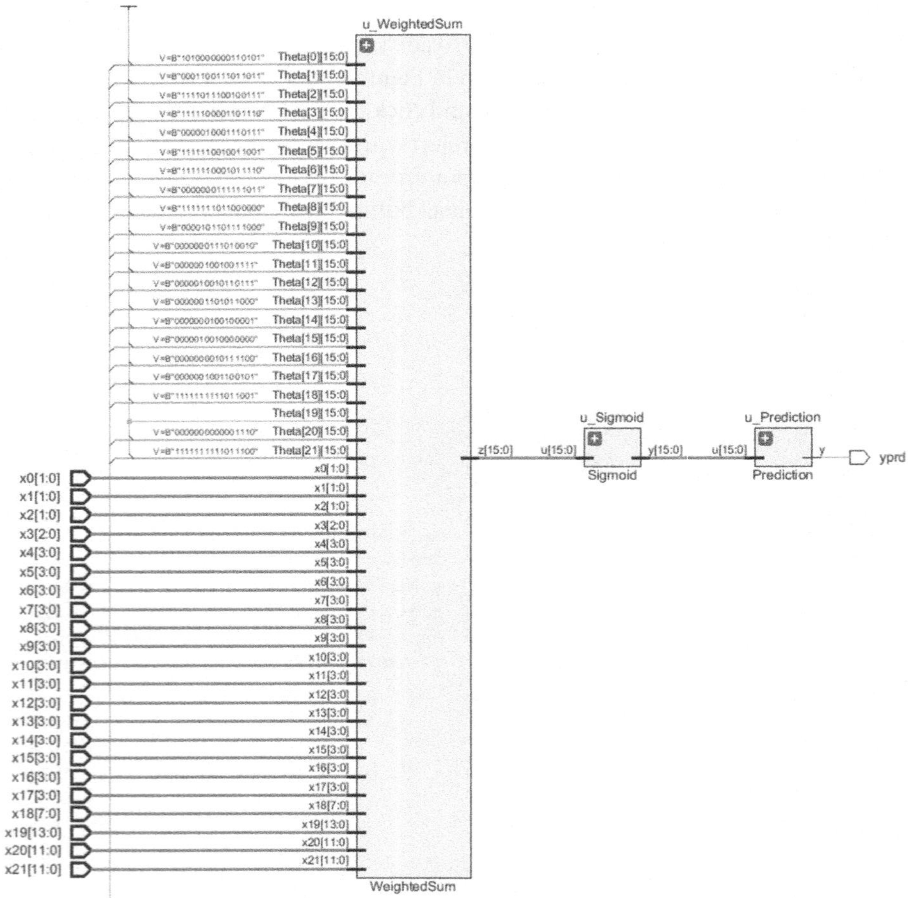

Figure 17.30 The schematic of "Bin_Class_Model".

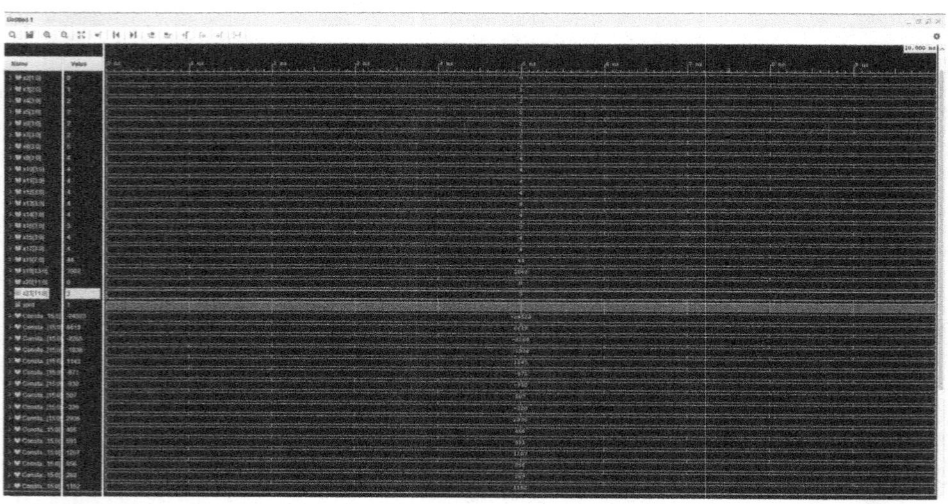

Figure 17.31 The simulated waveforms.

Now, we can simulate the project by selecting Flow → Run Simulation → Run Behavioral Simulation. In the opened simulation window, we click the Restart button. Then, in the waveforms window, we right-click on "x0", select "Force Constant..." option, and configure its settings, as illustrated in Figure 17.26. We repeat the same process for x1 = 1, x2 = 0, x3 = 1, x4 = 2, x5 = 2, x6 = 2, x7 = 2, x8 = 5, x9 = 4, x10 = 4, x11 = 4, x12 = 4, x13 = 4, x14 = 4, x15 = 3, x16 = 4, x17 = 4, x18 = 44, x19 = 3062, x20 = 0, and x21 = 3. Then, we click the Run button for 10 ns once and click on the Zoom Fit button in waveforms window to get the simulated waveforms, as illustrated in Figure 17.31.

This completes the process of achieving the same prediction with FPGA using the piecewise linear approximation of the sigmoid function.

18

Deploying a Deep Neural Network on FPGA

18.1 Introduction

In this section, we are going to deploy a deep neural network (DNN) model on an FPGA using Simulink and Vivado. First of all, we download a dataset from Kaggle that is suitable for deploying a DNN, and using MATLAB, we train the model based on minimizing the cost function. We use the diabetes dataset to predict whether the patient is diabetic or not based on various input parameters. Then, we implement the trained DNN model using HDL Coder and generate synthesizable VHDL code. Finally, using Vivado, we run synthesis analysis and simulation to verify the designed VHDL project.

18.2 The Kaggle Dataset

In this chapter, we will use the diabetes dataset from Kaggle to predict whether the patient is diabetic or not based on various input parameters. The weblink for downloading the diabetes dataset is as follows:

https://www.kaggle.com/datasets/mathchi/diabetes-data-set

An overview of the diabetes dataset in Excel is shown in Figure 18.1.

18.3 Introduction to Neural Networks

A neural network is a computational model inspired by the structure and function of the human brain. It is a fundamental component of deep learning, a subfield of artificial intelligence (AI). The neural networks are designed to recognize patterns and make intelligent decisions, much like the human brain does. Here are some key concepts and components of neural networks:

1. Neuron: The basic building blocks of a neural network are artificial neurons, also known as nodes or units. These neurons receive inputs, perform computations, and produce outputs.
2. Layers: Neurons are organized into layers within a neural network. There are typically three types of layers:
   a. Input Layer: This layer receives the initial data or features that are fed into the network. Each neuron in the input layer represents a feature or input variable.

Fundamentals of VHDL for FPGA Programming Using Vivado, First Edition. Majid Pakdel.

Companion Website: www.wiley.com/go/pakdelVHDL

	A	B	C	D	E	F	G	H	I
1	Pregnancies	Glucose	BloodPressure	SkinThickness	Insulin	BMI	DiabetesPedigreeFunction	Age	Outcome
2	6	148	72	35	0	33.6	0.627	50	1
3	1	85	66	29	0	26.6	0.351	31	0
4	8	183	64	0	0	23.3	0.672	32	1
5	1	89	66	23	94	28.1	0.167	21	0
6	0	137	40	35	168	43.1	2.288	33	1
7	5	116	74	0	0	25.6	0.201	30	0
8	3	78	50	32	88	31	0.248	26	1
9	10	115	0	0	0	35.3	0.134	29	0
10	2	197	70	45	543	30.5	0.158	53	1

Figure 18.1 Overview of the diabetes dataset.

   b. Hidden Layers: These layers lie between the input and output layers and are responsible for processing and learning patterns in the data. DNNs have multiple hidden layers, hence the term "deep learning."
   c. Output Layer: The final layer of neurons produces the network's output, which could be a classification, regression, or some other task-specific result.
3. Weights and Biases: Each connection between neurons has associated weights and biases. These parameters are learned during training and determine the strength of the connections between neurons.
4. Activation Function: An activation function is applied to the weighted sum of inputs and biases in each neuron. It introduces nonlinearity into the network, allowing it to learn complex relationships in the data. Common activation functions include sigmoid, ReLU (rectified linear unit), and tanh (hyperbolic tangent).
5. Feedforward Propagation: The process of passing data through the network from the input layer to the output layer is called feedforward propagation. Neurons in each layer compute their activations based on inputs and weights, and this process continues through the layers until the final output is generated.
6. Backpropagation: To train a neural network, you need a labeled dataset and a loss function that measures the difference between the predicted output and the true target values. Backpropagation is an algorithm used to update the weights and biases in the network to minimize this loss. It works by propagating the error from the output layer backward through the hidden layers, adjusting the parameters along the way.
7. Training: Training a neural network involves iteratively presenting the training data to the network, computing the loss, and using backpropagation to update the model's parameters. This process continues until the model's performance converges to an acceptable level.

Neural networks can be used for a wide range of tasks, including image and speech recognition, natural language processing, recommendation systems, and many others. They have demonstrated impressive performance in various fields and have contributed to significant advancements in AI and machine learning. A typical DNN is shown in Figure 18.2.

The mathematical model of a neuron is depicted in Figure 18.3.

In the term w_{ki}, k represents the layer number and i represents the instance number. We also have the following equation.

$$u = \sum_{i=1}^{m}(w_{ki}x_i) + b_k \tag{18.1}$$

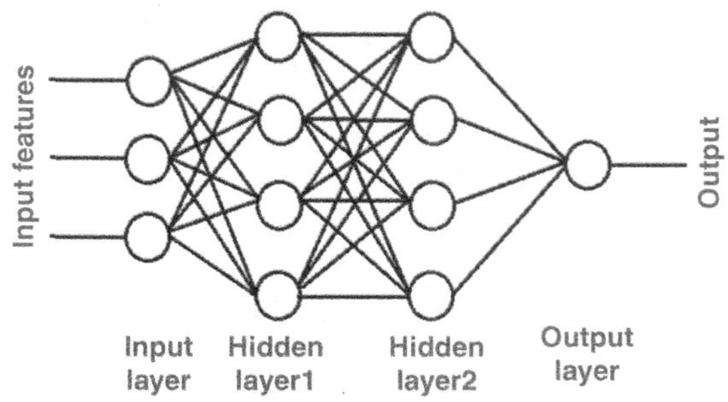

Figure 18.2 A typical deep neural network.

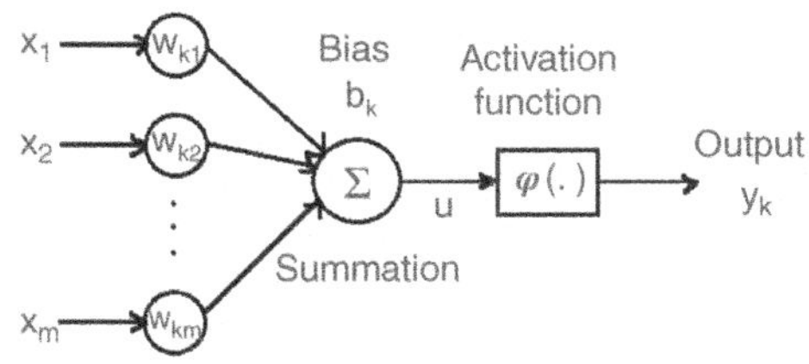

Figure 18.3 The neuron mathematical model.

So, we can rewrite Equation (18.1) as follows.

$$u = \begin{bmatrix} w_{k1} \; w_{k2} \; w_{k3} \ldots w_{km} \end{bmatrix} \begin{bmatrix} x_1 \\ x_2 \\ x_3 \\ \ldots \\ x_m \end{bmatrix} + b_k \tag{18.2}$$

Then, we will have the following equation.

$$u = (\mathbf{w_k})^T \mathbf{x} + b_k \tag{18.3}$$

Note: In the matrix representation of the weighted sum u, the weights of layers should be arranged as a row vector, while input instances should be arranged as a column vector. Once the weighted sum u is calculated, the neuron's output can be computed by passing u to the activation function, as demonstrated in the following equation.

$$y_k = \varphi(u) \tag{18.4}$$

18.4 Piecewise Linear Approximation of Activation Functions

The sigmoid, tanh, and ReLU are common activation functions used in artificial neural networks. Each of these functions introduces nonlinearity into the network's computations, allowing it to learn complex relationships in the data.

1. Sigmoid Activation Function: The sigmoid function is defined by the following equation:

$$\sigma(x) = \varphi(x) = \frac{1}{1 + e^{-x}} \tag{18.5}$$

The piecewise linear approximation of sigmoid function is given by the following equation:

$$\sigma(x) = \begin{cases} 0 & x < -3 \\ 0.072x + 0.22 & -3 \le x < -2 \\ 0.21x + 0.5 & -2 \le x < 2 \\ 0.072x + 0.78 & 2 \le x < 3 \\ 1 & x \ge 3 \end{cases} \tag{18.6}$$

We use the following MATLAB m-file for drawing actual and piecewise linear approximation of sigmoid function.

```
x = -5:0.1:5;
z = 1./(1+exp(-x));
y = zeros(1,length(x));
for i = 1:length(x)
   j = -5.1+0.1*i;
   y(i) = sigmoid(j);
end
plot(x,y,'r','LineWidth',2);
hold on;
grid on;
plot(x,z,'b','LineWidth',2);
legend('Approx','Actual');
title('Piecewise Linear Approximation of Sigmoid Function');
xlabel('x');
ylabel('y');
```

The function "sigmoid" is defined by the following codes.

```
function y = sigmoid(u)
if u < -3
    y = 0;
elseif (u >= -3 && u < -2)
    y = 0.072*u + 0.22;
elseif (u >= -2 && u < 2)
    y = 0.21*u + 0.5;
elseif (u >= 2 && u < 3)
    y = 0.072*u + 0.78;
else
    y = 1;
end
```

The actual and piecewise linear approximation plots for sigmoid function are illustrated in Figure 18.4.

2. Tanh Activation Function: The tanh function is defined by the following equation:

$$\tanh = \varphi(x) = \frac{e^x - e^{-x}}{e^x + e^{-x}} \tag{18.7}$$

The piecewise linear approximation of tanh function is given by the following equation:

$$\tanh = \begin{cases} -1 & x < -3 \\ 0.18x - 0.64 & -2 \le x < -1 \\ 0.82x & -1 \le x < 1 \\ 0.18x + 0.64 & 1 \le x < 2 \\ 1 & x \ge 2 \end{cases} \tag{18.8}$$

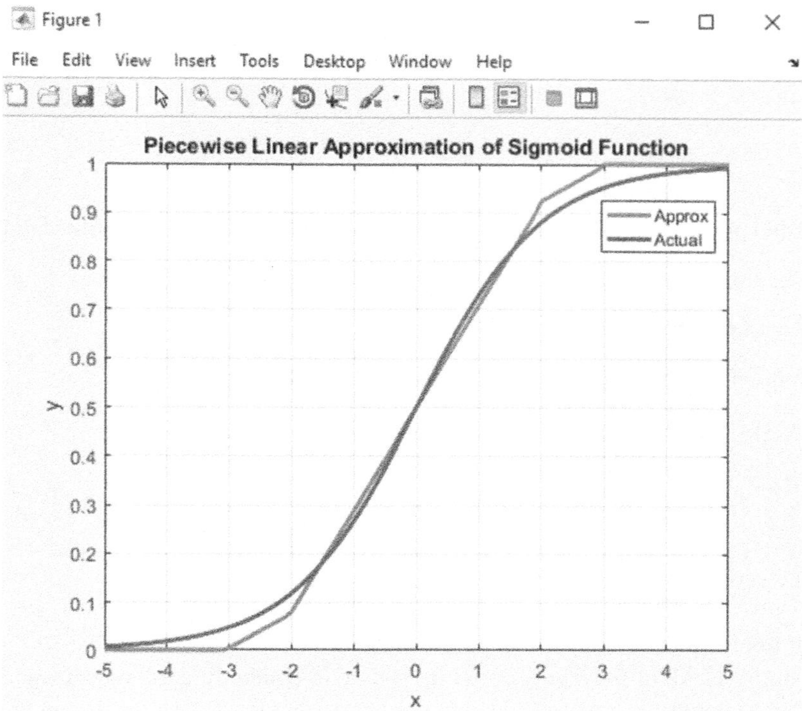

Figure 18.4 The actual and approximate plots for sigmoid function.

We use the following MATLAB m-file for drawing actual and piecewise linear approximation of tanh function.

```
x = -5:0.1:5;
z = (exp(x)-exp(-x))./(exp(x)+exp(-x));
y = zeros(1,length(x));
for i = 1:length(x)
   j = -5.1+0.1*i;
   y(i) = Tanh(j);
end
plot(x,y,'r','LineWidth',2);
hold on;
grid on;
plot(x,z,'b','LineWidth',2);
legend('Approx','Actual');
title('Piecewise Linear Approximation of Tanh Function');
xlabel('x');
ylabel('y');
```

The function "Tanh" is defined by the following codes.

```
function y = Tanh(u)
if u < -2
    y = -1;
elseif (u >= -2 && u < -1)
    y = 0.18*u - 0.64;
```

```
elseif (u >= -1 && u < 1)
    y = 0.82*u;
elseif (u >= 1 && u < 2)
    y = 0.18*u + 0.64;
else
    y = 1;
end
```

The actual and piecewise linear approximation plots for tanh function are illustrated in Figure 18.5.

1. ReLU: The ReLU function is defined by the following equation:

$$\text{ReLU} = \varphi(x) = \max\,(0,x) \tag{18.9}$$

We use the following MATLAB m-file for drawing the ReLU function:

```
x = -10:0.1:10;
z = max(0,x);
plot(x,z,'b','LineWidth',2);
title('ReLU = max(0,x)');
xlabel('x');
ylabel('y');
grid on;
```

The plot of ReLU function is shown in Figure 18.6.

The following MTLAB m-file (sig_approx1.m) is used to obtain the piecewise linear approximation of sigmoid function.

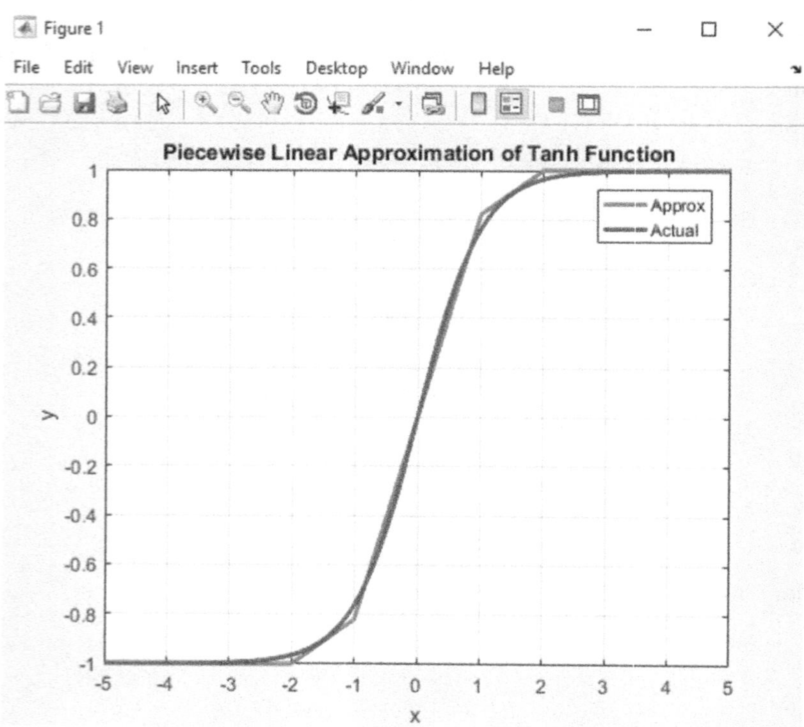

Figure 18.5 The actual and approximate plots for tanh function.

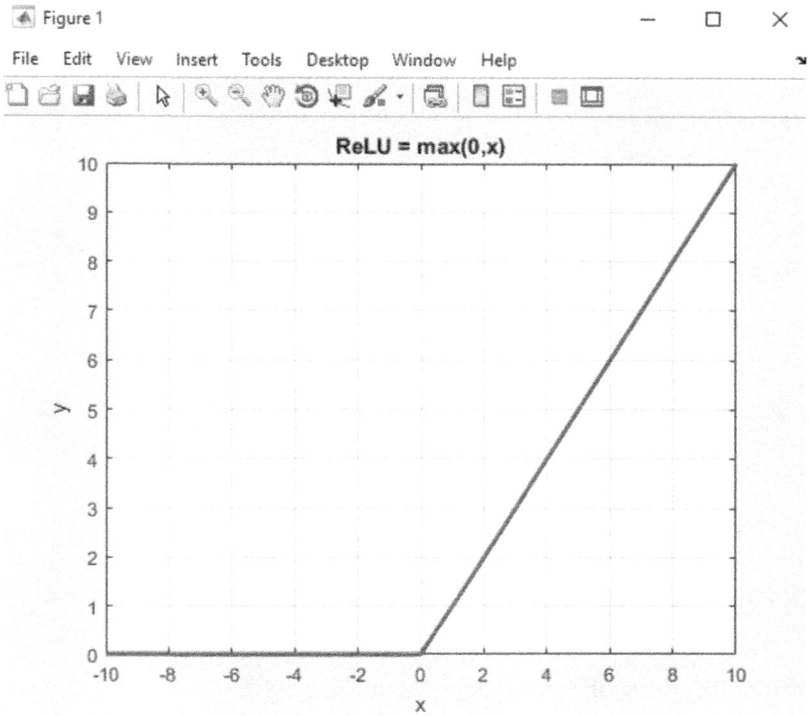

Figure 18.6 The plot of ReLU function.

```
clear all; close all; clc
degree = 1;
dx = 0.1;
f = @(x) 1./(1 + exp(-x));
x = -5:dx:5;
y =f(x);
plot(x,y);
grid on;
xlabel('x');

x2 = -3:dx:-2;
y2 = f(x2);
p2 = polyfit(x2, y2, degree);

x3 = -2:dx:2;
y3 = f(x3);
p3 = polyfit(x3, y3, degree);

x4 = 2:dx:3;
y4 = f(x4);
p4 = polyfit(x4, y4, degree);

zf2 = polyval(p2,x2);
zf3 = polyval(p3,x3);
zf4 = polyval(p4,x4);
```

```
x1 = -10:dx:-3;
zf1 = zeros(1,length(x1));

x5 = 3:dx:10;
zf5 = ones(1,length(x5));

x = -10:dx:10;
ya = f(x);

plot(x1,zf1,'LineWidth',2);
hold on;
plot(x2,zf2,'LineWidth',2);
plot(x3,zf3,'LineWidth',2);
plot(x4,zf4,'LineWidth',2);
plot(x5,zf5,'LineWidth',2);

plot(x,ya,'k','LineWidth',1);
grid on;
legend('y1','y2','y3','y4','y5','sig');
xlabel('x');
ylabel('y');
title('Piecewise Linear Approximation of Sigmoid');
xlim([-5 5]);
```

If we run that m-file, we get the plot as depicted in Figure 18.7.

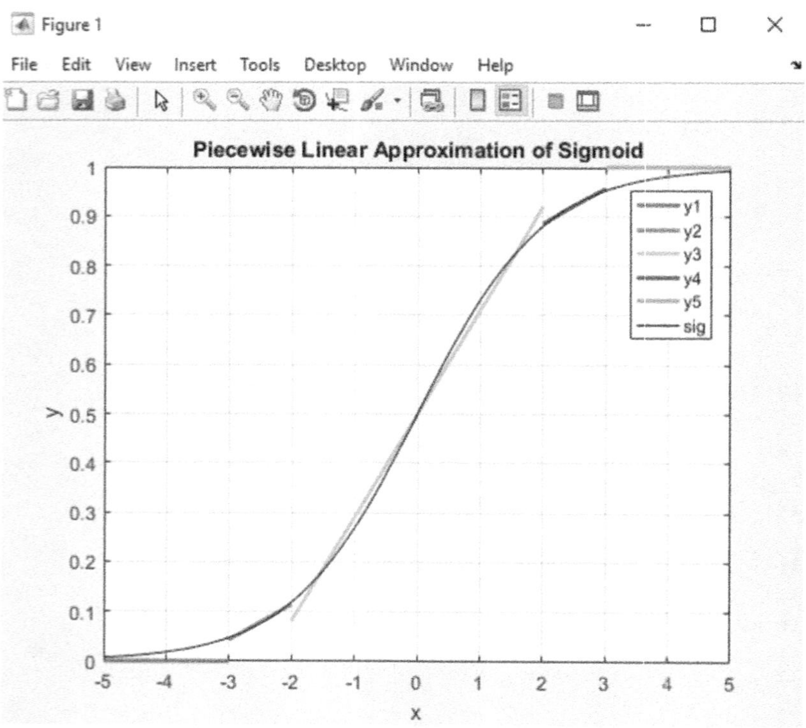

Figure 18.7 The piecewise linear approximation of sigmoid function.

If we type p2, p3, and p4 in the MATLAB command window and press the enter button, we get the following results.

```
>> p2
p2 =
    0.0713    0.2571
>> p3
p3 =
    0.2094    0.5000
>> p4
p4 =
    0.0713    0.7429
```

Here, the first element represents the slope, and the second element represents the offset of the line equation. We have 0.0713 ≈ 0.072, 0.2571 ≈ 0.22, 0.2094 ≈ 0.21, and 0.7429 ≈ 0.78 in Equation (18.6). Now, to implement the piecewise linear approximation of the sigmoid function in HDL Coder, we use the following m-file (sig_approx2.m).

```
clear all; close all; clc
dx = 0.01;
f = @(x) 1./(1 + exp(-x));
x = -5:dx:5;

for k = 1:length(x)
    y(k) = sig_pw_HDL(x(k));
end
plot(x,y,'LineWidth',2);
grid on;
hold on;

ya = f(x);
plot(x,ya,'LineWidth',2);
ylim([-0.1 1.1]);
xlabel('x');
ylabel('y');
legend('Approx','Actual');
title('Piecewise Linear Approximation of Sigmoid');
```

Here, the function sig_pw_HDL is defined by the following codes.

```
function ysig = sig_pw_HDL(x)
WL = 16;
FL = 10;

F = hdlfimath();
nt = numerictype(1,WL,FL);

xf = fi(x,nt);

m1 = fi(0.072,nt);
c1 = fi(0.22,nt);

m2 = fi(0.21,nt);
c2 = fi(0.5,nt);
```

```
m3 = fi(0.072,nt);
c3 = fi(0.78,nt);

if xf < fi(-3,nt)
    y = fi(0,nt,F);
elseif xf < fi(-2,nt)
    y = fi(m1*x+c1,nt,F);
elseif ((xf >= fi(-2,nt)) && (xf <= fi(2,nt)))
    y = fi(m2*x+c2,nt,F);
elseif xf < fi(3,nt)
    y = fi(m3*x+c3,nt,F);
else
    y = fi(1,nt,F);
end

if y <= fi(0,nt)
    ysig = fi(0,nt,F);
elseif y >= fi(1,nt)
    ysig = fi(1,nt,F);
else
    ysig = fi(y,nt,F);
end
```

If we run the m-file (sig_approx2.m), we get the plot as illustrated in Figure 18.8, which is exactly the same as the plot in Figure 18.4.

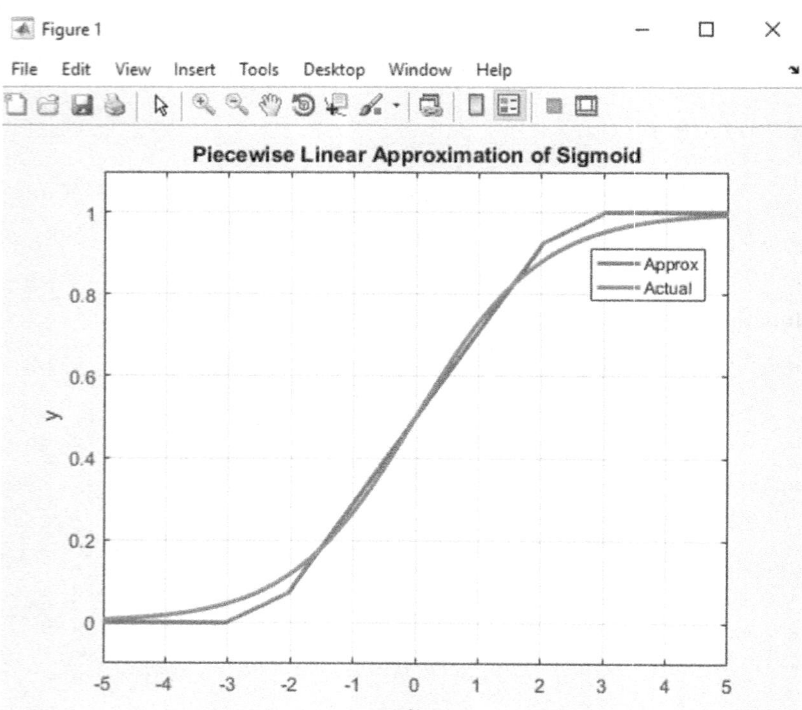

Figure 18.8 The HDL Coder piecewise linear approximation of sigmoid.

The same process is used for piecewise linear approximation of tanh function. The following MTLAB m-file (tanh_approx1.m) is used to obtain the piecewise linear approximation of tanh function.

```
clear all; close all; clc
degree = 1;
dx = 0.1;
f = @(x) (exp(x) - exp(-x))./(exp(x) + exp(-x));
x = -5:dx:5;
y =f(x);
plot(x,y);
grid on;
xlabel('x');

x2 = -2:dx:-1;
y2 = f(x2);
p2 = polyfit(x2, y2, degree);

x3 = -1:dx:1;
y3 = f(x3);
p3 = polyfit(x3, y3, degree);

x4 = 1:dx:2;
y4 = f(x4);
p4 = polyfit(x4, y4, degree);

zf2 = polyval(p2,x2);
zf3 = polyval(p3,x3);
zf4 = polyval(p4,x4);

x1 = -10:dx:-2;
zf1 = -1*ones(1,length(x1));

x5 = 2:dx:10;
zf5 = ones(1,length(x5));

x = -10:dx:10;
ya = f(x);

plot(x1,zf1,'LineWidth',2);
hold on;
plot(x2,zf2,'LineWidth',2);
plot(x3,zf3,'LineWidth',2);
plot(x4,zf4,'LineWidth',2);
plot(x5,zf5,'LineWidth',2);

plot(x,ya,'k','LineWidth',1);
grid on;
legend('y1','y2','y3','y4','y5','tanh');
xlabel('x');
ylabel('y');
title('Piecewise Linear Approximation of Tanh');
xlim([-5 5]);
```

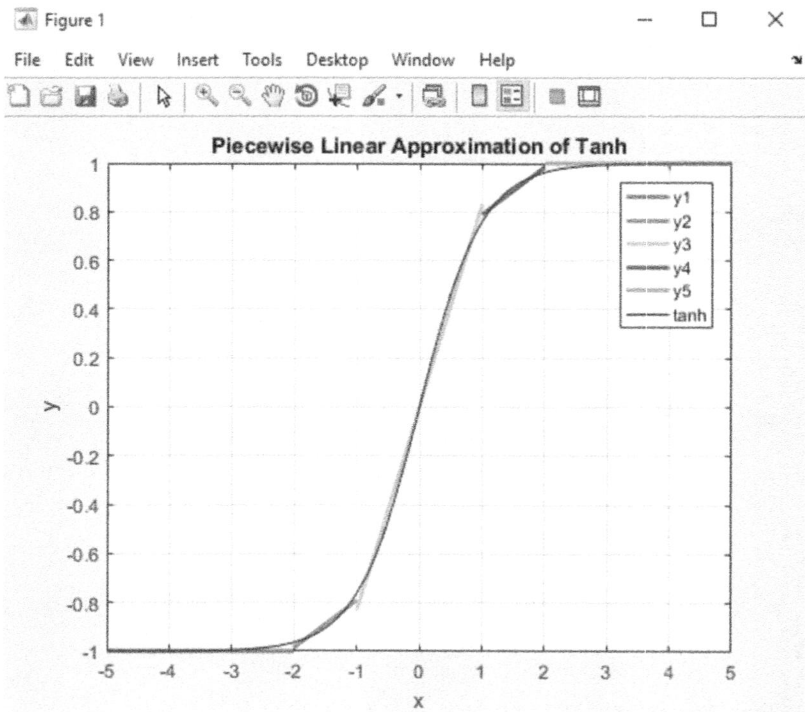

Figure 18.9 The piecewise linear approximation of tanh function.

If we run the m-file (tanh_approx1.m), we get the plot as shown in Figure 18.9.

If we type p2, p3, and p4 in the MATLAB command window and press the enter button, we get the following results.

```
>> p2
p2 =
    0.1962   -0.5941
>> p3
p3 =
    0.8319   -0.0000
>> p4
p4 =
    0.1962    0.5941
```

Here, the first element represents the slope, and the second element represents the offset of the line equation. We have $0.1962 \approx 0.18$, $-0.5941 \approx -0.64$, $0.8319 \approx 0.82$, and $0.5941 \approx 0.64$ in Equation (18.8). Now, for using piecewise linear approximation of tanh function in the HDL Coder, we use the following m-file (tanh_approx2.m).

```
clear all; close all; clc

f = @(x) (exp(x) - exp(-x))./(exp(x) + exp(-x));
x = linspace(-5,5,200);
```

```
for k = 1:length(x)
    y(k) = tanh_pw_HDL(x(k));
end
plot(x,y,'LineWidth',2);
grid on;
hold on;

ya = f(x);
plot(x,ya,'LineWidth',2);
ylim([-1.1 1.1]);
xlabel('x');
ylabel('y');
legend('Approx','Actual');
title('Piecewise Linear Approximation of Tanh');
```

Here, the function tanh_pw_HDL is defined by the following codes.

```
function ysig = tanh_pw_HDL(x)
WL = 16;
FL = 10;

F = hdlfimath();
nt = numerictype(1,WL,FL);

xf = fi(x,nt);

m1 = fi(0.18,nt);
c1 = fi(-0.64,nt);

m2 = fi(0.82,nt);
c2 = fi(0,nt);

m3 = fi(0.18,nt);
c3 = fi(0.64,nt);

if xf < fi(-2,nt)
    y = fi(-1,nt,F);
elseif xf < fi(-1,nt)
    y = fi(m1*x+c1,nt,F);
elseif ((xf >= fi(-1,nt)) && (xf < fi(1,nt)))
    y = fi(m2*x+c2,nt,F);
elseif xf < fi(2,nt)
    y = fi(m3*x+c3,nt,F);
else
    y = fi(1,nt,F);
end

if y <= fi(-1,nt)
    ysig = fi(-1,nt,F);
elseif y >= fi(1,nt)
    ysig = fi(1,nt,F);
else
    ysig = fi(y,nt,F);
end
```

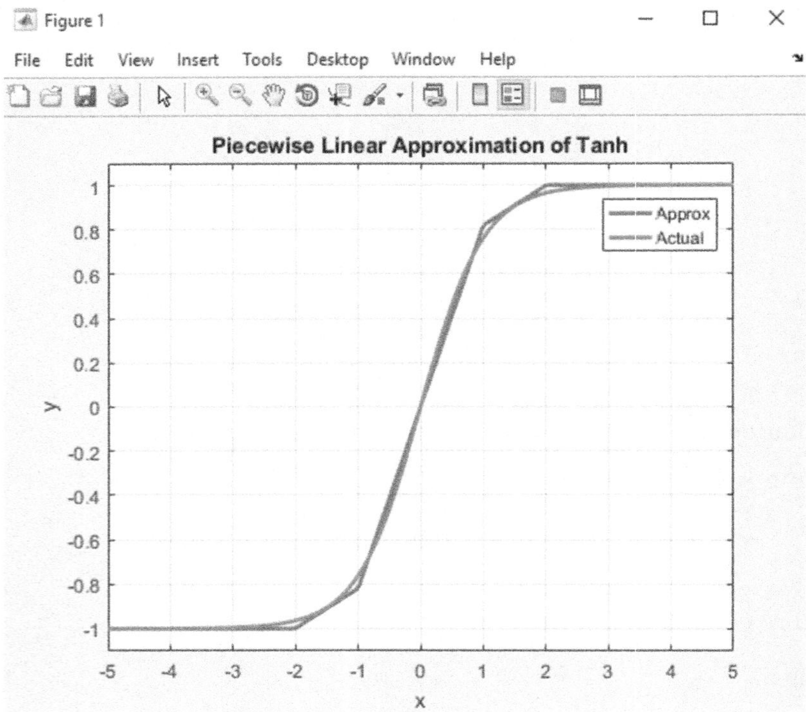

Figure 18.10 The HDL Coder piecewise linear approximation of tanh.

If we run the m-file (tanh_approx2.m), we get the plot as illustrated in Figure 18.10, which is exactly the same as the plot in Figure 18.5.

18.5 Training the Neural Network Model

The following MATLAB m-file (Train_NN_Model.m) is used for training the neural network model.

```
close all; clear all; clc
cd('H:\ML-FPGA\NN');
t = readtable('diabetes.csv');
%Histogram_Plots(t)

%Training and Testing Data
All_Data = table2array(t);
Xin = All_Data(:,1:end-1);
Yout = All_Data(:,end);

%Normalize the Data
X = mapminmax(Xin');
Y = mapminmax(Yout',0,1);
Len = length(Y);
```

```
%Training and Testing Data
Prob = 0.8;
NTr = round(Prob*Len);
XTrain = X(:,1:NTr);
YTrain = Y(:,1:NTr);

XTest = X(:,NTr+1:end);
YTest = Y(:,NTr+1:end);

XTest = XTest'; % Features Along Column
YTest = YTest';

%% Set Neural Network Parameters
% Define sizes
[R,C] = size(XTrain);
inputSize = R;
hiddenLayerSize1 = 3;
hiddenLayerSize2 = round(R/2);

% Define Neural Network
net = feedforwardnet([hiddenLayerSize1,hiddenLayerSize2]);

% Set Training Parameters
net.trainParam.epochs = 1000;
net.trainParam.goal = 1e-6;
net.trainParam.max_fail = 200;

% Specify the logsig transfer function for each layer
ActFun_Layaer1 = 'tansig'; %tanh
ActFun_Layaer2 = 'tansig'; %tanh
OutputLayer = 'logsig'; %sigmoid

net.layers{1}.transferFcn = ActFun_Layaer1; % 1st hidden layer
net.layers{2}.transferFcn = ActFun_Layaer2; % 2nd hidden layer
net.layers{3}.transferFcn = OutputLayer; % Output layer

% Train the Neural Network
size(XTrain)
size(YTrain)

net = train(net,XTrain,YTrain);

% Generate Simulink Model of Trained Neural Network
gensim(net);

%% Extract Weights and Biases
W1d = net.IW{1,1};  % Input Layer
b1d = net.b{1};

W2d = net.LW{2,1};  % Hidden Layer 1
b2d = net.b{2};
```

```
W3d = net.LW{3,2};  % Output Layer
b3d = net.b{3};

% Save W's and b's in .mat file (for later use)
save('Trained_NN_Model.mat','W1d','W2d','W3d','b1d','b2d','b3d','net');

%% Conversion to Fixed Point
% Convert XTest Data into Fixed Point
XTestf = Convert_Data_to_Fixed_Point(X, XTest);

% Convert W's and b's
[W1, b1, W2, b2, W3, b3] = Convert_Parameters_to_Fixed_Point();

save('Trained_fixed_NN_Model.mat','W1','W2','W3','b1','b2','b3');
```

The codes for the function "Histogram_Plots", which has been commented out in the m-file (Train_NN_Model.m), are shown as follows.

```
function Histogram_Plots(t)
Num_Disc_Var = [1 9];
Num_Cont_Var = [2 3 4 5 6 7 8];

NumD_Data = t(:,Num_Disc_Var);
NumD_Arr = table2array(NumD_Data);
NumD_str = string(NumD_Arr);
NumD_cell = cellstr(NumD_str);

% Histogram Plots
histogram(categorical(NumD_cell(:,1)),'BarWidth',0.3);
grid on;
ylabel('Frequency');
title('Pregnancies');

figure
histogram(categorical(NumD_cell(:,2)),'BarWidth',0.3);
grid on;
ylabel('Frequency');
title('Outcome');

NumC_Data = t(:,Num_Cont_Var);
NumC_Arr = table2array(NumC_Data);

figure
histogram(NumC_Arr(:,1));
grid on;
ylabel('Frequency');
title('Glucose');
xlabel('#');

figure
histogram(NumC_Arr(:,2));
grid on;
ylabel('Frequency');
```

```
title('Blood Pressure');
xlabel('Level');

figure
histogram(NumC_Arr(:,3));
grid on;
ylabel('Frequency');
title('Skin Thickness');

figure
histogram(NumC_Arr(:,4));
grid on;
ylabel('Frequency');
title('Insulin');
xlabel('mm');

figure
histogram(NumC_Arr(:,5));
grid on;
ylabel('Frequency');
title('BMI');
xlabel('Level');

figure
histogram(NumC_Arr(:,6));
grid on;
ylabel('Frequency');
title('Diabetes Pedigree Function');

figure
histogram(NumC_Arr(:,7));
grid on;
ylabel('Frequency');
title('Age');
xlabel('Years');
```

If we uncomment and run the m-file (Train_NN_Model.m), the histogram plots of the diabetes dataset will be displayed. Also, we have used 80% of the data for training and 20% for testing. The neural network consists of an input layer with the size equal to the number of input features (eight neurons), hidden layer 1 (layer 1) with 3 neurons, hidden layer 2 (layer 2) with four neurons, and an output layer (layer 3) with one neuron. The activation function for layer 1 and layer 2 is tanh ('tansig'); however, for the output layer (layer 3), the activation function is sigmoid ('logsig'). Also, we save the trained neural network parameters (weights and biases) in the file "Trained_NN_Model.mat." Now, we can run the MATLAB m-file (Train_NN_Model.m) and view the neural network training window, as shown in Figure 18.11.

Also, the Simulink model of the feedforward neural network is generated, as depicted in Figure 18.12.

If we type W1d, b1d, W2d, b2d, and W3d, b3d in the MATLAB command window, we get their values (W1d = 3×8, b1d = 3×1, W2d = 4×3, b2d = 4×1, and W3d = 1×4, b3d = 1×1). Also, if we write ">> view(net)" in the MATLAB command window, we can see our designed neural network as illustrated in Figure 18.13.

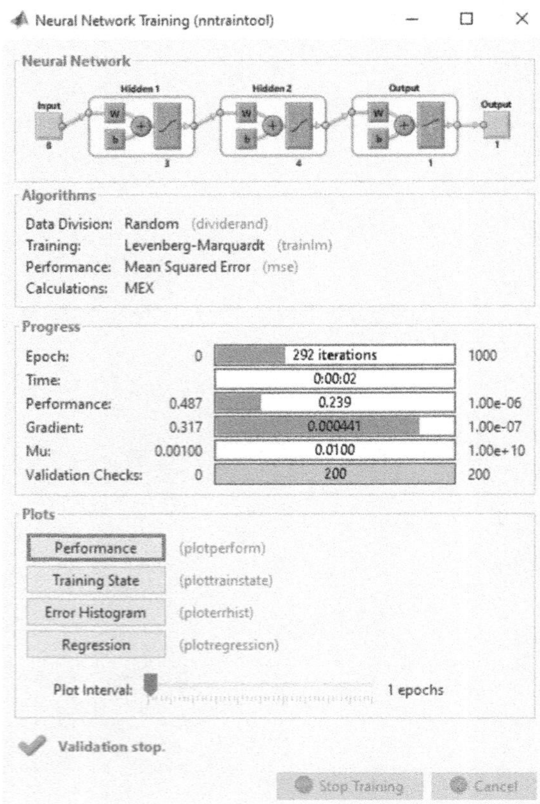

Figure 18.11 The neural network training window.

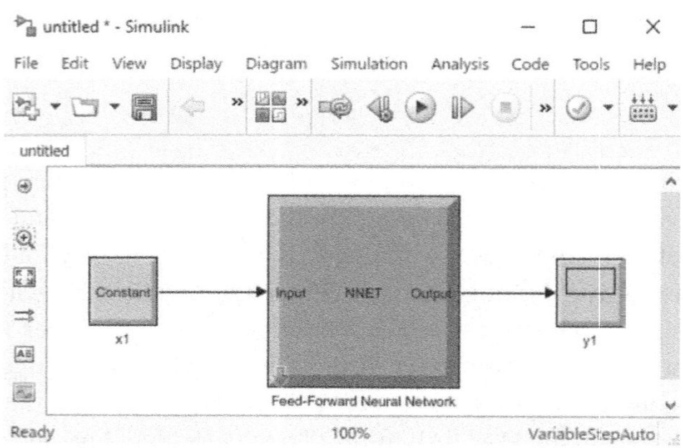

Figure 18.12 The Simulink model of the feedforward neural network.

If we click on the constant x1 in Figure 18.12, we can see its block parameters window as shown in Figure 18.14.

As you can see in Figure 18.14, the x1 is a normalized 8×1 matrix by default. If we run the Simulink model, we can view the y1 waveform on the scope, which is about 0.82, as depicted in Figure 18.15.

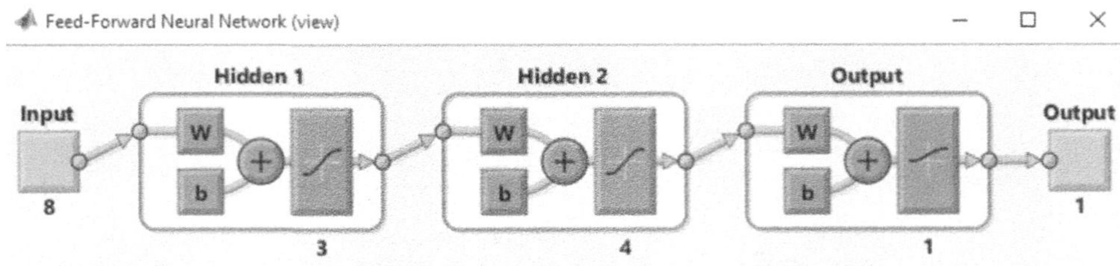

Figure 18.13 Viewing the designed neural network.

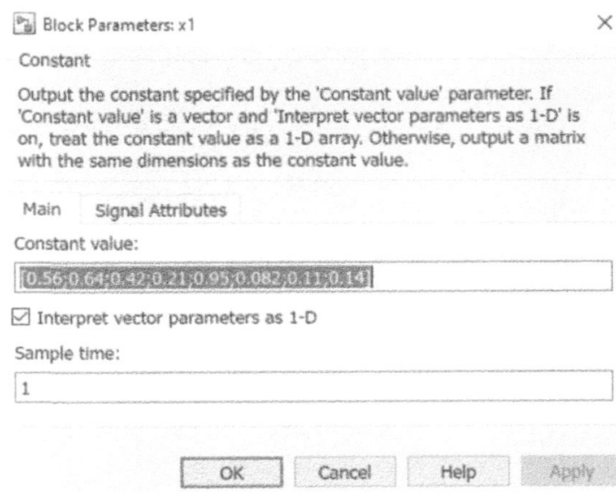

Figure 18.14 The block parameters window for constant x1.

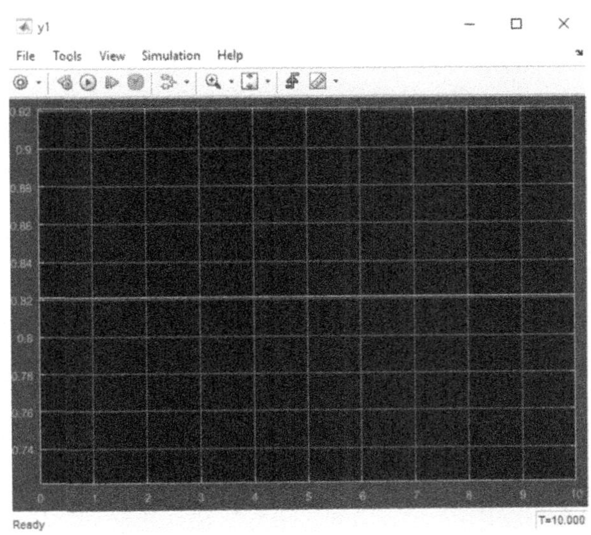

Figure 18.15 The y1 waveform on the scope.

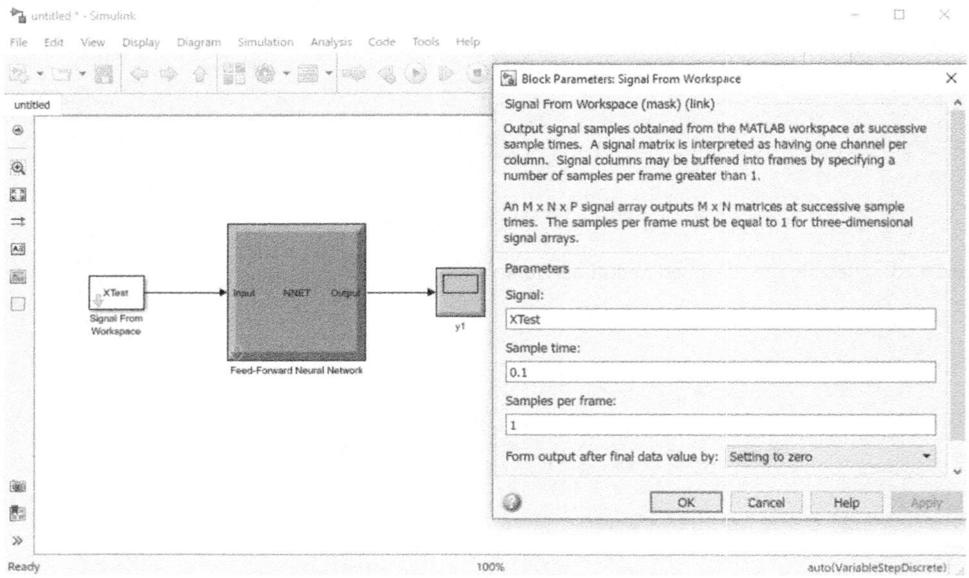

Figure 18.16 The Signal From Workspace block parameters.

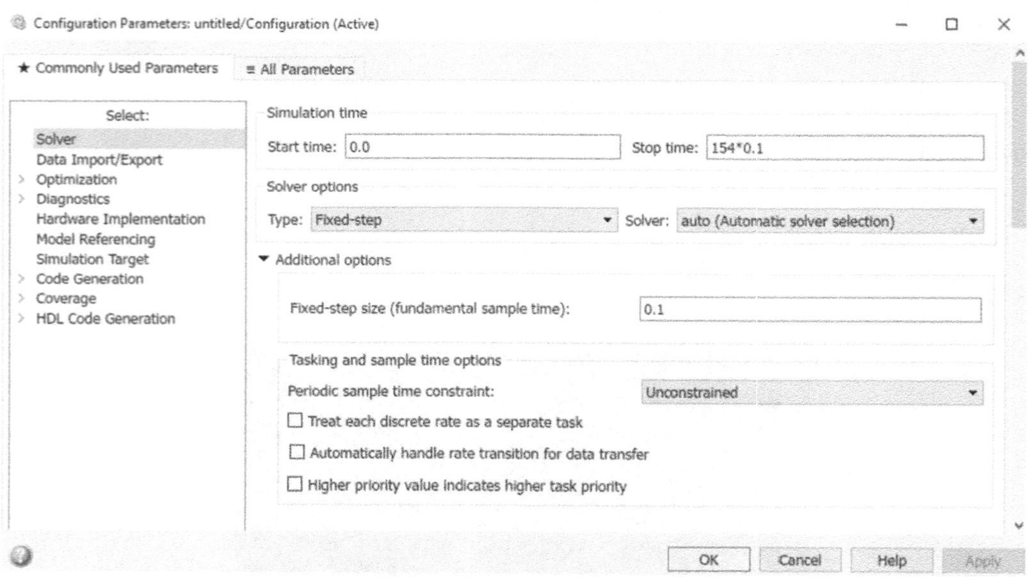

Figure 18.17 The Solver tab of configuration parameters.

Now, we can delete the constant x1 and bring the Signal From Workspace block into the Simulink model and set its parameters as illustrated in Figure 18.16.

Since XTest is a 154×8 matrix, we set the Solver tab of configuration parameters as shown in Figure 18.17.

```
>> size(XTest)
ans =
 154     8
```

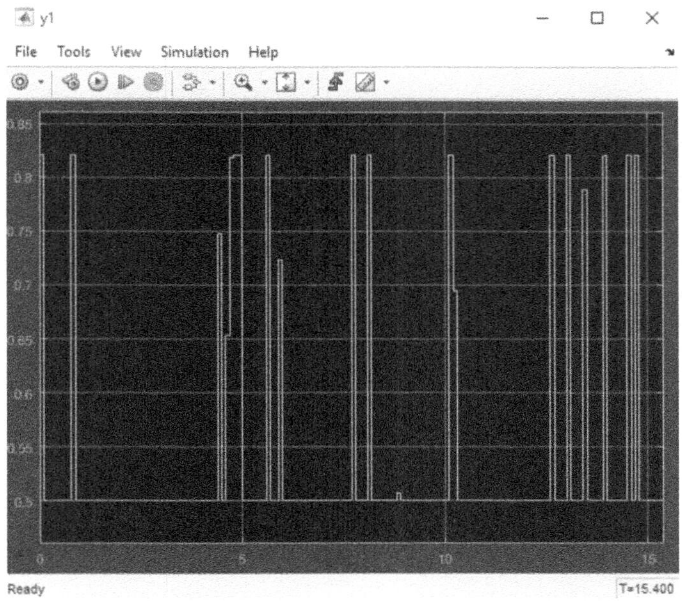

Figure 18.18 The waveform of y1 on scope.

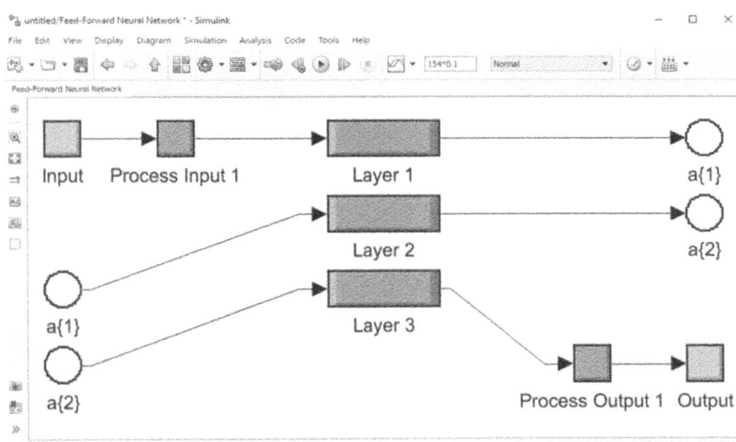

Figure 18.19 Deleting Process Input 1 and Process Output 1 blocks.

Now, if we run the Simulink model in Figure 18.16, we get the waveform of y1 on scope as depicted in Figure 18.18.

Inside the feedforward neural network block, there are Process Input 1 and Process Output 1 blocks, which are used for normalization, as illustrated in Figure 18.19. However, since we have already normalized the data inside the MATLAB m-file (Train_NN_Model.m), we can delete these blocks from the model and connect the input directly to layer 1, and layer 3 to the output. Now, if we run the simulation again, the waveform of y1 will remain unchanged.

We bring a MATLAB function block into the Simulink model and write the following code inside it.

```
function y = fcn(u)

if u > 0.5
    y = 1;
else
    y = 0;
end
```

Also, we bring a To Workspace block into the model with the block parameters as shown in Figure 18.20. The Simulink model is depicted in Figure 18.21. Now, if we run the simulation, the waveform of y2 on the scope is displayed, as illustrated in Figure 18.22.

We create an m-file titled "Cal_Accu.m" to calculate the accuracy of the simulation using the following codes.

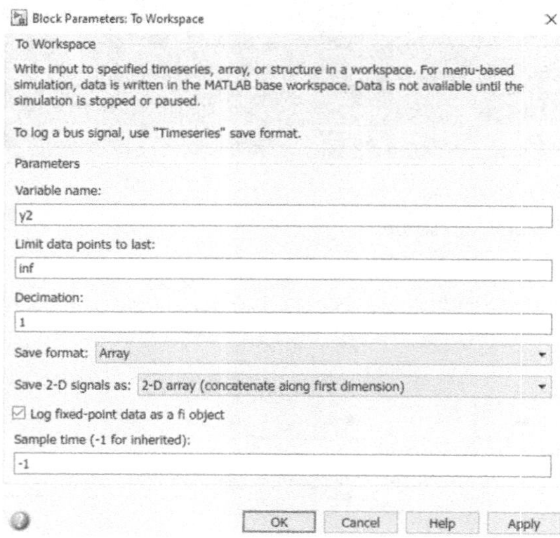

Figure 18.20 The block parameters of "To Workspace."

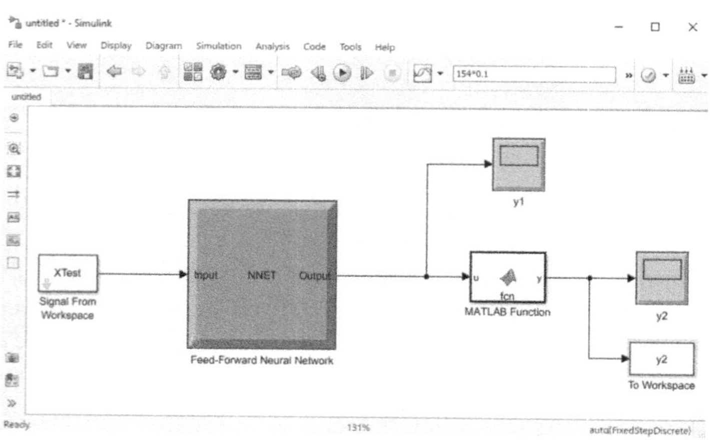

Figure 18.21 The Simulink model.

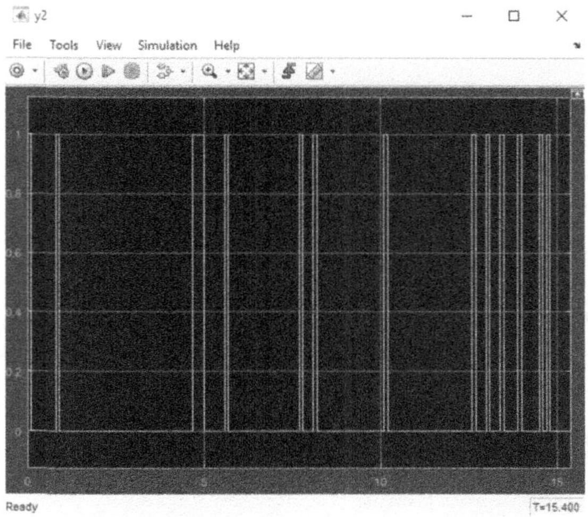

Figure 18.22 The waveform of y2 on the scope.

```
close all; clc;
ySim = y2;
ySim(end) = [];
Accuracy_Sim = 100*mean(ySim == YTest)
```

If we run that m-file, the following text is displayed in the MATLAB command window.

```
Accuracy_Sim =
 70.1299
```

So, the accuracy of the simulation is approximately 70.13%. We should note that the HDL Coder cannot generate code for the feedforward neural network block in Figure 18.21, and it is used only for simulation purposes. So, we should build our DNN manually. We define the function "NN_Layer_Tanh_HDL" by the following codes.

```
function ytan = NN_Layer_Tanh_HDL(Bias,Weight,X)
% This function implements NN_Layer using tanh function
% It is HDL supported, and can be used in HDL code generation
% Weight = they should be arranged as row vector
% X = Instances of input features should be arranged as column vector
% Bias = It can be row or column vector
% y = It is output vector arranged as column vector containing outputs
% From all neurons of given layer
% Note: All inputs of this function should be in fixed point format

Nodes = size(Weight,1);
WL = 16;
FL = 10;
PM = fi(zeros(Nodes,1),1,WL,FL);
WS = fi(zeros(Nodes,1),1,WL,FL);
ytan = fi(zeros(Nodes,1),1,WL,14);

for k = 1:Nodes
    PM(k) = Weight(k,:)*X(:,1);
    WS(k) = PM(k) + Bias(k);
```

```
    y = tanh_pw_HDL(WS(k));
    ytan(k) = y;
end
```

Then, in the Simulink model, we bring the HDL Coder MATLAB function and write the following code in it.

```
function ytan = NN_Layer_Tanh_HDL(Bias,Weight,X)
% Call Function
ytan = NN_Layer_Tanh_HDL(Bias,Weight,X);
```

The MATLAB function (Hidden_Layer1) is shown in Figure 18.23.

The codes for the function "Convert_Data_to_Fixed_Point" in MATLAB m-file (Train_NN_Model.m) are as follows.

```
function XTest_FixedPoint = Convert_Data_to_Fixed_Point(XFeatures, XTest)
% This function converts double type XTest (Test Data) into fixed point by
% considering the maximum and minimum values of XFeatures matrix along with
% fractional part.
% It is assumed that XTest is submatrix of XFeatures.
% Convert matrices to fixed-point representation based on statistics from
% XFeatures.

% 1. Find maximum and minimum values across all elements of XFeatures
max_val = max(XFeatures(:));
min_val = min(XFeatures(:));

% 2. Find maximum and minimum fractional values greather than zero across all
elements of XFeatures
fractional_values = abs(XFeatures - floor(XFeatures)); % Fractional parts of all
elements of XFeatures
fractional_values_non_zero = fractional_values(fractional_values > 0); %
Filter non-zero fractional parts of all elements of XFeatures
if isempty(fractional_values_non_zero) % If XFeatures has only positive integers
    FL = 0;
    max_fractional = 0;
    min_fractional = 0;
else
    max_fractional = max(fractional_values_non_zero);
    min_fractional = min(fractional_values_non_zero);
    % To determine FL, consider the smallest non-zero fractional value and
    % ensure
    FL = 1+ceil(-log2(min_fractional)); % 1 is added (manually)
end
```

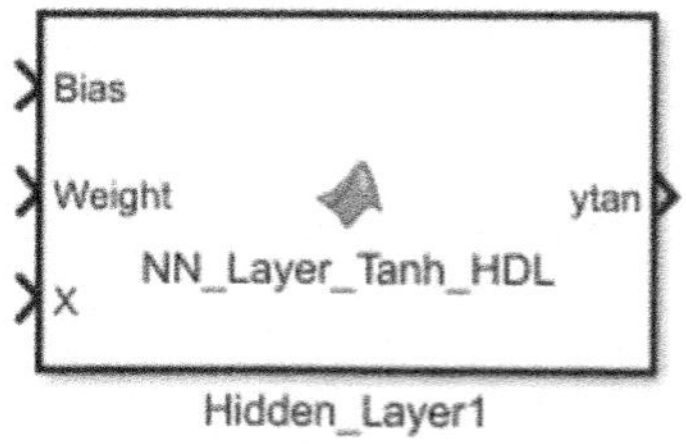

Figure 18.23 The MATLAB function (Hidden_Layer1).

```
% 3. Calculate WL
% WL = number of bits of the integer part of max value + 1 bit for sign +
% FL
WL = 1 + ceil(log2(max(abs([max_val, min_val])))) + FL;

% 4. Convert XTest to fixed-point using WL and FL
XTest_FixedPoint = fi(XTest, 1, WL, FL);

% Optional: Display the results (You can comment this out if not needed)
fprintf('Max value: %f\n', max_val);
fprintf('Min value: %f\n', min_val);
fprintf('Max fractional (non-zero): %f\n', max_fractional);
fprintf('Min fractional (non-zero): %f\n', min_fractional);
fprintf('\n------------- Data (Testing) Conversion');
fprintf('\nWord Length (WL): %d bits\n', WL);
fprintf('\nFraction Length (WL): %d bits\n\n\n', FL);
end
```

Also, the codes for the function "Convert_Parameters_to_Fixed_Point" in MATLAB m-file (Train_NN_Model.m) are as follows.

```
function [W1, b1, W2, b2, W3, b3] = Convert_Parameters_to_Fixed_Point()

data = load('Trained_NN_Model.mat');
W1d = data.W1d;
b1d = data.b1d;
W2d = data.W2d;
b2d = data.b2d;
W3d = data.W3d;
b3d = data.b3d;

maxW1d_val = max(W1d(:));
minW1d_val = min(W1d(:));

fractional_W1dvalues = abs(W1d - floor(W1d));
fractional_W1dvalues_non_zero = fractional_W1dvalues(fractional_W1dvalues > 0);
if isempty(fractional_W1dvalues_non_zero)
    FLW1d = 0;
else
    minW1d_fractional = min(fractional_W1dvalues_non_zero);
    FLW1d = 1+ceil(-log2(minW1d_fractional));
end

WLW1d = 1 + ceil(log2(max(abs([maxW1d_val, minW1d_val])))) + FLW1d;
W1 = fi(W1d, 1, WLW1d, FLW1d);
%%%%%%%%%%%%%%%%%%%%%%%%%%%%%%%%%%%%%%%%%%%%%%%%%%%%%%%%%%%%%%%%%%%%%%%%%%%%%%
maxb1d_val = max(b1d(:));
minb1d_val = min(b1d(:));

fractional_b1dvalues = abs(b1d - floor(b1d));
fractional_b1dvalues_non_zero = fractional_b1dvalues(fractional_b1dvalues > 0);
```

```
if isempty(fractional_b1dvalues_non_zero)
    FLb1d = 0;
else
    minb1d_fractional = min(fractional_b1dvalues_non_zero);
    FLb1d = 1+ceil(-log2(minb1d_fractional));
end

WLb1d = 1 + ceil(log2(max(abs([maxb1d_val, minb1d_val])))) + FLb1d;
b1 = fi(b1d, 1, WLb1d, FLb1d);
%%%%%%%%%%%%%%%%%%%%%%%%%%%%%%%%%%%%%%%%%%%%%%%%%%%%%%%%%%%%%%%%%%%%%%%%%%%%%%%%%%
maxW2d_val = max(W2d(:));
minW2d_val = min(W2d(:));

fractional_W2dvalues = abs(W2d - floor(W2d));
fractional_W2dvalues_non_zero = fractional_W2dvalues(fractional_W2dvalues > 0);
if isempty(fractional_W2dvalues_non_zero)
    FLW2d = 0;
else
    minW2d_fractional = min(fractional_W2dvalues_non_zero);
    FLW2d = 1+ceil(-log2(minW2d_fractional));
end

WLW2d = 1 + ceil(log2(max(abs([maxW2d_val, minW2d_val])))) + FLW2d;
W2 = fi(W2d, 1, WLW2d, FLW2d);
%%%%%%%%%%%%%%%%%%%%%%%%%%%%%%%%%%%%%%%%%%%%%%%%%%%%%%%%%%%%%%%%%%%%%%%%%%%%%%%%%%
maxb2d_val = max(b2d(:));
minb2d_val = min(b2d(:));

fractional_b2dvalues = abs(b2d - floor(b2d));
fractional_b2dvalues_non_zero = fractional_b2dvalues(fractional_b2dvalues > 0);
if isempty(fractional_b2dvalues_non_zero)
    FLb2d = 0;
else
    minb2d_fractional = min(fractional_b2dvalues_non_zero);
    FLb2d = 1+ceil(-log2(minb2d_fractional));
end

WLb2d = 1 + ceil(log2(max(abs([maxb2d_val, minb2d_val])))) + FLb2d;
b2 = fi(b2d, 1, WLb2d, FLb2d);
%%%%%%%%%%%%%%%%%%%%%%%%%%%%%%%%%%%%%%%%%%%%%%%%%%%%%%%%%%%%%%%%%%%%%%%%%%%%%%%%%%
maxW3d_val = max(W3d(:));
minW3d_val = min(W3d(:));

fractional_W3dvalues = abs(W3d - floor(W3d));
fractional_W3dvalues_non_zero = fractional_W3dvalues(fractional_W3dvalues > 0);
if isempty(fractional_W3dvalues_non_zero)
    FLW3d = 0;
else
    minW3d_fractional = min(fractional_W3dvalues_non_zero);
    FLW3d = 1+ceil(-log2(minW3d_fractional));
end

WLW3d = 1 + ceil(log2(max(abs([maxW3d_val, minW3d_val])))) + FLW3d;
```

```
W3 = fi(W3d, 1, WLW3d, FLW3d);
%%%%%%%%%%%%%%%%%%%%%%%%%%%%%%%%%%%%%%%%%%%%%%%%%%%%%%%%%%%%%%%%%%%%%%%%%%%%
maxb3d_val = max(b3d(:));
minb3d_val = min(b3d(:));

fractional_b3dvalues = abs(b3d - floor(b3d));
fractional_b3dvalues_non_zero = fractional_b3dvalues(fractional_b3dvalues > 0);
if isempty(fractional_b3dvalues_non_zero)
    FLb3d = 0;
else
    minb3d_fractional = min(fractional_b3dvalues_non_zero);
    FLb3d = 1+ceil(-log2(minb3d_fractional));
end

WLb3d = 1 + ceil(log2(max(abs([maxb3d_val, minb3d_val])))) + FLb3d;
b3 = fi(b3d, 1, WLb3d, FLb3d);
```

We design our DNN as depicted in Figure 18.24.

The MATLAB function "Hidden_Layer2" is the same as "Hidden_Layer1"; however, the MATLAB function "OutputLayer" code is as follows.

```
function ysig = NN_Layer_Sigmoid_HDL(Bias,Weight,X)
% Call Function
ysig = NN_Layer_Sigmoid_HDL(Bias,Weight,X);
```

Also, the function "NN_Layer_Sigmoid_HDL" is defined by the following codes, which is similar to the function "NN_Layer_Tanh_HDL."

```
function ysig = NN_Layer_Sigmoid_HDL(Bias,Weight,X)
%This function implements NN_Layer using sigmoid function
% It is HDL supported, and can be used in HDL code generation
% Weight = they should be arranged as row vector
% X = Instances of input features should be arranged as column vector
% Bias = It ca be row or column vactor
% y = It is output vector arranged as column vector containing outputs
```

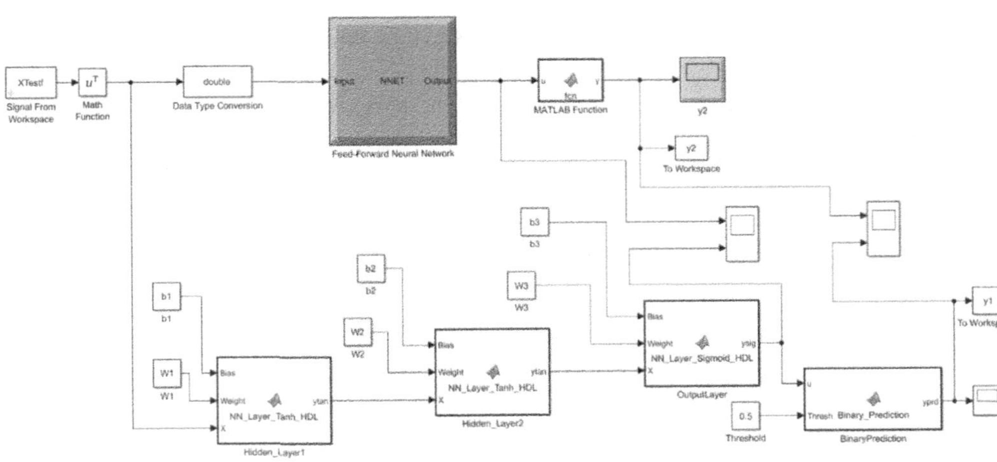

Figure 18.24 The deep neural network design.

```
% from all neurons of given layer
% Note: All inputs of this function should be in fixed point format

Nodes = size(Weight,1);
WL = 16;
FL = 10;
PM = fi(zeros(Nodes,1),1,WL,FL);
WS = fi(zeros(Nodes,1),1,WL,FL);
ysig = fi(zeros(Nodes,1),1,WL,14);

for k = 1:Nodes
    PM(k) = Weight(k,:)*X(:,1);
    WS(k) = PM(k) + Bias(k);  %u
    y = sig_pw_HDL(WS(k));    %y = sigmoid(u)
    ysig(k) = y;
end
```

The MATLAB function "BinaryPrediction" in Figure 18.24 is defined by the following codes.

```
function yprd = Binary_Prediction(u,Thresh)
% Call Function
yprd = Binary_Prediction(u,Thresh);
```

The codes for the function "Binary_Prediction" are as follows.

```
function yprd = Binary_Prediction(u,Thresh)
if u > Thresh
    yprd = fi(1,0,1,0);
else
    yprd = fi(0,0,1,0);
end
```

The block parameters for the Main tab and Signal Attributes tab of the Constant Threshold are illustrated in Figures 18.25 and 18.26, respectively. Also, the block parameters for the Main tab of Constant W1 is shown in Figure 18.27, and the same settings are used for other parameters (b1, W2, b2, W3, and b3) as well.

The output prediction using the feedforward neural network and the manually designed DNN are compared in Figure 18.28. Also, the inputs for the MATLAB function prediction blocks for both generated and manual DNN are compared in Figure 18.29.

Now, if we run the Calculation Accuracy m-file (Cal_Accu.m) with the following codes, we get the results in the MATLAB command window as follows.

```
close all; clc;
ySim = y2;
yHDL = y1;
ySim(end) = [];
yHDL(end) = [];
Accuracy_Sim = 100*mean(ySim == YTest)
Accuracy_HDL = 100*mean(yHDL == YTest)

Accuracy_Sim =
   70.1299
Accuracy_HDL =
   68.8312
```

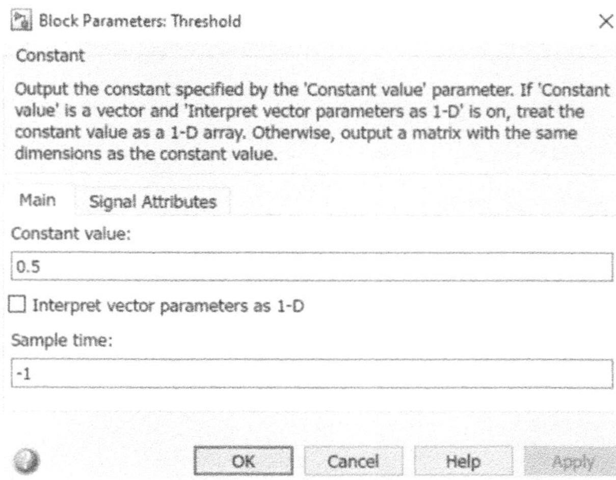

Figure 18.25 Block parameters for the Main tab of the threshold.

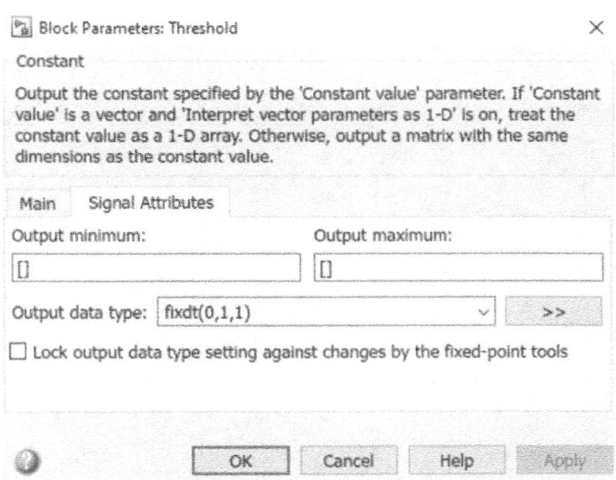

Figure 18.26 Block parameters for the Signal Attributes tab of the threshold.

So, the simulation accuracy is about 70.13%, and the HDL accuracy is about 68.83%. We select the Hidden_Layer1, Hidden_Layer1, OutputLayer, and BinaryPrediction blocks, then right-click and select the Create Subsystem from Selection option, as depicted in Figure 18.30.

The created subsystem (DNN) is illustrated in Figure 18.31. Also, the block design of the subsystem "DNN" is shown in Figure 18.32.

Now we can generate the VHDL code for the subsystem "DNN" in Figure 18.31 using HDL Coder. To do that, we write the following code in the MATLAB command window.

```
>> hdlsetup('simulation_DNN_sub')
```

The name of the Simulink design is simulation_DNN_sub. Then, we go to the Configuration Parameters window and set the Solver tab values as demonstrated in Figure 18.33. Next, we go to the HDL Code Generation tab and set the parameters as depicted in Figure 18.34.

Then we type the following code in the MATLAB command window.

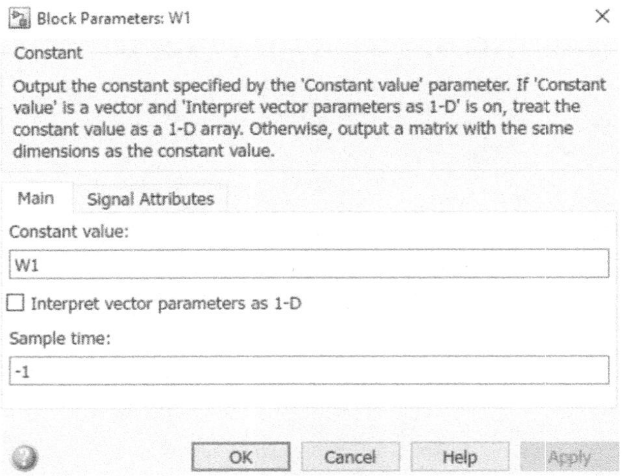

Figure 18.27 Block parameters for the Main tab of Constant W1.

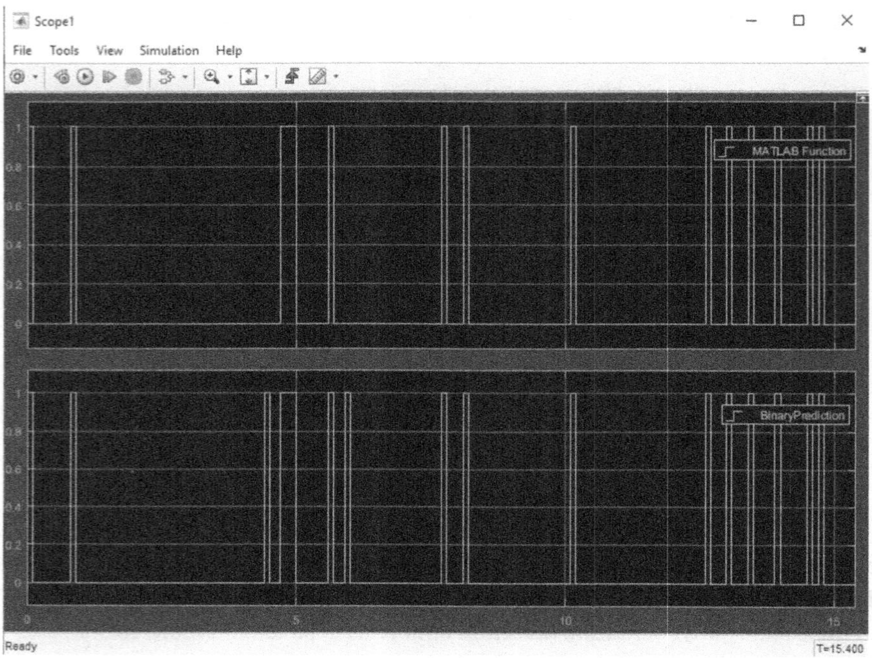

Figure 18.28 Comparison of the output prediction using generated and manual DNN.

```
>> checkhdl('simulation_DNN_sub/DNN')
```

The HDL Check Report window as a web browser pops up, as shown in Figure 18.35.

As shown in Figure 18.35, we got two errors and one warning, since the matrices (W1 and W2) are not supported at Simulink block interfaces in HDL Coder. We type W1 and press the enter button, and the result will be as follows.

```
>> W1
W1 =
   -3.4688    -6.5938    -1.9375    -0.2969    -4.0469    -4.9063    -3.1406    -3.4688
   -0.6406     0.6719     2.3125    -0.1719    -0.6406     0.0938    -1.2656    -1.6094
    1.7813     3.8750   -13.0938     0.5938     0.8750     3.7031     2.4219     1.8281
          DataTypeMode: Fixed-point: binary point scaling
            Signedness: Signed
            WordLength: 11
        FractionLength: 6
```

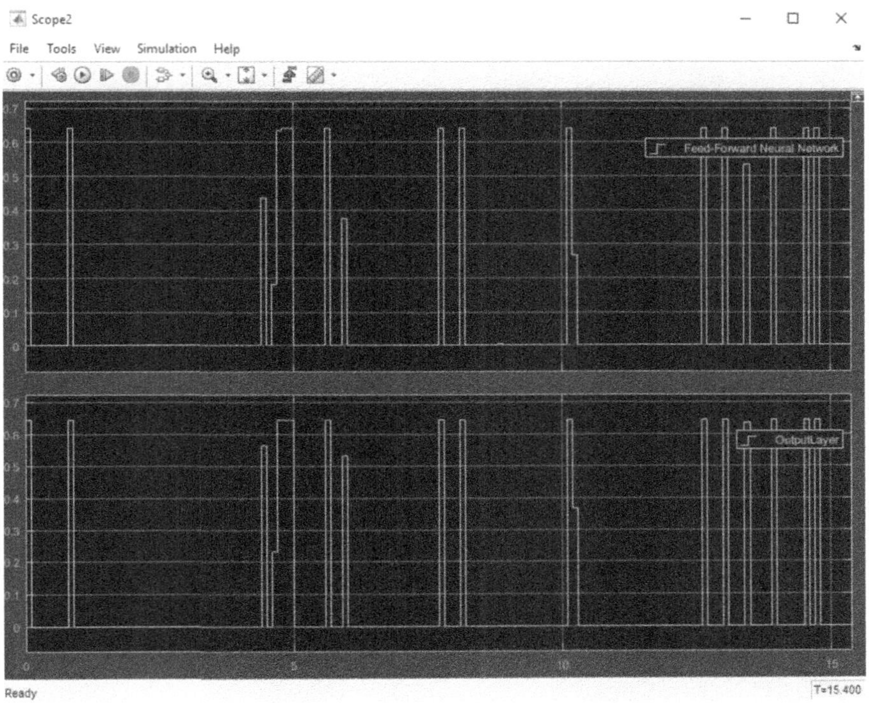

Figure 18.29 Comparison of the inputs to the MATLAB function prediction blocks.

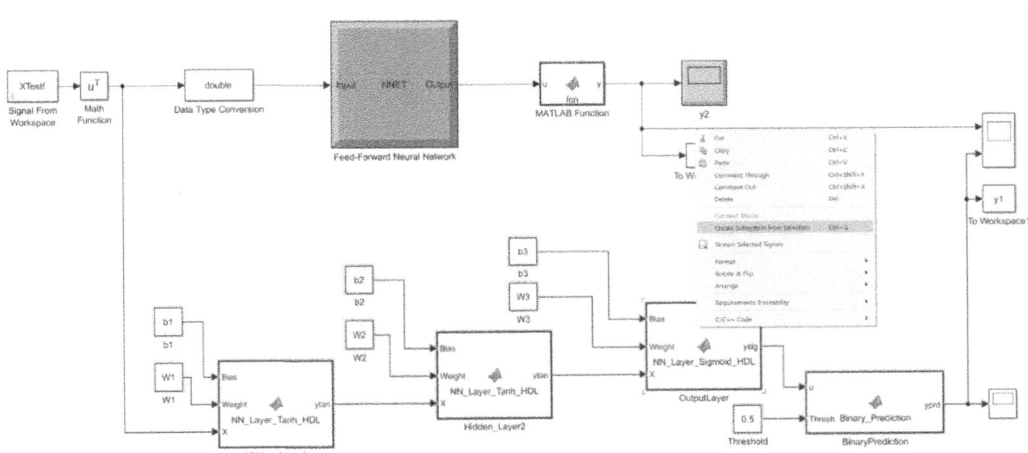

Figure 18.30 Selecting the Create Subsystem from Selection option.

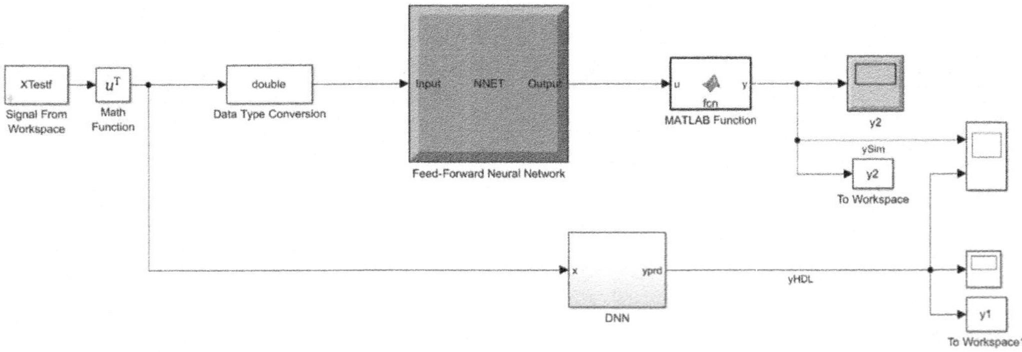

Figure 18.31 The created subsystem (DNN).

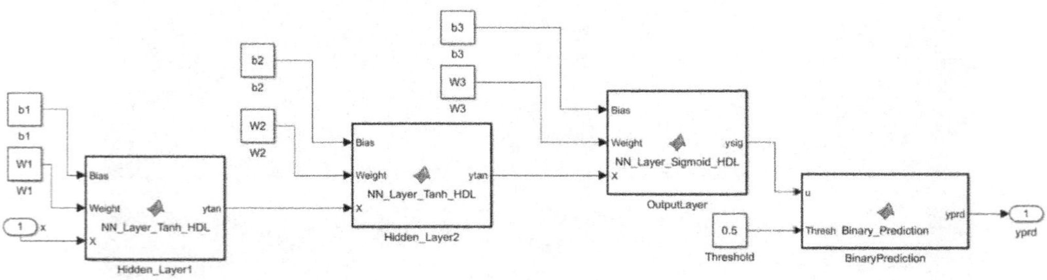

Figure 18.32 The block design of subsystem "DNN."

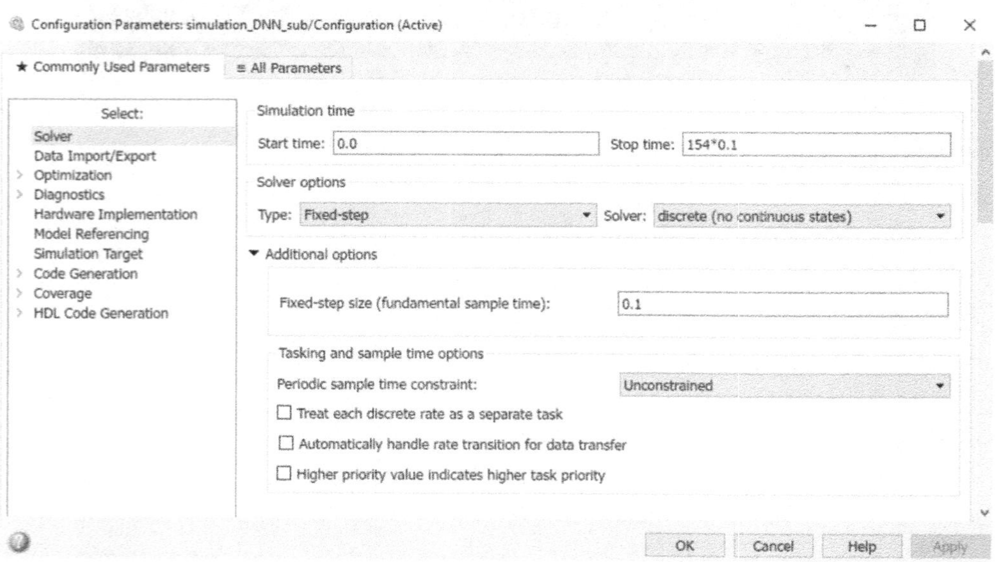

Figure 18.33 The configuration parameters for the Solver tab.

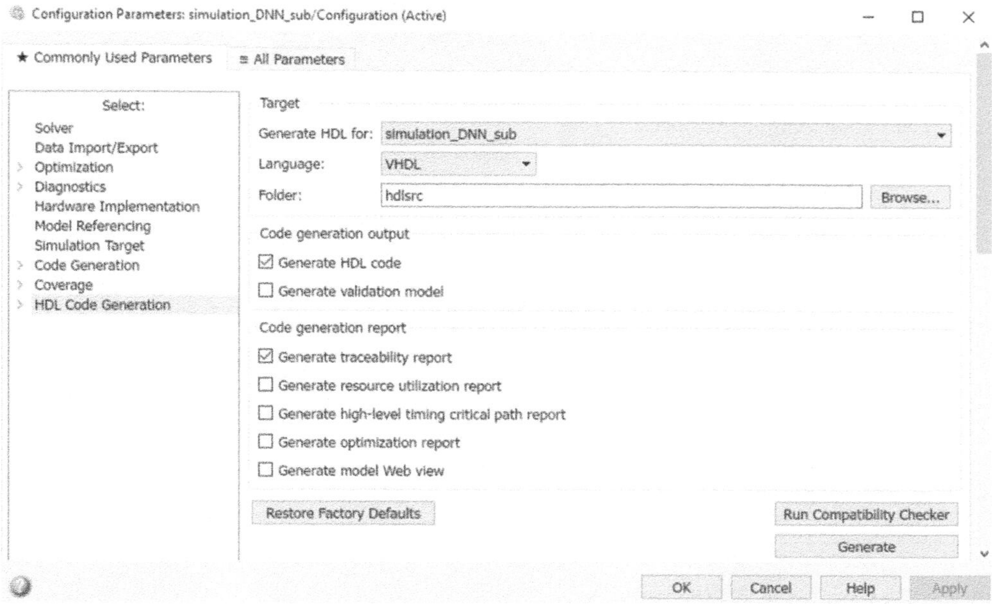

Figure 18.34 The HDL Code Generation tab of Configuration Parameters window.

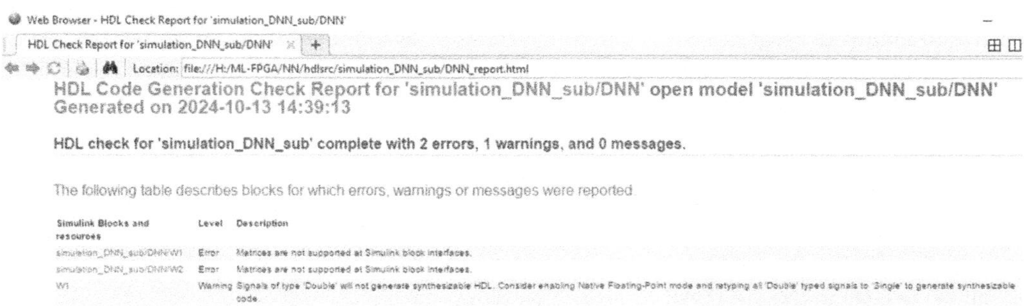

Figure 18.35 The HDL Check Report window.

Then, we click on the Hidden_Layer1 MATLAB function, and edit the codes as follows.

```
function ytan = NN_Layer_Tanh_HDL(Bias,X)
% Call Function
W1 = [-3.4688  -6.5938  -1.9375  -0.2969  -4.0469  -4.9063  -3.1406  -3.4688
   -0.6406   0.6719   2.3125  -0.1719  -0.6406   0.0938  -1.2656  -1.6094
    1.7813   3.8750  -13.0938   0.5938   0.8750   3.7031   2.4219   1.8281];
W1 = fi(W1,1,11,6);

ytan = NN_Layer_Tanh_HDL(Bias,W1,X);
```

We type W2 and press the enter button in the MATLAB command window, and the result will be as follows.

```
>> W2
W2 =
    1.9688    -1.5938    -0.1250
   -1.6563    -3.2813     1.0938
    4.2188    -0.2188     2.2813
   -9.6875    -4.7813     0.7813
          DataTypeMode: Fixed-point: binary point scaling
            Signedness: Signed
            WordLength: 10
        FractionLength: 5
```

Then, we click on the Hidden_Layer2 MATLAB function, and edit the codes as follows.

```
function ytan = NN_Layer_Tanh_HDL(Bias,X)
% Call Function
W2 = [1.9688    -1.5938    -0.1250
   -1.6563    -3.2813     1.0938
    4.2188    -0.2188     2.2813
   -9.6875    -4.7813     0.7813];
W2 = fi(W2,1,10,5);

ytan = NN_Layer_Tanh_HDL(Bias,W2,X);
```

As you can see in Figure 18.36, we delete constants W1 and W2, since they have been defined inside the MATLAB functions Hidden_Layer1 and Hidden_Layer2.

If we simulate the Simulink design again, we can view the output prediction waveforms as depicted in Figure 18.37.

Now, we can generate the VHDL code for the subsystem "DNN" in Figure 18.36 using HDL Coder. To do that, we write the following code in the MATLAB command window.

```
>> hdlsetup('simulation_DNN_sub')
```

The name of the Simulink design is simulation_DNN_sub. Then, we go again to the Configuration Parameters window and set the Solver tab values as demonstrated in Figure 18.33. Next, we go to the HDL Code Generation tab and set the parameters as depicted in Figure 18.34. Then, we type the following code in the MATLAB command window.

```
>> checkhdl('simulation_DNN_sub/DNN')
```

The HDL Check Report window as a web browser pops up, as shown in Figure 18.38.

Now, we use the following code for HDL code generation in the MATLAB command window.

```
>> makehdl('simulation_DNN_sub/DNN')
```

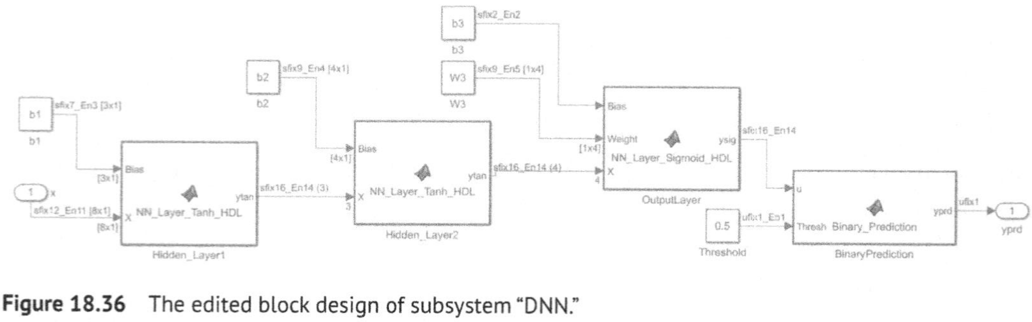

Figure 18.36 The edited block design of subsystem "DNN."

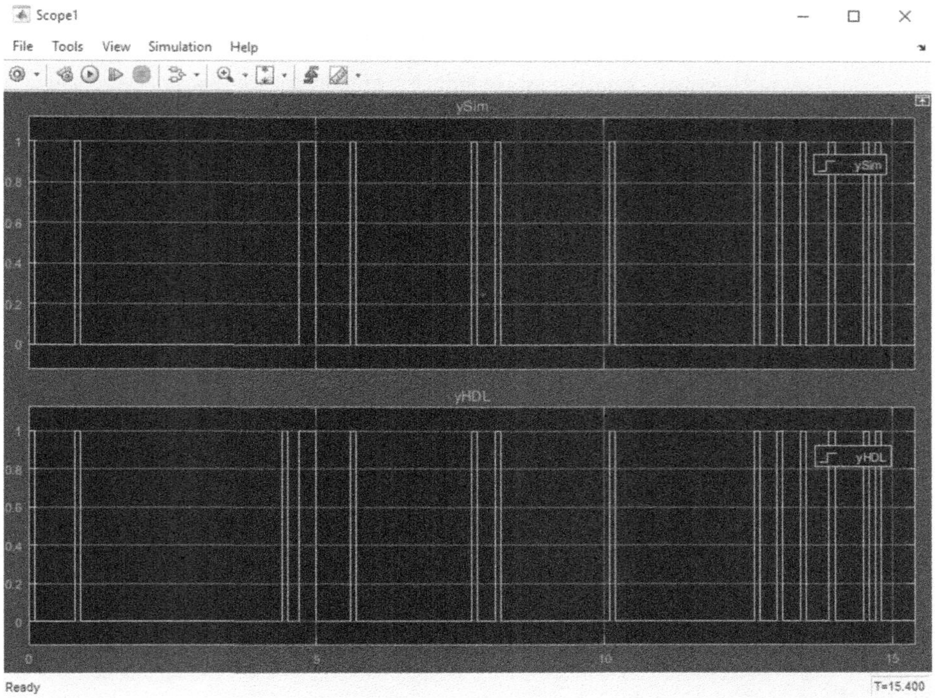

Figure 18.37 Comparison of the output predictions.

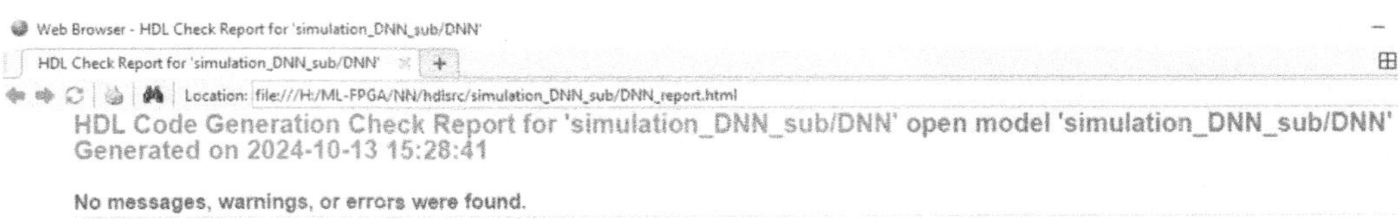
Web Browser - HDL Check Report for 'simulation_DNN_sub/DNN'
HDL Check Report for 'simulation_DNN_sub/DNN'
Location: file:///H:/ML-FPGA/NN/hdlsrc/simulation_DNN_sub/DNN_report.html
HDL Code Generation Check Report for 'simulation_DNN_sub/DNN' open model 'simulation_DNN_sub/DNN'
Generated on 2024-10-13 15:28:41

No messages, warnings, or errors were found.

Figure 18.38 The HDL Check Report window.

The VHDL codes are generated successfully, and the Code Generation Report window is displayed, as depicted in Figure 18.39.

18.6 Using Generated VHDL Code in Vivado

Now, we run Vivado and create a project named "DNN", then click the Next button. Next, we select Project Type as RTL Project and click the Next button. In the Add Sources window, before we add the generated VHDL files, we copy and paste the generated VHDL files from Figure 18.39 into the "DNN" folder of the Vivado project. Now, by clicking on the Add Files button, we can add them to the project, as shown in Figure 18.40.

Next, we select our Xilinx development board and click on the Next button. Finally, we click the Finish button to create the "DNN" project with the added generated VHDL file. Then, we click on Run Synthesis, as illustrated in Figure 18.41.

As demonstrated in Figure 18.41, the synthesis is successfully completed, and we can click on the Cancel button. Then, we open the schematic in the RTL Analysis tab as shown in Figure 18.42.

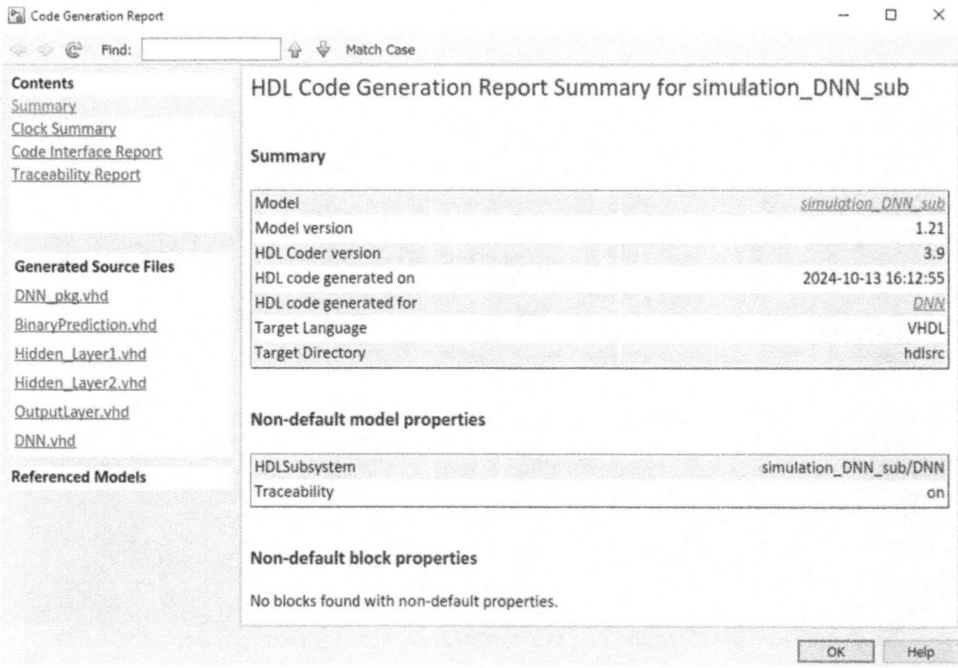

Figure 18.39 The Code Generation Report window.

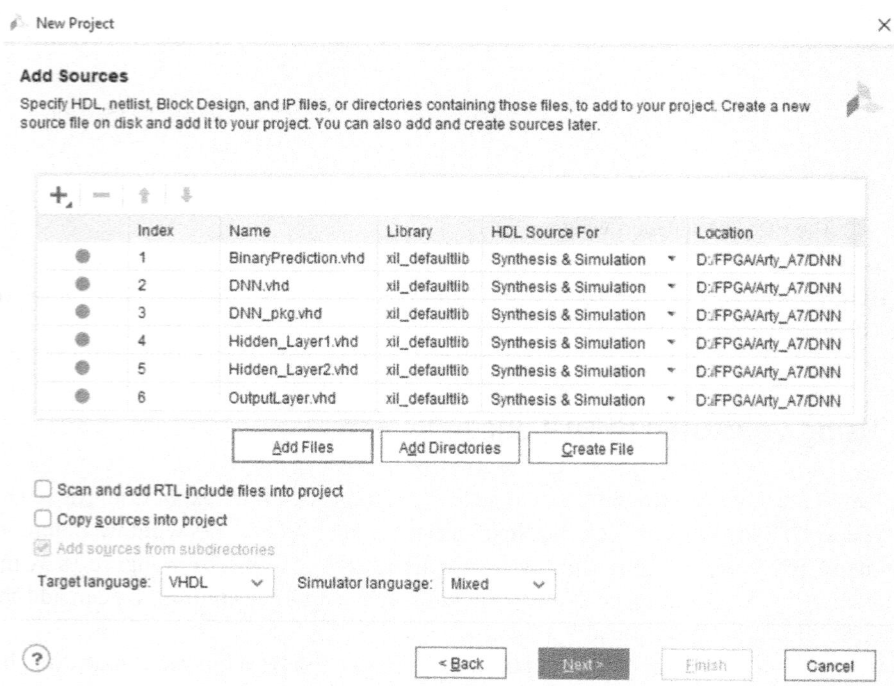

Figure 18.40 Adding generated VHDL files to the Sources folder.

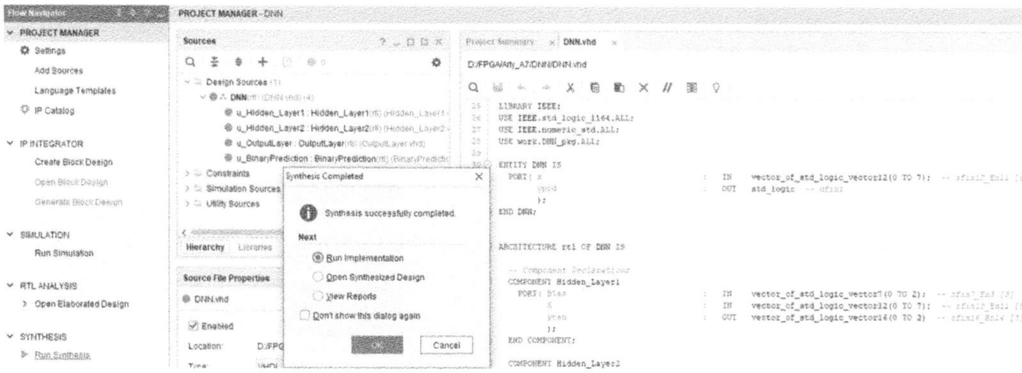

Figure 18.41 Successfully running the synthesis.

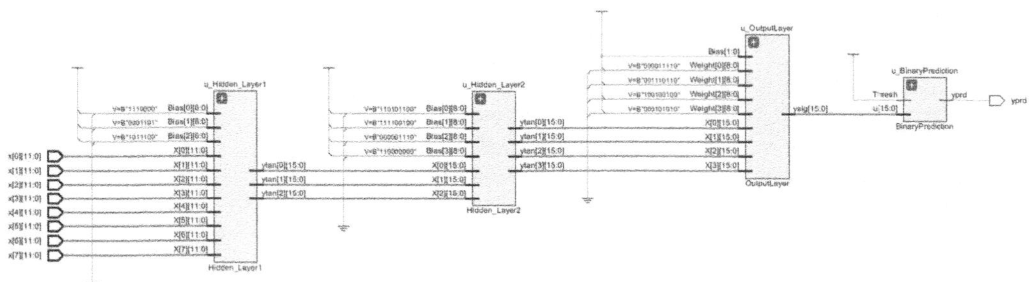

Figure 18.42 The schematic of "DNN."

Now, we can simulate the project by selecting Flow → Run Simulation → Run Behavioral Simulation. In the opened simulation window, we click on the Restart button. We write the following code in the MATLAB command window.

```
>> XTestf(50,:)
ans = 
    0.0586    0.4575    0.3115   -0.0708   -0.6929    0.1299   -0.5225   -0.3667
          DataTypeMode: Fixed-point: binary point scaling
            Signedness: Signed
            WordLength: 12
        FractionLength: 11
>> t50 = XTestf(50,:);
>> t50.bin
ans =
000001111000   001110101001   001001111110   111101101111   101001110101   000100
001010   101111010010   110100010001
>> y1(50)
ans =
     1
          DataTypeMode: Fixed-point: binary point scaling
            Signedness: Unsigned
            WordLength: 1
        FractionLength: 0
>> y2(50)
ans =
     1
```

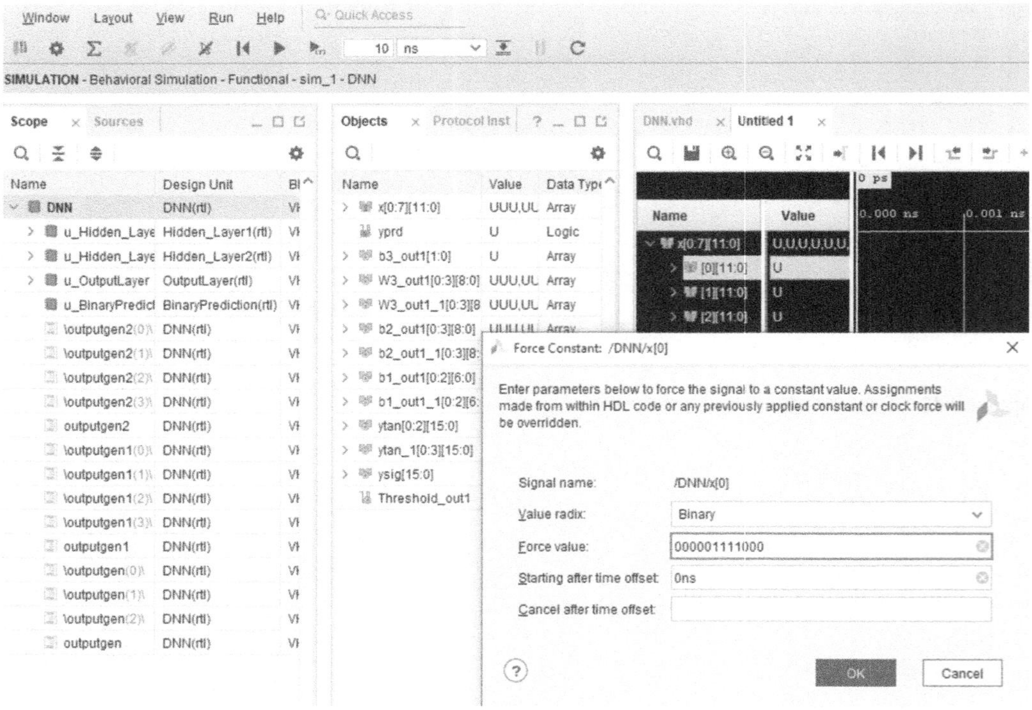

Figure 18.43 The Force Constant window for x[0].

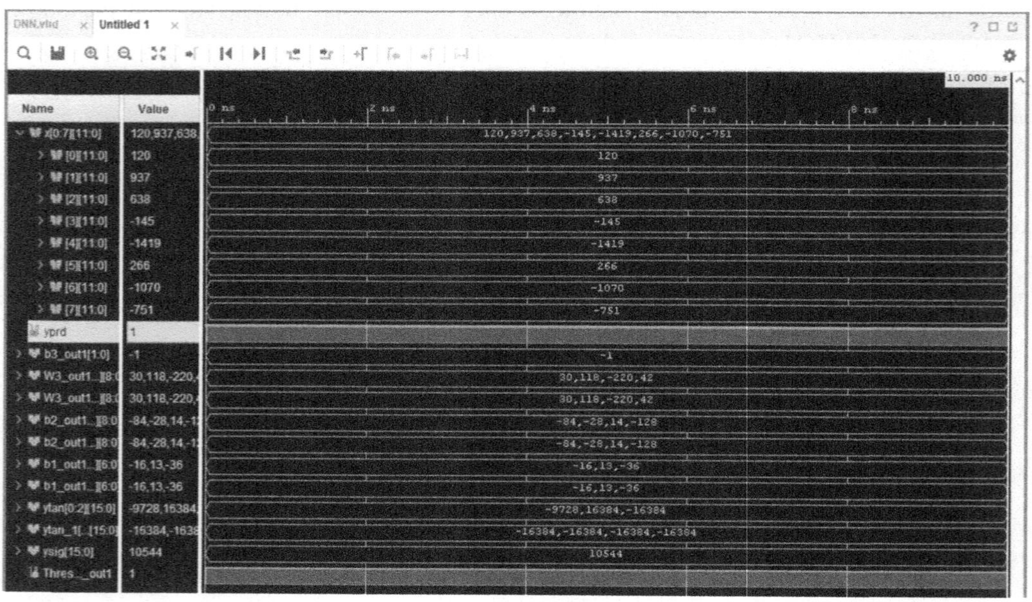

Figure 18.44 The simulated waveforms.

Then, in the waveform window, we right-click on "x[0]" and select the "Force Constant..." option, and configure its settings as illustrated in Figure 18.43.

We repeat the same process for x[1] = 001110101001, x[2] = 001001111110, x[3] = 111101101111, x[4] = 101001110101, x[5] = 000100001010, x[6] = 101111010010, and x[7] = 110100010001. Then, we click on the Run button for 10 ns once, and click on the Zoom Fit button in waveforms window to get the simulated waveforms as illustrated in Figure 18.44.

So, we achieve the same prediction using FPGA (yprd = 1) as well.

Index

Fundamentals of VHDL for FPGA Programming Using Vivado, First Edition. Majid Pakdel.

Companion Website: www.wiley.com/go/pakdelVHDL

d